Neural Surface Antigens

From Basic Biology Towards Biomedical Applications

Neural Surface Antigens
From Basic Biology Towards Biomedical Applications

Jan Pruszak
Emmy Noether-Group for Stem Cell Biology
Department of Molecular Embryology
Institute of Anatomy and Cell Biology
University of Freiburg, Freiburg im Breisgau, Germany

AMSTERDAM • BOSTON • HEIDELBERG • LONDON • NEW YORK • OXFORD
PARIS • SAN DIEGO • SAN FRANCISCO • SINGAPORE • SYDNEY • TOKYO
Academic Press is an imprint of Elsevier

Academic Press is an imprint of Elsevier
125 London Wall, London EC2Y 5AS, UK
525 B Street, Suite 1800, San Diego, CA 92101-4495, USA
225 Wyman Street, Waltham, MA 02451, USA
The Boulevard, Langford Lane, Kidlington, Oxford OX5 1GB, UK

Library of Congress Cataloging-in-Publication Data
A catalog record for this book is available from the Library of Congress

British Library Cataloguing-in-Publication Data
A catalogue record for this book is available from the British Library

ISBN: 978-0-12-800781-5

For information on all Academic Press publications
visit our website at http://store.elsevier.com/

Working together
to grow libraries in
developing countries

www.elsevier.com • www.bookaid.org

Publisher: Elsevier
Acquisition Editor: Christine Minihane
Editorial Project Manager: Shannon Stanton
Production Project Manager: Karen East and Kirsty Halterman
Designer: Greg Harris

Typeset by TNQ Books and Journals
www.tnq.co.in

Contents

9. NG2 (Cspg4): Cell Surface Proteoglycan on Oligodendrocyte Progenitor Cells in the Developing and Mature Nervous System

Akiko Nishiyama, Aaron Lee and Christopher B. Brunquell

10. Comprehensive Overview of CD133 Biology in Neural Tissues across Species

József Jászai, Denis Corbeil and Christine A. Fargeas

11. Fundamentals of NCAM Expression, Function, and Regulation of Alternative Splicing in Neuronal Differentiation

Ana Fiszbein, Ignacio E. Schor and Alberto R. Kornblihtt

Contributors

Robert Beattie Department of Biomedicine, University of Basel, Mattenstrasse, Basel, Switzerland

Nadège Bondurand INSERM U955, IMRB, Equipe 6, Créteil, France; Faculté de Médecine, Université Paris Est, Créteil, France

Hélène Boudin INSERM UMR913, IMAD, University of Nantes, Nantes, France

Christopher Boyce BD Biosciences, La Jolla, CA, USA

Florence Broders-Bondon Institut Curie/CNRS UMR144, Paris, France

Christopher B. Brunquell Department of Physiology and Neurobiology, University of Connecticut, Storrs, CT, USA

Krista D. Buono Department of Neurology and Neuroscience, New Jersey Medical School, Rutgers University-New Jersey Medical School, Newark, NJ, USA; ICON Central Laboratories, 123 Smith Street, Farmingdale, NY

Christian T. Carson BD Biosciences, La Jolla, CA, USA

Si Chen Division of Molecular Neurobiology, German Cancer Research Center (DKFZ), Heidelberg, Germany

Denis Corbeil Tissue Engineering Laboratories (BIOTEC), Medizinische Fakultät der Technischen Universität Dresden, Dresden, Germany

Mirko Corselli BD Biosciences, La Jolla, CA, USA

Sylvie Dufour Institut Curie/CNRS UMR144, Paris, France; INSERM U955, IMRB, Equipe 6, Créteil, France; Faculté de Médecine, Université Paris Est, Créteil, France

Nil Emre BD Biosciences, La Jolla, CA, USA

Christine A. Fargeas Tissue Engineering Laboratories (BIOTEC), Medizinische Fakultät der Technischen Universität Dresden, Dresden, Germany

Ana Fiszbein Laboratorio de Fisiología y Biología Molecular, Departamento de Fisiología, Biología Molecular y Celular, IFIBYNE-CONICET, Facultad de Ciencias Exactas y Naturales, Universidad de Buenos Aires, Buenos Aires, Argentina

Talita Glaser Departamento de Bioquímica, Instituto de Química, Universidade de São Paulo, S.P., Brazil

Isaias Glezer Departamento de Bioquímica, Escola Paulista de Medicina, Universidade Federal de São Paulo, São Paulo, Brazil

Matthew T. Goodus Department of Neurology and Neuroscience, New Jersey Medical School, Rutgers University-New Jersey Medical School, Newark, NJ, USA

Robert Hermann Division of Molecular Neurobiology, German Cancer Research Center (DKFZ), Heidelberg, Germany

Yutaka Itokazu Department of Neuroscience and Regenerative Medicine, Medical College of Georgia, Georgia Regents University, Augusta, GA, USA; Charlie Norwood VA Medical Center, Augusta, GA, USA

József Jászai Institute of Anatomy, Medizinische Fakultät der Technischen Universität Dresden, Dresden, Germany

Henry J. Klassen University of California, Irvine, CA, USA

Alberto R. Kornblihtt Laboratorio de Fisiología y Biología Molecular, Departamento de Fisiología, Biología Molecular y Celular, IFIBYNE-CONICET, Facultad de Ciencias Exactas y Naturales, Universidad de Buenos Aires, Buenos Aires, Argentina

Aaron Lee Department of Physiology and Neurobiology, University of Connecticut, Storrs, CT, USA

Steven W. Levison Department of Neurology and Neuroscience, New Jersey Medical School, Rutgers University-New Jersey Medical School, Newark, NJ, USA

Enric Llorens-Bobadilla Division of Molecular Neurobiology, German Cancer Research Center (DKFZ), Heidelberg, Germany

Antoine Louveau Neuroscience Department, Center for Brain Immunology and Glia, University of Virginia, Charlottesville, VA, USA

Sujeivan Mahendram McMaster Stem Cell and Cancer Research Institute, McMaster University, Hamilton, Ontario, Canada; Departments of Biomedical Sciences and Surgery, Faculty of Health Sciences, McMaster University, Hamilton, Ontario, Canada

Ana Martin-Villalba Division of Molecular Neurobiology, German Cancer Research Center (DKFZ), Heidelberg, Germany

Nicole McFarlane McMaster Stem Cell and Cancer Research Institute, McMaster University, Hamilton, Ontario, Canada; Departments of Biomedical Sciences and Surgery, Faculty of Health Sciences, McMaster University, Hamilton, Ontario, Canada

Lisamarie Moore Department of Neurology and Neuroscience, New Jersey Medical School, Rutgers University-New Jersey Medical School, Newark, NJ, USA

Tanzila Mukhtar Department of Biomedicine, University of Basel, Mattenstrasse, Basel, Switzerland

Akiko Nishiyama Department of Physiology and Neurobiology, University of Connecticut, Storrs, CT, USA

Ágatha Oliveira Departamento de Bioquímica, Instituto de Química, Universidade de São Paulo, S.P., Brazil

Geoffrey W. Osborne The University of Queensland, Queensland Brain Institute/The Australian Institute for Bioengineering and Nanotechnology, Queensland, Australia

Jan Pruszak Institute of Anatomy and Cell Biology, University of Freiburg, Freiburg im Breisgau, Germany

Serge Rivest Faculty of Medicine, Department of Molecular Medicine, Neuroscience Laboratory, CHU de Québec Research Center, Laval University, Quebec, Canada

Christiana Ruhrberg Department of Cell Biology, UCL Institute of Ophthalmology, London, UK

Laura Sardà-Arroyo Departamento de Bioquímica, Instituto de Química, Universidade de São Paulo, S.P., Brazil

Ignacio E. Schor Laboratorio de Fisiología y Biología Molecular, Departamento de Fisiología, Biología Molecular y Celular, IFIBYNE-CONICET, Facultad de Ciencias Exactas y Naturales, Universidad de Buenos Aires, Buenos Aires, Argentina; European Molecular Biology Laboratory, Heidelberg, Germany

Sheila K. Singh McMaster Stem Cell and Cancer Research Institute, McMaster University, Hamilton, Ontario, Canada; Departments of Biochemistry and Biomedical Sciences, Faculty of Health Sciences, McMaster University, Hamilton, Ontario, Canada; Departments of Biomedical Sciences and Surgery, Faculty of Health Sciences, McMaster University, Hamilton, Ontario, Canada

Minomi K. Subapanditha McMaster Stem Cell and Cancer Research Institute, McMaster University, Hamilton, Ontario, Canada; Departments of Biochemistry and Biomedical Sciences, Faculty of Health Sciences, McMaster University, Hamilton, Ontario, Canada

Mathew Tata Department of Cell Biology, UCL Institute of Ophthalmology, London, UK

Verdon Taylor Department of Biomedicine, University of Basel, Mattenstrasse, Basel, Switzerland

Miguel Tillo Department of Cell Biology, UCL Institute of Ophthalmology, London, UK

Henning Ulrich Departamento de Bioquímica, Instituto de Química, Universidade de São Paulo, S.P., Brazil

Chitra Venugopal McMaster Stem Cell and Cancer Research Institute, McMaster University, Hamilton, Ontario, Canada; Departments of Biomedical Sciences and Surgery, Faculty of Health Sciences, McMaster University, Hamilton, Ontario, Canada

Jason G. Vidal BD Biosciences, La Jolla, CA, USA

Tamra Werbowetski-Ogilvie Regenerative Medicine Program, Department of Biochemistry & Medical Genetics and Physiology, University of Manitoba, Winnipeg, MB, Canada

Lissette Wilensky BD Biosciences, La Jolla, CA, USA

André Machado Xavier Departamento de Bioquímica, Escola Paulista de Medicina, Universidade Federal de São Paulo, São Paulo, Brazil

Takeshi Yagi KOKORO-Biology Group, Laboratories for Integrated Biology, Graduate School of Frontier Biosciences, Osaka University, Suita, Osaka, Japan; Core Research for Evolutional Science and Technology (CREST), Japan Science and Technology Agency, Japan

Robert K. Yu Department of Neuroscience and Regenerative Medicine, Medical College of Georgia, Georgia Regents University, Augusta, GA, USA; Charlie Norwood VA Medical Center, Augusta, GA, USA

Amber N. Ziegler Department of Neurology and Neuroscience, New Jersey Medical School, Rutgers University-New Jersey Medical School, Newark, NJ, USA

Foreword

Although cell-based therapy for treating neurological disorders is in its infancy, recent advances in iPSC-based technology and our ability to make multiple kinds of neurons and regional specific glia suggest that this is likely to change. In addition, the ability to obtain large quantities of defined cell types from hundreds of individuals both normal and those afflicted by a particular genetic disease allows one to consider designing elegant screens.

In both of these types of applications it is critical that a defined population of cells that is homogenous in its characteristics is obtained. This has been difficult in many fields of stem cell biology as all our processes of differentiation lead to a mixed final population that is at best enriched for a desired phenotype. Much effort has gone into developing sorting and selection methods to accelerate both drug discovery and cell-based therapy.

This book *Neural Surface Antigens*, edited by Dr Jan Pruszak as one of the pioneers in this area, focuses on functionally characterizing and identifying cell surface antigens for biomedical applications. The articles by a knowledgeable panel of international authors have been carefully selected based on our understanding of nervous system development where cell surface antigens are used to segregate developing cell populations and as such are uniquely expressed both spatially and temporally. Covering neuronal as well as glial cell types, separate chapters are devoted to various surface antigens including adhesion molecules (e.g., NCAM, integrins), representatives of transmembrane receptor signaling (e.g., CD95, toll-like receptors, neurotrophins), semaphorins and other glycoproteins, proteoglycans as well as glycolipids. Additional chapters are devoted to the process of cell selection and the associated concepts and technologies required with a particular focus on flow cytometry.

I believe this book will serve as a valuable reference to the novice and expert alike. It provides a context to why and how surface antigens may be chosen as markers and also describes their biological function in regulating cellular interdependencies in neural development, cancer, and stem cell biology. While there are books on individual molecules and books on techniques, an integrated compilation such as this one is not available and may well set an example for other fields of translational stem cell biology. I hope the readers will find this collection as useful as I and my laboratory did.

Baltimore, December 2014
Mahendra Rao MBBS, PhD
V.P Strategic Affairs, Q therapeutics,
SLC, UT 84,108
&
VP Regenerative Medicine,
New York Stem Cell Foundation,
New York, NY 10,032

Preface

Recent progress in stem cell research has begun to transform concepts and applications in biology and medicine. Beyond instilling hope and high expectations with respect to cell therapeutic measures, personalized medicine, cancer eradication, and human cellular model systems in the near future, this rapidly developing field has begun to unveil the intricacies of phenotypic plasticity in development, tissue homeostasis, and disease.

In the context of our own research in neural stem cell biology and neuroregeneration, a major obstacle to translational progress has been the inability to precisely mimic in the dish the faithful development of cells exclusively toward the phenotype of interest: the equivalent of a particular physiological cell type in need of being replaced or of being studied in biomedical in vitro assays and screens. To eliminate confounding contaminants of unwanted cells and to isolate specific subsets of cells, stem cell scientists have begun to revert to flow cytometric and other cell isolation methods based on neural surface antigens. Along with that has come a quest for novel markers and marker combinations to better define the target population.

Parts of these efforts may yield a surface antigen marker "tree" for neuropoiesis, a definition of neural developmental stages and phenotypes by neural surface antigens, analogous to the well-established hematopoietic lineage analysis. As opposed to the "fishing approaches" of earlier times, today's high-throughput screening approaches imply an exhaustive, comprehensive analysis of surface molecules expressed on neural cell populations. In that context it becomes humbling to be made aware of the sheer complexity of possibilities that biology provides by the dynamics of posttranslational modifications, membrane trafficking, and conformational changes of these molecules and the introduction of numerous splice variants—features that may not only correlate with, but also contribute to explaining the complexity of the nervous system.

Beyond description, the real fun starts with the functional implications and effects of such differential surface antigen expression. While the implications are immediately apparent in fundamental neural cell biology, neural development, and neuro-oncology alike, what determines an individual cell's decision to develop in a microcontext appropriate manner has remained unanswered. Which mechanisms govern a cell's decision to grow or to differentiate? The improved understanding of surface antigens and their signaling pathways lies at the heart of this exciting and important challenge. "All" inputs to a particular cell are mediated by the molecules presented on its surface. A cell senses its position in the world via the differential composition of molecules expressed on its outer membrane. Surface molecules comprise growth factor receptors, adhesion molecules and cell–cell interaction proteins. Biochemically, they include glycoproteins and glycolipids, channels, and immunoglobulin superfamily members. They can be membrane-spanning, GPI-anchored or extrinsic and may themselves be cleaved off, secreted, and act as long-range signaling molecules. Some may be more prominent on different subsets of neurons, others on glia, and/or on transformed cells of either lineage. The selected expert contributions from leading authorities working on neural surface antigens in the fields of neural stem cell biology, neurodevelopment and cancer presented in this volume for the first time explore and cover this topic for the neural lineage. It is targeting researchers ranging from student-level to experienced investigators in cellular neurobiology and biomedicine.

The book is divided into three parts. The first (Chapters 1 and 2) covering fundamentals that may prepare the readership from various backgrounds and fields of specialization for the remainder of the volume. The second section (Chapters 3–13) dealing with particular subsets of surface antigens and family of molecules largely from a fundamental biological perspective. And the final part (Chapters 14–18) focusing in on biomedical applications when exploiting surface molecules as markers. The concluding Chapter 19 represents an attempt to synthesize and integrate these components and to provide an outlook on future challenges and opportunities in exploring neural surface antigens in basic biology and biomedical applications.

Unique to this book is its intention to serve as an integrator at multiple levels, across particular surface molecule families, encouraging to explore and to identify commonalities in between researchers working in disparate fields. It also demands and provides justification for an overview, bird's eye view perspective of neural surface antigens (transcriptome, proteome, "surfaceome"), and the development of analogous analytical tools for computational, large-scale readout of presence and cellular effects of neural surface antigens.

As the editor, I am indebted and thankful to all contributors, and I am incredibly pleased to witness such a diverse project come to fruition. I thank Christine Minihane and Shannon Stanton at Elsevier for proposing the book and for their overall editorial support from the publisher's side throughout the project. Together with my coauthors, I thank the readers for using this book, for applying its concepts and approaches to their own particular research questions and for continuous discourse toward refinement of an integrated functional understanding of neural surface antigen dynamics and signaling.

Jan Pruszak
Freiburg, 2015

Chapter 1

Fundamentals of Neurogenesis and Neural Stem Cell Development

Robert Beattie*, Tanzila Mukhtar* and Verdon Taylor

Department of Biomedicine, University of Basel, Mattenstrasse, Basel, Switzerland

1.1 NEURULATION: FORMATION OF THE CENTRAL NERVOUS SYSTEM ANLAGE

During the early stages of postgastrulation embryonic development, the ectoderm differentiates to form the epidermis and the neural ectoderm, the primordium of the nervous system (for review see Ref. [1]). In vertebrates, the central nervous system (CNS) begins as the neural plate, an ectodermal-derived structure that folds dorsally to form the neural tube through a process called neurulation. Neurulation is divided into the sequential phases of primary and secondary neurulation initiated through a combination of growth factors and inhibitory signals secreted by the underlying axial mesoderm (notochord), dorsal ectoderm, and Spemann organizer (Figure 1.1). The neural tube then differentiates rostrally into the future brain and caudally to form the spinal cord and most of the peripheral nervous system, which will not be covered here. The rostral part of the neural tube segregates into three swellings, establishing the forebrain, midbrain, and hindbrain. In parallel, the rostrocaudal tube is segmented into modules called neuromeres.

During neurulation, neural crest cells (NCCs) are formed at the neural plate border, a junction between the surface ectoderm and the most dorsal neurepithelium. NCCs are unique to vertebrates, and induction of NCCs begins in mammals during embryogenesis in the midbrain and continues caudally toward the tail [2,3]. Initially, NCCs are an integral part of the neurepithelium and are morphologically indistinguishable. Upon induction, NCCs delaminate from the lateral neural plate/dorsal neural tube and migrate throughout the embryo. Various classes of NCCs include cranial, cardiac, vagal, trunk, and sacral, all of which have unique migration patterns. NCCs give rise to the majority of the peripheral nervous system and the bone and cartilage of the head; they also generate smooth muscle cells and pigment cells. In avians, fish, and amphibians,

NCC delamination requires cytoskeletal and cytoadhesive changes brought on by key transcription factors from the Snail gene family. Snail1 and Snail2 directly repress E-cadherin, which facilitates cell migration [2]. So far no such correlation has been identified during mammalian embryogenesis. The transcription factor Smad-interacting protein 1 is known to downregulate E-cadherin expression and is required for correct delamination of NCCs [2,6]. Because NCCs have both multipotent and self-renewing capabilities, it is hypothesized that they comprise a heterogeneous population of progenitors, each of which specifies a distinct cell type in the body [7]. Alternatively, NCC differentiation could be guided by intrinsic cues or extrinsic signals emanating from the tissues they interact with during migration [2,6]. For example, the role of extrinsic fibroblast growth factor (FGF) signaling has been demonstrated in determining the specific fate of craniofacial mesenchyme [2]. Because NCCs have many of the hallmarks of early stem cell progenitors, they may be interesting candidates for studying tissue engineering and regenerative medicine in the future. For a detailed review, please refer to [2,3,6].

1.2 NEURULATION AND NEURAL TUBE FORMATION

The mammalian brain and most of the spinal cord are formed during the first phase of neurulation, which is commonly divided into four phases. In mice, neurulation begins at around embryonic day (E) 8 with the induction of the neural plate when the inhibitory signals chordin, noggin, and follistatin are secreted by the Spemann organizer. These factors block bone morphogenic protein 4 (BMP4) signaling, inducing dorsal epiblast cells and allowing the anteroposterior midline of the ectoderm to adopt a neuroectodermal fate. These neuroectodermal cells undergo an apicobasal thickening and generate the neural plate along the

---

* Equal contribution.

Neural Surface Antigens. http://dx.doi.org/10.1016/B978-0-12-800781-5.00001-3

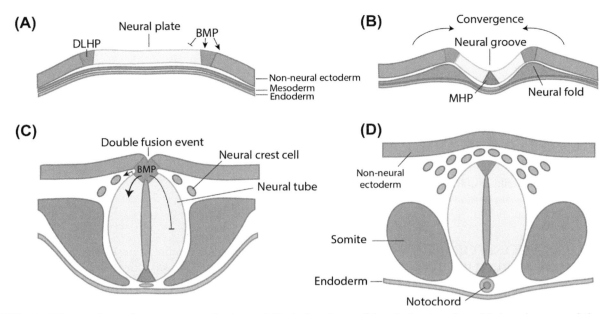

FIGURE 1.1 Schemes of central nervous system development. The brain and most of the spinal cord are formed during primary neurulation, which is commonly divided into four phases. (A) Epiblast cells are induced to a neuroectoderm fate, generating the neural plate. (B) The remodeling phase, in which the neural plate undergoes convergent extension and begins to fold along the median hinge point (MHP) and dorsolateral hinge points. (C) The two neural folds converge at the midpoint and then proceed to fuse, leading to the dorsal closure of the neural tube. During neurulation, neural crest cells (NCCs) are formed at the neural plate border, a junction between the surface ectoderm and the most dorsal neurepithelium. NCCs are unique to vertebrates, and induction of NCCs begins in mammals during embryogenesis in the midbrain and continues caudally toward the tail [2,3]. (D) By embryonic day 9 in the mouse, fusion is complete. BMP—bone morphogenic protein. *Adapted from Refs [4,5].*

dorsal midline of the embryo. Once committed, neuroecto-dermal cells no longer require inhibitory signals for neural plate formation to proceed (Figure 1.1) [8,9].

The neural plate undergoes a remodeling phase, whereby convergent extension increases the length (rostrocaudally) and narrows the width (transversely) simultaneously. During these processes, the neural plate continues to thicken apicobasally, generating cellular forces that begin to bend the neural plate and induce neural tube formation. As the lateral folds of the neural plate converge to the midline, the epidermal ectoderm delaminates from the neurepithelium of the neural plate, and fusion of both the ectoderm and the dorsal neural tube proceeds [8,9]. The neural tube zips closed posteriorly from the hindbrain and anteriorly from the midhindbrain junction, while remaining open over the future fourth ventricle posterior to the cerebellum. By E9 in the mouse, fusion is complete and the neural tube is closed, forming the primitive ventricles of the future brain regions.

Far less is known about secondary neurulation, which is the formation of the posterior region of the neural tube and caudalmost portion of the spinal cord. Secondary neurulation begins from a solid mass of cells forming from the tail bud. These cells form the medullary cord, which then cavitates to form multiple lumina. Finally, these lumina fuse into a single lumen, continuing the central canal of the neural tube in the most rostral aspects. In contrast to primary neurulation, here the process is more a hollowing

out of a mass of cells rather than tube formation from an ectodermal plate of cells [10].

1.3 REGIONALIZATION OF THE MAMMALIAN NEURAL TUBE

1.3.1 Molecular Basis of Regionalization

The neurepithelium of the neural tube follows a sequential series of overlapping and competing patterning steps during brain development. Timing is critical, particularly in structures such as the cerebral cortex, where even moderate changes in gene expression pattern can lead to serious developmental, motor, behavioral, psychological, and cognitive disorders. The best characterized morphogens and signaling pathways involved in regional identity include Sonic hedgehog (Shh), retinoic acid (RA), FGF, wingless (Wnt), and BMP signaling (Figure 1.2) [11,12]. Shh is secreted by the notochord (axial mesoderm) beneath the floor plate of the neural tube and controls neuronal cell fate in a concentration-dependent manner [13]. RA is secreted from the mesoderm and defines the posterior CNS, including the hindbrain and spinal cord. RA contributes to segmentation of the hindbrain into eight distinct compartments called rhombomeres, which later give rise to the medulla, pons, and cerebellum. FGF activity along with RA and Wnt leads to the caudalization of the neural tissue [14,15]. Wnt

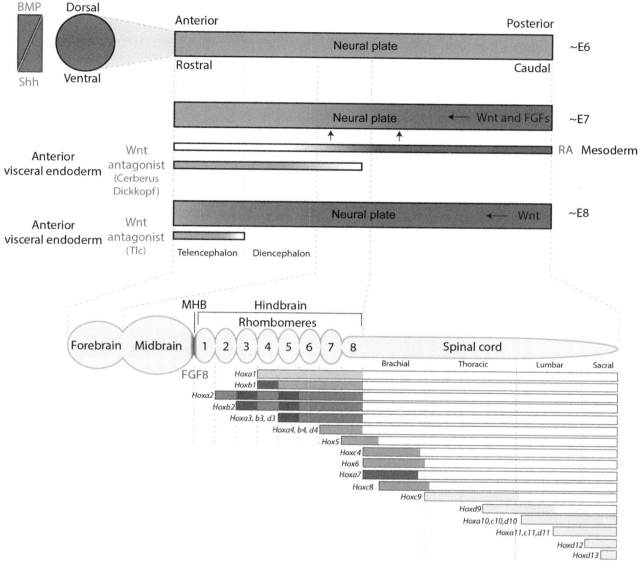

FIGURE 1.2 **Regionalization during neural tube formation is dependent on overlapping agonistic and antagonistic morphogen gradients.** Dorsoventral patterning of the neural tube is largely dependent on bone morphogenic protein (BMP) and Sonic hedgehog (Shh) signaling. Some of the key factors involved in patterning the anteroposterior axis include wingless (Wnt) and its antagonists (Cerberus, Dickkopf, Tlc), fibroblast growth factor (FGF), and retinoic acid. Distribution of these factors leads to the eventual segmentation of the neural tube into the forebrain, midbrain, hindbrain, and spinal cord. FGF8 expression delineates the MHB. Additionally, the Hox family of genes, located on four different chromosomes (HoxA, HoxB, HoxC, and HoxD), is crucial in spatiotemporal patterning of the neural tube. *Hox1–Hox5* are responsible for hindbrain segmentation, and *Hox4–Hox11* are involved in patterning of the spinal cord. MHB—midbrain–hindbrain boundary. *Adapted from Refs [11,21–25].*

signaling is crucial in the development of the neural tube, particularly in establishing anteroposterior polarity. Several Wnt antagonists, including Cerberus, Dickkopf, and Tlc, are important in patterning the dorsal telencephalon [16–20]. Diffusion of BMPs and their antagonists along the neural plate creates a gradient of high BMP activity dorsally to low activity ventrally. This leads to the specification of distinct pools of progenitors in the dorsal spinal cord [4,12].

Additionally, the Hox gene family of homeodomain-containing transcription factors is highly conserved across vertebrates and plays a key role in body patterning [22]. The majority of the 39 Hox genes found throughout vertebrates are expressed in the CNS where they play crucial roles in neuronal specification and selectivity. Hox genes are organized into clusters (HoxA, HoxB, HoxC, and HoxD) on four different chromosomes and exhibit a 3′–5′ gradient of sensitivity to RA. *Hox1–Hox5* (like RA) are involved in hindbrain segmentation into rhombomeres. *Hox4–Hox11* are expressed in the spinal cord and lead to rostrocaudal positioning of neuronal subtypes (Figure 1.2) [23,24].

1.3.2 Structural Organization of Cellular Compartments and Boundaries in the Developing Neural Tube

As the neural tube progressively becomes more regionalized, the organization of distinct structural domains arises. Segmentation of the neural tube in the mouse begins initially by assigning anterior–posterior identity along the neuraxis, dividing into the forebrain, midbrain, hindbrain, and spinal cord. The hindbrain (or rhombencephalon) is further divided into rhombomeres which give rise to the metencephalon (the pons and the cerebellum) as well as the myelencephalon (the medulla oblongata). The midbrain (or mesencephalon) is located caudal to the hindbrain and rostral to the forebrain. The forebrain (or prosencephalon) divides into the diencephalon (prethalamus, thalamus, hypothalamus, subthalamus, epithalamus, and pretectum) and the telencephalon (cerebrum) (Figure 1.2). The cerebrum can be further divided into the cerebral cortex, the basal ganglia, and the limbic system (Figure 1.2). For a full review of the cellular compartments and boundaries in vertebrate brain development see Kiecker and Lumsden [25].

1.4 ONSET OF NEUROGENESIS IN THE TELENCEPHALON

The mammalian neocortex modulates processing of sensory information and motor activity and mediates cognition. The isocortex formation of the cerebral cortex develops in an inside-out temporal fashion and comprises six histologically distinct neuronal layers. These layers differ in neuronal composition, connectivity, and density. The earliest born neurons populate the deep layers (VI and V), and the later born neurons migrate past the deep layer neurons to form the upper layers (IV, III, and II) of the future cerebral cortex (see later sections). Diverse neuronal subtypes that contribute to the complex neural circuitry are specified by a multitude of factors. Much progress has been made toward understanding the molecular pathways and mechanisms controlling neuronal cell-type diversity in the cortex. However, detailed mechanistic knowledge of the interplay between the transcriptional networks and upstream factors has yet to be elucidated [26].

1.5 THE TRANSITION OF THE NEUREPITHELIUM TO NEURAL STEM CELLS

Neurogenesis is composed of an orchestrated series of cellular events that include proliferation, fate commitment, differentiation, maturation, expansion, migration, and functional integration of newborn neurons into neuronal circuits. In the developing mouse CNS there are at least two distinct classes of progenitor cells, the apical progenitors (APs) and the basal progenitors (BPs) (Figure 1.3). The APs include neuroepithelial progenitors (NEPs), which generate radial glial cells (RGCs), and short neural precursors, all of which have stem cell character [27–30]. By E9, the neurepithelium is a single layer of NEPs, which form the pseudo-stratified neurepithelium. Owing to the displacement of the cell body (karyon) of the NEPs during the cell cycle, the ventricular zone resembles a multilayered structure but it is actually a pseudo-stratified single-cell epithelium. The migration of the nucleus (karyon)

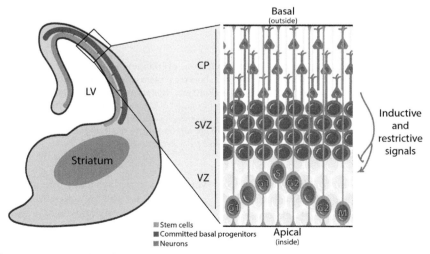

FIGURE 1.3 **Scheme of a coronal hemisection of the developing mouse telencephalon and the stem and progenitor populations.** As neurogenesis continues, neural stem cells (NSCs) retain contact with the outside of the neural tube and their apical end feet line the tube, resulting in long polarized processes. NSCs undergo interkinetic nuclear migration during the cell cycle. DNA replication (S phase) always takes place when the cell body reaches the ventricular (VZ)–subventricular zone (SVZ) boundary, mitosis (M) and karyokinesis take place at the luminal surface (apical) of the neural tube. Committed progeny of the NSCs, basal progenitors, migrate to the SVZ where they may divide before differentiating into immature neurons that migrate to the superficial layers of the forming cortical plate (CP) and future cerebral cortex.

along the apicobasal process during the cell cycle is referred to as interkinetic nuclear migration and is cell cycle dependent. Mitosis occurs at the apical side of the cell at the lumen of the neural tube, whereas S phase takes place at the basal boundary of the ventricular zone, and G1 and G2 occur during directed migration of the nucleus (Figure 1.3) [31,32]. As NEPs and RGCs transition from symmetric proliferation to asymmetric neurogenic divisions during neurogenesis their cell cycle lengthens almost entirely due to lengthening of the G1 phase.

NSCs in the ventricular zone (VZ) of the neural tube connect with one another through tight and adherens junctions at their apical ends. The maintenance of cell polarity is dependent upon the adherens junctions and polarity is critical for NSC function [27,33]. Between E9 and E10 (before the onset of neurogenesis) NEPs maintain their radial morphology, but begin to exhibit astroglial hallmarks and downregulate tight junctions and other epithelial markers, ultimately

transforming into a more restricted distinct cell type called RGCs [28,34]. The nuclei of RGCs continue to migrate along the apical–basal axis during the cell cycle, but interkinetic nuclear movement becomes continually more restricted to the apical end of the extending basal process (Figure 1.3). By the time neurogenesis begins in the forebrain, between E10 and E11 in the mouse, RGCs start to upregulate markers characteristic of astroglia, including glutamate transporter, brain–lipid-binding protein (BLBP), glial fibrillary acidic protein (GFAP), and vimentin. Apical end feet of the RGCs remain anchored to one another through adherens junctions [35,36].

As development continues, a class of intermediate progenitors called BPs is formed. Unlike NEPs and RGCs, BPs do not have apical connections to the lumen of the neural tube but instead undergo a limited number of cell divisions in the subventricular zone (SVZ), a region basal and adjacent to the VZ (Figure 1.4) [37,38]. BPs in the SVZ upregulate

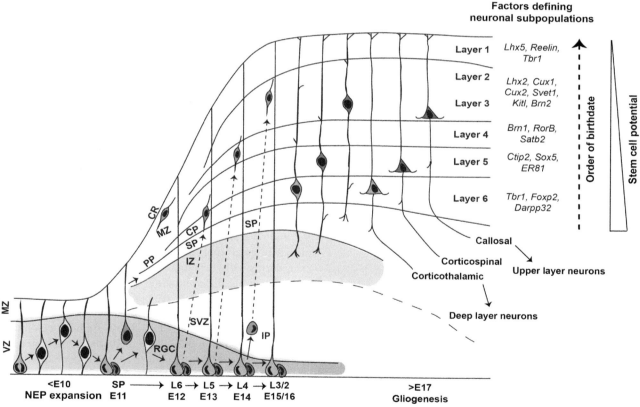

FIGURE 1.4 Neurogenesis and migration of neurons in the mouse cortex. Neural epithelial progenitors (NEPs) in the ventricular zone (VZ) of the developing telencephalon generate the many neuronal subtypes of the six-layered cerebral cortex, potentially starting as a homogeneous multipotent cell population that becomes fate restricted over time during neurogenesis. Before neurogenesis commences, NEPs undergo a series of symmetric divisions in the VZ, expanding the stem cell pool. As neurogenesis proceeds, the VZ NEPs transform into radial glial cells (RGCs) and generate basal progenitors (BPs), which populate the subventricular zone (SVZ). Newly formed neurons derived directly either from NSCs or from the BPs migrate radially outward forming the various cortical layers in an inside-out fashion. The first projection neurons populate the preplate (PP) forming the nascent cortical plate (CP). The CP later becomes layers 2 to 6 of the neocortex. CP neurons split the PP into the marginal zone (MZ) and subplate (SP). Each layer of the cerebral cortex is composed of different neuronal subtypes, which are generated sequentially throughout neurogenesis. Toward the end of neurogenesis the radial scaffolding of the RGCs is dismantled and RGCs become gliogenic, generating cortical and subependymal zone astrocytes and a sheet of ependymal cells lining the ventricles. Some of the key transcription factors used in defining neuronal subtypes are listed adjacent to their respective cortical layer. *Adapted from [42].*

the transcription factors cut-like homeobox 1 (Cux1), Cux2, and Tbr2, and although limited self-renewing divisions have been shown, they subsequently undergo symmetric differentiating cell divisions to generate two neurons [39–41].

1.5.1 Asymmetric versus Symmetric Cell Divisions

During cortical development, neural progenitors can undergo three modes of cell division. Before neurogenesis begins NEPs divide symmetrically, giving rise to two NEP daughter cells, allowing for rapid expansion of the progenitor pool. Later, NSCs can undergo asymmetric divisions, allowing for both self-renewal of the NSC and generation of a differentiated daughter cell [43,44]. The committed daughter cells are either a single neuron or a BP, which can undergo further cell divisions. RGCs act as a scaffold for the newborn neurons to migrate into the forming cerebral cortex. The third mode of cell division involves an amplification step at the BP stage, increasing the progenitor pool before finally differentiating into neurons. Because a single RGC can give rise to multiple BPs, and a single BP can give rise to two or more neurons, the SVZ is generally recognized as one of the main sites of amplification during neurogenesis [29,45,46]. Regulation of the number of RGCs that divide to give rise directly to neurons or BPs is crucial in controlling neurogenesis. Too many daughter cells differentiating directly into neurons results in overall neurogenesis being severely reduced owing to a lack of BP amplification. Although mitotic spindle orientation is not the only determinant, it has been shown to play a direct role in RGC daughter cell fate.

1.6 PROGENITOR FATE COMMITMENT AND RESTRICTION

A detailed understanding of the mechanisms that lead to the formation of multiple neuronal subtypes from a single population of neocortical stem cells is still lacking [47]. Two alternative models have been proposed to explain the process of temporal expansion and differentiation in the cortex. The "common progenitor" model proposes that NSCs restrict their fate temporally as neurogenesis progresses, sequentially generating neurons unique to each layer of the cerebral cortex. Alternatively, the "multiple progenitor" model proposes that NSCs are a heterogeneous pool at the outset, in which each NSC subtype would be guided by intrinsic and extrinsic signals to generate specific neuronal subtypes or astrocytes. Currently, there is evidence supporting both models [48].

1.6.1 The Common Progenitor Model

Heterochronic transplantation experiments performed in ferrets by McConnell and colleagues revealed that the

potential of NSCs is restricted over time. With age, NSCs become more defined in their fate, eventually losing the ability to generate deep-layer neurons [49,50]. Further supporting the common progenitor model, clonal analysis showed that neocortical NSCs generate deep- and upper-layer neurons in vitro in a sequential and temporal manner [51]. Additionally, retroviral lineage tracing experiments labeling NSCs in vivo support fate restriction of NSCs during development and NSC multipotency [52]. Fezf2, a transcription factor enriched in cortical layer 5 and important in fate specification and connectivity of subcerebral projection neurons, is expressed by NSCs throughout cortical neurogenesis [26,53,54]. Fate mapping experiments demonstrated that these Fezf2+ NSCs could sequentially generate both deep- and upper-layer neurons while becoming fate restricted over time [53]. Ectopic expression of Fezf2 directed the late cortical progenitors to differentiate into deeper-layer projection-like neurons, emphasizing its instructive role. Moreover, Fezf2 is expressed by NSCs as early as E8.5 in the pallial neurepithelium, suggesting its impact on fate determination [42,47,48].

1.6.2 Multiple Progenitor Model

Early evidence showed that several transcription factors are responsible for the fate determination of various neuronal subtypes. These factors and the onset of their expression during development imply different subsets of progenitors, which are predetermined and committed to generate specific neuronal subtypes [48]. These fate-restricted NSCs in the developing telencephalon express the transcription factors Cux1 and Cux2, both of which have been associated with differentiated and specific neuron subtypes in the cerebral cortex [55]. Cux1 and Cux2 are expressed in the VZ and SVZ abundantly during upper-layer neurogenesis, primarily specifying callosal projection neurons [55]. However, during early development Cux2+ NSCs proliferate and expand without differentiating. Later, when neurons of the superficial cortical layers are being generated, these NSCs and progenitors switch to a neurogenic mode and generate Cux2+ upper-layer neurons. These findings challenged the existing common progenitor model but left many questions unanswered. Subsequent lineage tracing experiments confirmed the presence of Cux2+ NSCs but suggested they generate both upper- and deep-layer neurons as well as interneurons derived from the ventral telencephalon [53]. The presence of multipotent NSCs expressing Fezf2 and Cux2 does not negate the possibility of the existence of fate-restricted progenitors, but additional single-cell analysis of fate and lineage will be required [50,53].

Other models in the field emphasize the presence of stem cells that are multipotent and switch their fate over the course of sequential rounds of cortical neurogenesis. This would suggest that NSCs would be initially committed to one fate during development and then switch to an alternate fate as corticogenesis proceeds. Multipotent NSCs could then generate multiple neuronal subtypes while still restricting their potential and eventually becoming unipotent. Further investigation of the mechanisms driving neurogenesis is crucial to understand NSC cell regulation [48].

1.7 MOLECULAR MECHANISMS OF NEURAL STEM CELL MAINTENANCE

1.7.1 Notch Signaling as a Key Regulator in Maintenance of NSCs

To maintain neurogenesis from the developing embryo into adulthood, NSCs must be able to self-renew. One of the best-studied signaling pathways shown to be involved in NSC maintenance, proliferation, quiescence, and survival is the Notch pathway [56–62]. Notch receptors are type 1 transmembrane proteins, which can be activated through extracellular protein–protein interactions with ligands of either the Delta or the Serrate (Delta-like and cluster of differentiation antigen CD339 or Jagged, respectively, in mammals) family on adjacent cells. Upon activation receptors undergo sequential proteolytic cleavage, first by a disintegrin and metalloprotease and then by a presenilin containing γ-secretase, releasing the intracellular domain of Notch (NICD) [63,64]. Canonical Notch signaling is mediated by the interaction of nuclear-translocated NICD with the CSL transcriptional complex (RBP-J in mice) (Figure 1.5). This interaction disrupts the preformed repressor complex and switches it to an activator by recruiting Mastermind and chromatin-modifying agents (i.e., histone acetyltransferase) to induce target gene expression [65–70]. The best-studied targets of the Notch pathway in mammals are the orthologs of hairy/enhancer of split (Hes/Hey). The direct canonical Notch targets, *Hes1* and *Hes5,* are two of these basic helix–loop–helix (bHLH) transcriptional regulators and are critical for neural development [71]. Hes1 and Hes5 directly repress transcription of proneural genes including *Ascl1* (Mash1), *Atoh1* (Math1), and *Neurog2* (Ngn2), thereby maintaining NSCs in a progenitor state [62,71]. Conversely, inactivation of Notch results in upregulation of the proneural genes and neural progenitor differentiation [61,72,73]. Manipulating the Notch signaling pathway using γ-secretase inhibitors, by ablating *RBP-J,* by knocking out individual members of the Notch family, or by expressing an activated NICD showed that Notch is key in modulating progenitor cell proliferation and neurogenesis during embryonic development [61,72,73].

The classic "lateral inhibition" model of Notch signaling in NSCs proposes that all early progenitors express similar levels of proneural genes and Notch ligands. Then through stochastic variations, the levels of receptors, ligands, and proneural genes fluctuate between adjacent cells, resulting in a "salt-and-pepper" pattern of Notch component gene expression. Cells with slightly higher ligand levels activate receptors in neighboring cells, causing an inhibition of proneural genes in those cells. Differences in the gene expression profiles of neighboring cells continue to be exacerbated and eventually lead to the lineage commitment of the high proneural gene-expressing cell. Real-time imaging in Hes reporter mice showed that negative feed-forward and feedback loops exist, resulting in oscillatory expression of downstream Notch signaling components and their targets over time, which are independent and not linked to cell cycle phases [60,74,75]. Therefore, a cell with high proneural gene expression at one time point may revert to a low proneural gene expression state shortly thereafter. Oscillations of Notch signaling in progenitors of the nervous system are analogous but not identical to the waves of Notch activity seen during somite formation. Oscillations in components of Notch may further alter the ability of NSCs to respond to external differentiation cues and be critical for regulating NSC potential [76]. Notch1 has been proposed to play a role in the maintenance of actively dividing NSCs in the adult neurogenic niche [77–81]. In the SVZ of the lateral ventricle wall and dentate gyrus (DG) of adult mice Notch activity promotes NSC survival and maintenance and stem cell self-renewal [77,78,80,82–84]. However, both the preservation of and the transition from a quiescent NSC state to an activated state appear to be RBP-J dependent [61,79,81,83].

Great efforts have been made over the years to identify molecular markers that discriminate populations of quiescent and activated NSCs from niche astrocytes; however, none have proven to be ideal [85,86]. Epidermal growth factor receptors have been associated with active SVZ NSCs that maintain astrocytic (BLBP) and glial (GFAP) markers, and Prominin-1 expression associates with NSCs, distinguishing them from parenchymal astrocytes [81,85]. In the adult DG horizontal, nonradial cells with active Notch signaling include a population of actively dividing NSCs [84]. However, there is also a population of quiescent horizontal DG NSCs that currently cannot be discerned based on molecular marker alone [84,87]. New genetic tools will need to be generated and markers identified that allow for independent and simultaneous lineage tracing of these two NSC populations. For an in-depth analysis of the role of Notch in quiescence and active NSC populations see Giachino and Taylor [88].

In the adult forebrain SVZ niche, NSCs receive inductive cues directing them to specific fates and restrictive signals, which limit their potential and prevent differentiation [89]. Some of these inductive cues most likely work in tandem with Notch [90]. Noncanonical activation of Notch through pigment epithelium-derived factor (PEDF) secreted by vascular endothelial cells within the adult lateral ventricle SVZ can bias cell fate toward RGC-like states. By activating nuclear factor κB, PEDF exports nuclear receptor corepressor, which

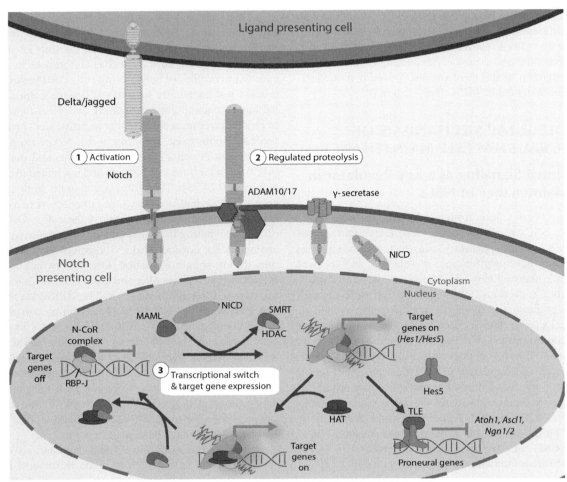

FIGURE 1.5 Canonical notch receptor signaling in the control of neurogenesis. Notch receptors and their ligands are type 1 transmembrane proteins. Notch receptor activation is triggered when either Delta or Jagged presented by neighboring cells binds to the ectodomain, resulting in regulated intramembrane proteolysis in which first a disintegrin and metalloprotease (ADAM10 or 17) and then a presenilin containing γ-secretase cleave the receptor, releasing a soluble intracellular domain (NICD). The NICD translocates to the nucleus where it interacts with the CSL (CBF1, Su(H), and Lag1—RBP-J in mice) protein complex including the DNA-binding protein recombining binding protein suppressor of hairless (RBP-J). The binding of NICD releases the nuclear receptor corepressor complex (N-CoR), which includes silencing mediator of retinoid and thyroid receptors (SMRT) and histone deacetylases (HDAC). The NICD-bound CSL complex is a positive regulator of Notch target genes including *Hes1* and *Hes5*. Hes5 is a basic helix–loop–helix transcription factor that, together with a zinc finger protein of the transducing-like enhancer of split (TLE) family, represses the proneural genes (*Atoh1*, *Ascl1*, and *Ngn1/2*) in NSCs and thereby inhibits neuronal differentiation. The NICD complex also interacts with a histone acetyltransferase (HAT), leading to epigenetic marking of target genes and transcriptional activation.

acts as a transcriptional inhibitor of the Notch target genes *Hes1* and *Egfr*, allowing for NSCs to undergo asymmetric, self-renewing divisions [91]. Other inductive cues include hypoxia-inducible factor 1α, which under hypoxic conditions is stabilized and cooperates with Notch signaling to promote expression of target genes by NSCs.

1.8 INTERNEURON GENERATION FROM THE VENTRAL TELENCEPHALON

Neuronal subtypes can be defined according to the neurotransmitters they secrete, which include γ-aminobutyric acid (GABA), acetylcholine, dopamine, and glutamine. In the

cerebral cortex, the excitatory glutamatergic cortical neurons and the inhibitory interneurons (i.e., GABAergic interneurons) mediate excitation and inhibition, respectively. Alterations in either population of neurons results in neurological and psychiatric disorders. The GABAergic interneurons are morphologically, physiologically, and neurochemically distinct from the glutamatergic excitatory neurons. As embryonic development proceeds, excitatory and inhibitory neurons mature and form synapses with one another to establish a complex cortical network. The distinct properties of the neuronal subtypes also aid in modulating the cortical output and plasticity of the cortex by creating local inhibitory networks. Disruption of the excitatory and inhibitory

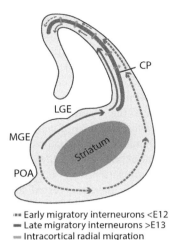

■■ Early migratory interneurons <E12
— Late migratory interneurons >E13
— Intracortical radial migration

FIGURE 1.6 Generation of interneurons in the mammalian forebrain. Interneurons (inhibitory neurons) are generated from the ventral telencephalon during embryonic development and migrate tangentially from the subpallium and integrate to the cortex. The subpallian sources of interneurons are the lateral ganglionic eminence (LGE), the medial ganglionic eminence (MGE), and the preoptic area/anterior endopeduncular (POA). Interneurons are divided into the early- and late-born populations, which adopt different tangential migratory pathways to the cerebral cortex. Upon reaching their mediolateral location in the dorsal telencephalon, the migratory interneuron neuroblasts migrate radially into the cortical plate (CP). *Adapted from Ref. [96].*

neuronal balance is implicated in neurological disorders in humans, such as schizophrenia, epilepsy, and autism [92].

Interneurons in a mouse are produced mainly between E11 and E17 [92]. In mice as well as in primates and humans interneurons originate from the ventral NSCs and progenitors residing in the medial ganglionic eminence (MGE), caudal ganglionic eminence, preoptic area, and anterior entopeduncular area of the subpallium (ventral telencephalon) (Figure 1.6) [92]. Following dorsal migration from the subpallium, the interneurons integrate into the cerebral cortex, in a sequential and temporal order similar to that seen in corticogenesis (Figure 1.6) [42]. The subpallial region expresses the transcription factors Dlx1 and Dlx2, which are essential for interneuron production, migration, and differentiation [92]. Other factors, including Nkx2.1 and Sox6, are expressed in the MGE and affect interneuron differentiation by controlling downstream transcriptional programs in NSCs and postmitotic neurons [92]. For a detailed review please refer to [93–95]. Because of the lack of markers or lineage tracing studies, little is known at the time of this writing about the molecular variations of single interneurons in different functional regions [26,92].

1.9 FORMATION OF THE CEREBRAL ISOCORTEX AND CORTICAL LAYERING

Upon induction of neurogenesis, numerous neocortical determinants are expressed along the dorsolateral wall of the telencephalon. Key factors, including LIM homeobox 2, forkhead box G1, empty spiracles homolog 2 (Emx2), and paired box 6 (Pax6), control the neocortical progenitor domains along the dorsal and ventral axis [26]. When the dorsal determinants Pax6 and Emx2 are ablated the ventral domain of the telencephalon is expanded [97]. It is speculated that Pax6 and Tlx regulate cell fate decisions in VZ NSCs, and loss-of-function studies show defects in the thickness of the superficial cortex [98–100]. Fezf2, Satb2, Ctip2, and Tbr1 have been identified as key NSC subtype markers, directing neuronal subtype specification (Figure 1.4) [26,48]. As neurogenesis proceeds, mature deep-layer neurons exhibit progressive postmitotic refinement of subtype identity, with layer-specific patterns of gene expression (Figure 1.4). At E13.5, postmitotic deep-layer neurons coexpress Ctip2 and Satb2. Later, neurons express either Ctip2 or Satb2 and become fate-restricted to subcerebral projection neurons and corticothalamic projection neurons, respectively [101,102]. This may indicate the presence of a more plastic state in which the neuronal cell fate is determined. Also, neuronal diversification and regional specialization result in increased sophistication of neural circuitry and determine functionally distinct areas. As of 2014, data suggest a progressive recruitment of transcription factors during cortical development, leading to a continually restricted neuronal fate [103]. Several inductive and repressive cues that regulate the regimental corticogenesis influence these factors that are expressed dynamically throughout development. In the future, lineage tracing of cell populations facilitated by surface molecular markers and cell-sorting approaches, single-cell analysis, and genome sequencing methods may provide additional insights into cortical patterning and specification. Additionally, assessing cortical plasticity and determining clear boundaries within subtypes during neurogenesis could pave the way for future therapeutic interventions [26].

1.10 OLIGODENDROGENESIS AND ASTROGENESIS

Glia cells carry out a diverse range of critical functions in the brain, including ensuring adequate nutrient supplies to neurons, providing scaffolding and support, insulating axons, removing cellular debris, and destroying pathogens. Oligodendrocytes compose one of the major types of glia cells and have the ability not only to provide support, but, when mature, also to myelinate and insulate axons, thereby ensuring proper signal transduction (reviewed in Ref. [104]). Differentiation of oligodendrocytes from early oligodendrocytic precursor cells (OPCs) into mature myelinating cells is a process termed oligodendrogenesis. Much like neurogenesis, oligodendrogenesis requires a complex network of morphogens, transcription factors, and signaling pathway cross talk for maturation to occur correctly. Some of the key signaling pathways involved include Wnt, Shh, BMP, and Notch [105].

Oligodendrocytes in the developing forebrain are produced in three sequential and competitive phases beginning in the embryo and continuing into early postnatal development [106]. Interestingly, oligodendrocytes from the first wave of oligodendrogenesis in the forebrain are almost completely absent from the postnatal brain, potentially eliminated or out-competed by later waves [106]. Oligodendrocytes begin in the developing neural tube as multipotent NEPs. Through sequential rounds of asymmetric cell divisions morphogen gradients of BMP and Shh restrict NEPs first to RGCs and then to OPCs [105]. OPCs then proceed to receive chemoattractant and mitogen signals instructing them to proliferate and migrate from the SVZ into the developing forebrain [105]. Once OLPs reach their destination, they proceed to integrate and differentiate, forming myelin and ensheathing axons.

Depending on the stage of oligodendrocyte maturation, oligodendrocytes express a range of markers. Key markers displayed by OPCs are platelet-derived growth factor receptor α (PDGFRα), Nkx2.2, and NG2, as well as the transcription factors Olig2 and Sox10 [107]. However, a variety of OPC populations exist, with some having limited or no expression of PDGFRα [108,109]. Conversely, all OPCs seem to express Sox10, and therefore Sox10 is generally accepted as an identifier of OPCs [106]. Later, mature myelinating oligodendrocytes express common markers such as myelin basic protein, myelin oligodendrocyte glycoprotein, and 2′,3′-cyclic nucleotide 3′-phosphodiesterase [107].

Another population of glial cells in the brain is the astrocytes that are important for neuronal function and are generated from the same pool of NSCs that gives rise to the neurons. Astrocytes are not just bystanders in brain function, they are also involved in the synaptic transmission and processing of neural circuits and modulate synapses and synaptic connectivity [110]. Additionally, astrocytes maintain a steady state in the CNS by modulating the ions, pH, neurotransmitter metabolism, and flow of blood [111]. At the onset of astrogliogenesis, stem cells can generate either BPs or astrocytes. BPs require the bHLH factor neurogenin-1 to inhibit the formation of astrocytes, whereas astrogliogenesis is promoted by the JAK/STAT signaling pathway [110].

1.10.1 Human Neocorticogenesis

The human cerebral cortex has expanded dramatically during phylogeny. Rodents have a smaller neocortex that lacks folding (lissencephalic), presenting limitations for studying the larger and highly folded (gyrencephalic) human neocortex. Human corticogenesis is characterized by the appearance of an enlarged SVZ that is split into an inner

SVZ (iSVZ) and an outer SVZ (oSVZ) by a thin fiber layer [112]. The increased neocortical surface area and volume in humans is associated with an expanded pool of progenitor cells in the oSVZ [113]. It has been proposed that the developing cerebral cortex has a columnar organization in which the newly born neurons migrate basally on a continuous radial fiber to the superficial layers. This results in the formation of radial columns of cells with related function [31,114]. The "radial unit hypothesis" integrates these concepts of cortical expansion and thalamic cues affecting size and cellular composition with neuronal function [103]. However, other studies have suggested that some lateral dispersion of clonally associated neurons exists, contrasting with the columnar organization model [31,115].

In the human brain, an increase in the number of neurons is achieved through three stages of extensive cellular expansion. In humans, cortical neuron production begins by gestational week (GW) 6. Subsequently the oSVZ develops, after GW11, and expands dramatically to become the main germinal region of the neocortex [113]. Compared to the NSCs and BPs in the rodent telencephalon, humans have additional progenitor pools including outer radial glia in the oSVZ that are defined by morphology and location. Outer radial glia numbers increase as they undergo multiple cell divisions and add to the BP pool. The cells in the oSVZ, like their VZ counterparts, also express nestin, vimentin, Pax6, and GFAP, with Tbr2 (a marker for BPs) being selective for oSVZ cells [116,117]. The oSVZ in the human dorsal cerebral cortex also contains a class of proliferating cells that express markers relevant to inhibitory neurons, including ASCL1, DLX2, NKX2-1, and calretinin, suggesting an expansion of immature interneurons migrating to their final location in the dorsal cortex [31,113]. Thymidine labeling experiments in primates show a relationship between the proliferation phases within the oSVZ and peaks of corticogenesis. This supports the hypothesis that the oSVZ and not the VZ is the main domain expanding during primate cortical development [103,118]. Experiments in ferrets, cats, and humans also revealed that with an increase in brain gyrification, there are more proliferating cells associated with the oSVZ than there are in the VZ/iSVZ.

Considering the dramatic differences between humans and rodents, various in vitro methods that recapitulate human brain development have been employed with great success. Birthdating studies by Gaspard et al., 2008, and Eiraku et al., 2008, demonstrated that mouse embryonic stem cells (ESCs) could be induced to undergo neurogenesis in vitro in a fashion similar to what is observed in the developing cerebral cortex [119–121]. These results were recapitulated in human ESCs, although with a slower induction of the neurogenic program, which may reflect the ontogenetic time frame of our species [119,120]. Subsequently, a fully stratified three-dimensional (3D) system for the generation of retinal tissue from mouse ESCs has been developed, which provides a valuable resource for developing therapeutics in cases of retinal degeneration [122]. With new technologies such as patient-derived iPS cells, the molecular basis and genetic mutations involved in neurodegenerative disorders may begin to be unraveled. Indeed, several studies have been able to model human corticogenesis in vitro using 2D culture systems, although a 3D system still remains a challenge [123–125].

As great as our advances in understanding the molecular biology of the developing brain have been in recent decades, there is still much that must be addressed before NSCs and iPS cells can be considered for therapeutic intervention. A deeper understanding of population-specific molecular markers, lineage relationships, and transcriptional profiles will contribute to developing new methods for the field of regenerative medicine.

REFERENCES

[1] Tam PP, Loebel DA. Gene function in mouse embryogenesis: get set for gastrulation. Nat Rev Genet 2007;8(5):368–81.

[2] Bhatt S, Diaz R, Trainor PA. Signals and switches in Mammalian neural crest cell differentiation. Cold Spring Harb Perspect Biol 2013;5(2).

[3] Green SA, Bronner ME. Gene duplications and the early evolution of neural crest development. Semin Cell Dev Biol 2013;24(2):95–100.

[4] Liu A, Niswander LA. Bone morphogenetic protein signalling and vertebrate nervous system development. Nat Rev Neurosci 2005;6(12):945–54.

[5] Jessell TM, Sanes JR. In: Kandel ER, Schwartz JH, Jessell TM, editors. Principles of neural science: the induction and patterning of the nervous system. 4th ed, New York: McGraw-Hill, Health Professions Division; 2000.

[6] Sauka-Spengler T, Bronner M. Snapshot: neural crest. Cell 2010; 143(3):486. 486 e1.

[7] Van de Putte T, et al. Mice lacking Zfhx1b, the gene that codes for Smad- interacting protein-1, reveal a role for multiple neural crest cell defects in the etiology of Hirschsprung disease-mental retardation syndrome. Am J Hum Genet 2003;72(2):465–70.

[8] Copp AJ, Greene ND, Murdoch JN. The genetic basis of mammalian neurulation. Nat Rev Genet 2003;4(10):784–93.

[9] Copp AJ. Neurulation in the cranial region–normal and abnormal. J Anat 2005;207(5):623–35.

[10] Shimokita E, Takahashi Y. Secondary neurulation: fate-mapping and gene manipulation of the neural tube in tail bud. Dev Growth Differ 2011;53(3):401–10.

[11] Rallu M, Corbin JG, Fishell G. Parsing the prosencephalon. Nat Rev Neurosci 2002;3(12):943–51.

[12] Lupo G, Harris WA, Lewis KE. Mechanisms of ventral patterning in the vertebrate nervous system. Nat Rev Neurosci 2006;7(2):103–14.

[13] Cohen M, Briscoe J, Blassberg R. Morphogen interpretation: the transcriptional logic of neural tube patterning. Curr Opin Genet Dev 2013;23(4):423–8.

[14] Tiberi L, Vanderhaeghen P, van den Ameele J. Cortical neurogenesis and morphogens: diversity of cues, sources and functions. Curr Opin Cell Biol 2012;24(2):269–76.

[15] Sansom SN, Livesey FJ. Gradients in the brain: the control of the development of form and function in the cerebral cortex. Cold Spring Harb Perspect Biol 2009;1(2):a002519.

[16] Ciani L, Salinas PC. WNTS in the vertebrate nervous system: from patterning to neuronal connectivity. Nat Rev Neurosci 2005; 6(5):351–62.

[17] Hebert JM, Fishell G. The genetics of early telencephalon patterning: some assembly required. Nat Rev Neurosci 2008;9(9):678–85.

[18] Houart C, et al. Establishment of the telencephalon during gastrulation by local antagonism of wnt signaling. Neuron 2002;35(2):255–65.

[19] Piccolo S, et al. The head inducer Cerberus is a multifunctional antagonist of Nodal, BMP and Wnt signals. Nature 1999;397(6721):707–10.

[20] Bafico A, et al. Novel mechanism of Wnt signalling inhibition mediated by Dickkopf-1 interaction with LRP6/Arrow. Nat Cell biol 2001;3(7):683–6.

[21] Maden M. Retinoic acid in the development, regeneration and maintenance of the nervous system. Nat Rev Neurosci 2007;8(10):755–65.

[22] Pearson JC, Lemons D, McGinnis W. Modulating Hox gene functions during animal body patterning. Nat Rev Genet 2005;6(12): 893–904.

[23] Akin ZN, Nazarali AJ. Hox genes and their candidate downstream targets in the developing central nervous system. Cell Mol Neurobiol 2005;25(3–4):697–741.

[24] Philippidou P, Dasen JS. Hox genes: choreographers in neural development, architects of circuit organization. Neuron 2013;80(1):12–34.

[25] Kiecker C, Lumsden A. Compartments and their boundaries in vertebrate brain development. Nat Rev Neurosci 2005;6(7):553–64.

[26] Molyneaux BJ, et al. Neuronal subtype specification in the cerebral cortex. Nat Rev Neurosci 2007;8(6):427–37.

[27] Gotz M, Huttner WB. The cell biology of neurogenesis. Nat Rev Mol Cell Biol 2005;6(10):777–88.

[28] Kriegstein AR, Gotz M. Radial glia diversity: a matter of cell fate. Glia 2003;43(1):37–43.

[29] Fishell G, Kriegstein AR. Neurons from radial glia: the consequences of asymmetric inheritance. Curr Opin Neurobiol 2003;13(1):34–41.

[30] Gal JS, et al. Molecular and morphological heterogeneity of neural precursors in the mouse neocortical proliferative zones. J Neurosci 2006;26(3):1045–56.

[31] Lui JH, Hansen DV, Kriegstein AR. Development and evolution of the human neocortex. Cell 2011;146(1):18–36.

[32] Taverna E, Huttner WB. Neural progenitor nuclei IN motion. Neuron 2010;67(6):906–14.

[33] Franco SJ, Muller U. Shaping our minds: stem and progenitor cell diversity in the mammalian neocortex. Neuron 2013;77(1):19–34.

[34] Malatesta P, Hartfuss E, Gotz M. Isolation of radial glial cells by fluorescent-activated cell sorting reveals a neuronal lineage. Development 2000;127(24):5253–63.

[35] Rasin MR, et al. Numb and Numbl are required for maintenance of cadherin-based adhesion and polarity of neural progenitors. Nat Neurosci 2007;10(7):819–27.

[36] Kuo CT, et al. Postnatal deletion of numb/numblike reveals repair and remodeling capacity in the subventricular neurogenic niche. Cell 2006;127(6):1253–64.

[37] Haubensak W, et al. Neurons arise in the basal neurepithelium of the early mammalian telencephalon: a major site of neurogenesis. Proc Natl Acad Sci U.S.A 2004;101(9):3196–201.

[38] Noctor SC, et al. Cortical neurons arise in symmetric and asymmetric division zones and migrate through specific phases. Nat Neurosci 2004;7(2):136–44.

[39] Nieto M, et al. Expression of Cux-1 and Cux-2 in the subventricular zone and upper layers II-IV of the cerebral cortex. J Comp Neurol 2004;479(2):168–80.

[40] Cubelos B, et al. Cux-2 controls the proliferation of neuronal intermediate precursors of the cortical subventricular zone. Cereb Cortex 2008;18(8):1758–70.

[41] Englund C, et al. Pax6, Tbr2, and Tbr1 are expressed sequentially by radial glia, intermediate progenitor cells, and postmitotic neurons in developing neocortex. J Neurosci 2005;25(1):247–51.

[42] Kwan KY, Sestan N, Anton ES. Transcriptional co-regulation of neuronal migration and laminar identity in the neocortex. Development 2012;139(9):1535–46.

[43] Noctor SC, et al. Neurons derived from radial glial cells establish radial units in neocortex. Nature 2001;409(6821):714–20.

[44] Miyata T, et al. Asymmetric inheritance of radial glial fibers by cortical neurons. Neuron 2001;31(5):727–41.

[45] Knoblich JA. Mechanisms of asymmetric stem cell division. Cell 2008;132(4):583–97.

[46] Kriegstein A, Alvarez-Buylla A. The glial nature of embryonic and adult neural stem cells. Annu Rev Neurosci 2009;32:149–84.

[47] Leone DP, et al. The determination of projection neuron identity in the developing cerebral cortex. Curr Opin Neurobiol 2008;18(1):28–35.

[48] Greig LC, et al. Molecular logic of neocortical projection neuron specification, development and diversity. Nat Rev Neurosci 2013; 14(11):755–69.

[49] Desai AR, McConnell SK. Progressive restriction in fate potential by neural progenitors during cerebral cortical development. Development 2000;127(13):2863–72.

[50] Han W, Sestan N. Cortical projection neurons: sprung from the same root. Neuron 2013;80(5):1103–5.

[51] Shen Q, et al. The timing of cortical neurogenesis is encoded within lineages of individual progenitor cells. Nat Neurosci 2006;9(6):743–51.

[52] Tan SS, Breen S. Radial mosaicism and tangential cell dispersion both contribute to mouse neocortical development. Nature 1993; 362(6421):638–40.

[53] Guo C, et al. Fezf2 expression identifies a multipotent progenitor for neocortical projection neurons, astrocytes, and oligodendrocytes. Neuron 2013;80(5):1167–74.

[54] Chen B, Schaevitz LR, McConnell SK. Fezl regulates the differentiation and axon targeting of layer 5 subcortical projection neurons in cerebral cortex. Proc Natl Acad Sci U.S.A 2005;102(47):17184–9.

[55] Franco SJ, et al. Fate-restricted neural progenitors in the mammalian cerebral cortex. Science 2012;337(6095):746–9.

[56] Androutsellis-Theotokis A, et al. Notch signalling regulates stem cell numbers in vitro and in vivo. Nature 2006;442(7104):823–6.

[57] Mizutani K, et al. Differential notch signalling distinguishes neural stem cells from intermediate progenitors. Nature 2007;449(7160):351–5.

[58] Ohtsuka T, et al. Hes1 and Hes5 as notch effectors in mammalian neuronal differentiation. EMBO J 1999;18(8):2196–207.

[59] Hitoshi S, et al. Notch pathway molecules are essential for the maintenance, but not the generation, of mammalian neural stem cells. Genes Dev 2002;16(7):846–58.

[60] Basak O, Taylor V. Identification of self-replicating multipotent progenitors in the embryonic nervous system by high notch activity and Hes5 expression. Eur J Neurosci 2007;25(4):1006–22.

[61] Imayoshi I, et al. Essential roles of notch signaling in maintenance of neural stem cells in developing and adult brains. J Neurosci 2010; 30(9):3489–98.

[62] Louvi A, Artavanis-Tsakonas S. Notch signalling in vertebrate neural development. Nat Rev Neurosci 2006;7(2):93–102.

[63] Brou C, et al. A novel proteolytic cleavage involved in notch signaling: the role of the disintegrin-metalloprotease TACE. Mol Cell 2000;5(2):207–16.

[64] Mumm JS, et al. A ligand-induced extracellular cleavage regulates gamma-secretase-like proteolytic activation of Notch1. Mol Cell 2000;5(2):197–206.

[65] Kurooka H, Kuroda K, Honjo T. Roles of the ankyrin repeats and C-terminal region of the mouse notch1 intracellular region. Nucleic Acids Res 1998;26(23):5448–55.

[66] Tamura K, et al. Physical interaction between a novel domain of the receptor notch and the transcription factor RBP-J kappa/Su(H). Curr Biol 1995;5(12):1416–23.

[67] Kato H, et al. Involvement of RBP-J in biological functions of mouse Notch1 and its derivatives. Development 1997;124(20):4133–41.

[68] Hsieh JJ, Hayward SD. Masking of the CBF1/RBPJ kappa transcriptional repression domain by Epstein-Barr virus EBNA2. Science 1995;268(5210):560–3.

[69] Dou S, et al. The recombination signal sequence-binding protein RBP-2N functions as a transcriptional repressor. Mol Cell Biol 1994;14(5):3310–9.

[70] Fiuza UM, Arias AM. Cell and molecular biology of Notch. J Endocrinol 2007;194(3):459–74.

[71] Hatakeyama J, et al. Hes genes regulate size, shape and histogenesis of the nervous system by control of the timing of neural stem cell differentiation. Development 2004;131(22):5539–50.

[72] Lütolf S, et al. Notch1 is required for neuronal and glial differentiation in the cerebellum. Development 2002;129(2):373–85.

[73] Gaiano N, Nye JS, Fishell G. Radial glial identity is promoted by Notch1 signaling in the murine forebrain. Neuron 2000;26(2): 395–404.

[74] Kageyama R, Ohtsuka T, Kobayashi T. The Hes gene family: repressors and oscillators that orchestrate embryogenesis. Development 2007;134(7):1243–51.

[75] Shimojo H, Ohtsuka T, Kageyama R. Oscillations in notch signaling regulate maintenance of neural progenitors. Neuron 58(1):52–64.

[76] Kageyama R, et al. Dynamic notch signaling in neural progenitor cells and a revised view of lateral inhibition. Nat Neurosci 2008; 11(11):1247–51.

[77] Ables JL, et al. Notch1 is required for maintenance of the reservoir of adult hippocampal stem cells. J Neurosci 2010;30(31):10484–92.

[78] Aguirre A, Rubio ME, Gallo V. Notch and EGFR pathway interaction regulates neural stem cell number and self-renewal. Nature 2010;467(7313):323–7.

[79] Basak O, et al. Neurogenic subventricular zone stem/progenitor cells are Notch1-dependent in their active but not quiescent state. J Neurosci 2012;32(16):5654–66.

[80] Nyfeler Y, et al. Jagged1 signals in the postnatal subventricular zone are required for neural stem cell self-renewal. EMBO J 2005;24(19):3504.

[81] Giachino C, et al. Molecular diversity subdivides the adult forebrain neural stem cell population. Stem Cells 2014;32(1):70–84.

[82] Basak O, Taylor V. Stem cells of the adult mammalian brain and their niche. Cell Mol Life Sci 2009;66(6):1057–72.

[83] Ehm O, et al. RBPJkappa-dependent signaling is essential for long-term maintenance of neural stem cells in the adult hippocampus. J Neurosci 2010;30(41):13794–807.

[84] Lugert S, et al. Quiescent and active hippocampal neural stem cells with distinct morphologies respond selectively to physiological and pathological stimuli and aging. Cell Stem Cell 2010;6(5):445–56.

[85] Pastrana E, Cheng LC, Doetsch F. Simultaneous prospective purification of adult subventricular zone neural stem cells and their progeny. Proc Natl Acad Sci U.S.A 2009;106(15):6387–92.

[86] Beckervordersandforth R, et al. In vivo fate mapping and expression analysis reveals molecular hallmarks of prospectively isolated adult neural stem cells. Cell Stem Cell 2010;7(6):744–58.

[87] Suh H, et al. In vivo fate analysis reveals the multipotent and self-renewal capacities of Sox2+ neural stem cells in the adult hippocampus. Cell Stem Cell 2007;1(5):515–28.

[88] Giachino C, Taylor V. Notching up neural stem cell homogeneity in homeostasis and disease. Front Neurosci 2014;8:32.

[89] Doetsch F. A niche for adult neural stem cells. Curr Opin Genet Dev 2003;13(5):543–50.

[90] Shen Q, et al. Endothelial cells stimulate self-renewal and expand neurogenesis of neural stem cells. Science 2004;304(5675):1338–40.

[91] Andreu-Agulló C, et al. Vascular niche factor PEDF modulates notch-dependent stemness in the adult subependymal zone. Nat Neurosci 2009;12(12):1514–23.

[92] Sultan KT, Brown KN, Shi SH. Production and organization of neocortical interneurons. Front Cell Neurosci 2013;7:221.

[93] Bartolini G, Ciceri G, Marin O. Integration of GABAergic interneurons into cortical cell assemblies: lessons from embryos and adults. Neuron 2013;79(5):849–64.

[94] Caputi A, et al. The long and short of GABAergic neurons. Curr Opin Neurobiol 2013;23(2):179–86.

[95] Southwell DG, et al. Interneurons from embryonic development to cell-based therapy. Science 2014;344(6180):1240622.

[96] Gao P, et al. Lineage-dependent circuit assembly in the neocortex. Development 2013;140(13):2645–55.

[97] Muzio L, et al. Conversion of cerebral cortex into basal ganglia in Emx2(-/-) Pax6(Sey/Sey) double-mutant mice. Nat Neurosci 2002;5(8):737–45.

[98] Caric D, et al. Determination of the migratory capacity of embryonic cortical cells lacking the transcription factor Pax-6. Development 1997;124(24):5087–96.

[99] Tarabykin V, et al. Cortical upper layer neurons derive from the subventricular zone as indicated by Svet1 gene expression. Development 2001;128(11):1983–93.

[100] Zimmer C, et al. Dynamics of Cux2 expression suggests that an early pool of SVZ precursors is fated to become upper cortical layer neurons. Cereb Cortex 2004;14(12):1408–20.

[101] Alcamo EA, et al. Satb2 regulates callosal projection neuron identity in the developing cerebral cortex. Neuron 2008;57(3):364–77.

[102] Srinivasan K, et al. A network of genetic repression and derepression specifies projection fates in the developing neocortex. Proc Natl Acad Sci U.S.A 2012;109(47):19071–8.

[103] Rakic P. Evolution of the neocortex: a perspective from developmental biology. Nat Rev Neurosci 2009;10(10):724–35.

[104] Pfeiffer SE, Warrington AE, Bansal R. The oligodendrocyte and its many cellular processes. Trends Cell Biol 1993;3(6):191–7.

[105] Li H, et al. Two-tier transcriptional control of oligodendrocyte differentiation. Curr Opin Neurobiol 2009;19(5):479–85.

[106] Kessaris N, et al. Competing waves of oligodendrocytes in the forebrain and postnatal elimination of an embryonic lineage. Nat Neurosci 2006;9(2):173–9.

[107] Fancy SP, et al. Myelin regeneration: a recapitulation of development? Annu Rev Neurosci 2011;34:21–43.

[108] Spassky N, et al. Sonic hedgehog-dependent emergence of oligodendrocytes in the telencephalon: evidence for a source of oligodendrocytes in the olfactory bulb that is independent of PDGFRα signaling. Development 2001;128(24):4993–5004.

[109] Spassky N, et al. Multiple restricted origin of oligodendrocytes. J Neurosci 1998;18(20):8331–43.

[110] Kanski R, et al. A star is born: new insights into the mechanism of astrogenesis. Cell Mol Life Sci 2014;71(3):433–47.

[111] Sofroniew MV, Vinters HV. Astrocytes: biology and pathology. Acta Neuropathol 2010;119(1):7–35.

[112] Smart IH, et al. Unique morphological features of the proliferative zones and postmitotic compartments of the neural epithelium giving rise to striate and extrastriate cortex in the monkey. Cereb Cortex 2002;12(1):37–53.

[113] Hansen DV, Rubenstein JL, Kriegstein AR. Deriving excitatory neurons of the neocortex from pluripotent stem cells. Neuron 2011;70(4):645–60.

[114] Rakic P. Neurons in rhesus monkey visual cortex: systematic relation between time of origin and eventual disposition. Science 1974;183(4123):425–7.

[115] Kaas JH. The evolution of brains from early mammals to humans. Wiley Interdiscip Rev Cogn Sci 2013;4(1):33–45.

[116] Zecevic N, Chen Y, Filipovic R. Contributions of cortical subventricular zone to the development of the human cerebral cortex. J Comp Neurol 2005;491(2):109–22.

[117] Bayatti N, et al. A molecular neuroanatomical study of the developing human neocortex from 8 to 17 postconceptional weeks revealing the early differentiation of the subplate and subventricular zone. Cereb Cortex 2008;18(7):1536–48.

[118] Lukaszewicz A, et al. G1 phase regulation, area-specific cell cycle control, and cytoarchitectonics in the primate cortex. Neuron 2005;47(3):353–64.

[119] Eiraku M, et al. Self-organized formation of polarized cortical tissues from ESCs and its active manipulation by extrinsic signals. Cell Stem Cell 2008;3(5):519–32.

[120] Gaspard N, et al. An intrinsic mechanism of corticogenesis from embryonic stem cells. Nature 2008;455(7211):351–7.

[121] Eiraku M, Sasai Y. Self-formation of layered neural structures in three-dimensional culture of ES cells. Curr Opin Neurobiol 2012;22(5):768–77.

[122] Eiraku M, Sasai Y. Mouse embryonic stem cell culture for generation of three-dimensional retinal and cortical tissues. Nat Protoc 2012;7(1):69–79.

[123] Espuny-Camacho I, et al. Pyramidal neurons derived from human pluripotent stem cells integrate efficiently into mouse brain circuits in vivo. Neuron 2013;77(3):440–56.

[124] Bershteyn M, Kriegstein AR. Cerebral organoids in a dish: progress and prospects. Cell 2013;155(1):19–20.

[125] Lancaster MA, et al. Cerebral organoids model human brain development and microcephaly. Nature 2013;501(7467):373–9.

Chapter 2

A Brief Introduction to Neural Flow Cytometry from a Practical Perspective

Geoffrey W. Osborne

The University of Queensland, Queensland Brain Institute / The Australian Institute for Bioengineering and Nanotechnology, Queensland, Australia

2.1 INTRODUCTION

Flow cytometry encompasses a range of standard techniques that, since their inception, have found widespread uptake in various fields of biology and medicine. The objective of this introductory chapter is to provide a brief summary of fundamental techniques and underlying principles of flow cytometry, and to outline the role it has played and can have in the investigation of neural cell types. This summative overview aims to encourage readers new to flow cytometry to exploit this powerful methodology in the pursuit of scientific inquiries in neurobiology.

2.2 WHAT IS FLOW CYTOMETRY?

Flow cytometry was initially developed in the 1960s [1,2] and 1970s, and excellent background material is available should the reader be interested in a historical perspective [3]. After flow cytometric instruments and antibodies became commercially available, the technology began to be widely used in a range of fields including immunology, hematology, and oncology. This process has continued until today with increasing refinement of the development of antibody probes and molecular techniques. Its implementation in neurobiology, however, has lagged behind, as traditional methods for the investigation of neural cells chiefly relied on histological techniques. Yet, flow cytometry offers a range of advantages; it should be regarded as a powerful additional tool in the neurobiologist's toolbox, and ways of integrating it with complementary technologies should be carefully considered.

Flow cytometers are analytical instruments that combine fluidics, optical components, electronics, and computer technologies (Figure 2.1) to provide information about the intracellular and extracellular characteristics of cells. The flow cytometer operates by forcing cell suspensions (sample) to travel in single file within a larger volume of fluid (sheath) through a process known as *hydrodynamic focusing*. These aligned cells then pass through an optical interrogation point, normally a laser beam spot of defined dimensions, where light signals are generated from each passing cell. These signals, which are subsequently collected by sensitive optical detectors and digitized, can have different characteristics:

- scattered light from the cell surface or from intracellular components;
- fluorescence light resulting from intrinsic intracellular or surface components (including autofluorescence or genetic reporter proteins);
- fluorescence resulting from fluorescent dyes or fluorochromes either binding directly and specifically to their target or directly coupled to antibodies or oligonucleotide probes.

The accumulated signal detected per individual cell reflects the sum total of the fluorescence emitted from the fluorophore, the intrinsic fluorescence of the cell (autofluorescence), and inherent systemic noise.

While the running of samples through a flow cytometer can be a simple task, the analysis and interpretation of the resulting data can be challenging for the novice. A simple structured approach can be helpful in this context.

Before running the experiment, clearly define the objective. For example, "I want to find the number of single cells that express marker X and are not dividing." Set up appropriate control samples that will aid in meeting the objective and generate interpretable data. Controls that are required are:

- Cells untreated (can be unstained or without treatment)
- Cells separately stained with each fluorophore/antibody of interest
- If cells are to be labeled with multiple fluorophores simultaneously, prepare controls that lack each of the fluorophores, i.e., fluorescence minus one (FMO) to characterize the impact of the presence or absence of each fluorophore in the mix.
- Use of isotype controls is optional

Neural Surface Antigens. http://dx.doi.org/10.1016/B978-0-12-800781-5.00002-5

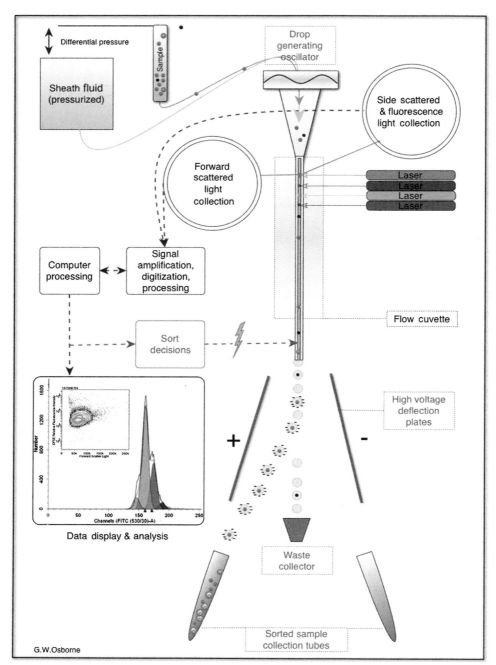

FIGURE 2.1 Schematic overview of the parts that are required to make a functional flow cytometer (black text) and additional requirements for sorting cells (blue text). Cells or particles in suspension (pink) travel in sheath fluid (blue), past lasers, where generated light signals are collected by detectors for processing and data display. Subsequently in cell sorters, sections of the sheath containing cells of interest are broken up and selectively defected and collected. See text for a complete explanation.

Use the controls to set the instrument, and apply compensation for spectral overlap of fluorophores (a topic comprehensively covered here: http://www.drmr.com); then on experimental samples. Carefully consider the total number of cells that need to be recorded [4] and ensure that subpopulations of cells contain sufficient numbers so that calculations performed are not overly biased due to low event numbers.

Analysis of flow cytometry data is becoming increasingly complex due to the large number of parameters recorded and the volumes of data now routinely generated. Do not become daunted by this, as there are many excellent publications [5–8] available to guide the researcher starting out in this area. In addition, a range of introductory articles on flow cytometric data analysis can be found online. In future, a range of automated data analysis programs [9–12] will pave the way for nonsubjective experiments analysis paradigms.

2.3 CHALLENGES AND OPPORTUNITIES OF NEURAL FLOW CYTOMETRY

To neurobiologists, the previous paragraphs illustrate one of the limitations of flow cytometry: the need for cells being in suspension. While blood cells naturally occur in suspension and are subsequently very well suited and rapidly prepared for cytometry assays, this is rarely the case for neural cell types under physiological conditions. In flow cytometry, it is required that the neural cells be dissociated from tissue samples [13] or must be detached from the plasticware in the case of neural cells growing in adherent culture. Thus, for most experiments, the selection of the correct dissociation method is absolutely critical for successful, reproducible, flow cytometric analysis of neural samples [14].

Once a cell suspension has been achieved, the strengths of flow cytometry compared to other similar techniques, such as microscopy, can be utilized. The key advantage that flow cytometry has over microscopy is its speed. Modern flow instruments can analyze conservatively at a rate of well beyond 10,000 cells per second, on a per-cell basis, so that the measurement is not a bulk measurement, as occurs, for example, with cells in a spectrophotometer cuvette. The immediate ramification of high sample throughput rate is twofold: (1) it allows the identification of infrequent cellular events, and (2) it provides enough of these rare events to render them statistically reliable. When analyzing cells derived from primary neural tissue, the samples are often characterized by large amounts of cellular debris resulting from tissue preparation procedures. High throughput speed in this scenario is important, as sufficient material can be analyzed within a reasonable timeframe to find the cells of interest among the bulk of debris. Other attributes of flow cytometry that make it attractive for use in the study of neural cells are that specific fluorescent signals, associated with molecules of interest in or on the cells, can be measured both qualitatively and quantitatively.

When coupled with fluorescence measurements, the speed of flow cytometry underlines another beneficial feature. In flow, excitation, emission, and detection of fluorescence occur over relatively short periods of time. Given that the total transit time through the laser is normally less than $10\,\mu s$, cells in flow are not normally subjected to photobleaching as frequently as in microscopy, due to the breaking of chemical bonds within the fluorochrome. Cells can then be identified, separated, and reanalyzed for confirmation of purity, based on the knowledge that usually the fluorochrome has not had time to photobleach during the short transit time through the instrument.

The ability to separate cells defined according to characteristic markers at the aforementioned rates in a highly specific manner is possibly the greatest advantage that flow cytometry has over other competing technologies, such as immunopanning [15] or magnetic separation [16,17]. While these latter methods may be capable of greater throughput, they lack the selectivity of flow cytometric sorting, for example, in cases where cells with an intermediate expression level of a fluorescent protein and an intermediate level of an antibody bound to the surface of the cell can be selected and separated ("low" vs. "high" expression of a surface antigen). When this selectivity is combined with the ability to multiplex numerous fluorophores—provided that spectral overlap can be controlled—the possibilities are unrivaled by any other technology.

Separation of cells by flow cytometry is normally known by the acronym FACS (fluorescence-activated cell sorting). Of note, cells can be sorted based on other criteria apart from fluorescence, such as scattered light, as previously mentioned. A précis of the sorting process is as follows: The liquid stream that carries the cells through the flow cytometer is subjected to mechanical vibrations such that the stream breaks up very precisely into droplets at a known distance from the laser interrogation point. Cells with desired characteristics are identified and cause an electrical charge to be applied to the liquid stream at the precise moment that a droplet containing the cell of interest breaks from the stream, resulting in a charged droplet traveling through the air. This charged droplet then passes between two metal plates that hold a fixed high voltage of opposing polarities. The charged particle is drawn toward the plate of the opposite polarity as it travels in the air, and is deflected into a collection vessel below the plates for further experimentation. Note that the charge is not applied directly to the cell; rather, the electrical charge is carried on the outside of the drop containing that cell, and cells can be sorted and remain viable. It is worth noting, however, that neural cells are often particularly fragile following dissociation, thus are more susceptible to the shear forces that occur with this electrostatic cell sorting process [18] than to the lower shear forces that may occur with other methods of separation. In spite of these caveats, FACS opens up many unique possibilities such as the ability to perform single-cell PCR on cells with particular known characteristics, and is increasingly used beyond hematology in numerous fields ranging from basic cell biology to clinical applications in biomedicine [19,20].

To successfully perform flow cytometry experiments on neural cells derived from culture or from primary tissue, careful consideration must be given to the cell preparation method. The selection of the appropriate dissociation buffer is critical for a successful experiment. For example, experience in the author's laboratory and that of others [14] indicates that certain cell-surface markers, such as the cluster of differentiation (CD) antigen CD24 are detrimentally affected by dissociation with papain, a common dissociation agent. Papain, trypsin [21] (and trypsin replacement),

Liberase-1 and Accutase are the most widely used solutions for dissociating cells of a neural type regardless of their source, and Panchision et al. provide a key manuscript in this area with detailed comparative analysis of dissociation properties, cell viability, and maintenance of surface epitopes [14,22]. A publication [23] at the time of writing indicates that Liberase removes markers for certain subsets on leukocytes, and this is worth considering in the context of related neural markers.

Other solutions are available from various commercial sources with different purported advantageous attributes that may well be worth trying if staining for the antigen of interest is unsuccessful using one of the more widely used agents. In essence, selection of the appropriate dissociation solution needs to be made on a case-by-case basis, with different surface molecules being susceptible to different agents depending on the tissue source and the antibody. In addition, it is worth remembering that surface antigen expression is highly dynamic, with some antigens continually trafficking to and from the cell surface as a normal part of maintaining cellular homeostasis. Other receptors may only be present during certain developmental stages, particularly when considering neurogenic regions such as the subventricular zone (SVZ) or olfactory bulb progenitor or neural stem cells (NSCs) [24]. Once the sample preparation has been empirically optimized for a given cell type, careful consideration must be given to the flow cytometry parameters that are to be used, particularly when sorting cells. Sorting flows cytometers (FACS) instruments frequently offer a range of adjustable settings. The recommended approach is to consider the parts of the cytometer that the sample will interact with as a "timeline" or experimental pipeline, and then set the instrument settings to provide the optimal conditions at each point in that timeline.

2.4 CELL SORTING OF NEURAL CELLS— STEP BY STEP

1. Initially, consider the sample tube that contains a carefully prepared sample. As a general rule of thumb, samples resuspended in polypropylene tubes show lower levels of sample loss due to adhering to plastic then those resuspended in polystyrene sample tubes. http://www.biocompare.com/Application-Notes/43253-Chemical-And-Thermal-Resistance-of-Polypropylene-Polystyrene-LDPE-HDPE-EVA-And-UV-Star/

2. Next, consider the medium in which your sample is resuspended, particularly if you are undertaking a cell sorting experiment for a relatively minor population of cells. The medium needs to provide an environment that maintains the cells in a "nonstressful" manner for what may be a period of hours while the sorting process occurs. Typically for flow cytometry sorting experiments that use electrostatic sorting, a phosphate-buffered saline solution containing 1–2% fetal calf serum would be considered a sample medium of choice for avoiding a refractive index mismatch between the sample media and the sheath (usually buffered saline) when interrogated by the laser excitation source. However, this may not be the best medium for maintaining your cells, and can be monitored in real time by the flow cytometer. This is done by using a viability indicator dye in the sample and calculating the number of nonviable cells at the start of the sorting experiment and then continually throughout the experiment. There is a wide range of viability dyes available, either DNA intercalating (propidium iodide, 7AAD, DRAQ7) or amine reactive (Zombie™). Choose one that is spectrally discrete from other fluorochromes used in the experiment to will make instrument setup and data interpretation much easier. An appreciable increase in the number of nonviable cells occurring over the course the experiment indicates that other suitable media, such as Roswell Park Memorial Institute (RPMI) media or Hank's buffered salt solution (HBSS) [25] should be tried.

3. Another factor is the sample temperature during the sorting process, as this can dramatically impact long-term cell survival following sorting. In our experience (but this is handled differently in different facilities), both sample and sort collection tubes give best results for neural samples when maintained at room temperature rather than at 4°C or 37°C. The original hypothesis behind keeping a sample at 4°C is that receptors that may rapidly cycle from the cell surface into the cytosol (internalization) will be limited in this action and that the surface receptor "snapshot" that our labeling with specific for recently tagged anybody's has captured would not be disrupted. In neural sample preparations, we see little or no evidence of this.

4. The next point in our virtual timeline to consider is the sheath pressure and sample differential pressure, the latter of which causes the sample to be pushed through the instrument. On analytic flow cytometers, the sheath pressure is normally fixed, with the possibility of slight variations in sample differential pressure. A general recommendation is to use the lowest possible differential pressure, as this leads to less variability of particle position in the laser focal spot and less heterogeneity in the collected resulting data. On cell sorting flow cytometers, again use the lowest sample differential pressure that can be used to move the sample through in a reasonably timely fashion. The downside of this is that cells sit for longer periods in the sample tube, where pH changes can occur, a phenomenon that can at least in part be offset by the addition of HEPES to the sample tube. This

can be optimized by resuspending the cells so the final volume does not exceed a concentration of 2×10^7 cells/mL. Concentrations greater than this tend to cause excessive sample clumping that perturbs the sample stream or clogs the system's sample intake lines. On cell sorters, low sheath pressures are also beneficial in increasing the survival of sorted neural cells. Low sheath pressures are those in the range 10–12 psi, (velocities of ~10 m/s, the equivalent of 60 kph). As the instrumentation is often optimized for the selection and separation of lymphoid cells, which are hardy and can withstand higher pressures, often the standard sheath pressures that FACS instruments are set to greatly exceed these levels—sometimes as high as 60 psi (25 m/s). Empirically, optimizing sheath pressures for sorting neural tissue derived from the murine brain, based on greatest numbers of viable cells yielded from sorting experiments and subsequent growth in the neurosphere assay [26], shows that at pressures greater than 28 psi there are measurable decreases in the number of viable cells. Additionally, it has been shown that lower system pressures are beneficial to survival of human stem-cell derived neurons [22]. For novices who are looking for some initial guidelines to be conveyed to a core facility engineer, the recommendation for initial parameters for a "jet-in-air" cell sorter may be in the range of 12 psi, with a 100- or 120-μm nozzle tip and a sample differential pressure of 0.5 psi as a starting point. For so-called cuvette-based cell sorters, the minimum pressure at which the systems can operate is often greater than jet-in-air cell sorters, thus the starting recommendation would be an operating pressure of around 20 psi with a 100-μm nozzle tip and a low sample differential setting of 10% or 20% of the absolute scale.

5. Coupled with the previously mentioned optimization of sheath pressure—the nozzle diameter, through which the sheath and sample pass prior to cells being interrogated at the laser interrogation point—is another important factor when trying to obtain healthy viable cells following FACS. Traditionally, 70-μm diameter nozzles were used, as this was the standard way that instruments were configured. However, more recently it has been shown that using nozzle diameters in the range of 85–90 μm (or preferably 100 μm) is beneficial when sorting fragile neural material. In the author's laboratory, a combination of a maximum pressure of 28 psi with 100-μm nozzle and low sample differential pressure is the standard approach for neural cell sorting experiments. Should the viability of the sort product be less than expected, the default position is to lower the sheath pressure to the range of 10–12 psi and repeat the experiment. Should viability still remain low, the next approach should be to increase the nozzle

diameter, the effect of which is to make larger droplets containing the cell of interest. In many instances, the combination of lower speed and lower drop volume result in better long-term viability, particularly when the cells are growing in tissue culture. On the latter point, post-cell sorting viability testing is often not indicative of how well or how poorly sorted cells will grow in culture.

6. Last, but by no means least on the cell's timeline as it passes through the flow cytometer, are the sort collection vessels. Cells may be sorted directly into the medium for tissue culture, into RNA or DNA collection buffers such as Trizol, or into multiwell plates such as 96-well or 384-well plates with single cell deposition. In all instances, having an appropriate medium in the collection container that effectively does the appropriate job—for example, lysing the cells to release RNA—should be carefully considered. In addition, sort collection tubes, whenever possible, should be made of polypropylene to increase the recovery of sorted cells.

7. Post-cell sorting, there are a number of other points that should be at least considered. First, before restaining and possible reanalysis, anecdotal evidence suggests that cell membrane epitope levels take varying amounts of time to return to normal after sorting. The time for this is cell type-dependent, and in the authors laboratory a minimum time of 3 h post-sorting is allowed before restaining of neural cells. Related to this point is the often immediate reanalysis of sorted populations of cells for the assessment of sort purity. Wherever sufficient sorted material is available, a reanalysis of the sorted sample utilizing the existing sorting gates can and should be performed. However, a couple of caveats apply to help interpret the results from resorting:

   a. Fluorescence can be quenched slightly by passing through the laser beam, therefore "positive" cells may fall slightly lower than the defined sort gate.
   b. Viability can be affected due to shear forces, particle acceleration and deceleration, impact, and pH difference, to name a few; thus, scattered light profiles may alter and no longer meet the sort gate.

With these caveats in mind, when performing cell sorts of positively labeled populations, wherever possible draw an additional sorting region encompassing an unlabeled cell population in the positively labeled sample, and sort this population solely for the purpose of evaluating sort purity. With no fluorophores quench, this resorted population should show a purity of close to 100% on a correctly configured and performing instrument.

We now consider the role that flow cytometry and sorting have played in the characterization of the main types of neural cells, namely NSCs, neurons, astrocytes, and oligodendrocytes. In tissue-based neurobiology, neural cells have historically been categorized based on morphology and staining using dyes [27] and histological techniques. In more recent times, with the advent of modern antibody techniques, fluorescently labeling antibodies or fluorescent proteins have become the norm.

2.5 FLOW CYTOMETRIC ANALYSIS OF NSCs

While the brain was long thought to be a post mitotic organ, it is now generally widely accepted that new neural cells are generated in at least two distinct regions of the adult mammalian brain: the SVZ in the wall of the lateral ventricles and the subgranular zone (SGZ) of the dentate gyrus in the hippocampus. These neurogenic niches harbor NSCs that can differentiate into all neural cell types. SVZ NSCs supply olfactory bulb neurons involved in olfactory processing, while hippocampal neurogenesis is important for learning and memory. This process is due to the maturation of NSCs, which at least in part have been identified by their surface antigen expression profiles. Here, we briefly consider examples of markers identified on NSCs found in primary tissue. A range of these and other markers will be dealt with in greater detail in subsequent chapters of this book.

One of the first and most significant reports in this area was the immunoselection against Notch1 surface antigen as reported by Johannson in 1999 [28]. This triggered a great deal of experimental work in looking for surface markers for NSCs, with other marker combinations such as CD133+ CD34− CD45− cells [29], or CD24 (also known as heat-stable antigen, HSA) being applied. Other work by Rietze et al. [33] used combinations of CD24, (which also stains neuronal progenitors and ependymal cells) and peanut agglutinin [30,31] showing a promising enrichment of NSCs. These authors demonstrated that the PNA-low HSA-low population was largely negative for the surface antigens CD34, CD90.2, CD117, CD135, and CD31, markers at the time commonly used to identify hematopoietic cells. This early work, while mainly focused on the surface antigen expression, also showed that cell size as assessed by forward scattered light was important (NSCs > 12 μm), which emphasizes the importance of scattered light as an analytic parameter in its own right.

More recently, we [32] utilized scattered light measurements as one of the main criteria for the enrichment of neuronal progenitors, and in many ways it is an under-utilized feature of flow cytometry. It can provide additional information that is essentially free, is normally being recorded anyway, and only requires that the researcher take the time to interpret what changes in light scatter may actually mean. Indeed, there are now protocols available [33] showing that for round cells, side scatter pulse-width signals provide accurate indications of cell size. Thus, when trying to identify and separate cells by flow cytometry, side scatter width measurements combined with surface marker expression is something that is worth monitoring during normal sample data acquisition.

Of the numerous strategies now available, by far the most prevalent CD marker used in the identification of putative NSCs is CD133. In 1997, CD133 (also known as Prominin-1) was found to be a novel marker for human hematopoietic stem cells [34,35], and while little is known about the biological function of CD133, it is thought to be a regulator of plasma membrane organization based on its cellular localization [35,36]. The initial observation, that CD133 expression was rapidly down-regulated as human epithelial cells [36] differentiate, provided a clue to the significant role that CD133 could play in identifying NSCs [37]. CD133 combined with forward and side scattered light parameters [38] have been the most widely used for identifying stem cells and their niches from primary tissue samples. At the same time, in vitro assays identified CD133 positive cells capable of forming self-renewing neurospheres that could differentiate into neurons and glial cells [29]. In addition to CD133, CD15 (also known as Lexis X or SSEA1 and widely used in stem biology studies), used either singly [39] or in combination with CD184 [40], leads to enrichment. Following on from this work, CD133 positive stem/progenitor-like cells were identified in the murine cerebellum [41], prostate [42], kidney [43,44], and liver [45]. However, the situation when using CD133 is not always completely straightforward, as there is substantial controversy when it is been shown in particular assays, such as the cell survival neurosphere assay, that conversely, CD133 negative cells may also give rise to neurospheres that provide an indirect indication of NSC numbers [46].

Flow cytometry of culture-derived NSCs is covered extensively in various chapters of this volume where their importance is given a thorough coverage. Flow cytometry has been a key technology in the elucidation and characterization of neural cells generated from embryonic stem (ES) cells, induced pluripotent stem (iPS) cells, or other sources. A quick PubMed search based on terms such as "flow cytometry" and "neural stem cells" will yield many hundreds of manuscripts that are relevant in this area. Readers are encouraged to clearly define their particular area of interest for the application of flow cytometry before performing online searches, to avoid having to sort through excessive irrelevant publications.

2.6 FLOW CYTOMETRY OF NEURONS

Neurons are specialized cells composed of distinct morphological parts: the cell body, also known as the soma, the dendrites, an axon, and presynaptic terminals, which together allow the transmission and electrochemical modulation of information. Given their important role, neurons have been of intense interest from the early onset of flow cytometry [47–51]. The search has been on to find markers and antibodies that show specificity to neurons [52]. Notably, there are few markers or combinations thereof that are specific for live unfixed neurons from *primary tissue*. This situation differs from that of cultured cells, or those derived from stem and progenitor cells differentiated toward neuronal phenotypes, an area that will be discussed later in this chapter. The following section highlights some neuronal markers and alternative labeling approaches commonly used in the application of flow cytometry of neural cell types.

Tubulin is a major component of microtubules. β-III-tubulin considered neuron specific, and is therefore one of the most widely used markers to identify neurons [53] Its intracellular location requires cells to be fixed and permeabilized to allow antibody entering and egress. Tuj1 is an anti-β-III-tubulin antibody (named after the mouse hybridoma cell line from which it was derived) that frequently appears in the literature of flow cytometry and neuroscience applications. Figure 2.2 displays this neuronal marker in conjunction with the astroglial marker glial fibrillary acidic protein (GFAP).

The microtubule-associated protein 2 (MAP2) is a ubiquitous neuronal cytoskeletal protein that binds to tubulin and stabilizes microtubules [54], and can be targeted by specific antibodies following fixation. It tends to be associated with more mature neuronal cells, and dendritic processes more specifically.

It would be remiss not to mention the important role that NeuN has had in the identification of neuronal populations by flow cytometry. NeuN is a nuclear marker first reported by Mullen et al. in 1992 [55], that shows specificity for neuronal

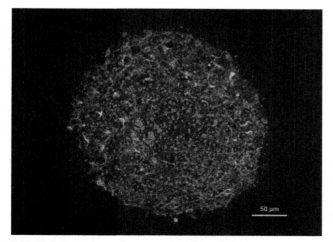

FIGURE 2.2 A neurosphere generated from flo cytomertically sorted cells from the sub ventricular zone of the murine brain shows the specificity of the neuron specific marker Tuj1 (red), the astroglial marker GFAP (green) with nuclei counterstained with DAPI (blue).

cell types throughout the central and peripheral nervous systems in mice. It has been used on both dissociated nuclei and on fixed and permeabilized cells in immunohistochemistry and flow cytometry studies. However, there is some controversy regarding NeuN staining in some tissues indicating a lack of specificity in certain neural cell types [56], and the use of NeuN needs to be assessed on a case-by-case basis as to whether it meets the requirements of the research project.

An alternative approach is to use retrograde labeling techniques for the identification of neurons. Neurons from the ventral mesencephalon have been fluorescently retrograde labeled using DiI and the labeled cells dissociated from brain tissue and separated by flow cytometry [57]. It is also possible to retrograde-label by injecting green fluorescent microspheres into the axonal projection fields of the pons and the cervical spinal cord [58] to specifically label corticospinal motor neurons. These cells were then successfully purified by FACS based on forward scatter and surface characteristics, combined with the fluorescence from the retrograde label. Interestingly, while these cells are fragile when taken at E18, P3, P6, P14 from the murine neocortex, the cells retain short proximal dendritic and/or axonal processes.

While these markers and labeling approaches showed great specificity, it would be beneficial to be able to find appropriate cell-surface markers for neurons that did not require the fixation step. One approach that was explored [59] is the labeling of dopaminergic neurons with a monoclonal antibody against the neural-specific protein 4 (NSP4) that was reported to be present on approximately 30% of mesencephalic cells in the mouse brain, and analyzed and separated by flow cytometry. These sorted cells, when grown in culture, show enrichment in dopaminergic neurons. Dopaminergic neurons were identified by the presence of tyrosine hydroxylase (TH) using a protocol that requires fixation and permeabilization of cells, then staining TH with a specific antibody. The NSP4 approach was promising because it demonstrated that characteristic markers for dopaminergic neurons could be identified and that the traditional methods [60,61] of analyzing dopaminergic neurons by flow cytometry, which involved staining for TH, could at least partially be replaced by a surface labeling protocol. More recently, Ganat et al. [62] utilized fluorescent protein-based approaches to specifically label dopaminergic neurons, sorted the cells, and then showed engraftment of specific subsets in the mouse brain. While this work was based on intracellular labeling, others [63,64] have combined intracellular and surface labeling in a novel approach to identify and enrich midbrain dopaminergic specific subsets of cells.

Another flow cytometry marker of importance is the neural cell adhesion molecule (NCAM; CD56) is that binds either the carbohydrate and polypeptide form of the NCAM on the surface of neural cells. CD56 can be found on the surface of neural material in a development-specific manner [56], whereby the levels of CD56 vary with the maturity of the neuronal cell, and has been used as a separation basis for cell sorting [22]. Interestingly, anti-CD56 antibodies are routinely used in hematological studies to identify relatively rare natural killer cells present in peripheral blood. In spite of this, CD56-expressing neural cells can be usefully combined with CD45 staining for leukocytes in primary samples where blood and primary neural tissue may be mixed. This allows the exclusion of leukocytes from further analysis.

2.7 FLOW CYTOMETRIC ANALYSIS OF GLIAL CELL TYPES

Astrocytes are the most abundant neural cell type, and share large morphological and functional variability that remains to be further resolved. Traditionally, the most widely used marker for the identification of astrocytes via immunofluorescence and immunohistochemical methods has been the presence of the intracellular GFAP. First identified in 1972 [65], there is now a plethora of monoclonal antibodies available from commercial vendors that, following permeabilization of the cell membrane, will bind to the GFAP present in cells of the astrocytic lineage. However, as with other neural cell types, flow cytometry-based experimental research has been hampered by a lack of suitable antibodies that will bind to common antigens present on the surface of astrocytes. Recently (2014), the situation improved with the development of an antibody that targets the extracellular epitope of the astrocyte-specific L-glutamate/L-aspartate transporter GLAST [66] labeled ACSA-1 (astrocyte cell surface antigen-1). Representing a major advance in identifying this important neural cell type, GLAST has been shown by immunohistochemistry, immunocytochemistry, and flow cytometry to label virtually all astrocytes that are identified as positive by other markers such as GFAP or nestin (Figures 2.3 and 2.4). Importantly, other cells such as oligodendrocytes, neurons, or microglia are not recognized by the GLAST antibody. One caveat to be aware of is that the GLAST epitope detected by the ACSA-1 antibody shows papain sensitivity, therefore it is critical to use a trypsin-based enzymatic tissue dissociation [66] (Figures 2.5).

Also, the use of the alternative glutamate transporter GLT has been described for the identification of astrocytes [67] used in association with Thy1 (CD90), that labels some neuronal populations, and CD11b, a widely

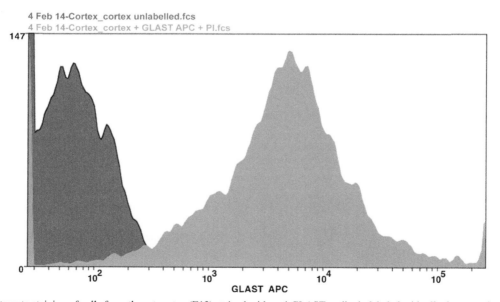

FIGURE 2.3 Astrocyte staining of cells from the rat cortex (E12) stained with anti-GLAST antibody labeled with allophycocyanin (APC) compared with unlabeled control.

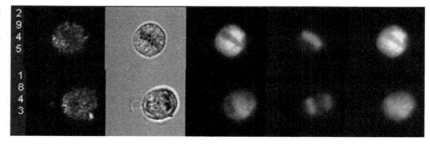

FIGURE 2.4 AMNIS imaging cytometry showing the images associated with two cells (numbered 2945 top row, and 1843 lower row) from a glioblastoma cell line. Left to right for each row, side scattered light signal, brightfield, GFAP (green(white in print versions)), nucleus stained with DRAQ5 (pink(gray in print versions)) with composite GFAP/DRAQ5 in last column. DRAQ5 staining shows chromatin condensation (top row) and then separation of DNA prior to cell division (bottom row).

used marker for the identification of microglia. Other antibody panels that can be used for flow cytometry of astrocytes rely on surface integrin expression. Brain astrocytes express αvβ3 integrin heterodimers on the cell surface [68,69] in vivo, where they play an important role in the regulation of neuronal process (neurite) outgrowth, and which are known to be up-regulated in neurological disease.

For the identification of oligodendroctyes from cells that are astrocytes, the cell-surface antigen O4 can be used, and cells may be sorted and cultured on this basis [70]. Oligodendrocyte precursors can be labeled with anti-NG2, a chondroitin sulfate proteoglycan present on the surface of many oligodendroglial progenitor cells (OPCs) in various species [71]. Cells that are initially NG2 positive show increasing levels of O4 expression as they mature. The importance of NG2 is detailed in one of the following chapters in this book.

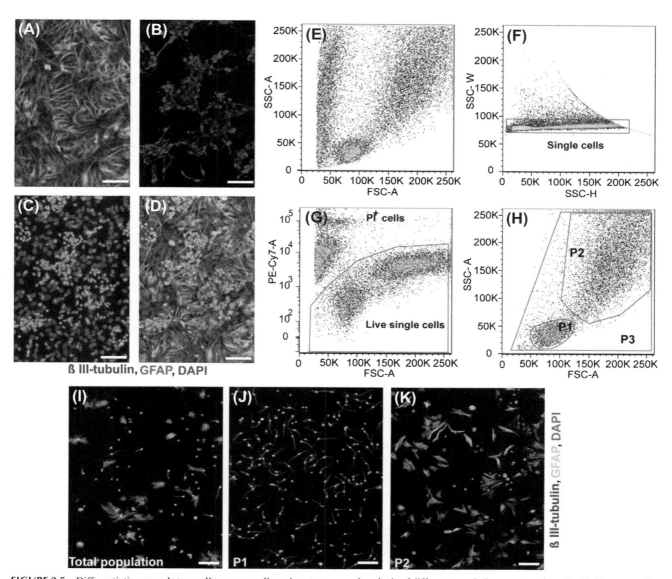

FIGURE 2.5 Differentiating neural stem cell progeny, cell-sorting strategy, and analysis of different populations post-sorting. (A–D): Representative micrographs of the neuroblast assay culture, 4 days after switching to growth factor free medium and stained for the astrocyte marker; GFAP (A), neuronal marker; β-III-tubulin (B) and counterstained with DAPI (C). Notice the colonies of β-III-tubulin-positive neuronal cells on top of the astrocyte monolayer (D) and the nucleus size difference. (E–H): Cells were first plotted based on FSC versus SSC (E) and then side scatter pulse width (SSC-W), versus side scatter pulse height (SSC-H) (F) to exclude doublets and clumps. After excluding dead or damaged cells based on propidium iodide (PI) uptake (G), cells were plotted based on FSC and SSC (H) and gates drawn to define three cell populations; P1 (FSC^{low} SSC^{low}), P2 (FSC^{high} SSC^{high}) and P3 (total cells). (I–K): Micrographs of sorted cells from P3 (I), P1 (J), and P2 (K) population that were stained (1 day after plating) for β-III-tubulin, GFAP, and DAPI.

2.8 CONCLUSION

This chapter is meant to provide some insights into the depth and breadth of the effect that flow cytometry has had, and will continue to have, in the area of neural cell characterization. Some practical advice has been shared regarding the specifics of utilizing flow cytometry in this scenario. Necessarily, many procedures and protocols will need to be optimized based on the particular cell of interest. Importantly, many of the challenges related to flow cytometry in neural cell analysis are intimately linked with sample preparation and the thorough and effective dissociation of cells into viable single-cell suspensions. Once this often significant hurdle is overcome, then a whole range of assays are available to investigators, enabling them to ask and answer novel scientific questions.

Novel techniques are continually being developed for flow cytometry. Two examples are the ability to detect, quantify, and track protein aggregation and mislocalization in individual cells and then separate particular cells based on these intracellular localizations [72]. Another recent advance that is revolutionizing flow cytometry assays are *nanoprobes* or *nanoflares*, oligonucleotide gold nanoparticle conjugates that can be used for detection of intracellular levels of mRNA [73] for cultured cells. These nanoprobes, when combined with surface markers, allow the detection of genes and possibly their downstream product in live cells.

Finally, it is hoped that this chapter encourages the reader to delve further into more detailed explanations of applicability in a later chapter. In addition, researchers are encouraged to consider flow cytometric methods as another essential tool in their laboratory toolbox, rather than an adjunct technology supporting something such as microscopy. When used appropriately, as with all tools, the results obtained can be truly beneficial to your research endeavors.

REFERENCES

[1] Fulwyler MJ. An electronic particle separator with potential biological application. Science 1965;150:371.

[2] Fulwyler MJ. Electronic separation of biological cells by volume. Science 1965;150:910.

[3] Shapiro HM. Practical flow cytometry. 3rd ed; 1995. pp. xxxviii+542p–xxxviii+542p.

[4] Le Meur N, et al. Data quality assessment of ungated flow cytometry data in high throughput experiments. Cytometry A 2007;71A:393.

[5] Lugli E, Roederer M, Cossarizza A. Data analysis in flow cytometry: the future just started. Cytometry A 2010;77:705.

[6] Herzenberg LA, Tung J, Moore WA, Herzenberg LA, Parks DR. Interpreting flow cytometry data: a guide for the perplexed. Nat Immunol 2006;7:681.

[7] Alvarez DF, Helm K, Degregori J, Roederer M, Majka S. Publishing flow cytometry data. Am J Physiol Lung Cell Mol Physiol 2010;298:L127.

[8] Pedreira CE, Costa ES, Lecrevisse Q, van Dongen JJ, Orfao A. Overview of clinical flow cytometry data analysis: recent advances and future challenges. Trends Biotechnol 2013;31:415.

[9] Meehan S, et al. AutoGate: automating analysis of flow cytometry data. Immunol Res 2014;58:218.

[10] Malek M, et al. flowDensity: reproducing manual gating of flow cytometry data by automated density-based cell population identification. Bioinformatics 2014;1–2, http://dx.doi.org/10.1093/bioinformatics/btu677.

[11] Finak G, et al. OpenCyto: an open source infrastructure for scalable, robust, reproducible, and automated, end-to-end flow cytometry data analysis. PLoS Comput Biol 2014;10:e1003806.

[12] Shih MC, Huang SH, Donohue R, Chang CC, Zu Y. Automatic B cell lymphoma detection using flow cytometry data. BMC Genomics 2013;14(Suppl. 7):S1.

[13] Meyer R, Zaruba M, McKhann G. Flow cytometry of isolated cells from the brain. Anal Quant Cytol 1980;2(66).

[14] Panchision DM, et al. Optimized flow cytometric analysis of central nervous system tissue reveals novel functional relationships among cells expressing CD133, CD15, and CD24. Stem Cells 2007;25:1560.

[15] Dugas JC, et al. A novel purification method for CNS projection neurons leads to the identification of brain vascular cells as a source of trophic support for corticospinal motor neurons. J Neurosci 2008;28:8294.

[16] Murphy SJ, Watt DJ, Jones GE. An evaluation of cell separation techniques in a model mixed cell population. J Cell Sci 1992;102(Pt 4):789.

[17] Kim DS, et al. Highly pure and expandable PSA-NCAM-positive neural precursors from human ESC and iPSC-derived neural rosettes. PLoS One 2012;7:e39715.

[18] Foo LC. Purification of astrocytes from transgenic rodents by fluorescence-activated cell sorting. Cold Spring Harb Protoc 2013; 2013. pdb.prot074229.

[19] Lovatt D, et al. The transcriptome and metabolic gene signature of protoplasmic astrocytes in the adult murine cortex. J Neurosci 2007;27:12255.

[20] Sergent-Tanguy S, Chagneau C, Neveu I, Naveilhan P. Fluorescent activated cell sorting (FACS): a rapid and reliable method to estimate the number of neurons in a mixed population. J Neurosci Methods 2003;129:73.

[21] Azari H, Millette S, Ansari S, Rahman M, Deleyrolle LP, Reynolds BA, Isolation and expansion of human glioblastoma multiforme tumor cells using the neurosphere assay. J Vis Exp 2011;56, e3633. http://dx.doi.org/10.3791/3633.

[22] Pruszak J, Sonntag K-C, Aung MH, Sanchez-Pernaute R, Isacson O. Markers and methods for cell sorting of human embryonic stem cell-derived neural cell populations. Stem Cells 2007;25(2257).

[23] Hagman DK, et al. Characterizing and quantifying leukocyte populations in human adipose tissue: impact of enzymatic tissue processing. J Immunol Methods 2012;386:50.

[24] Mamber C, Kozareva DA, Kamphuis W, Hol EM. Shades of gray: the delineation of marker expression within the adult rodent subventricular zone. Prog Neurobiol 2013;111:1.

[25] Hedlund E, et al. Embryonic stem cell-derived Pitx3-enhanced green fluorescent protein midbrain dopamine neurons survive enrichment by fluorescence-activated cell sorting and function in an animal model of Parkinson's disease. Stem Cells 2008;26:1526.

[26] Reynolds BA, Weiss S. Generation of neurons and astrocytes from isolated cells of the adult mammalian central nervous system. Science 1992;255:1707.

[27] Delaney PA. Reliable methods for the fixation and staining of Nissl substance. Anat Rec 1927;36:111.

[28] Johansson CB, et al. Identification of a neural stem cell in the adult mammalian central nervous system. Cell 1999;96:25.

[29] Uchida N, et al. Direct isolation of human central nervous system stem cells. Proc Natl Acad Sci USA 2000;97:14720.

[30] Eldi P, Rietze RL. Flow cytometric characterization of neural precursor cells and their progeny. Methods Mol Biol 2009;549:77.

[31] Rietze RL, et al. Purification of a pluripotent neural stem cell from the adult mouse brain. Nature 2001;412:736.

[32] Azari H, Osborne GW, Yasuda T, Golmohammadi MG, Rahman M, et al. Purification of immature neuronal cells from neural stem cell progeny. PLoS ONE 2011;6(6):e20941. http://dx.doi.org/10.1371/journal.pone.0020941.

[33] Hoffman RA. Pulse width for particle sizing. Curr Protoc Cytom 2009. [chapter 1], Unit 1 23.

[34] Yin AH, et al. AC133, a novel marker for human hematopoietic stem and progenitor cells. Blood 1997;90:5002.

[35] Weigmann A, Corbeil D, Hellwig A, Huttner WB. Prominin, a novel microvilli-specific polytopic membrane protein of the apical surface of epithelial cells, is targeted to plasmalemmal protrusions of non-epithelial cells. Proc Natl Acad Sci USA 1997;94:12425.

[36] Corbeil D, et al. The human AC133 hematopoietic stem cell antigen is also expressed in epithelial cells and targeted to plasma membrane protrusions. J Biol Chem 2000;275:5512.

[37] Walker TL, et al. Prominin-1 allows prospective isolation of neural stem cells from the adult murine hippocampus. J Neurosci 2013;33(3010).

[38] Murayama A, Matsuzaki Y, Kawaguchi A, Shimazaki T, Okano H. Flow cytometric analysis of neural stem cells in the developing and adult mouse brain. J Neurosci Res 2002;69:837.

[39] Capela A, Temple S. LeX is expressed by principle progenitor cells in the embryonic nervous system, is secreted into their environment and binds Wnt-1. Dev Biol 2006;291:300.

[40] Corti S, et al. Multipotentiality, homing properties, and pyramidal neurogenesis of CNS-derived LeX(ssea-1)+/CXCR4+ stem cells. FASEB J 2005.

[41] Lee A, et al. Isolation of neural stem cells from the postnatal cerebellum. Nat Neurosci 2005;8:723.

[42] Shepherd CJ, et al. Expression profiling of CD133+ and CD133− epithelial cells from human prostate. Prostate 2008;68:1007.

[43] Bussolati B, et al. Isolation of renal progenitor cells from adult human kidney. Am J Pathol 2005;166:545.

[44] Florek M, et al. Prominin-1/CD133, a neural and hematopoietic stem cell marker, is expressed in adult human differentiated cells and certain types of kidney cancer. Cell Tissue Res 2005;319:15.

[45] Schmelzer E, et al. Human hepatic stem cells from fetal and postnatal donors. J Exp Med 2007;204:1973.

[46] Sun Y, et al. CD133 (prominin) negative human neural stem cells are clonogenic and tripotent. PLoS One 2009;4:e5498.

[47] Lee EHY, Geyer MA. Selective effects of apomorphine on dorsal raphe neurons: a cytofluorimetric study. Brain Res Bull 1982;9:719.

[48] Calof AL, Reichardt LF. Motoneurons purified by cell sorting respond to two distinct activities in myotube-conditioned medium. Dev Biol 1984;106:194.

[49] Dyer SA, Derby MA, Cole GJ, Glaser L. Identification of subpopulations of chick neural retinal cells by monoclonal antibodies: a fluorescence activated cell sorter screening technique. Brain Res 1983;285:197.

[50] Hobi R, Studer M, Ruch F, Kuenzle CC. The DNA content of cerebral cortex neurons. Determination by cytophotometry and high performance liquid chromatography. Brain Res 1984;305:209.

[51] Moskal JR, Schaffner AE. Monoclonal antibodies to the dentate gyrus: immunocytochemical characterization and flow cytometric analysis of hippocampal neurons bearing a unique cell-surface antigen. J Neurosci 1986;6:2045.

[52] Yuan SH, et al. Cell-surface marker signatures for the isolation of neural stem cells, glia and neurons derived from human pluripotent stem cells. PLoS One 2011;6:e17540.

[53] Ferreira A, Caceres A. Expression of the class III beta-WW isotype in developing neurons in culture. J Neurosci Res 1992;32:516.

[54] Herzog W, Weber K. Fractionation of brain microtubule-associated proteins. Isolation of two different proteins which stimulate tubulin polymerization in vitro. Eur J Biochem/FEBS 1978;92:1.

[55] Mullen RJ, Buck CR, Smith AM. NeuN, a neuronal specific nuclear protein in vertebrates. Development 1992;116:201.

[56] Cannon JR, Greenamyre JT. NeuN is not a reliable marker of dopamine neurons in rat substantia nigra. Neurosci Lett 2009;464:14.

[57] Lopez Lozano JJ, Notter MF, Gash DM, Leary JF. Selective flow cytometric sorting of viable dopamine neurons. Brain Res 1989;486:351.

[58] Arlotta P, et al. Neuronal subtype-specific genes that control corticospinal motor neuron development in vivo. Neuron 2005;45:207.

[59] di Porzio U, Rougon G, Novotny EA, Barker JL. Dopaminergic neurons from embryonic mouse mesencephalon are enriched in culture through immunoreaction with monoclonal antibody to neural specific protein 4 and flow cytometry. Proc Natl Acad Sci USA 1987;84:7334.

[60] Joh TH, Reis DJ. Different forms of tyrosine hydroxylase in central dopaminergic and noradrenergic neurons and sympathetic ganglia. Brain Res 1975;85:146.

[61] Joh TH, Shikimi T, Pickel VM, Reis DJ. Brain tryptophan hydroxylase: purification of, production of antibodies to, and cellular and ultrastructural localization in serotonergic neurons of rat midbrain. Proc Natl Acad Sci USA 1975;72:3575.

[62] Ganat YM, et al. Identification of embryonic stem cell-derived midbrain dopaminergic neurons for engraftment. J Clin Invest 2012;122:2928.

[63] Turac G, et al. Combined flow cytometric analysis of surface and intracellular antigens reveals surface molecule markers of human neuropoiesis. PLoS One 2013;8:e68519.

[64] Doi D, et al. Isolation of human induced pluripotent stem cell-derived dopaminergic progenitors by cell sorting for successful transplantation. Stem Cell Rep 2014;2:337.

[65] Bignami A, Eng LF, Dahl D, Uyeda CT. Localization of the glial fibrillary acidic protein in astrocytes by immunofluorescence. Brain Res 1972;43:429.

[66] Jungblut M, et al. Isolation and characterization of living primary astroglial cells using the new GLAST-specific monoclonal antibody ACSA-1. Glia 2012;60:894.

[67] Schwarz JM, Smith SH, Bilbo SD. FACS analysis of neuronal-glial interactions in the nucleus accumbens following morphine administration. Psychopharmacology 2013;230:525.

[68] Milner R, et al. Distinct roles for astrocyte alphavbeta5 and alphav-beta8 integrins in adhesion and migration. J Cell Sci 1999;112 (Pt 23):4271.

[69] Milner R, Relvas JB, Fawcett J, ffrench-Constant C. Developmental regulation of alphav integrins produces functional changes in astrocyte behavior. Mol Cell Neurosci 2001;18:108.

[70] Trotter J, Schachner M. Cells positive for the O4 surface antigen isolated by cell sorting are able to differentiate into astrocytes or oligodendrocytes. Dev Brain Res 1989;46:115.

[71] Horiuchi M, Lindsten T, Pleasure D, Itoh T. Differing in vitro survival dependency of mouse and rat NG2[+] oligodendroglial progenitor cells. J Neurosci Res 2010;88:957.

[72] Ramdzan YM, et al. Tracking protein aggregation and mislocalization in cells with flow cytometry. Nat Methods 2012;9:467.

[73] Prigodich AE, et al. Nano-flares for mRNA regulation and detection. ACS Nano 2009;3:2147.

Chapter 3

CD36, CD44, and CD83 Expression and Putative Functions in Neural Tissues

Isaias Glezer[1], Serge Rivest[2] and André Machado Xavier[1]

[1]Departamento de Bioquímica, Escola Paulista de Medicina, Universidade Federal de São Paulo, São Paulo, Brazil; [2]Faculty of Medicine, Department of Molecular Medicine, Neuroscience Laboratory, CHU de Québec Research Center, Laval University, Quebec, Canada

3.1 INTRODUCTION

The identification of specific surface antigens in various leukocyte populations supported the routine use of well-known phenotyping techniques in immunological research, diagnostics, and therapy. For instance, over 300 clusters of differentiation (CD) molecules have been officially characterized so far. Obviously, surface antigens did not evolve naturally to provide the necessary resources for the technical development of the field of immunobiology. Surface molecules are key players in cellular communication and cell–cell and cell–extracellular matrix (ECM) contacts. Remarkably, a discrete glial or neuronal population may express a gene that encodes a specific CD molecule polypeptide chain, whereas others do not. In addition to gene expression, the antigen presence on the extracellular face of the cell membrane will ultimately depend on tight control over membrane trafficking and surface expression dynamics. The challenge of revealing surface molecules that specify neural populations is a first step toward understanding the possible complex neurologic functions associated with these antigens. Because a large-scale analysis of all CD molecules in the brain is not available, starting points such as the one presented in this chapter are necessary for this emerging subject.

Although the motivations for selecting the scavenger receptor CD36, the cell adhesion molecule CD44, and the immunoglobulin superfamily (IgSF) member CD83 can be circumstantial, the comparative study involving these three glycoproteins strongly argues in favor of the promising perspectives of this field. A closer inspection of the anatomical distribution of these transcripts was primarily motivated by the differential regulation of gene expression in the central nervous system (CNS), as detailed below. The genes encoding these three molecules present different expression patterns in the brain, including signaling requirements for their induction in mice exposed to restraint stress or their response to systemic administration of the proinflammatory trigger molecule lipopolysaccharide (LPS) from

gram-negative bacteria [1,2]. CD36 antigen, also known as fatty acid translocase (FAT), platelet glycoprotein 4 (GPIV), glycoprotein IIIb (GPIIIB), glycoprotein PAS-4, and scavenger receptor class B member 3 (SCARB3), seems to be multifunctional in terms of signaling, phagocytosis, and lipid metabolism [3]. In contrast, the function of CD44 antigen, also known as extracellular matrix receptor III (ECMR-III), Hermes antigen, hyaluronate receptor, lymphocyte antigen 24, phagocytic glycoprotein I (Pgp-1), GP90 lymphocyte homing/adhesion receptor, epican, heparan sulfate proteoglycan, HUTCH-I, chondroitin sulfate proteoglycan 8, hematopoietic cell E- and L-selectin ligand, and homing function and Indian blood group system, is mostly related to adhesion [4], and its expression has been associated with microglia, astrocytes, and neurons. On the other hand, CD83 antigen (alternative names: B-cell activation protein, BL11, and cell surface protein HB1) is less characterized than CD36 or CD44 and appears to be specific to mature dendritic cells involved in antigen presentation to lymphocytes, because these CD83-positive cells express the highest levels of major histocompatibility complex class II [5]. Hence, the three CD molecules are sharply different in their molecular functions and gene regulation. Here, we review the remarkably distinct neuroanatomical patterns of the previously reported brain expression of these three genes/molecules and discuss whether their putative roles in the brain relate to their diverse structure and function.

3.2 THE PUTATIVE CD36 FUNCTIONS IN THE CNS: A MULTIFUNCTIONAL SCAVENGER RECEPTOR AND LIPID SENSOR

3.2.1 CD36 Structure and General Functions

CD36 is a glycoprotein expressed at the cell surface of several differentiated cells. It possesses two transmembrane

Neural Surface Antigens. http://dx.doi.org/10.1016/B978-0-12-800781-5.00003-7

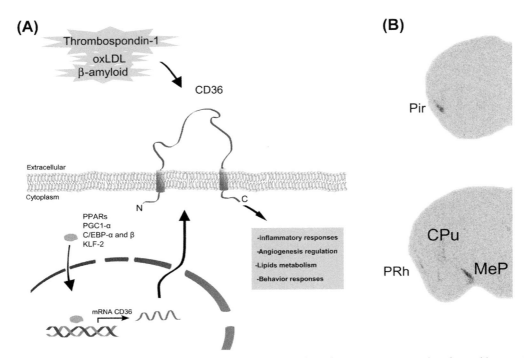

FIGURE 3.1 CD36 general structure/functions and brain gene expression. (A) Schematic receptor representation along with transcription inducers, ligands, and coupled cellular events. (B) Representative autoradiography photographs depicting *Cd36* mRNA expression in the CNS. Abbreviations: CPu, caudate putamen; MeP, medial amygdaloid nucleus (posterior); Pir, piriform cortex; PRh, perirhinal cortex.

segments leading to two short cytoplasmic tails (N- and C-terminal; Figure 3.1), and in humans the glycoprotein is encoded by the *CD36* gene located on chromosome 7 [6,7]. The predicted size of the CD36 polypeptide chain is 53 kDa according to the nucleotide sequence (cDNA) [8]. In addition, CD36 is heavily modified posttranslationally by N-linked glycosylation. Most of the sites are located in the large extracellular loop of the protein, but no individual site was found to be essential for proper trafficking of CD36 [7,9]. The amino acid identity is highly conserved comparing murine and human proteins, and in *Drosophila*, Croquemort proteins represent a well-established family homolog [10,11]. The *CD36* gene promoter is responsive to peroxisome proliferator-activated receptors, peroxisome proliferator-activated receptor γ coactivator 1α, CCAAT-enhancer-binding proteins α and β, and Krüppel-like factor 2 transcription factors [12–15]. However, most of these studies focused on lipid metabolism, and owing to *CD36/Cd36* promoter complexity, it is very likely that tissue-specific effects may influence gene expression regulation.

The CD36 receptor belongs to the class B scavenger receptor (SR) family, which also includes the SR-BI membrane receptor and lysosomal integral membrane protein II. SRs, including CD36 itself, contribute to atherosclerosis progression via modified LDL phagocytosis, disturbed macrophage migration, and the formation of so-called foam cells [16,17]. Monocytes/macrophages, platelets, hepatocytes, and endothelial cells express CD36, which promotes varied functions related to the inflammatory process, negative regulation of microvascular angiogenesis, oxidized lipoprotein or phospholipid signaling, and microorganism or apoptotic cell engulfment. CD36 is also one of the receptors for the matrix glycoprotein and inhibitor of neovascularization thrombospondin-1, explaining its involvement in angiogenesis and modulation of vascular endothelial growth factor signaling [3,18]. In adipocytes and muscle cells, CD36 plays an important role in lipid homeostasis through the transport of long-chain fatty acids [19]. A conserved structure of the receptor seems to associate with lipid signaling because the expression of the *Drosophila melanogaster* CD36 homolog sensory neuron membrane protein in a population of sensorial olfactory neurons is important for proper responses to the lipid pheromone *cis*-vaccenyl acetate, which is an ester derived from fatty alcohol [20]. In rodent gustatory cells, CD36 mediates fatty acid detection and preference for lipid ingestion [21,22], strongly suggesting a conserved role for this receptor in lipid sensorial recognition. In addition, diacylglyceride detection in immune cells via toll-like receptors (TLRs) depends on CD36 [23]. In hepatocytes, the receptor is also a crucial target of parkin, a protein associated with Parkinson's disease, mediating a regulatory mechanism in lipid metabolism [24]. All together, it is evident that CD36 is an extraordinarily versatile molecule.

3.2.2 *Cd36* Gene Expression Maps to Olfactory Relays and Reproductive Behavior-Associated Brain Regions

Investigation of the constitutive expression of *Cd36* in the murine CNS revealed surprising findings, including colocalization of the transcript with the neuronal marker NeuN [2]. According to in situ hybridization (ISH) histochemistry, most of the brain is negative for the transcript, and low to moderate expression levels map to main and accessory olfactory bulb (MOB and AOB) relays also associated with gonadotropin-releasing hormone neurons [25–28]. These regions include the dorsal taenia tecta, some segments of the piriform cortex, the bed nucleus of the stria terminalis, the nucleus of the lateral olfactory tract, the medial amygdaloid nucleus (especially the posterodorsal part), the posterolateral cortical amygdaloid nucleus, and variable signals in the bed nucleus of the accessory olfactory tract [2] (Figure 3.1). It is worth noting that the transcript expression was variable among individuals, and some regions could not be reproducibly detected, including the ventromedial hypothalamic nucleus (VMH). Although the reason for this variability remains obscure, certain hypothalamic regions are indeed consistent, including the medial preoptic nucleus and ventral premammillary nucleus, which are involved in sexual behavior [29–31] and could functionally relate to *Cd36* expression in MOB and AOB relays. In addition to these sites, pyramidal CA1 field neurons in the ventral hippocampal formation show strong labeling of the riboprobe. At the time of writing, no plausible specific relation can be tracked between these neurons and the other mentioned brain regions. Robust *Cd36* (mRNA) labeling is present in ependymal cells of the fourth ventricle. The role of gene expression in this site, as well as in weakly labeled regions in the mesencephalon, needs further investigation to establish a possible functional link. Nevertheless, a strong neuroanatomical appeal can be found in terms of a putative olfactory *Cd36* circuitry. No systematic neuroanatomical mapping has been conducted so far regarding CD36 protein expression, which will provide important information regarding cellular distribution of the surface antigen, such as in cell bodies, dendrites, etc. In addition, the development of *Cd36* gene reporter mice will also reveal important cell expression data in histological analysis.

3.2.3 CD36 Functions in the Nervous System

Most of the prior CD36 investigation conducted in the CNS aimed to target the immune functions associated with microglia/macrophage expression. Indeed, CD36 expression in these cells is important for Aβ-amyloid phagocytosis and reactive oxygen species generation [32,33]. Another study concluded that CD36 expression correlates with

β-amyloid deposits rather than with Alzheimer pathology [34,35]. These observations suggest that CD36 induction in microglial cells is associated with pathological conditions. A study showed that mouse embryonic stem cell-derived neural stem cells (NSCs) could fuse with cocultured neurons, which is a process dependent upon microglial CD36 presence [36]. It still needs to be verified if there is relevance for this phenomenon in vivo and during pathological states. In contrast to microglial findings, another group reported CD36 expression in subsets of astrocytes. Interference with CD36 expression or gene deficiency provoked diminished astrocyte glial fibrillary acidic protein (GFAP) expression and proliferation, in addition to reduced glial scar formation in a stroke model [37,38]. The expression of the CD36 molecule in glial cells is highly correlated with brain lesions and other pathological features. However, expression of the scavenger receptor in various cells makes it difficult to ascertain specific roles of CD36 in cellular events. For instance, CD36-null mice present decreased proinflammatory signaling in the brain after middle cerebral artery occlusion [39], but no information regarding a particular cell type response can be inferred from this or other studies in general.

CD36 is also involved in CNS drug response according to one report. The receptor was identified as a cause of a lack of response to valproic acid treatment in a subset of spinal muscular atrophy patients. The study focused on survival motor neuron 2 (*SMN2*) gene induction caused by the drug owing to its effects on histone acetylation and made use of induced pluripotent stem (iPS) cell technology to generate GABAergic neurons from patients' fibroblasts to conduct extensive validation [40]. According to gene expression profiling, *Cd36* overexpression correlates with valproic acid nonresponder cells and renders various cells nonresponsive to the drug. Thus, it is possible that different levels of CD36 expression influence neuronal function at the transcriptional level, which awaits future investigation.

Regarding the putative neuronal roles of CD36, two studies deserve particular attention. First, mice deficient in the *Cd36* gene show memory impairment [41]. Although several factors could influence behavioral tests, it is reasonable to attribute the reported effects to neuronal cells. This is plausible in light of the constitutive gene expression seen in CNS neuronal nuclei (see above), which does not seem to be the case for glial cells. Second, one study demonstrated fatty acid sensing (excitability) in a subpopulation of dissociated VMH glucose-responsive hypothalamic neurons, which was partially reduced by treatment with the established CD36 inhibitor succinimidyl oleate [42]. It should be noted that *Cd36* (mRNA) is not reliably detected in VMH in every sample (see above). This indicates that the VMH could be a site of important modulation of *Cd36* gene expression subject to individual variation or that only a small population of neurons in this nucleus expresses the transcript. It is also of note that succinimidyl oleate inhibits the mitochondrial respiratory chain (complex III) and should not be regarded as a specific CD36 inhibitor [43]. Nevertheless, data confirmed the involvement of CD36 in these hypothalamic neurons using an RNA interference (RNAi) approach in vivo by the use of adeno-associated virus vectors designed to express short-hairpin RNA against *Cd36* mRNA. The results show that in addition to modifying the long-chain fatty acid response, CD36 depletion in the ventromedial hypothalamus modifies distribution of adiposity and glucose metabolism in rats [44]. Hence, it is plausible to conclude that neuronal CD36 probably plays important neurophysiological roles.

3.3 THE EXPRESSION OF CD44 ADHESION MOLECULE IN NEURAL CELLS

3.3.1 CD44 Structure and General Functions

CD44 is an integral membrane glycoprotein, initially reported with other nomenclature, such as Pgp-1, ECMR-III. or Hermes antigen [45–49]. Although early studies related CD44 to ECM interaction, it was only in 1990 that the identification of hyaluronan (hyaluronic acid; HA) as a major ligand could provide a substantial advance in terms of its biological relevance [50]. HA is a nonsulfated glycosaminoglycan polymer produced by integral plasma membrane enzymes called hyaluronan synthases, and it is found in the ECM quite abundantly in some tissues, especially in soft tissue [51]. Other ligands for CD44 include laminin, fibronectin, osteopontin (OPN), serglycin, and collagen. CD44 is polymorphic and encoded by a single gene that transcribes several variants of alternative splicing. One of the CD44 variants is ubiquitously expressed, whereas the overexpression and the presence of multiple variants are often associated with malignant neoplasia [52].

The human *CD44* gene structure comprises the common exons 1–5 and 16–20, which spliced together yield the ubiquitous CD44s ("s" for "standard"; also called CD44H, "hematopoietic") polypeptide chain. Exons 6–15 are subject to alternative splicing, encoding the variants designated by CD44v [52–54]. Exon 19 is usually spliced out, yielding a 73-amino-acid cytoplasmic tail, whereas inclusion of this exon determines a 9-amino-acid short tail. It should be noted that an exon numbering alternative exists, comprising exons 1–10 for the standard isoform and the variant exons v1–v10. There is a reasonable degree of conservation (47–93% amino acid identity) between orthologs comparing databases that include 11 mammals and two avian species; and the transmembrane domain is essentially invariant. The N-terminal globulin protein domain extracellular region is determined by the first five exons (1–5) and interacts with

CD44 ligands in general, acting as docking sites for several components of the ECM. The membrane-proximal portion (stem structure) is variable in size and glycosylation depending on alternative splicing. For example, inclusion of alternatively spliced variant exon 3 (v3) encodes a heparan sulfate site that promotes attachment of heparin-binding proteins (Figure 3.2). The major posttranslation modifications include phosphorylation, palmitoylation, and proteolytic cleavage. In addition, several N- and O-linked glycosylation sites promote drastic changes in molecular mass. The 37 kDa estimated molecular weight of CD44s turns out to be an observed band of ~80 kDa, as estimated by gel electrophoresis. Certain types of glycosylation diminish HA-binding, providing a clear mechanism of CD44 activity modulation (for review see Refs [55–57]). According to gene reporter studies conducted with the *CD44* promoter, transcription factors that promote transcription include the activating protein-1 heterodimer composed of Fos/c-Jun and early growth response 1 (also known as transcription factor Zif268 or nerve growth factor-induced protein A), which could link proinflammatory and growth factor signaling to the gene expression [58–62].

CD44 functions vary from lymphocyte activation, mostly through adhesion, to presentation of chemical factors and hormones, which in this case CD44 functions as a coreceptor or acts as a platform for growth factors. However, cell adhesion and tissue architecture maintenance are the most considerable of all. Other functions include signal transduction through association with cytoskeleton actins and monitoring changes in the ECM related with growth [56]. During embryonic development, the CD44 receptor intermediates HA interactions with cells, which in turn induces the activation of ERK and promotes cell proliferation [63,64]. Alternatively, CD44 cleavage by γ-secretase activity generates an intracellular domain fragment that regulates transcriptional events and CD44 function, in this case, via a dominant-negative effect [65,66]. Although the involvement of this receptor in immunity has long been known, it is noteworthy that its modulatory role on the immune response has also been explored with *Cd44*-deficient mice, which develop normally [67]. For instance, one study demonstrated a negative regulation on innate signaling engagement through TLRs [68], and several other studies reported a similar or more complex interplay upon immune challenge (please refer to Chapter 5 for more information on TLRs). The fact that *Cd44*-deficient mice do not show the expected obvious phenotype led to assumptions that another gene replaces *Cd44* during development, implying that conditional and cell-specific knockout mice are much needed to evaluate CD44 roles (for discussion see Ponta et al. [56]). Beyond neuroimmunological and neurodegenerative paradigms, a myriad of studies have focused on the role of CD44 relevant to various neoplasia models, but it is beyond the scope of this chapter to review them.

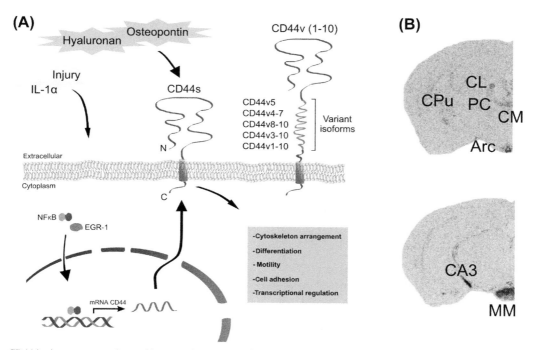

FIGURE 3.2 CD44 brain gene expression and its general structure and cellular functions. (A) Schematic receptor representation along with transcription inducers, ligands, and coupled cellular events. (B) Representative autoradiography photographs depicting *Cd44* transcript localization in the CNS. Abbreviations: Arc, arcuate hypothalamic nucleus; CA3, field CA3 of the hippocampus; CL, centrolateral thalamic nucleus; CM, centromedial thalamic nucleus; CPu, caudate putamen; MM, mammillary nucleus; PC, paracentral thalamic nucleus.

3.3.2 Constitutive Murine *Cd44* Gene Expression Maps to Intralaminar Nuclei of the Thalamus and Few Other Limited Nuclei of the CNS

In adult mice, isotopic ISH in combination with NeuN immunohistochemistry (IHC) reveals a distinct neuronal expression pattern of *Cd44* (mRNA). Moderate signals were detected in thalamic nuclei that underlie the intralaminar formation, such as the paracentral thalamic nucleus, centromedial thalamic nucleus, centrolateral thalamic nucleus, intermediodorsal thalamic nucleus, and parafascicular thalamic nucleus (Figure 3.2). Of note, the habenular commissure also stained positive for the transcript in a substantial manner [2]. Relevant *Cd44* expression is also spotted in some caudate putamen cells, lightly noticeable in few cortical regions and central amygdala, and highly flagrant in the medial mammillary nucleus, pyramidal cell layer of CA1 ventral hippocampus, arcuate nucleus (Arc), and median eminence (ME) of the hypothalamus. At the protein level, Jones and colleagues reported similar regions, but not all of them, by means of IHC with anti-CD44 antibody [69]. Importantly, isotopic ISH signal specificity was verified against the respective sense probes and by the use of two different riboprobes, one for the first five exons and another for the last four exons, both regions present in virtually all transcripts [2]. Although ISH and IHC are mostly equivalent, the few discrepancies

could be explained by the particularities of the different methodologies and by the incursion of the protein through projecting axons, as strongly suggested by the presence of CD44 in axons. The thalamostriatal system has received increasing attention, especially because of pathological findings in progressive supranuclear palsy (PSP) and Parkinson's disease (PD) and the described neurophysiological roles in awareness and motor control [70,71]. It will be very interesting in the future to address the role of CD44 in the thalamostriatal system. The fact that selective regions express *Cd44* prompted the verification of a relationship between the expression sites. The literature suggests that prominent *Cd44* expression sites are associated with awareness and cognition. In addition, another appealing feature of *Cd44* expression relies on the fact that the central amygdala conveys cortical information to autonomic nuclei [31,72–74]. In this sense, it is interesting to speculate a role for Arc and ME as neurosecretory outputs from a putative *Cd44*+ circuitry. Whether CD44+ neurons interact with other CD44-expressing cells, through, for example HA, or if CD44 acts as coreceptor for important molecules involved in neuronal function, awaits investigation. There is no current evidence that CD44-positive neurons are involved in cognition or that CD44 itself plays a role in cell network plasticity or stability. In fact, it is also possible that CD44 expression in circuitry-associated regions is a remnant of common gene expression activated during development. It is nevertheless important to point out that selective analysis of neurons that express *CD44* gene is a possible way to extract regional information about neuronal function.

3.3.3 CD44 Glycoprotein Expression and Functions in the Nervous System

Attempts to purify brain HA-binding proteins date back to the early 1980s [75], and other isolated proteins (proteoglycans) that share this chemical attribute should not be confounded with CD44. Curiously, the description of CD44 in the brain by means of immunoblotting occurred about the same period, one decade earlier than the characterization of its ability to also bind to HA [76]. Before focusing on the roles of CD44 in neuronal cells, it is interesting to verify that glia cells express this receptor in varied settings. In addition, tumors affecting the CNS can express CD44 variants [77], and endothelial cell CD44 influences *Cryptococcus neoformans* brain infection [78]. Hence, several cell types associated with the CNS can express CD44.

During nervous system development, CD44 expression precedes that of the astroglial marker GFAP. The CD44+ precursor cell population is distinct from the populations recognized by A2B5 ganglioside (expressed in type II astrocytes and cells committed to oligodendrocyte lineage), homeobox protein Nkx-2.2 (oligodendrocyte progenitor marker), NG2 proteoglycan (or chondroitin sulfate proteoglycan 4, an oligodendrocyte and possibly some astrocyte progenitors marker), and embryonic neural cell adhesion molecule (or NCAM-H or PSA-NCAM, a neuronal progenitor marker) antibodies, and in general, the percentage of CD44+/GFAP+ cells increases postnatally [79,80]. Intriguingly, a new study describes different populations of CD44-positive astrocytes in human brain. Astrocytes that express this surface antigen presented long processes and were consistently observed in the subpial layer, deep layers of the cortex and hippocampus [81]. The authors also observed variable numbers and shapes of CD44+ astrocytes without long processes in the cortex, depending on age of the subject, suggesting the existence of acquired phenotypes. Astrocytes with different morphologies display CD44 expression in brains from Alzheimer's disease (AD) and PSP patients, implying that CD44 must be useful for identification of astroglial cells relevant to the neuropathological progression of some neurodegenerative diseases [82–84]. Of note, as mentioned above, CD44 is processed by γ-secretase activity, which is of relevance to AD pathology [85].

The expression of CD44 antigen is also useful for characterizing neural stem/progenitor cells. The presence of CD44 in the majority of human embryonic stem cells (hESCs), NSC contaminants, and neural ectoderm-depleted embryoid body makes feasible the use of this marker in a negative selection strategy. For instance, it is possible to sort CD184+/CD271−/CD44−/CD24+ NSCs derived from the H9 hESC line without the undesired cells. In contrast, the isolated CD184+/CD44+ progenitor cells seem to differentiate into mitotic capable astrocytes both in vitro and in vivo [86]. It should be noted that cellular processes are retracted and/or cleaved to a certain extent during preparation for fluorescence-activated cell sorting, implying that surface expression in cell bodies is the index analyzed in these studies. An independent report suggests the use of CD44 as a marker of astrocyte precursor cells in developing postnatal murine cerebellum [87]. However, the same group reported later wide CD44 expression in varied precursor cell types, which becomes restricted to granular cells in adult mouse cerebellum [88]. Oligodendrocyte progenitor cells (OPCs), which are rapidly recruited for repair and subjected to inflammatory modulation [89–91], depend on CD44 to migrate to demyelinated sites in the spinal cord, as suggested by the use of blocking antibodies [92,93]. In a 2013 study, induced overexpression of CD44 in neural precursor cells enabled an improved *trans*-endothelial migration and invasion of perivascular tissues [94]. According to the authors, the strategy is relevant for the development of cell-based therapy to increase migration of intravenously delivered cells into target tissues. Altogether, the evidence clearly shows that subsets of macroglial progenitors express CD44, which is important for migratory cells such as OPCs.

Regarding neuronal expression of CD44, some of the first studies were conducted in the developing nervous system. For example, in the developing chick embryo, CD44 initially appears on cephalic neural fold cells. Later, subpopulations of pre- and migratory cranial neural crest cells also express the receptor [95]. A pioneering study conducted by Stretavan and colleagues described the surface antigen expression on mouse embryonic chiasma neurons [96]. The report not only characterized the neuronal presence of 85 kDa sialylated glycoprotein CD44s by immunoblot and PCR, but also made use of antibodies and a mammalian cell expression system to show its inhibitory effect on retinal axon growth. CD44 also seems to regulate axon crossing at the midline of the chiasma and axon divergence [97]. Interestingly, CD44 protein in the retina is known to be present in Müller cells, but not in retinal neurons [98–100]. However, amacrine cells seem to express the glycoprotein as well, according to observations in retina from *CD44*-reporter transgenic mice [101]. In addition, the use of the CD44 ligands laminin and osteopontin as substrates for culture revealed a role for CD44 in axonal growth of retinal ganglion cells [102]. Noticeably, CD44 expression is to a great extent associated with the visual system and it will be important to prove the role of this molecule in the development and maintenance of this system.

The levels of CD44 ligands HA and OPN change upon neuronal injury. OPN, which also binds certain integrins, is a proinflammatory cytokine released by macrophages and T lymphocytes and an ECM protein synthesized in various tissues, including neural [103,104]. Increased OPN levels are associated with neurodegenerative diseases including multiple sclerosis (MS), AD, and PD [105]. OPN rarely exacerbates such diseases, except for MS, and protective effects have been observed under certain conditions [103]. Since the levels of CD44 ligands undergo critical changes in the injured neural tissue, a potential role for the surface receptor has been investigated. Wang and colleagues reported that CD44 deficiency is neuroprotective in an ischemic injury model [106]. In contrast, OPN-deficient mice displayed increased thalamic lesions when submitted to cortical ischemic stroke [107]. Interestingly, the thalamus is an important site of CD44 neuronal constitutive expression [2]. However, lesions can induce persistent nonneuronal expression of the receptor for up to 2 months, at least in a stab-injury model [108]. Therefore, the development and use of cell-specific *Cd44*-deficient mouse models is necessary to establish a precise CD44 role and rule out the aforementioned possible developmental compensation. In addition, this would help to evaluate the contradictory results, especially regarding OPN, which can bind receptors other than CD44. For example, other studies showed that brain mechanical injury upregulates OPN, which signals typically via integrin $\alpha_v\beta_3$ and CD44; similar associations were reported in the substantia nigra [109,110]. Although it remains to be determined whether it is OPN or HA that masters CD44 signaling after injury, several reports showed CD44 induction in response to brain lesions, for example, in the hippocampal molecular layer, neuropil of thalamic nuclei, polymorphic nucleated cells, infiltrated monocytes, axons and dendrites of motor neurons, demyelinating Schwann cells, and astrocytes [69,111,112]. Because CD44 induction is very consistent in lesion models, it can be used to evaluate axonal regeneration owing to its involvement in neurite outgrowth [113]. Therefore, CD44 induction in various cells, including neurons, is part of an innate response to lesions in the nervous system and seems to organize regenerative mechanisms. CD44 is also expressed in neural crest cells and dorsal root ganglia (neurons and Schwann cells) [114,115]. A study conducted with dorsal root ganglion neurons showed a transduction mechanism involving CD44 and tyrosine kinases that inhibited the plasma membrane Ca^{2+} pump [116]. Engagement of this signaling pathway may influence the excitability of sensory neurons after injury, providing an elegant link between CD44 induction and adaptive neuronal functions. In light of these advances, it will be interesting to explore the functional properties of neural CD44+ cells in animal models and differentiated iPS cells from neurodegenerative disease patients. In comparison with CD36, the roles of CD44 seem to be more complex in the nervous system because of its expression during development and induction in various cell types after injury. In the next section, it will be evident that CD83 is much less studied and characterized, and its inclusion in this chapter

provides an opportune contrast that illustrates the unexplored diversity of surface antigens in the brain.

3.4 THE GLYCOPROTEIN CD83

3.4.1 CD83 Structure and General Functions

CD83 is an integral membrane protein that belongs to the immunoglobulin superfamily. In immunology, this antigen has been extensively used for detecting activated/mature dendritic cells (DCs) since the original description of its expression selectivity toward these very efficient antigen-presenting cells [5]. DCs are not the only cells that express CD83, which is also found transiently on a wide range of leukocytes. The human gene located on chromosome 6 encodes an ~45 kDa membrane protein that possesses a single extracellular V-type immunoglobulin domain (Figure 3.3) and three N-glycosylation sites. The congruence of amino acids between human and murine CD83 is 63%, and in general, the protein is quite conserved in vertebrates. Intriguingly, very little is known for sure regarding CD83 ligands owing to variable results in the literature. Gene expression yields either a membrane (mCD83) or a soluble protein (sCD83) [117–120]. There is evidence that the receptor forms dimers by means of a fifth free cysteine residue present in the polypeptide chain [121]. The human gene promoter has been characterized, disclosing that multiple NF-κB and interferon regulatory sites cooperate for maturation-specific CD83 expression in DCs [122]. Four different splice variants have been reported, one yielding the full-length transmembrane CD83, whereas the others translate to additional shorter fragments, which potentially provide sCD83 proteins [123]. A particular regulatory feature of CD83 expression includes the unusual utilization of a chromosomal region maintenance 1-mediated nuclear export pathway of the transcript, instead of Tap (also termed NXF member 1)-mediated export used by the majority of mRNAs [124]. In addition, the protein HuR is required for *CD83* transcript accumulation in the cytoplasm through binding to a *cis*-regulatory coding sequence and not the canonical AU-rich elements in the 3′ untranslated region [125]. Overall, CD83 is an uncommon cell surface antigen in many aspects.

Since the early 2000s, CD83 function could be inferred by the use of genetic and recombinant protein approaches. The generation of CD83-deficient mice revealed that the antigen expression on thymic stromal cells is important for CD4+ T cell generation [126,127]. Also, observations of *Cd83* promoter activity have been investigated with a knock-in approach that expresses enhanced green fluorescent protein under *Cd83* promoter control, confirming gene activation in B, T, and dendritic cell populations in vivo [128]. Experiments using RNAi demonstrated that CD83 is indeed involved in T cell activation [118]. Conversely, sCD83 seems

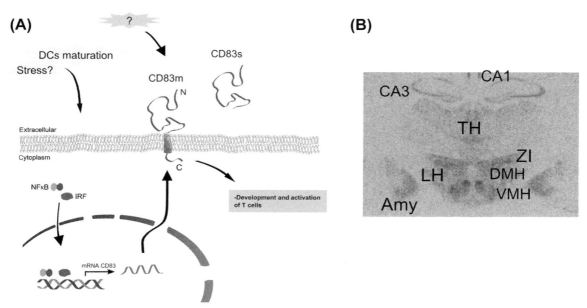

FIGURE 3.3 Widespread CD83 expression in the CNS and its general structure and cellular functions. (A) Schematic receptor representation along with transcription inducers, ligands, and coupled cellular events. (B) Representative autoradiography photographs depicting *Cd83* mRNA expression in the brain. Abbreviations: Amy, amygdala; CA1, field CA1 of the hippocampus; CA3, field CA3 of the hippocampus; DMH, dorsomedial hypothalamic nucleus; LH, lateral hypothalamic area; TH, thalamus; VMH, ventromedial hypothalamic nucleus; ZI, zona incerta.

to be immunosuppressive in different models [129,130]. Despite the important insights into CD83 immune functions, relevant data that could help to understand the role of the surface antigen in neural cells are still missing.

3.4.2 Neuronal *Cd83* Gene Expression is Widespread in Murine Brain, albeit Not Ubiquitous

The fact that robust CD83 expression is in general restricted to a few immune cell populations contrasts sharply with the extensive mRNA expression throughout the CNS. The first report of the mouse *Cd83* gene characterization in 1998 showed relatively abundant expression in brain tissue compared to lung, heart, muscle, testis, and others. In fact, brain levels were quite comparable to those of spleen, which presented the highest transcript enrichment as judged by Northern blot analysis [131]. ISH assay using riboprobes from two independent cloning vectors revealed that *Cd83* (mRNA) is present in the entire rostrocaudal extension of the brain, but not in white matter [2]. Signals were also absent in dorsal and ventral striatum, globus pallidus, dorsal peduncular cortex, and the dorsal part of the lateral septal nucleus and were faint in the substantia nigra. This fact suggests little association of Cd83+ neurons with the classical basal ganglia movement control and certain limbic circuits. The hypothalamus is highly positive for the transcript with the exception of the suprachiasmatic nucleus, whereas the thalamus is also an important expression site, excluding the reticular thalamic nucleus (Figure 3.3). It is quite

remarkable that specific diencephalic nuclei do not express *Cd83*. Lack of expression is also observed in the cerebellar cortex, central nucleus of the inferior colliculus, paramedian reticular nucleus, olive nuclei, lateral lemniscus, and ventral cochlear nuclei.

Brain *Cd83* (mRNA) expression levels are also variable in signal-positive regions. For example, the locus coeruleus is a prominent site of expression, which along with several *Cd83+* thalamic nuclei, reticular formation, and cortical regions, suggests association of the transcript with brain circuits involved in sleep, arousal, consciousness, and psychiatric disorders [132,133]. At the level of the anterior thalamic nuclei, the anterodorsal nucleus presents very high signal intensity. This nucleus may contribute to various aspects of cognitive and memory functions [134]. Other numerous neuroanatomical features could be highlighted regarding *Cd83* gene expression, including high transcript levels in the periaqueductal gray. However, functional characterization of Cd83+ neurons seems to be crucial to establish a meaningful identity of these cells.

3.4.3 Possible CD83 Functions in the Nervous System

Unfortunately, only a small number of reports have described brain CD83 expression relevant to CNS diseases, and all of them focused on DC-like cells, which are referred to as brain DCs because it is unknown how much these cells resemble typical DCs [135–139]. It is quite puzzling that neuronal expression is not reported in their results, raising

TABLE 3.1 Summary of Transcript and Protein Expression in Adult and Developing Nervous Tissue

	Adult NS	Developing Brain	Evidence
Cd36 mRNA	CNS neurons (limited nuclei)	ND	Combined ISH/IHC (NeuN)
CD36 protein	Brain tissue Hypothalamic neurons Neurons (rare)—infarct area Astrocytes Microglia/Macrophage	Not detected in NSCs by FACS Microglia/Macrophage	Western blot Functional Double-IF (βIII-tubulin) Double-IF (GFAP)/Functional Double-IF (CD11b)/Functional
Cd44 mRNA	Brain tissue Dorsal root ganglion (DRG) CNS neurons (some nuclei) Astrocyte (culture) Microglia/Macrophage	Embryonic ventral diencephalon Migrating neural crest cells Retina Immature Purkinje neurons Cerebellar progenitors	RT-PCR Combined ISH/IHC (NeuN) ISH Northern blot Combined ISH/IHC (Iba-1)
CD44 protein	Brain tissue CNS neurons DRG neurons Amacrine cells (retina) Müller cell (retina) Schwann cell Astrocytes Microglia/macrophage	Embryonic ventral diencephalon Immature Purkinje neurons Migrating neural crest cells Cerebellar neural progenitors Optic chiasma neurons (embryo) Retina cell glial progenitor Astrocyte precursor OPCs/OPC (cell line or hESC-derived) Fibroblast-like stem cell	Western Blot/Functional IHC IHC/Functional IHC/Cytometry Double-IF (βIII-tubulin) Knock-in transgenic mice IHC/Double-IF Double-IF (GFAP)/Cytometry Double-IF (NG2)/functional Functional (in vitro)/ Cytometry IHC
Cd83 mRNA	Brain tissue Neurons (various nuclei) Microglia/Macrophage	ND	Northern blot Combined ISH/IHC (NeuN) Combined ISH/IHC (Iba-1)
CD83 protein	Microglia/Macrophage/DC	Not detected in NSCs by FACS	IHC

Please note that CD44 is found in various cell types during adulthood and development. *Cd83* and *Cd36* genes appear to be expressed only in terminally differentiated cells. *Cd83* mRNA is detected in several brain nuclei, but only microglial expression has been characterized at the protein level. This contrasts with reported CD36 glycoprotein presence in neurons and astrocytes. Abbreviations: IF, immunofluorescence; IHC, immunohistochemistry; ISH, in situ hybridization histochemistry. Examples of cell markers: βIII-tubulin (TUBB3), neuronal cells; CD11b (Mac-1 or ITGAM), microglia/macrophages; GFAP, astrocytes in general; Iba-1 (AIF1), microglia/macrophages; NeuN (Fox-3 or RBFOX3), neuronal nucleus; ND, not determined. For further details, please see main text.

the possibility that antibodies are cell-specific owing to different glycosylation patterns. The CD83 protein has not been studied during development, except for its negative expression in NSCs [86]. It is also true that *Cd83* transcript translation into membrane protein has not been verified in neuronal cells in all brain extensions through IHC yet. Interestingly, a genome-wide [140] study that investigated transgenerational effects in autism found significant association with the *CD83* gene. Therapeutically, sCD83 has been shown to interfere with the progression of experimental acute encephalitis, an MS model, revealing important immunomodulatory potential for CD83 [129]. Despite these clinical perspectives, no clear association of CD83 with neuronal function or neuronal cells in disease exists at present.

It is worth mentioning that restraint stress of the animal increases global *Cd83* brain transcript levels. This result implies that the *Cd83* gene promoter in neuronal cells is responsive to this stimulus. The rise in glucocorticoid (GC) levels caused by the stressor does not seem to explain the intensified signals, because systemic LPS, which also increases GCs, fails to increase the transcript [2]. In addition, LPS infused in the brain provokes *Cd83* induction in microglia/macrophages, or even possibly brain DCs, only when GC receptor signaling is blocked [1]. Thus, GCs do not seem to take part in the stress effect on gene expression, or their effect is context dependent. More investigation is required to understand the signaling implicated in *Cd83* neuronal transcription in the presence of stressors.

Functionally, CD83 persists as a perplexing molecule in the CNS regarding its possible neural functions.

3.5 CONCLUDING REMARKS

The structural and functional diversity of CD molecules is proportionate to the existence of unique combinations of expression profiles in neural cells. Depending on the circumstances, CD44 can select different cells ranging from myeloid to macroglial and neuronal, in addition to their progenitor or fully differentiated states. The usefulness of this marker, which is highly involved in cell adhesion, depends on its association with other neural surface antigens or genetically engineered reporter proteins under the control of specific promoters. CD36 seems to be useful for identifying fatty-acid-responsive hypothalamic neurons, and further research is needed to determine if some olfactory circuitry neurons express this antigen on their cell surface as well. Less is known regarding CD83, but in the case of protein translation being coupled to transcript levels, tissue dissection combined with high- and low-signal cytometer analysis might be able to sort unique neurons such as those of the anterodorsal thalamic nucleus. The function of CD83 in neuronal cells remains undetermined. Table 3.1 provides a summary of the transcript and protein expression of these three molecules in the adult and developing nervous tissue.

The probing of neuronal cells, specifically, with antibodies against surface antigens faces the problem of antibody characterization and antigen enrichment in the cell body. Because some antigens such as CD44 can be present in cellular projections lost during cell sorting procedures, the use of mRNA provides a plausible solution. A similar situation applies in the case of CD83, because it is possible that the cell secretes the protein as sCD83. Fortunately, live-cell sorting can be performed with mRNA probing using molecular beacons [141] or reagents such as nano-flare probes [142]. The use of this strategy along with ISH analysis and conventional sorting will represent a powerful association for cell-based large-scale profiling. The advances in robotic sample processing and single-cell analysis promise huge data acquisition, which will allow an extraordinary growth of knowledge about antigen cell markers and neural population phenotyping.

REFERENCES

[1] Glezer I, Chernomoretz A, David S, Plante MM, Rivest S. Genes involved in the balance between neuronal survival and death during inflammation. PLoS One 2007;2:e310.

[2] Glezer I, Bittencourt JC, Rivest S. Neuronal expression of Cd36, Cd44, and Cd83 antigen transcripts maps to distinct and specific murine brain circuits. J Comp Neurol 2009;517:906–24.

[3] Silverstein RL, Febbraio M. CD36, a scavenger receptor involved in immunity, metabolism, angiogenesis, and behavior. Sci Signal 2009;2:re3.

[4] Johnson P, Ruffell B. CD44 and its role in inflammation and inflammatory diseases. Inflamm Allergy Drug Targets 2009;8:208–20.

[5] Zhou LJ, Tedder TF. Human blood dendritic cells selectively express CD83, a member of the immunoglobulin superfamily. J Immunol 1995;154:3821–35.

[6] Fernández-Ruiz E, Armesilla AL, Sánchez-Madrid F, Vega MA. Gene encoding the collagen type I and thrombospondin receptor CD36 is located on chromosome 7q11.2. Genomics 1993;17:759–61.

[7] Silverstein RL. Type 2 scavenger receptor CD36 in platelet activation: the role of hyperlipemia and oxidative stress. Clin Lipidol 2009;4:767–79.

[8] Oquendo P, Hundt E, Lawler J, Seed B. CD36 directly mediates cytoadherence of *Plasmodium falciparum* parasitized erythrocytes. Cell 1989;58:95–101.

[9] Hoosdally SJ, Andress EJ, Wooding C, Martin CA, Linton KJ. The human scavenger receptor CD36: glycosylation status and its role in trafficking and function. J Biol Chem 2009;284:16277–88.

[10] Franc NC, Dimarcq JL, Lagueux M, Hoffmann J, Ezekowitz RAB. Croquemort, a novel *Drosophila* hemocyte/macrophage receptor that recognizes apoptotic cells. Immunity 1996;4:431–43.

[11] Hart K, Wilcox M. A *Drosophila* gene encoding an epithelial membrane protein with homology to CD36/LIMP II. J Mol Biol 1993;234:249–53.

[12] Teboul L, Febbraio M, Gaillard D, Amri EZ, Silverstein R, Grimaldi PA. Structural and functional characterization of the mouse fatty acid translocase promoter: activation during adipose differentiation. Biochem J 2001;360:305–12.

[13] Benton CR, Nickerson JG, Lally J, Han X-X, Holloway GP, Glatz JFC, et al. Modest PGC-1alpha overexpression in muscle in vivo is sufficient to increase insulin sensitivity and palmitate oxidation in subsarcolemmal, not intermyofibrillar, mitochondria. J Biol Chem 2008;283:4228–40.

[14] Qiao L, Zou C, Shao P, Schaack J, Johnson PF, Shao J. Transcriptional regulation of fatty acid translocase/CD36 expression by CCAAT/enhancer-binding protein alpha. J Biol Chem 2008;283:8788–95.

[15] Jin-Lian C, Xiao-Jie L, Kai-Lin Z, Kun Y. Kruppel-like factor 2 promotes liver steatosis through upregulation of CD36. J Lipid Res 2013;55(1):32–40.

[16] Nicholson AC, Han J, Febbraio M, Silversterin RL, Hajjar DP. Role of CD36, the macrophage class B scavenger receptor, in atherosclerosis. Ann N Y Acad Sci 2001;947:224–8.

[17] Park YM, Febbraio M, Silverstein RL. CD36 modulates migration of mouse and human macrophages in response to oxidized LDL and may contribute to macrophage trapping in the arterial intima. J Clin Invest 2009;119:136–45.

[18] Chu LY, Ramakrishnan DP, Silverstein RL. Thrombospondin-1 modulates VEGF signaling via CD36 by recruiting SHP-1 to VEGFR2 complex in microvascular endothelial cells. Blood 2013;122:1822–32.

[19] Abumrad NA, el-Maghrabi MR, Amri EZ, Lopez E, Grimaldi PA. Cloning of a rat adipocyte membrane protein implicated in binding or transport of long-chain fatty acids that is induced during preadipocyte differentiation. Homology with human CD36. J Biol Chem 1993;268:17665–8.

[20] Benton R, Vannice KS, Vosshall LB. An essential role for a CD36-related receptor in pheromone detection in *Drosophila*. Nature 2007;450:289–93.

[21] Gaillard D, Passilly-Degrace P, Besnard P. Molecular mechanisms of fat preference and overeating. Ann N Y Acad Sci 2008;1141: 163–75.

[22] Laugerette F, Passilly-Degrace P, Patris B, Niot I, Febbraio M, Montmayeur J-P, et al. CD36 involvement in orosensory detection of dietary lipids, spontaneous fat preference, and digestive secretions. J Clin Invest 2005;115:3177–84.

[23] Hoebe K, Georgel P, Rutschmann S, Du X, Mudd S, Crozat K, et al. CD36 is a sensor of diacylglycerides. Nature 2005;433:523–7.

[24] Kim K-Y, Stevens MV, Akter MH, Rusk SE, Huang RJ, Cohen A, et al. Parkin is a lipid-responsive regulator of fat uptake in mice and mutant human cells. J Clin Invest 2011;121:3701–12.

[25] Boehm U, Zou Z, Buck LB. Feedback loops link odor and pheromone signaling with reproduction. Cell 2005;123:683–95.

[26] Guillamon A, Segovia S. Sex differences in the vomeronasal system. Brain Res Bull 1997;44:377–82.

[27] Scalia F, Winans SS. The differential projections of the olfactory bulb and accessory olfactory bulb in mammals. J Comp Neurol 1975;161:31–55.

[28] Yoon H, Enquist LW, Dulac C. Olfactory inputs to hypothalamic neurons controlling reproduction and fertility. Cell 2005;123:669–82.

[29] Simerly RB. Organization and regulation of sexually dimorphic neuroendocrine pathways. Behav Brain Res 1998;92:195–203.

[30] Simerly RB, Swanson LW. Projections of the medial preoptic nucleus: a *Phaseolus vulgaris* leucoagglutinin anterograde tract-tracing study in the rat. J Comp Neurol 1988;270:209–42.

[31] Swanson LW. Cerebral hemisphere regulation of motivated behavior. Brain Res 2000;886:113–64.

[32] Coraci IS, Husemann J, Berman JW, Hulette C, Dufour JH, Campanella GK, et al. CD36, a class B scavenger receptor, is expressed on microglia in Alzheimer's disease brains and can mediate production of reactive oxygen species in response to beta-amyloid fibrils. Am J Pathol 2002;160:101–12.

[33] Yamanaka M, Ishikawa T, Griep A, Axt D, Kummer MP, Heneka MT. PPARγ/RXRα-induced and CD36-mediated microglial amyloid-β phagocytosis results in cognitive improvement in amyloid precursor protein/presenilin 1 mice. J Neurosci 2012;32:17321–31.

[34] Ricciarelli R, D'Abramo C, Zingg J-M, Giliberto L, Markesbery W, Azzi A, et al. CD36 overexpression in human brain correlates with beta-amyloid deposition but not with Alzheimer's disease. Free Radic Biol Med 2004;36:1018–24.

[35] El Khoury JB, Moore KJ, Means TK, Leung J, Terada K, Toft M, et al. CD36 mediates the innate host response to beta-amyloid. J Exp Med 2003;197:1657–66.

[36] Cusulin C, Monni E, Ahlenius H, Wood J, Brune JC, Lindvall O, et al. Embryonic stem cell-derived neural stem cells fuse with microglia and mature neurons. Stem Cells 2012;30:2657–71.

[37] Cho S, Park E, Febbraio M, Anrather J, Park L, Racchumi G, et al. The class B scavenger receptor CD36 mediates free radical production and tissue injury in cerebral ischemia. J Neurosci 2004;25:2504–12.

[38] Bao Y, Qin L, Kim E, Bhosle S, Guo H, Febbraio M, et al. CD36 is involved in astrocyte activation and astroglial scar formation. J Cereb Blood Flow Metab 2012;32:1567–77.

[39] Kunz A, Abe T, Hochrainer K, Shimamura M, Anrather J, Racchumi G, et al. Nuclear factor-kappaB activation and postischemic inflammation are suppressed in CD36-null mice after middle cerebral artery occlusion. J Neurosci 2008;28:1649–58.

[40] Garbes L, Heesen L, Hölker I, Bauer T, Schreml J, Zimmermann K, et al. VPA response in SMA is suppressed by the fatty acid translocase CD36. Hum Mol Genet 2013;22:398–407.

[41] Abumrad NA, Ajmal M, Pothakos K, Robinson JK. CD36 expression and brain function: does CD36 deficiency impact learning ability? Prostagl Other Lipid Mediat 2005;77(1–4):77–83.

[42] Le Foll C, Irani BG, Magnan C, Dunn-Meynell AA, Levin BE. Characteristics and mechanisms of hypothalamic neuronal fatty acid sensing. Am J Physiol Regul Integr Comp Physiol 2009;297:R655–64.

[43] Drahota Z, Vrbacký M, Nusková H, Kazdová L, Zídek V, Landa V, et al. Succinimidyl oleate, established inhibitor of CD36/FAT translocase inhibits complex III of mitochondrial respiratory chain. Biochem. Biophys Res Commun 2010;391:1348–51.

[44] Le Foll C, Dunn-Meynell A, Musatov S, Magnan C, Levin BE. FAT/CD36: a major regulator of neuronal fatty acid sensing and energy homeostasis in rats and mice. Diabetes 2013;62:2709–16.

[45] Hughes EN, Mengod G, August JT. Murine cell surface glycoproteins. Characterization of a major component of 80,000 daltons as a polymorphic differentiation antigen of mesenchymal cells. J Biol Chem 1981;256:7023–7.

[46] Trowbridge IS, Lesley J, Schulte R, Hyman R, Trotter J. Biochemical characterization and cellular distribution of a polymorphic, murine cell-surface glycoprotein expressed on lymphoid tissues. Immunogenetics 1982;15:299–312.

[47] Carter WG, Wayner EA. Characterization of the class III collagen receptor, a phosphorylated, transmembrane glycoprotein expressed in nucleated human cells. J Biol Chem 1988;263:4193–201.

[48] Gallatin WM, Wayner EA, Hoffman PA, St John T, Butcher EC, Carter WG. Structural homology between lymphocyte receptors for high endothelium and class III extracellular matrix receptor. Proc Natl Acad Sci USA 1989;86:4654–8.

[49] Jalkanen ST, Bargatze RF, Herron LR, Butcher EC. A lymphoid cell surface glycoprotein involved in endothelial cell recognition and lymphocyte homing in man. Eur J Immunol 1986;16:1195–202.

[50] Aruffo A, Stamenkovic I, Melnick M, Underhill CB, Seed B. CD44 is the principal cell surface receptor for hyaluronate. Cell 1990;61:1303–13.

[51] Fraser JRE, Laurent TC, Laurent UBG. Hyaluronan: its nature, distribution, functions and turnover. J Intern Med 1997;242:27–33.

[52] Goodison S, Urquidi V, Tarin D. CD44 cell adhesion molecules. Mol Pathol 1999;52:189–96.

[53] Goodfellow PN, Banting G, Wiles MV, Tunnacliffe A, Parkar M, Solomon E, et al. The gene, MIC4, which controls expression of the antigen defined by monoclonal antibody F10.44.2, is on human chromosome 11. Eur J Immunol 1982;12:659–63.

[54] Screaton GR, Bell MV, Jackson DG, Cornelis FB, Gerth U, Bell JI. Genomic structure of DNA encoding the lymphocyte homing receptor CD44 reveals at least 12 alternatively spliced exons. Proc Natl Acad Sci USA 1992;89:12160–4.

[55] Bajorath J. Molecular organization, structural features, and ligand binding characteristics of CD44, a highly variable cell surface glycoprotein with multiple functions. Proteins 2000;39:103–11.

[56] Ponta H, Sherman L, Herrlich PA. CD44: from adhesion molecules to signalling regulators. Nat Rev Mol Cell Biol 2003;4:33–45.

[57] Thorne RF, Legg JW, Isacke CM. The role of the CD44 transmembrane and cytoplasmic domains in co-ordinating adhesive and signalling events. J Cell Sci 2004;117:373–80.

[58] Fitzgerald KA, O'Neill LA. Characterization of CD44 induction by IL-1: a critical role for Egr-1. J Immunol 1999;162:4920–7.

[59] Foster LC, Wiesel P, Huggins GS, Pañares R, Chin MT, Pellacani A, et al. Role of activating protein-1 and high mobility group-I(Y) protein in the induction of CD44 gene expression by interleukin-1beta in vascular smooth muscle cells. FASEB J 2000;14:368–78.

[60] Zhang M, Wang MH, Singh RK, Wells A, Siegal GP. Epidermal growth factor induces CD44 gene expression through a novel regulatory element in mouse fibroblasts. J Biol Chem 1997;272:14139–46.

[61] Vendrov AE, Madamanchi NR, Niu X-L, Molnar KC, Runge M, Szyndralewiez C, et al. NADPH oxidases regulate CD44 and hyaluronic acid expression in thrombin-treated vascular smooth muscle cells and in atherosclerosis. J Biol Chem 2010;285:26545–57.

[62] Maltzman JS, Carman JA, Monroe JG. Role of EGR1 in regulation of stimulus-dependent CD44 transcription in B lymphocytes. Mol Cell Biol 1996;16:2283–94.

[63] Hatano H, Shigeishi H, Kudo Y, Higashikawa K, Tobiume K, Takata T, et al. RHAMM/ERK interaction induces proliferative activities of cementifying fibroma cells through a mechanism based on the CD44-EGFR. Lab Invest 2011;91:379–91.

[64] Matrosova VY, Orlovskaya IA, Serobyan N, Khaldoyanidi SK. Hyaluronic acid facilitates the recovery of hematopoiesis following 5-fluorouracil administration. Stem Cells 2004;22:544–55.

[65] Lammich S, Okochi M, Takeda M, Kaether C, Capell A, Zimmer A-K, et al. Presenilin-dependent intramembrane proteolysis of CD44 leads to the liberation of its intracellular domain and the secretion of an Abeta-like peptide. J Biol Chem 2002;277:44754–9.

[66] Mellor L, Knudson CB, Hida D, Askew EB, Knudson W. Intracellular domain fragment of CD44 alters CD44 function in chondrocytes. J Biol Chem 2013;288:25838–50.

[67] Protin U, Schweighoffer T, Jochum W, Hilberg F. CD44-deficient mice develop normally with changes in subpopulations and recirculation of lymphocyte subsets. J Immunol 1999;163:4917–23.

[68] Kawana H, Karaki H, Higashi M, Miyazaki M, Hilberg F, Kitagawa M, et al. CD44 suppresses TLR-mediated inflammation. J Immunol 2008;180:4235–45.

[69] Jones LL, Liu Z, Shen J, Werner A, Kreutzberg GW, Raivich G. Regulation of the cell adhesion molecule CD44 after nerve transection and direct trauma to the mouse brain. J Comp Neurol 2000;426:468–92.

[70] Henderson JM, Carpenter K, Cartwright H, Halliday GM. Loss of thalamic intralaminar nuclei in progressive supranuclear palsy and Parkinson's disease: clinical and therapeutic implications. Brain 2000;123(Pt 7):1410–21.

[71] Smith Y, Galvan A, Ellender TJ, Doig N, Villalba RM, Huerta-Ocampo I, et al. The thalamostriatal system in normal and diseased states. Front Syst Neurosci 2014;8:5.

[72] Berendse HW, Groenewegen HJ. Organization of the thalamostriatal projections in the rat, with special emphasis on the ventral striatum. J Comp Neurol 1990;299:187–228.

[73] Smith Y, Raju DV, Pare JF, Sidibe M. The thalamostriatal system: a highly specific network of the basal ganglia circuitry. Trends Neurosci 2004;27:520–7.

[74] van der Werf YD, Witter MP, Groenewegen HJ. The intralaminar and midline nuclei of the thalamus. Anatomical and functional evidence for participation in processes of arousal and awareness. Brain Res Rev 2002;39:107–40.

[75] Delpech B, Halavent C. Characterization and purification from human brain of a hyaluronic acid-binding glycoprotein, hyaluronectin. J Neurochem 1981;36:855–9.

[76] Dalchau R, Kirkley J, Fabre JW. Monoclonal antibody to a human leukocyte-specific membrane glycoprotein probably homologous to the leukocyte-common (L-C) antigen of the rat. Eur J Immunol 1980;10:737–44.

[77] Resnick DK, Resnick NM, Welch WC, Cooper DL. Differential expressions of CD44 variants in tumors affecting the central nervous system. Mol Diagn 1999;4:219–32.

[78] Jong A, Wu C-H, Gonzales-Gomez I, Kwon-Chung KJ, Chang YC, Tseng H-K, et al. Hyaluronic acid receptor CD44 deficiency is associated with decreased cryptococcus neoformans brain infection. J Biol Chem 2012;287:15298–306.

[79] Liu Y, Wu Y, Lee JC, Xue H, Pevny LH, Kaprielian Z, et al. Oligodendrocyte and astrocyte development in rodents: an in situ and immunohistological analysis during embryonic development. Glia 2002;40:25–43.

[80] Alfei L, Aita M, Caronti B, De Vita R, Margotta V, Medolago Albani L, et al. Hyaluronate receptor CD44 is expressed by astrocytes in the adult chicken and in astrocyte cell precursors in early development of the chick spinal cord. Eur J Histochem 1999;43:29–38.

[81] Sosunov AA, Wu X, Tsankova NM, Guilfoyle E, McKhann GM, Goldman JE. Phenotypic heterogeneity and plasticity of isocortical and hippocampal astrocytes in the human brain. J Neurosci 2014;34:2285–98.

[82] Akiyama H, Tooyama I, Kawamata T, Ikeda K, McGeer PL. Morphological diversities of CD44 positive astrocytes in the cerebral cortex of normal subjects and patients with Alzheimer's disease. Brain Res 1993;632:249–59.

[83] Yamada T, Calne DB, Akiyama H, McGeer EG, McGeer PL. Further observations on tau-positive glia in the brains with progressive supranuclear palsy. Acta Neuropathologica 1993;85(3):308–15.

[84] Haegel H, Tolg C, Hofmann M, Ceredig R. Activated mouse Astrocytes and T Cells express similar CD44 variants. Role of CD44 in Astrocyte/T cell binding. J Cell Biol 1993;122:1067–77.

[85] Koo EH, Kopan R. Potential role of presenilin-regulated signaling pathways in sporadic neurodegeneration. Nat Med 2004;10(Suppl):S26–33.

[86] Yuan SH, Martin J, Elia J, Flippin J, Paramban RI, Hefferan MP, et al. Cell-surface marker signatures for the isolation of neural stem cells, glia and neurons derived from human pluripotent stem cells. PLoS One 2011;6:e17540.

[87] Cai N, Kurachi M, Shibasaki K, Okano-Uchida T, Ishizaki Y. CD44-positive cells are candidates for astrocyte precursor cells in developing mouse cerebellum. Cerebellum 2012;11:181–93.

[88] Naruse M, Shibasaki K, Yokoyama S, Kurachi M, Ishizaki Y. Dynamic changes of CD44 expression from progenitors to subpopulations of astrocytes and neurons in developing cerebellum. PLoS One 2013;8:e53109.

[89] Crawford AH, Chambers C, Franklin RJM. Remyelination: the true regeneration of the central nervous system. J Comp Pathol 2013;149:242–54.

[90] Glezer I, Lapointe A, Rivest S. Innate immunity triggers oligodendrocyte progenitor reactivity and confines damages to brain injuries. FASEB J 2006;20:750–2.

[91] Glezer I, Rivest S. Oncostatin M is a novel glucocorticoid-dependent neuroinflammatory factor that enhances oligodendrocyte precursor cell activity in demyelinated sites. Brain Behav Immun 2010;24:695–704.

[92] Piao J-H, Wang Y, Duncan ID. CD44 is required for the migration of transplanted oligodendrocyte progenitor cells to focal inflammatory demyelinating lesions in the spinal cord. Glia 2013;61:361–7.

[93] Sundberg M, Skottman H, Suuronen R, Narkilahti S. Production and isolation of NG2+ oligodendrocyte precursors from human embryonic stem cells in defined serum-free medium. Stem Cell Res 2010;5:91–103.

[94] Deboux C, Ladraa S, Cazaubon S, Ghribi-Mallah S, Weiss N, Chaverot N, et al. Overexpression of CD44 in neural precursor cells improves trans- endothelial migration and facilitates their invasion of perivascular tissues in vivo. PLoS One 2013;8:e57430.

[95] Corbel C, Lehmann A, Davison F. Expression of CD44 during early development of the chick embryo. Mech Dev 2000;96:111–4.

[96] Sretavan DW, Feng L, Puré E, Reichardt LF. Embryonic neurons of the developing optic chiasm express L1 and CD44, cell surface molecules with opposing effects on retinal axon growth. Neuron 1994;12:957–75.

[97] Lin L, Chan S-O. Perturbation of CD44 function affects chiasmatic routing of retinal axons in brain slice preparations of the mouse retinofugal pathway. Eur J Neurosci 2003;17:2299–312.

[98] Chaitin MH, Wortham HS, Brun-Zinkernagel AM. Immunocytochemical localization of CD44 in the mouse retina. Exp Eye Res 1994;58:359–65.

[99] Nishina S, Hirakata A, Hida T, Sawa H, Azuma N. CD44 expression in the developing human retina. Graefe's Arch Clin Exp Ophthalmol 1997;235:92–6.

[100] Shinoe T, Kuribayashi H, Saya H, Seiki M, Aburatani H, Watanabe S. Identification of CD44 as a cell surface marker for Müller glia precursor cells. J Neurochem 2010;115:1633–42.

[101] Sarthy V, Hoshi H, Mills S, Dudley VJ. Characterization of green fluorescent protein-expressing retinal cells in CD 44-transgenic mice. Neuroscience 2007;144:1087–93.

[102] Ries A, Goldberg JL, Grimpe B. A novel biological function for CD44 in axon growth of retinal ganglion cells identified by a bioinformatics approach. J Neurochem 2007;103:1491–505.

[103] Brown A. Osteopontin: a key link between immunity, inflammation and the central nervous system. Transl Neurosci 2012;3:288–93.

[104] Al'Qteishat A, Gaffney J, Krupinski J, Rubio F, West D, Kumar S, et al. Changes in hyaluronan production and metabolism following ischaemic stroke in man. Brain 2006;129:2158–76.

[105] Carecchio M, Comi C. The role of osteopontin in neurodegenerative diseases. J Alzheimer's Dis 2011;25:179–85.

[106] Wang X, Xu L, Wang H, Zhan Y, Puré E, Feuerstein GZ. CD44 deficiency in mice protects brain from cerebral ischemia injury. J Neurochem 2002;83:1172–9.

[107] Schroeter M, Zickler P, Denhardt DT, Hartung H-P, Jander S. Increased thalamic neurodegeneration following ischaemic cortical stroke in osteopontin-deficient mice. Brain 2006;129:1426–37.

[108] Stylli SS, Kaye AH, Novak U. Induction of CD44 expression in stab wounds of the brain: long term persistence of CD44 expression. J Clin Neurosci 2000;7:137–40.

[109] Plantman S. Osteopontin is upregulated after mechanical brain injury and stimulates neurite growth from hippocampal neurons through β1 integrin and CD44. Neuroreport 2012;23:647–52.

[110] Ailane S, Long P, Jenner P, Rose S. Expression of integrin and CD44 receptors recognising osteopontin in the normal and LPS-lesioned rat substantia nigra. Eur J Neurosci 2013:1–9.

[111] Bausch SB. Potential roles for hyaluronan and CD44 in kainic acid-induced mossy fiber sprouting in organotypic hippocampal slice cultures. Neuroscience 2006;143:339–50.

[112] Borges K, McDermott DL, Dingledine R. Reciprocal changes of CD44 and GAP-43 expression in the dentate gyrus inner molecular layer after status epilepticus in mice. Exp Neurol 2004;188:1–10.

[113] Ruff CA, Staak N, Patodia S, Kaswich M, Rocha-Ferreira E, Da Costa C, et al. Neuronal c-Jun is required for successful axonal regeneration, but the effects of phosphorylation of its N-terminus are moderate. J Neurochem 2012;121:607–18.

[114] Ikeda K, Nakao J, Asou H, Toya S, Shinoda J, Uyemura K. Expression of CD44H in the cells of neural crest origin in peripheral nervous system. Neuroreport 1996;7:1713–6.

[115] Casini P, Nardi I, Ori M. Hyaluronan is required for cranial neural crest cells migration and craniofacial development. Dev Dyn 2012;241(2):294–302.

[116] Ghosh B, Li Y, Thayer SA. Inhibition of the plasma membrane Ca^{2+} pump by CD44 receptor activation of tyrosine kinases increases the action potential afterhyperpolarization in sensory neurons. J Neurosci 2011;31:2361–70.

[117] Zhou LJ, Schwarting R, Smith HM, Tedder TF. A novel cell-surface molecule expressed by human interdigitating reticulum cells, langerhans cells, and activated lymphocytes is a new member of the Ig superfamily. J Immunol 1992;149:735–42.

[118] Prechtel AT, Turza NM, Theodoridis AA, Steinkasserer A. CD83 knockdown in monocyte-derived dendritic cells by small interfering RNA leads to a diminished T cell stimulation. J Immunol 2007;178:5454–64.

[119] Prazma CM, Tedder TF. Dendritic cell CD83: a therapeutic target or innocent bystander? Immunol Lett 2008;115:1–8.

[120] Wang X, Wei MQ, Liu X. Targeting CD83 for the treatment of graft-versus-host disease. Exp Ther Med 2013;5:1545–50.

[121] Lechmann M, Kotzor N, Zinser E, Prechtel AT, Sticht H, Steinkasserer A. CD83 is a dimer: comparative analysis of monomeric and dimeric isoforms. Biochem Biophys Res Commun 2005;329:132–9.

[122] Stein MF, Lang S, Winkler TH, Deinzer A, Erber S, Nettelbeck DM, et al. Multiple interferon regulatory factor and NF-κB sites cooperate in mediating cell-type- and maturation-specific activation of the human CD83 promoter in dendritic cells. Mol Cell Biol 2013;33:1331–44.

[123] Dudziak D, Nimmerjahn F, Bornkamm GW, Laux G. Alternative splicing generates putative soluble CD83 proteins that inhibit T cell proliferation. J Immunol 2005;174:6672–6.

[124] Rodriguez MS, Dargemont C, Stutz F. Nuclear export of RNA. Biol Cell/Under Auspices Eur Cell Biol Organ 2002;96:639–55.

[125] Prechtel AT, Chemnitz J, Schirmer S, Ehlers C, Langbein-Detsch I, Stülke J, et al. Expression of CD83 is regulated by HuR via a novel cis-active coding region RNA element. J Biol Chem 2006;281:10912–25.

[126] Fujimoto Y, Tu L, Miller AS, Bock C, Fujimoto M, Doyle C, et al. CD83 expression influences CD4+ T cell development in the thymus. Cell 2002;108:755–67.

[127] García-Martínez LF, Appleby MW, Staehling-Hampton K, Andrews DM, Chen Y, McEuen M, et al. A novel mutation in CD83 results in the development of a unique population of CD4+ T cells. J Immunol 2004;173:2995–3001.

[128] Lechmann M, Shuman N, Wakeham A, Mak TW. The CD83 reporter mouse elucidates the activity of the CD83 promoter in B, T, and dendritic cell populations in vivo. Proc Natl Acad Sci USA 2008;105:11887–92.

[129] Zinser E, Lechmann M, Golka A, Lutz MB, Steinkasserer A. Prevention and treatment of experimental autoimmune encephalomyelitis by soluble CD83. J Exp Med 2004;200:345–51.

[130] Ge W, Arp J, Lian D, Liu W, Baroja ML, Jiang J, et al. Immunosuppression involving soluble CD83 induces tolerogenic dendritic cells that prevent cardiac allograft rejection. Transplantation 2010;90:1145–56.

[131] Twist CJ, Beier DR, Disteche CM, Edelhoff S, Tedder TF. The mouse Cd83 gene: structure, domain organization, and chromosome localization. Immunogenetics 1998;48:383–93.

[132] Jones BE. Paradoxical sleep and its chemical/structural substrates in the brain. Neuroscience 1991;40:637–56.

[133] Parvizi J, Damasio A. Consciousness and the brainstem. Cognition 2001;79:135–60.

[134] Shibata H, Honda Y. Thalamocortical projections of the anterodorsal thalamic nucleus in the rabbit. J Comp Neurol 2012;520:2647–56.

[135] Kaneko T, Kaneko M, Chokechanachaisakul U, Kawamura J, Kaneko R, Sunakawa M, et al. Artificial dental pulp exposure injury up-regulates antigen-presenting cell-related molecules in rat central nervous system. J Endod 2010;36:459–64.

[136] Hollenbach R, Sagar D, Khan ZK, Callen S, Yao H, Shirazi J, et al. Effect of morphine and SIV on dendritic cell trafficking into the central nervous system of rhesus macaques. J Neurovirol 2014;20:175–83.

[137] Dietel B, Cicha I, Kallmünzer B, Tauchi M, Yilmaz A, Daniel WG, et al. Suppression of dendritic cell functions contributes to the anti-inflammatory action of granulocyte-colony stimulating factor in experimental stroke. Exp Neurol 2012;237:379–87.

[138] Henkel JS, Engelhardt JI, Siklós L, Simpson EP, Kim SH, Pan T, et al. Presence of dendritic cells, MCP-1, and activated microglia/macrophages in amyotrophic lateral sclerosis spinal cord tissue. Ann Neurol 2004;55:221–35.

[139] Plumb J, Armstrong MA, Duddy M, Mirakhur M, McQuaid S. CD83-positive dendritic cells are present in occasional perivascular cuffs in multiple sclerosis lesions. Mult Scler 2003;9:142–7.

[140] Tsang KM, Croen LA, Torres AR, Kharrazi M, Delorenze GN, Windham GC, et al. A genome-wide survey of transgenerational genetic effects in autism. PLoS One 2013;8:e76978.

[141] Larsson HM, Lee ST, Roccio M, Velluto D, Lutolf MP, Frey P, et al. Sorting live stem cells based on Sox2 mRNA expression. PLoS One 2012;7:e49874.

[142] Seferos DS, Giljohann DA, Hill HD, Prigodich AE, Mirkin CA. Nano-flares: probes for transfection and mRNA detection in living cells. J Am Chem Soc 2007;129:15477–9.

Chapter 4

Life and Death in the CNS: The Role of CD95

Si Chen*, Robert Hermann*, Enric Llorens-Bobadilla* and Ana Martin-Villalba
Division of Molecular Neurobiology, German Cancer Research Center (DKFZ), Heidelberg, Germany

4.1 INTRODUCTION

CD95, also known as Fas or Apo-1, is a type 1 transmembrane receptor that belongs to the tumor necrosis factor (TNF) receptor superfamily. Members of this family, perhaps best known for their role as inducers of apoptosis, are sensors of the extracellular environment of the cell and signal through a variety of intracellular effectors. Proteins in this group are characterized by the presence of well-conserved cysteine-rich domains at the extracellular receptor–ligand interaction site [1]. CD95, one of the best-studied receptors in the group, is in addition a member of the so-called "death receptor" subfamily, defined by the presence of an 80-amino-acid "death domain" (DD) in the cytoplasmic segment of the protein. The DD functions as a docking site for the adaptor protein FADD (Fas-associated protein with death domain). This domain also contains a YXXL motif that, once phosphorylated, can interact with SH2-containing proteins [2,3]. CD95 also harbors a less-characterized membrane-proximal domain involved in the interaction with other adaptor proteins [4]. The ligand for CD95, CD95L, is a type 2 transmembrane protein, and a member of the TNF superfamily of cytokines. Like in the other members of the family, trimerization of the ligand molecules through well-conserved internal residues exposes a less-conserved external surface that allows receptor-specific binding [1]. Binding of the trimerized ligand to CD95 induces the trimerization of the receptor and triggers downstream signaling. CD95L at the membrane can undergo metalloproteinase-mediated cleavage to generate a soluble form of the ligand [5].

CD95 apoptotic signaling is mediated upon binding of the trimerized CD95L, leading to clustering of the CD95 receptor and stabilization of the open conformation of the intracellular DD [6]. This allows FADD to be bound to DD and the subsequent recruitment of procaspase-8 through the death effector domain (DED) of FADD. Once the number of procaspase-8 molecules exceeds the number of FADD molecules at the receptor, the caspase-activating chain can cluster and form the death-inducing signaling complex (DISC) [7,8]. Caspase-8 is then activated through self-cleavage, resulting in activation of caspase-3 either directly or indirectly via the mitochondrial apoptotic pathway and, finally, apoptotic cell death [9] (Figure 4.1).

However, there is also clear evidence suggesting that other cell fates may be mediated by CD95 activation that have nonapoptotic, even opposite, outcomes. CD95 ligation has been shown to activate (1) the three main MAPK (mitogen-activated protein kinase) pathways, namely p38, c-Jun N-terminal kinase (JNK) 1/2, and extracellular signal-regulated kinase (ERK) 1/2; (2) the phosphoinositide 3-kinase (PI3K) pathway via binding of SH2 domains to phosphorylated DD; and (3) the transcription factor nuclear factor κB (NF-κB). Activation of these pathways results in cell proliferation, migration, and inflammation (Figure 4.1) [10,11].

The mechanisms that lead to differential outcomes of CD95 activation are not fully elucidated. Similar to other TNF receptor superfamily members, CD95 has multifaceted functions under different conditions. One possible explanation is that components of both apoptotic and nonapoptotic signaling are activated simultaneously, but it is the balance of apoptosis-inhibitory molecules, such as inhibitors of apoptosis and members of the Bcl-2 family, and the availability of cell-growth-inducing kinases that determine the ultimate cell fate. Moreover, it has been suggested that the membrane-bound CD95L is responsible for eliciting apoptosis, whereas other cellular functions are mediated through circulating soluble CD95L, although this remains controversial [10].

In the following sections, we summarize both apoptotic and nonapoptotic functions of CD95 in the healthy and diseased brain and give an overview of the differential

---

* Equal contribution.

Neural Surface Antigens. http://dx.doi.org/10.1016/B978-0-12-800781-5.00004-9

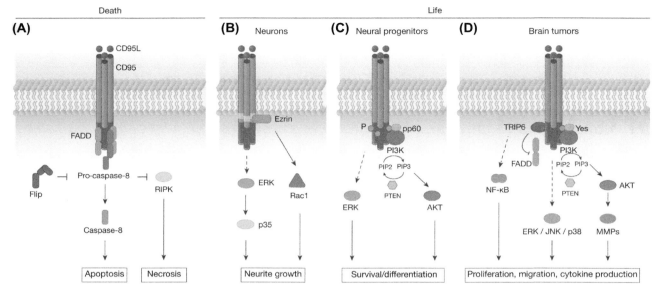

FIGURE 4.1 CD95 signaling in life and death. (A) Death signaling by CD95. Upon binding of the trimerized CD95L, FADD is recruited to the DD of CD95. Procaspase-8 subsequently binds FADD through its DED to form the DISC, which is inhibited in the presence of Flip. Activation of procaspase-8 leads to apoptosis. In addition, DISC formation prevents RIPK activation and programmed necrosis. (B–D) Life signaling by CD95. (B) CD95 signaling in neurons. Trimerized CD95L binding results in activation of ERK and subsequent upregulation of p35 expression that causes neurite growth in neurons. Alternatively, the adaptor protein ezrin is recruited to the MPD of CD95 and the small GTPase Rac1 is activated to signal neurite growth. (C) CD95 signaling in neural progenitors. ERK or AKT activation induces survival and differentiation of neural progenitors upon CD95L binding. AKT in this case is activated through phosphorylation of the DD by Src family kinase pp60 and subsequent binding of PI3K and is inhibited by PTEN. (D) CD95 signaling in brain tumors. Upon CD95L binding, the Src family kinase Yes phosphorylates the DD that serves as the docking site for the p85 subunit (not shown) of PI3K. PI3K-dependent activation of AKT induces migration via the induction of MMPs. Moreover, the adaptor protein TRIP6 inhibits FADD association of CD95 by binding to the intracellular domain of CD95 and promotes NF-κB activation and migration. Last, the ERK, JNK, and p38 pathways are activated upon CD95L binding and are responsible for survival, proliferation, and the induction of cytokine expression. Abbreviations: CD95L, CD95 ligand; FADD, Fas-associated protein with death domain; FLIP, FLICE-like inhibitory protein; RIPK, receptor-interacting protein kinase; ERK, extracellular signal-regulated kinase; MPD, membrane proximal domain; PTEN, phosphatase and tensin homolog; P, phosphotyrosine; PI3K, phosphoinositide 3-kinase; PIP2, phosphatidylinositol 4,5-bisphosphate; PIP3, phosphatidylinositol (3,4,5)-trisphosphate; MMP, metalloprotease; TRIP6, thyroid hormone receptor interactor 6; JNK, c-Jun N-terminal kinase; NF-κB, nuclear factor κB. *Adapted from Ref. [10].*

signaling pathways that are involved in the diverse and fascinating outcomes of CD95 activation.

4.2 CD95 EXPRESSION IN THE HEALTHY AND DISEASED BRAIN

4.2.1 CD95 Expression in the Developing and Healthy Adult Brain

For many years little attention was given to the CD95L/CD95 system in neural tissues because the most apparent defects of mutant mice were immune system related. However, this view has changed dramatically, as it has become clear that various cells express CD95L and CD95 during normal neural development, in restricted areas of the adult brain and in the diseased state.

During brain development CD95 appears to be widely expressed. In rodents, CD95 expression starts at embryonic day (E)14–E15, continues until birth, peaks at early postnatal stages, and declines thereafter [12,13]. Areas expressing CD95 include the ventricular and subventricular zones, hippocampus, cerebellum, and cortex. CD95L seems to follow

a similar expression pattern [14,15]. Although expression remains poorly characterized in vivo, it is clear that embryonic neural progenitors and developing neurons express CD95 in vitro and that CD95 expression is conserved in human embryonic neural progenitors [16–18].

After completion of neural development, CD95 and CD95L expression levels are downregulated and become almost undetectable in the healthy adult brain. It is remarkable, however, that expression persists in the neurogenic regions. In both the subventricular zone (SVZ) and the subgranular zone of the hippocampal dentate gyrus (DG), CD95 appears to be restricted to neural stem cells (NSCs) and it is downregulated during lineage differentiation [19]. Somatic stem cells in other organs also express CD95 and thus CD95 expression might be a common feature of adult tissue stem cells [20–22]. Of note, mouse and human embryonic stem cells do not seem to express CD95, or express it at very low levels. Here, expression increases as differentiation proceeds toward tissue-restricted progenitors [23,24].

In the adult human brain, expression probably remains at locations proximal to blood vessels, presumably to ensure peripheral deletion of autoimmune cells [25].

4.2.2 CD95 Expression after Brain Injury

Injury to the brain increases CD95 expression in several cell populations, including neurons, astrocytes, oligodendrocytes, and microglia (reviewed in Ref. [26]). No study, however, directly addressed whether this upregulation also occurs in NSCs. CD95 and CD95L (re)expression or upregulation in neurons and glial cells of compromised brains seems to be a broadly used, general response in mouse models for a variety of neurologic disorders such as inflammatory diseases (e.g., multiple sclerosis/experimental autoimmune encephalomyelitis), trauma (traumatic brain injury, spinal cord injury), neurodegenerative diseases (Alzheimer disease, Parkinson disease, amyotrophic lateral sclerosis), infectious diseases (viral neurodegeneration in Moloney murine leukemia virus, human immunodeficiency virus), and cerebral ischemia and other insults (seizure, hydrocephalus, ethanol). In addition this increased expression was also reported to occur in human neurological disorders [19,27,28]. A detailed summary of CD95L and CD95 expression regulation in the compromised brain can be found in a review by Choi et al. [26].

4.2.3 CD95 Expression in Brain Tumors

Studies of CD95 expression and function in human brain tumors have been mainly carried out in gliomas. The expression of CD95 was initially identified in human glioma cell lines and ex vivo glioma cells [29] and was later confirmed in tumor biopsies of various grades of astrocytomas on the mRNA as well as the protein level [30,31]. Expression of CD95 occurs in a subset of tumors and tends to positively correlate with the degree of malignancy, with WHO Grade III and IV tumors having the highest levels of CD95 compared to lower grade gliomas. The CD95 protein was found to be predominantly expressed in association with large areas of ischemic necrosis within tumor tissue [30,32,33].

The ligand for the CD95 receptor is also coexpressed in many astrocytomas, mainly as the membrane-bound form [34]. The expression does not correlate with the grade of malignancy as opposed to CD95 and was initially found to be more dispersed [32,35]. Later, in a syngeneic mouse glioma model, CD95L was found to be expressed mainly at the tumor/host interface and was detected in tumor cells, neurons, and macrophages as well as close to vessels within the tumor [36].

A few studies have also begun to characterize CD95 expression in glioma cancer stem cells (CSCs). The mRNA expression of CD95 and CD95L in glioma CSCs that are grown as tumor spheres in basic fibroblast growth factor and epidermal growth factor has been reported to be lower compared to that in primary adherent glioma cell cultures in serum [37,38]. The mechanistic and functional implications of this observation are still poorly studied.

In other types of brain tumors, such as oligodendrogliomas, ependymomas, and pediatric medulloblastomas, a subset of tumors was also found to express CD95 as well as the CD95L [39–42]. In these tumors, the functions of CD95 remain largely unknown. In summary, CD95 is widely expressed among various kinds of tumors of the central nervous system (CNS) and may represent a general feature of tumorigenesis.

4.2.4 Controversies Regarding CD95 Detection

Despite being the subject of intensive research for many years, many CD95 reagents, most prominently antibodies, were apparently not tested thoroughly enough or used with the necessary caution. In fact some antibodies were found to be nonspecific long after being used in published work [43]. Moreover, expression of CD95 seems to be more easily and frequently detectable in (primary) cultured cells or cell lines compared to their in vivo counterparts. This might be a direct consequence of a function of the CD95 system, i.e., responding to external insults and/or changed microenvironment with its subsequent upregulation. Thus, in vitro data need to be interpreted with caution. Consequently, the precise spatiotemporal expression patterns of CD95L and CD95 in the healthy and pathologic brain remain insufficiently characterized.

4.3 FUNCTIONS AND SIGNALING OF CD95 IN THE CNS

4.3.1 CD95-Mediated Neurite Growth in Neurons

Early reports on the role of the CD95L/CD95 system in developing neurons focused on establishing a link to the apoptotic phases of immature neurons and often inferred an apoptotic function of CD95 activation [13,26,44]. Other death-inducing phenotypes of activated CD95 in developing neurons were found using in vitro assays with cell lines and primary cultures, which, however, might not reflect the actual requirement for the receptor in the developing brain in vivo [44,45]. In contrast to these findings, later studies revealed that nonapoptotic signaling of CD95 is indeed crucial for neuron maturation. For instance, activation of the CD95 receptor in developing or regenerating neurons promotes their neurite growth. In this case, activation of CD95 leads to downstream activation of ERK and subsequent expression of p35 (see below). Accordingly, neurite growth after sciatic nerve injury is delayed in *lpr* mice and can be accelerated by local administration of activating CD95 antibodies [46]. CD95 also modulated branching of neuronal processes in developing cortical and hippocampal neurons [15]. Cultured hippocampal neurons increase their number of branches after CD95 activation and mice with no or reduced levels of functional CD95 or CD95L display

---

Box 4.1 Glossary of Mouse Models

lpr: Mice with strongly decreased CD95 expression due to an endogenous retrovirus insertion into intron 2 of the *fas* gene.

MRL-*lpr*: Mouse model of systemic lupus erythematosus, in which *lpr* was crossed on the Murphy Roths Large background.

gld: Mice bearing a point mutation near the 3′ end of the coding sequence of the *fasl* gene, causing expression of a nonfunctional form of CD95L.

EAE: Experimental autoimmune encephalomyelitis mouse model presenting inflammatory demyelinating disease in the CNS; model of multiple sclerosis.

MOG-cre:CD95$^{f/f}$: A cross between mice in which Cre recombinase is cloned into the first exon of the oligodendrocyte-specific myelin oligodendrocyte glycoprotein (*mog*) gene and mice harboring floxed alleles of CD95 exon 9.

LysM-cre:CD95$^{f/f}$: Mouse model of specific CD95 deletion in peripheral myeloid.

SOD^{G93A} and SOD^{G85R}: Mutations of the SOD1 gene inducing amyotrophic lateral sclerosis.

MPTP: 1-Methyl-4-phenyl-1,2,3,6-tetrahydropyridine-induced model of Parkinson disease.

---

abnormal branching phenotypes: pyramidal neurons of the CA1 region have less complex dendritic trees in *lpr* mice compared to their wild-type counterparts, and *gld* mice exhibit similar reductions in branching (see Box 4.1 for explanation of mouse models) [15].

Downstream components of the apoptotic branch of CD95 signaling have been extensively studied and were briefly described in the introduction to this chapter. Nonapoptotic signaling of CD95 seems to be similarly diverse and is the subject of ongoing investigations. Enhanced branching of developing neurons after CD95 activation was shown to be caspase-independent and to involve dephosphorylation of Tau, a protein promoting assembly and stabilization of microtubules and thus crucial for the establishment of neuronal morphology [15]. On the other hand, stimulation of CD95 in regenerating neurons activates ERK and subsequently upregulates expression of p35 (Figure 4.1), a neuron-specific activator of cyclin-dependent kinase 5 [46]. The ERK/p35 pathway has already been described as the molecular mechanism underlying nerve growth factor (NGF)-induced axon regeneration, suggesting that NGF and CD95 can converge on the ERK pathway, presumably via Raf-1 [47]. Moreover, a previously unknown functional region of CD95, the cytoplasmic membrane proximal domain, was found to recruit the adaptor protein ezrin and subsequently activate the small GTPase Rac1 to signal neurite growth in cortical neurons (Figure 4.1) [4].

4.3.2 Functional Relevance of CD95 in the Injured CNS

CD95 activity changes drastically after various insults to the brain. In addition to CD95- and CD95L-expressing immune cells infiltrating the brain, many cells within the brain, including neurons, (re)express CD95 and/or CD95L. The functions of CD95 signaling during these phases can be diverse, including both apoptotic and nonapoptotic proregenerative signaling. Thus, the CD95/CD95L system plays crucial roles in many neuropathological conditions, which are briefly summarized here (see Ref. [48] for an extensive review).

The CD95L/CD95 system was shown to be of fundamental importance during cerebral adult focal ischemia through middle cerebral artery or bilateral common carotid artery occlusion (MCAo and BCCAo). Inhibition of CD95 activity (using CD95 *lpr* mice) decreased the infarct volume in several studies and improved locomotor recovery [49–51].

After spinal cord injury (SCI), activation of CD95 leads to apoptosis of cells in the affected area, including injured motor neurons. Several studies have shown that blocking CD95 signaling after SCI, either by using genetic models (CD95 *lpr* mice) or through injection of anti-CD95L antibody or soluble CD95 receptor, reduces apoptotic events, increases axonal regeneration, and improves locomotor activity [52–55]. These studies relied on mouse models and experiments with general CD95 deficiency or inhibition and thus the contribution of CD95 to injury regarding the specific cell type/compartment was not characterized. Later Letellier et al. showed that SCI triggers CD95L expression in peripheral blood myeloid cells, enhancing their migration to the injury site via activation of a CD95/PI3K/matrix metalloprotease (MMP)-9 signaling pathway in these cells [56]. Deletion of CD95L in myeloid cells, but not of CD95 in neural cells, led to the functional recovery of spinal-injured animals, suggesting that CD95L acts on peripheral myeloid cells to induce tissue damage.

Neurodegenerative diseases are widespread, multicausal, and complex pathologies characterized by progressive neuronal death and with poor prognosis for patients. Implication of the CD95 system in death and survival decisions in neurons has triggered interest in studying this pathway in many of the neurodegenerative diseases. For instance, in superoxide dismutase 1 (SOD1) mutation-driven models of amyotrophic lateral sclerosis (ALS) experimental interference with CD95 signaling increased survival, decreased motor neuron loss, and improved motor function [57–59]. Similarly, both Aβ25–35/1–40- and Aβ1–42-induced neurotoxicity in Alzheimer disease was reduced in cortical neurons of CD95 *lpr* and CD95 *gld* mice [60,61]. The role of CD95 in Parkinson disease remains more controversial: both increased and decreased dopaminergic neuron death in the substantia nigra were reported in CD95-deficient mice after exposure to the neurotoxin MPTP [62–64]. In experimental autoimmune encephalomyelitis (EAE) models of multiple sclerosis genetic deletion of CD95 (*lpr* mice) or CD95L function (*gld* mice) decreased apoptotic cell death (including that of oligodendrocytes), inflammation, demyelination, and severity of symptoms [65–69].

The involvement of CD95 in the induction of neuronal death has been implicated by many studies in the examples above. However, many of these studies were only correlational and it was assumed that upregulation of CD95 will lead to cell death. Because it has become clear that the CD95L/CD95 system has nonapoptotic functions, these studies have to be taken with caution. The best demonstrations of a causal link between CD95 function and neuronal cell death are found under experimental conditions including oxidative stress, such as ischemia [47]. Reduced infarct volume in *lpr* mice or mice treated with CD95L antibody established a causal link of CD95 activity and neuronal death. Similarly, in motor neurons of mouse models for ALS, reactive oxygen species (ROS) accumulation renders these neurons sensitive to CD95-mediated apoptosis. In fact, motor neurons seem to be the only neurons spontaneously sensitive to CD95-induced death [70]. On the other hand, in many other studied insults to the brain, such as trauma or neurodegenerative disease, causative links are weak or missing, whereas more recent evidence for a nonapoptotic function of CD95 in these conditions is emerging. Thus, upregulation of CD95 in neurons and other cells under those conditions might rather represent a protective mechanism designed to increase proregenerative signaling to the cell [47].

4.3.3 Role of CD95 in Neural Progenitors

As mentioned above, both embryonic and adult neural progenitors express CD95. Initial reports studying mechanisms of cell death in neural progenitors found that although neural progenitors are sensitive to mitochondrial cell death, they are insensitive to CD95-induced apoptosis [16,71,72]. When C17.2 cells, an immortalized mouse embryonic neural progenitor cell line, or primary adult neural progenitors were treated with DMNQ, a drug that increases mitochondrial ROS production, cells readily underwent apoptosis. Induction of apoptosis was blocked by the caspase inhibitor zVAD-fmk. These cells also underwent apoptosis when treated with the caspase-dependent apoptosis inducer staurosporin. However, when the cells were treated with an activating CD95 monoclonal antibody no induction of apoptosis could be observed. Ricci-Vittiani et al. also reached similar conclusions in their work with primary embryonic and adult human neural progenitors [17]. When treated with agonistic antibodies to CD95, and to other death receptors, neural progenitor cells did not undergo apoptosis. This was initially hypothesized to be due to the low expression of caspase-8, a critical downstream effector of the extrinsic apoptotic pathway.

However, pretreatment of these cells with proinflammatory cytokines that strongly upregulate caspase-8 expression did not render the cells sensitive to CD95-mediated death. In addition, exogenous overexpression of caspase-8 to levels similar to those observed in apoptosis-sensitive cell lines, such as Jurkat, was also insufficient to engage CD95 activation in apoptotic signaling. In contrast, parallel experiments conducted in neurons showed that they are extremely sensitive to CD95-mediated apoptosis when exposed to the same inflammatory environment in vitro. This suggests that even if neural progenitors possess a competent apoptotic machinery, they are selectively protected from CD95-induced death.

In the absence of an apoptosis-inducing function, the fact that CD95 is present in neural progenitors in homeostasis and in pathological conditions, and that this expression is conserved across species, suggested an alternative role.

Initial evidence showed that CD95 activation in neural progenitor cells could activate the growth pathway ERK [16], which was in line with the growing body of evidence of nonapoptotic roles for CD95 in the nervous system [15,46,48]. The role of CD95 in adult neural stem cells was later clarified in an extensive study by Corsini et al. [19]. Using neural progenitor cells from the adult SVZ it was observed that CD95L stimulation increased their survival in serial sphere assays. This finding was later corroborated in an independent study using neonatal neural progenitors [73]. Importantly, only CD95-expressing cells from the SVZ and the DG have neurosphere-forming capacity, suggesting that CD95 expression is a defining feature of neurosphere-forming cells. The relationship between CD95 and other surface markers enriched in neurosphere-initiating cells (i.e., CD133, LeX, CD24) remains to be studied.

In addition, CD95L stimulation in NSCs increased their differentiation toward the neuronal lineage [19]. This proneurogenic effect of CD95 was shown to act through the activation of the PI3K/AKT (protein kinase B)/mammalian target of rapamycin (mTOR) signaling axis. CD95L stimulation activates PI3K signaling through the recruitment of the Src family kinase member pp60 (c-Src). Subsequently, PI3K activation leads to the activation of AKT and mTOR, which initiate the neurogenic program. In vivo, conditional deletion of CD95 in adult NSCs and their progeny led to reduced generation of newborn neurons. Interestingly, these mice, in which CD95 was exclusively removed from the adult neurogenic compartment, showed reduced performance in memory tasks. The deficits in neurogenesis and working memory could be rescued by running, a powerful neurogenic stimulus, suggesting either that CD95 signaling is not required for exercise-induced neurogenesis or that alternative stem cell pools exist and responded to the stimulus [19]. Thus, CD95 acts as a survival and neurogenic signal in the adult stem cell niches through the activation of nonapoptotic pathways and is required to maintain homeostatic neurogenesis.

4.3.3.1 Potential of CD95 in Neural Progenitor-Based Therapies

CD95/CD95L activity is strongly increased in the brain in several pathological conditions. Because CD95-expressing neural progenitors interpret CD95 activation as a survival signal, it is attractive to harness this system to direct NSCs to migrate to the injury site and contribute to the repair process. Indeed, in a mouse model of global ischemia that induces CA1 neuronal damage, transplanted wild-type but not CD95-deficient neural progenitors were able to generate neurons that integrated into the injured CA1 and contributed to functional recovery [19].

Another exciting possibility is to exploit CD95 activity in neural progenitors not only in cis, activating their survival and differentiation, but also in trans, inducing the death of apoptosis-sensitive immune cells. This strategy would be of particular interest to prevent immune rejection of grafted cells or in the treatment of neurological disorders with an autoimmune component. Pluchino et al. originally tested this possibility in a mouse model of EAE [74]. Neural progenitors from the SVZ of adult mice injected intravenously in EAE mice selectively entered the CNS, where they exerted a neuroprotective effect. The grafted progenitor cells increased the apoptosis of CNS-infiltrating lymphocytes, an effect that was abrogated upon CD95L blockade. Other reports, in contrast, have shown that transplanted wild-type and CD95-deficient neural progenitors do not behave differently [75]. Interestingly, an analogous strategy has also proven successful with bone marrow-derived mesenchymal cell transplants in mouse models of colitis and systemic sclerosis [20]. In summary, CD95 in stem cells can have a dual activity, on one side increasing stem cell survival and differentiation capacity and on the other side providing immunosuppression.

4.3.4 Multifaceted Functions of CD95 in Brain Tumors

4.3.4.1 CD95-Mediated Apoptosis in Glioma

CD95 was early recognized as an apoptosis receptor that triggers cell death in malignant cells, resulting in tumor regression [76]. Owing to its killing effect and the low expression level of CD95 in normal brain [77], CD95 activation was soon regarded as a possible strategy for specific glioma therapy. Indeed, in various glioma cell lines, treatment with an agonistic anti-CD95 antibody (CH-11 monoclonal) induces apoptosis [29]. Resistance to CD95-induced apoptosis was also observed in some glioma lines (e.g., LN-319, LN-405, LN-308); however, this was explained by their low level of CD95 expression. Upon transfection of the CD95 cDNA, susceptibility to apoptosis can be induced [78,79]. Additionally, CD95 apoptosis-resistant glioma lines have low levels of caspase-8 owing to hypermethylation of its regulatory elements. Restoring the function of caspase-8, which is essential for apoptotic signaling, by global demethylation using 5-aza-2′-deoxycytidine or caspase-8 gene transfer renders cells sensitive to CD95-induced apoptosis and also chemotherapy-mediated cell death [80].

CD95-induced apoptosis is not restricted in vitro and could also be observed in vivo in a syngeneic mouse glioma model (SMA560) [81] and a syngeneic rat glioma model (36B10) [79] after local administration of CD95 antibody.

Interestingly, treatment with the CD95L could even potentiate CD95-mediated cytotoxicity in glioma cells. Freshly resected high-grade gliomas grown ex vivo are only moderately sensitive to anti-CD95 antibody, but soluble CD95L increased apoptosis significantly and could largely inhibit clonal tumor cell growth [82].

As discussed above, studies with agonistic anti-CD95 antibodies point toward an apoptosis-inducing effect. However, this has to be taken with caution, because it has been shown that the Fc portion of various agonistic antibodies alone, including that of anti-CD95 antibodies, can induce cytotoxicity [83,84].

Based on the finding that two adjacent trimeric CD95 ligands are required for CD95 signaling and formation of the DISC [85], the recombinant CD95L APO010, a hexameric protein containing two trimers of CD95L fused to the collagen domain of ACRP30/adiponectin, was developed. This molecule triggers caspase-dependent glioma cell death in both cell lines and primary glioma cells, although the effectiveness depends on the level of CD95 expression. U87MG xenografts receiving intracranial administration of APO010 showed 80% increased survival. However, systemic administration of APO010 showed only a mild delay in tumor growth, due to insufficient delivery. In combination with temozolomide, the current standard of care for high-grade gliomas, a synergistic effect could also be shown in LNT-229 cells [86]. Based on the effects of APO010 in these preclinical studies, a phase I nonrandomized clinical trial using APO010 in patients with solid tumors was launched in 2007 (NCT00437736). However, this trial was discontinued (for unknown reasons).

Several studies have shown that CD95L, when administered together with chemotherapeutics such as camptothecin [87], etoposide [88], doxorubicin, and vincristine [89], could enhance glioma apoptotic death and increase sensitivity to CD95 or chemotherapeutic-mediated apoptosis through chemotherapy-induced oxidative stress [87]. Thus, combining CD95L with chemotherapy was also suggested as a possible treatment strategy.

Moreover, CD133+ glioma CSCs isolated from U-87 cells express monomeric CD95 and are resistant to CD95-mediated apoptosis, whereas more mature cells express oligomeric CD95 and are sensitive to CD95-mediated apoptosis [37]. This indicates that glioma CSCs may not be entirely eradicated with CD95-inducing therapy.

4.3.4.2 CD95-Induced Proliferation, Migration, and Immune Functions in Glioma

The overwhelming majority of studies of CD95/CD95L interactions have concentrated on cell death as a primary readout, overshadowing the role of CD95 engagement as a driver of other outcomes. The realization that CD95 cannot have only apoptotic functions in malignancies came from the notion that a significant degree of resistance to CD95-mediated apoptosis exists in various tumor types despite moderate to high expression of CD95 [29,41,82,90–93]. Moreover, in gliomas, the apoptotic index is very low, with up to 0.1% in low-grade astrocytomas and less than 2% in high-grade glioblastoma multiforme (GBM), which makes it unlikely that the presence of CD95 and CD95L is sufficient to trigger extensive apoptosis in vivo [32]. The first indication that CD95 may have additional, nonapoptotic functions came from studies showing that CD95 ligation augmented the proliferation of CD3-activated primary T cells [94].

Subsequent molecular studies of CD95 in different systems unveiled manifold nonapoptotic functions of the CD95 signaling complex involving MAPKs (p38, JNK, ERK), NF-κB, and PI3K downstream signaling that have implications in proliferation, migration, and immune cell recruitment in various human malignancies (Figure 4.1) [10,11,91].

In the case of glioma, CD95 stimulation with CH-11 was also shown to enhance proliferation and cell cycle progression in several cell lines (A-172, T98G, YKG-1). Conversely, the anti-CD95L antibody NOK-2 inhibited proliferation. The enhanced proliferation is mediated by activation of the MAPK/ERK pathway upon CD95 activation (Figure 4.1) [95].

Likewise, in many other types of cancers, e.g., ovarian and liver cancers, CD95 has a growth-promoting role that is mediated by JNK and ERK activation downstream of CD95 resulting in phosphorylation of Jun and increased expression of the transcription factors Egr1 and Fos. Reducing CD95 or CD95L expression thus reduced tumor cell growth [96].

In 2014, inhibition of CD95 was shown to induce a new form of cell death termed "death induced by CD95 or CD95L elimination." CD95 is required for cancer cell survival in various cancer types, including gliomas, and its elimination leads to death that resembles a necrotic form of mitotic catastrophe [97].

Thus, CD95-mediated proliferation seems to be a general tumorigenic mechanism in various types of tumors, and inhibition, instead of enhancing CD95 activity, should be considered as a strategy for therapy.

Another protumorigenic function of CD95 in glioma is its role as a key driver of basal invasion [36]. In both apoptosis-resistant and apoptosis-sensitive glioma cell lines, as well as patient-derived glioma lines, activated

CD95 was shown to recruit both the Src family kinase Yes and the p85 subunit of PI3K, causing PI3K and AKT activation and the subsequent expression of MMPs, including MMP-2 and MMP-9 (Figure 4.1). In a syngeneic mouse model of intracranial GBM (SMA560), neutralization of CD95 dramatically reduced the number of invading cells. Interestingly, the intracranially transplanted glioma cells increased their expression of CD95 and CD95L, opening the possibility that the microenvironment promotes malignancy by inducing CD95/CD95L expression [36].

Similarly, Wisniewski et al. observed impaired motility and invasiveness of C6 rat glioma cells after blockade of CD95 signaling with a FasL-interfering protein. This was associated with a downregulation of MAPK signaling, AP-1- and NF-κB-driven transcription, and TIMP2 and MMP-2 activity [98].

Moreover, the adaptor protein thyroid hormone receptor interactor 6 (TRIP6) is implicated in blocking CD95-mediated apoptosis and in inducing migration. TRIP6 is expressed at high levels in apoptosis-resistant glioma cell lines and inhibits FADD association with CD95 by directly binding to the intracellular domain of CD95, thus blocking the proapoptotic pathway. This effect is enhanced by the growth-factor-like phospholipid lysophosphatidic acid. TRIP6 activates NF-κB to further block apoptosis and is phosphorylated via Src family kinases to induce a migratory phenotype in glioma cells (Figure 4.1) [99].

With regard to the highly invasive phenotype of GBM that complicates treatment and worsens outcome, targeting the CD95/CD95L system, which contributes to the invasion-prone phenotype of malignant glioma cells, seems to be a promising strategy.

Based on this idea, a clinical trial for GBM was launched with APG101 (Apocept™), a fusion protein consisting of the extracellular domain of the CD95 receptor fused to the Fc domain of IgG. APG101 binds CD95L and thereby prevents activation of the CD95 signaling pathway. In the first phase I trial in healthy volunteers and two GBM patients with the primary endpoints being safety and tolerability, no adverse events were reported [100]. The study was continued in a randomized controlled phase II clinical trial in which a combination of APG101 with radiotherapy was compared to stand-alone radiotherapy in relapsed GBM patients (NCT01071837). The study objective of increasing the percentage of patients reaching 6 months of progression-free survival by 100% in the experimental arm was substantially exceeded [101]. The success of this clinical trial demonstrates that the strategy of blocking rather than stimulating CD95 should be further pursued.

When looking at tumor development, the influence of the microenvironment should be taken into account.

TABLE 4.1 Overview of CD95 Expression and Function in the Central Nervous System

Species	Cell Type/Tissue	Manipulation/Model	Phenotype	References
Embryonic and Adult Neural Progenitors				
Mouse	C17.2 cells, immortalized neural progenitors	Fas mAb treatment	Resistance to apoptosis	[16]
	Adult SVZ and DG neural progenitors	CD95L treatment	Increased survival and neuronal differentiation	[19]
	Neonatal neural progenitors	CD95L treatment	Increased survival	[73]
		CD95-deficiency (*lpr*)	Decreased proliferation, increased survival, neuronal and oligodendrocyte differentiation	[110]
Human	Embryonic and adult neural progenitors	Fas mAb (CH-11) treatment	Resistance to apoptosis	[17]
Mouse	Subventricular zone progenitors	CD95L-overexpression	Increased generation of neuroblasts	[19]
	Subgranular zone progenitors	Deletion of CD95	Decreased generation of neuroblasts	[19]
	Subventricular zone progenitors	Transplantation of CD95-deficient cells in ischemic brain	Reduced integration at injury site	[19]
	Mouse neonatal neural progenitors	Transplantation of CD95-deficient cells in EAE mice	Equal neuroprotection	[75]
Regeneration & Neurite Growth				
Mouse	Sciatic neurons (58 weeks of age)	Sciatic nerve crush injury in *lpr* mice	Recovery delayed in *lpr* mice, accelerated by CD95 activating antibody	[46]
	Primary cortical neurons (from E15)	Ezrin-derived inhibitory peptide	Prevented CD95-induced neurite growth in primary neurons	[4]
	Hippocampus, CA1 pyramidal neurons (P5), Purkinje cells (P10), cultured hippocampal neurons (from E15 and P1)	*lpr* and *gld* mice, CD95L treatment in vitro	Reduced branching in vivo in *lpr* and *gld* mice, no increase in apoptosis, but branching in vitro	[15]
Ischemia				
Mouse	Brain (adult)	Cerebral ischemia: BCCAo or MCAo in *lpr* mice, *gld* mice, or anti-CD95L antibody treatment	Reduced infarct volumes 24 h postinjury, CD95 mRNA induction 6 h postinjury, improved locomotor activity	[49–51,111]
Human	Astrocytes (fetal and adult), pyramidal neurons, CA1 region	IL-1, IL-6, TNF-α treatment, ischemia patient	Increased expression of CD95 and CD95L, increased apoptosis	[19,27]
Trauma				
Mouse	Spinal cord	SCI by transection or clip compression; MRL/*lpr* mice, anti-CD95L antibody, sCD95	Reduced apoptosis, enhanced functional neurological outcome, better locomotor recovery, damaged area smaller	[52–55]
		SCI by transection; LysM-cre:CD95L$^{-/-}$ mice	Decreased number of neutrophils and macrophages infiltrating the injured spinal cord, better functional recovery	[56]
	Motor cortex, hippocampus (6–8weeks of age)	Controlled cortical impact in MRL/*lpr* mice	Impaired improvement of the spatial memory function, fewer TUNEL-positive hippocampal neurons	[112]

Species	Cell Type/Tissue	Manipulation/Model	Phenotype	References
Rat	Cortex (adult)	Cortical impact injury, stereotactic lesion	Upregulated expression of CD95 in neurons and astrocytes and of CD95L in neurons, astrocytes, and microglia	[113,114]
Neurodegenerative Disease				
Mouse	Dopaminergic neurons (8–10 and 17–18 weeks of age)	MPTP-induced model for PD in *lpr* and *gld* mice	Damage of the midbrain dopaminergic system, dopaminergic neurons rendered highly susceptible to degeneration in response to MPTP, exacerbated neurotoxicity or attenuated MPTP-induced death	[62–64]
	Brain (adult)	EAE in *lpr* and *gld* or MOG-cre:CD95$^{-/-}$ mice	Decreased apoptosis, inflammation and severity of symptoms, partial protection from development of EAE	[65–69]
		ALS (SOD1 mutations), *gld* mice, CD95-Fc, CD95 siRNA	Reduced loss of motor neurons, improved motor function and survival	[57–59]
	Cortical neurons	Alzheimer disease, *lpr* and *gld* mice	Decreased neurotoxicity	[60,61]
Brain Tumors: Apoptosis				
Human	TN98G, LN-215, LN-18	CD95 agonistic CH-11 antibody	Apoptosis	[29]
	LN-319, LN-405, LN-308	Transfection of CD95 cDNA	Apoptosis of initially resistant cells	[78]
Mouse	SMA560	Jo-2 antibody (CD95 agonist)	Apoptosis, increased survival in vivo of tumor-bearing mice	[81]
Human	Patient GBM-derived cultures	Soluble murine CD95L	Apoptosis, higher than with CD95 antibody	[82]
Rat	F98, 36B10 glioma cells and orthotopic xenograft	CD95L overexpression (rAd-CMV-FasL); CD95 overexpression (CD95 cDNA)	Apoptosis, increased survival of tumor-bearing rats	[79,92]
Human	U87, U373, LN-229, LN-308, T98G glioma cells and orthotopic xenograft	CD95L with chemotherapeutics (camptothecin, adriamycin, etoposide, doxirubicin, carmustina, vincristine, teniposide, 5-fluorouracil, cytarabine)	Increase in sensitivity to apoptosis, reduction of tumor growth	[87–89]
Brain Tumors: Resistance to Apoptosis				
Human	LN-405, LN308, LN-319, LN-229 glioma cells	CH-11 antibody (CD95 agonist)	Resistance to apoptosis	[29]
	Ex vivo medulloblastomas and primitive neuroectodermal tumors	FasL stimulation		[41]
	T98G glioma cells	Knockdown of CD95 or CD95L	Death induced by CD95 R/L elimination (swelling, ROS, DNA DSBs followed by reduction in cell growth, activation of caspase-2 resulting in MOMP and ultimately cell death)	[97]

Continued

TABLE 4.1 Overview of CD95 Expression and Function in the Central Nervous System—cont'd

Species	Cell Type/Tissue	Manipulation/Model	Phenotype	References
Brain Tumors: Proliferation				
Human	A-172, T98G, YKG-1 glioma cells	CH-11 antibody (CD95 agonist)	Enhanced proliferation, cell cycle progression; apoptotic pathway also activated	[95]
		NOK-2 antibody (CD95L neutralizing)	Inhibition of proliferation	[95]
Brain Tumors: Migration/Invasion				
Human	T98G glioma cells; patient GBM-derived cultures	CD95L-T4 (CD95 activating)	Resistance to apoptosis; no change in proliferation rate; increase in migration	[36]
	U87-MG, U373-MG glioma cells	Crosslinked CD95L (CD95 activating)	Increase in migration	[99]
Mouse	SMA560 syngeneic model	MFL3 antibody (CD95L neutralizing)	Reduced migration	[36]
Rat	C6 glioma cells	FasL interfering protein (CD95L neutralizing)		[98]
Brain Tumors: Immune Recruitment, Immune Evasion				
Human	CRT-MG, U373-MG, U87-MG glioma cells	CH-11 (CD95 agonist)	Increase in MCP-1, IL-8, and ICAM-1 expression	[90,107,108]
Rat	9L, C6, F98 glioma cells	Coculture with Jurkat A3 T-lymphocytes	T cell death by glioma cell secreted CD95L (inhibited by MLF4 or CD95L shRNA)	[105]
	9L cells injected intracranially or subcutaneously injected into immunocompetent Fischer 344 rats	Knockdown of CD95L	shCD95L inhibited T cell death, increased T cell infiltration in tumors, and reduced tumor growth	[105]

mAb, monoclonal antibody.

Because CD95 also plays an important role in cells of the immune system, its functions were also assessed in tumor–immune interactions. The effect of CD95/CD95L system on the immune system in the context of a tumor is again very controversial. Because gliomas express CD95L, and CD95L is implicated in providing immune privilege, such as in the brain, by inducing cell death in CD95-expressing T cells, it was hypothesized that CD95L was a way for tumor cells to mount a "counterattack" against CD95+ tumor-infiltrating lymphocytes to escape from an anti-tumor immune response [34,102,103]. A first hint of the "tumor strikes back" model in glioma was found in GBM tumor specimens in which apoptotic T cells expressing CD95 were localized in the vicinity of or in direct contact with CD95L-expressing tumor cells, suggesting that they can induce apoptosis in invading T cells [104]. Furthermore, coculturing CD95L-expressing rat 9L glioma cells with Jurkat T lymphocytes led to T cell death, and conversely, downregulation of 9L-derived CD95L reduced the tumor growth of subcutaneous and intracranial transplanted 9L cells significantly owing to enhanced T-cell infiltration [105].

Later studies suggested, however, that the immune counterattack model might not entirely apply. CD95L-overexpressing grafted transplants were rejected and this was associated with massive infiltration of neutrophils and other innate immune cells [106]. Instead, several studies have linked CD95 to inflammation by showing the induction of proinflammatory mediators upon CD95 ligation, also in malignant glioma cells [90,107–109]. Upon stimulation of glioma cell lines with the CH-11 antibody or human recombinant soluble CD95L, the expression of the chemokines MCP-1, IL-8, and IL6 was induced via the ERK1/2 and p38 MAPK pathway [90,107] and intercellular adhesion molecule-1 in a caspase-dependent manner. MCP-1 can attract macrophages, T cells, and natural killer cells, and IL-8 can influence the microvascular endothelial morphogenesis, thus promoting tumor growth. Therefore, the CD95/CD95L system may actually enhance immune response and lead to a tumor-promoting inflammatory environment (Table 4.1).

In summary, early efforts were focused on the apoptotic functions of CD95 and concentrated on sensitizing glioma cells to CD95-induced apoptosis. However, it is now known that CD95 can also have protumorigenic functions that cannot be overlooked. The controversies over the opposite outcomes of CD95 activation may be driven by the use of different experimental settings (e.g., different apoptosis-sensitive/resistant cells, in vitro/in vivo experiments) and insufficient methods of detecting apoptosis in the early studies of CD95-mediated death in glioma cells. Thus, the ultimate outcome of CD95 activation may be highly context-dependent.

4.4 CONCLUSIONS

CD95, once believed to be only a mediator of cell death, has emerged as a receptor with pleiotropic functions in the CNS. These functions are revealed during brain development, response to injury, and tumorigenesis. CD95-elicited phenotypes are highly context-dependent and require careful dissection of the role of CD95 in the various cell populations involved.

In addition, CD95 expression defines specific cell populations during development and adult homeostasis and its upregulation is a general feature of brain injury or inflammation.

CD95 is thus a functional marker that offers great therapeutic opportunities that are just beginning to be exploited.

REFERENCES

[1] Locksley RM, Killeen N, Lenardo MJ. The TNF and TNF receptor superfamilies: integrating mammalian biology. Cell 2001;104:487–501.

[2] Atkinson EA, Ostergaard H, Kane K, Pinkoski MJ, Caputo A, Olszowy MW, et al. A physical interaction between the cell death protein Fas and the tyrosine kinase p59fynT. J Biol Chem 1996;271:5968–71.

[3] Sancho-Martinez I, Martin-Villalba A. Tyrosine phosphorylation and CD95: a FAScinating switch. Cell Cycle 2009;8:838–42.

[4] Ruan W, Lee CT, Desbarats J. A novel juxtamembrane domain in tumor necrosis factor receptor superfamily molecules activates Rac1 and controls neurite growth. Mol Biol Cell 2008;19:3192–202.

[5] Kayagaki N, Kawasaki A, Ebata T, Ohmoto H, Ikeda S, Inoue S, et al. Metalloproteinase-mediated release of human Fas ligand. J Exp Med 1995;182:1777–83.

[6] Scott FL, Stec B, Pop C, Dobaczewska MK, Lee JJ, Monosov E, et al. FADD death domain complex structure unravels signalling by receptor clustering. Nature 2009;457:1019–22.

[7] Schleich K, Warnken U, Fricker N, Öztürk S, Richter P, Kammerer K, et al. Stoichiometry of the CD95 death-inducing signaling complex: experimental and modeling evidence for a death effector domain chain model. Mol Cell 2012;47:306–19.

[8] Dickens LS, Boyd RS, Jukes-Jones R, Hughes MA, Robinson GL, Fairall L, et al. A death effector domain chain DISC model reveals a crucial role for caspase-8 chain assembly in mediating apoptotic cell death. Mol Cell 2012;47:291–305.

[9] Kischkel F, Hellbardt S, Behrmann I, Germer M, Pawlita M, Krammer P, et al. Cytotoxicity-dependent APO-1 (Fas/CD95)-associated proteins form a death-inducing signaling complex (DISC) with the receptor. EMBO J 1995;14:5579.

[10] Martin-Villalba A, Llorens-Bobadilla E, Wollny D. CD95 in cancer: tool or target? Trends Mol Med 2013;19(6):329–35.

[11] Brint E, O'Callaghan G, Houston A. Life in the Fas lane: differential outcomes of Fas signaling. Cell Mol Life Sci 2013;70:4085–99.

[12] Park C, Sakamaki K, Tachibana O, Yamashima T, Yamashita J, Yonehara S. Expression of fas antigen in the normal mouse brain. Biochem Biophys Res Commun 1998;252:623–8.

[13] Cheema ZF, Wade SB, Sata M, Walsh K, Sohrabji F, Miranda RC. Fas/Apo [apoptosis]-1 and associated proteins in the differentiating cerebral cortex: induction of caspase-dependent cell death and activation of NF-κB. J Neurosci 1999;19:1754–70.

[14] French LE, Hahne M, Viard I, Radlgruber G, Zanone R, Becker K, et al. Fas and Fas ligand in embryos and adult mice: ligand expression in several immune-privileged tissues and coexpression in adult tissues characterized by apoptotic cell turnover. J Cell Biol 1996;133:335–43.

[15] Zuliani C, Kleber S, Klussmann S, Wenger T, Kenzelmann M, Schreglmann N, et al. Control of neuronal branching by the death receptor CD95 (Fas/Apo-1). Cell Death Differ 2006;13:31–40.

[16] Tamm C, Robertson JD, Sleeper E, Enoksson M, Emgård M, Orrenius S, et al. Differential regulation of the mitochondrial and death receptor pathways in neural stem cells. Eur J Neurosci 2004;19:2613–21.

[17] Ricci-Vitiani L, Pedini F, Mollinari C, Condorelli G, Bonci D, Bez A, et al. Absence of caspase 8 and high expression of PED protect primitive neural cells from cell death. J Exp Med 2004;200:1257–66.

[18] Klassen H, Schwartz MR, Bailey AH, Young MJ. Surface markers expressed by multipotent human and mouse neural progenitor cells include tetraspanins and non-protein epitopes. Neurosci Lett 2001;312:180–2.

[19] Corsini NS, Sancho-Martinez I, Laudenklos S, Glagow D, Kumar S, Letellier E, et al. The death receptor CD95 activates adult neural stem cells for working memory formation and brain repair. Cell Stem Cell 2009;5:178–90.

[20] Akiyama K, Chen C, Wang D, Xu X, Qu C, Yamaza T, et al. Mesenchymal-stem-cell-induced immunoregulation involves FAS-ligand-/FAS-mediated T cell apoptosis. Cell Stem Cell 2012;10:544–55.

[21] Dybedal I, Tamm C, Yang L, Robertson JD, Bryder D, Sleeper E, et al. Human reconstituting hematopoietic stem cells up-regulate Fas expression upon active cell cycling but remain resistant to Fas-induced suppression. Blood 2003;102:118–26.

[22] Reinehr R, Clynes RA, Sommerfeld A, Towers TL, Häussinger D, Presta LG, et al. CD95 ligand is a proliferative and antiapoptotic signal in quiescent hepatic stellate cells. Gastroenterology 2008;134:1494–7.

[23] Brunlid G, Pruszak J, Holmes B, Isacson O, Sonntag K-C. Immature and neurally differentiated mouse embryonic stem cells do not express a functional Fas/Fas ligand system. Stem Cells 2007;25:2551–8.

[24] Hirano K, Sasaki N, Ichimiya T, Miura T, Van Kuppevelt TH, Nishihara S. 3-O-sulfated heparan sulfate recognized by the antibody HS4C3 contribute to the differentiation of mouse embryonic stem cells via Fas signaling. PLoS ONE 2012;7:e43440.

[25] Bechmann I, Mor G, Nilsen J, Eliza M, Nitsch R, Naftolin F. FasL (CD95L, Apo1L) is expressed in the normal rat and human brain: evidence for the existence of an immunological brain barrier. Glia 1999;27:62–74.

[26] Choi C, Benveniste EN. Fas ligand/Fas system in the brain: regulator of immune and apoptotic responses. Brain Res Rev 2004;44 :65–81.

[27] Choi C, Park JY, Lee J, Lim JH, Shin EC, Ahn YS, et al. Fas ligand and Fas are expressed constitutively in human astrocytes and the expression increases with IL-1, IL-6, TNF-alpha, or IFN-gamma. J Immunol 1999;162:1889–95.

[28] Sairanen T, Karjalainen-Lindsberg M-L, Paetau A, Ijäs P, Lindsberg PJ. Apoptosis dominant in the periinfarct area of human ischaemic stroke–a possible target of antiapoptotic treatments. Brain 2006;129:189–99.

[29] Weller M, Frei K, Groscurth P, Krammer PH, Yonekawa Y, Fontana A. Anti-Fas/APO-1 antibody-mediated apoptosis of cultured human glioma cells. Induction and modulation of sensitivity by cytokines. J Clin Invest 1994;94:954–64.

[30] Tachibana O, Nakazawa H, Lampe J, Watanabe K, Kleihues P, Ohgaki H. Expression of Fas/APO-1 during the progression of astrocytomas. Cancer Res 1995;55:5528–30.

[31] Tachibana O, Lampe J, Kleihues P, Ohgaki H. Preferential expression of Fas/APO1 (CD95) and apoptotic cell death in perinecrotic cells of glioblastoma multiforme. Acta Neuropathol 1996;92:431–4.

[32] Gratas C, Tohma Y, Van Meir EG, Klein M, Tenan M, Ishii N, et al. Fas ligand expression in glioblastoma cell lines and primary astrocytic brain tumors. Brain Pathol 1997;7:863–9.

[33] Tohma Y, Gratas C, Van Meir EG, Desbaillets I, Tenan M, Tachibana O, et al. Necrogenesis and Fas/APO-1 (CD95) expression in primary (de novo) and secondary glioblastomas. J Neuropathol Exp Neurol 1998;57:239–45.

[34] Saas P, Walker PR, Hahne M, Quiquerez AL, Schnuriger V, Perrin G, et al. Fas ligand expression by astrocytoma in vivo: maintaining immune privilege in the brain? J Clin Invest 1997;99:1173–8.

[35] Husain N, Chiocca EA, Rainov N, Louis DN, Zervas NT. Co-expression of Fas and Fas ligand in malignant glial tumors and cell lines. Acta Neuropathol 1998;95:287–90.

[36] Kleber S, Sancho-Martinez I, Wiestler B, Beisel A, Gieffers C, Hill O, et al. Yes and PI3K bind CD95 to signal invasion of glioblastoma. Cancer Cell 2008;13:235–48.

[37] Bertrand J, Begaud-Grimaud G, Bessette B, Verdier M, Battu S, Jauberteau M-O. Cancer stem cells from human glioma cell line are resistant to Fas-induced apoptosis. Int J Oncol 2009;34:717–27.

[38] Tao J. Expression levels of Fas/Fas-L mRNA in human brain glioma stem cells. Mol Med Rep 2012;5(5):1202–6.

[39] Frankel B, Longo SL, Ryken TC. Co-expression of Fas and Fas ligand in human non-astrocytic glial tumors. Acta Neuropathol 1999;98:363–6.

[40] Weller M, Schuster M, Pietsch T, Schabet M. CD95 ligand-induced apoptosis of human medulloblastoma cells. Cancer Lett 1998;128:121–6.

[41] Riffkin CD, Gray AZ, Hawkins CJ, Chow CW, Ashley DM. Ex vivo pediatric brain tumors express Fas (CD95) and FasL (CD95L) and are resistant to apoptosis induction. Neuro-Oncology 2001;3:229–40.

[42] Bodey B, Siegel SE, Kaiser HE. Fas (APO-1, CD95) receptor expression and new options for immunotherapy in childhood medulloblastomas. Anticancer Res 1999;19:3293–314.

[43] Restifo NP. Not so Fas: re-evaluating the mechanisms of immune privilege and tumor escape. Nat Med 2000;6:493–5.

[44] Cheema ZF, Santillano DR, Wade SB, Newman JM, Miranda RC. The extracellular matrix, p53 and estrogen compete to regulate cell-surface Fas/Apo-1 suicide receptor expression in proliferating embryonic cerebral cortical precursors, and reciprocally, Fas-ligand modifies estrogen control of cell-cycle proteins. BMC Neurosci 2004;5:11.

[45] Le-Niculescu H, Bonfoco E, Kasuya Y, Claret FX, Green DR, Karin M. Withdrawal of survival factors results in activation of the JNK pathway in neuronal cells leading to Fas ligand induction and cell death. Mol Cell Biol 1999;19:751–63.

[46] Desbarats J, Birge RB, Mimouni-Rongy M, Weinstein DE, Palerme J-S, Newell MK. Fas engagement induces neurite growth through ERK activation and p35 upregulation. Nat Cell Biol 2003;5:118–25.

[47] Lambert C, Landau AM, Desbarats J. Fas-beyond death: a regenerative role for Fas in the nervous system. Apoptosis 2003;8:551–62.

[48] Reich A, Spering C, Schulz JB. Death receptor Fas (CD95) signaling in the central nervous system: tuning neuroplasticity? Trends Neurosci 2008;31:478–86.

[49] Martin-Villalba A, Herr I, Jeremias I, Hahne M, Brandt R, Vogel J, et al. CD95 ligand (Fas-L/APO-1L) and tumor necrosis factor-related apoptosis-inducing ligand mediate ischemia-induced apoptosis in neurons. J Neurosci 1999;19:3809–17.

[50] Rosenbaum DM, Gupta G, D'Amore J, Singh M, Weidenheim K, Zhang H, et al. Fas (CD95/APO-1) plays a role in the pathophysiology of focal cerebral ischemia. J Neurosci Res 2000;61:686–92.

[51] Martin-Villalba A, Hahne M, Kleber S, Vogel J, Falk W, Schenkel J, et al. Therapeutic neutralization of CD95-ligand and TNF attenuates brain damage in stroke. Cell Death Differ 2001;8:679–86.

[52] Yoshino O, Matsuno H, Nakamura H, Yudoh K, Abe Y, Sawai T, et al. The role of Fas-mediated apoptosis after traumatic spinal cord injury. Spine 2004;29:1394–404.

[53] Demjen D, Klussmann S, Kleber S, Zuliani C, Stieltjes B, Metzger C, et al. Neutralization of CD95 ligand promotes regeneration and functional recovery after spinal cord injury. Nat Med 2004;10:389–95.

[54] Casha S, Yu WR, Fehlings MG. FAS deficiency reduces apoptosis, spares axons and improves function after spinal cord injury. Exp Neurol 2005;196:390–400.

[55] Ackery A, Robins S, Fehlings MG. Inhibition of Fas-mediated apoptosis through administration of soluble Fas receptor improves functional outcome and reduces posttraumatic axonal degeneration after acute spinal cord injury. J Neurotrauma 2006;23:604–16.

[56] Letellier E, Kumar S, Sancho-Martinez I, Krauth S, Funke-Kaiser A, Laudenklos S, et al. CD95-ligand on peripheral myeloid cells activates Syk kinase to trigger their recruitment to the inflammatory site. Immunity 2010;32:240–52.

[57] Raoul C, Buhler E, Sadeghi C, Jacquier A, Aebischer P, Pettmann B, et al. Chronic activation in presymptomatic amyotrophic lateral sclerosis (ALS) mice of a feedback loop involving Fas, Daxx, and FasL. Proc Natl Acad Sci USA 2006;103:6007–12.

[58] Petri S, Kiaei M, Wille E, Calingasan NY. Flint Beal M. Loss of Fas ligand-function improves survival in G93A-transgenic ALS mice. J Neurol Sci 2006;251:44–9.

[59] Locatelli F, Corti S, Papadimitriou D, Fortunato F, Del Bo R, Donadoni C, et al. Fas small interfering RNA reduces motoneuron death in amyotrophic lateral sclerosis mice. Ann Neurol 2007;62:81–92.

[60] Morishima Y, Gotoh Y, Zieg J, Barrett T, Takano H, Flavell R, et al. β-amyloid induces neuronal apoptosis via a mechanism that involves the c-Jun N-terminal kinase pathway and the induction of Fas ligand. J Neurosci 2001;21:7551–60.

[61] Su JH, Anderson AJ, Cribbs DH, Tu C, Tong L, Kesslack P, et al. Fas and Fas ligand are associated with neuritic degeneration in the AD brain and participate in beta-amyloid-induced neuronal death. Neurobiol Dis 2003;12:182–93.

[62] Hayley S, Crocker SJ, Smith PD, Shree T, Jackson-Lewis V, Przedborski S, et al. Regulation of dopaminergic loss by Fas in a 1-methyl-4-phenyl-1,2,3,6-tetrahydropyridine model of Parkinson's disease. J Neurosci 2004;24:2045–53.

[63] Ballok DA, Earls AM, Krasnik C, Hoffman SA, Sakic B. Autoimmune-induced damage of the midbrain dopaminergic system in lupus-prone mice. J Neuroimmunol 2004;152:83–97.

[64] Landau AM, Luk KC, Jones M-L, Siegrist-Johnstone R, Young YK, Kouassi E, et al. Defective Fas expression exacerbates neurotoxicity in a model of Parkinson's disease. J Exp Med 2005;202:575–81.

[65] Sabelko KA, Kelly KA, Nahm MH, Cross AH, Russell JH. Fas and Fas ligand enhance the pathogenesis of experimental allergic encephalomyelitis, but are not essential for immune privilege in the central nervous system. J Immunol 1997;159:3096–9.

[66] Waldner H, Sobel RA, Howard E, Kuchroo VK. Fas- and FasL-deficient mice are resistant to induction of autoimmune encephalomyelitis. J Immunol 1997;159:3100–3.

[67] Sabelko-Downes KA, Russell JH, Cross AH. Role of Fas–FasL interactions in the pathogenesis and regulation of autoimmune demyelinating disease. J Neuroimmunol 1999;100:42–52.

[68] Dittel BN, Merchant RM, Janeway CA. Evidence for Fas-dependent and Fas-independent mechanisms in the pathogenesis of experimental autoimmune encephalomyelitis. J Immunol 1999;162:6392–400.

[69] Hövelmeyer N, Hao Z, Kranidioti K, Kassiotis G, Buch T, Frommer F, et al. Apoptosis of oligodendrocytes via Fas and TNF-R1 is a key event in the induction of experimental autoimmune encephalomyelitis. J Immunol 2005;175:5875–84.

[70] Raoul C, Estévez AG, Nishimune H, Cleveland DW, deLapeyrière O, Henderson CE, et al. Motoneuron death triggered by a specific pathway downstream of Fas. potentiation by ALS-linked SOD1 mutations. Neuron 2002;35:1067–83.

[71] Sleeper E, Tamm C, Frisén J, et al. Cell death in adult neural stem cells. Cell Death Differ 2002;9:1377–8.

[72] Ceccatelli S. Neural stem cells and cell death. Toxicol Lett 2004;149:59–66.

[73] Knight JC, Scharf EL, Mao-Draayer Y. Fas activation increases neural progenitor cell survival. J Neurosci Res 2010;88(4):746–57.

[74] Pluchino S, Zanotti L, Rossi B, Brambilla E, Ottoboni L, Salani G, et al. Neurosphere-derived multipotent precursors promote neuroprotection by an immunomodulatory mechanism. Nat Cell Biol 2005;436:266–71.

[75] Hackett C, Tamm C, Knight J, Robertson JD, Mao-Draayer Y, Sleeper E, et al. Transplantation of Fas-deficient or wild-type neural stem/progenitor cells (NPCs) is equally efficient in treating experimental autoimmune encephalomyelitis (EAE). Am J Transl Res 2014;6:119.

[76] Trauth BC, Klas C, Peters AM, Matzku S, Möller P, Falk W, et al. Monoclonal antibody-mediated tumor regression by induction of apoptosis. Science 1989;245:301–5.

[77] Rensing-Ehl A, Frei K, Flury R, Matiba B, Mariani SM, Weller M, et al. Local Fas/APO-1 (CD95) ligand-mediated tumor cell killing in vivo. Eur J Immunol 1995;25:2253–8.

[78] Weller M, Malipiero U, Rensing-Ehl A, Barr PJ, Fontana A. Fas/APO-1 gene transfer for human malignant glioma. Cancer Res 1995;55:2936–44.

[79] Frankel B, Longo SL, Kyle M, Canute GW, Ryken TC. Tumor Fas (APO-1/CD95) up-regulation results in increased apoptosis and survival times for rats with intracranial malignant gliomas. Neurosurgery 2001;49:168–75. discussion 175–6.

[80] Fulda S, Küfer MU, Meyer E, van Valen F, Dockhorn-Dworniczak B, Debatin KM. Sensitization for death receptor- or drug-induced apoptosis by re-expression of caspase-8 through demethylation or gene transfer. Oncogene 2001;20:5865–77.

[81] Ashley DM, Ashley DM, Kong FM, Kong FM, Bigner DD, Bigner DD, et al. Endogenous expression of transforming growth factor beta1 inhibits growth and tumorigenicity and enhances Fas-mediated apoptosis in a murine high-grade glioma model. Cancer Res 1998;58:302–9.

[82] Frei K, Ambar B, Adachi N, Yonekawa Y, Fontana A. Ex vivo malignant glioma cells are sensitive to Fas (CD95/APO-1) ligand-mediated apoptosis. J Neuroimmunol 1998;87:105–13.

[83] Clynes RA, Towers TL, Presta LG, Ravetch JV. Inhibitory Fc receptors modulate in vivo cytotoxicity against tumor targets. Nat Med 2000;6:443–6.

[84] Xu Y, Szalai AJ, Zhou T, Zinn KR, Chaudhuri TR, Li X, et al. Fc gamma Rs modulate cytotoxicity of anti-Fas antibodies: implications for agonistic antibody-based therapeutics. J Immunol 2003;171:562–8.

[85] Holler N, Tardivel A, Kovacsovics-Bankowski M, Hertig S, Gaide O, Martinon F, et al. Two adjacent trimeric Fas ligands are required for Fas signaling and formation of a death-inducing signaling complex. Mol Cell Biol 2003;23:1428–40.

[86] Eisele G, Roth P, Hasenbach K, Aulwurm S, Wolpert F, Tabatabai G, et al. APO010, a synthetic hexameric CD95 ligand, induces human glioma cell death in vitro and in vivo. Neuro Oncol 2011;13:155–64.

[87] Xia S, Rosen EM, Laterra J. Sensitization of glioma cells to Fas-dependent apoptosis by chemotherapy-induced oxidative stress. Cancer Res 2005;65:5248–55.

[88] Giraud S, Bessette B, Boda C, Lalloue F, Petit D, Mathonnet M, et al. In vitro apoptotic induction of human glioblastoma cells by Fas ligand plus etoposide and in vivo antitumour activity of combined drugs in xenografted nude rats. Int J Oncol 2007;30:273–81.

[89] Roth W, Fontana A, Trepel M, Reed JC, Dichgans J, Weller M. Immunochemotherapy of malignant glioma: synergistic activity of CD95 ligand and chemotherapeutics. Cancer Immunol Immunother 1997;44:55–63.

[90] Choi C, Xu X, Oh JW, Lee SJ, Gillespie GY, Park H, et al. Fas-induced expression of chemokines in human glioma cells: involvement of extracellular signal-regulated kinase 1/2 and p38 mitogen-activated protein kinase. Cancer Res 2001;61:3084–91.

[91] Barnhart BC, Legembre P, Pietras E, Bubici C, Franzoso G, Peter ME. CD95 ligand induces motility and invasiveness of apoptosis-resistant tumor cells. EMBO J 2004;23:3175–85.

[92] Ambar BB, Frei K, Malipiero U, Morelli AE, Castro MG, Lowenstein PR, et al. Treatment of experimental glioma by administration of adenoviral vectors expressing Fas ligand. Hum Gene Ther 1999;10:1641–8.

[93] Algeciras-Schimnich A, Barnhart BC, Peter ME. Apoptosis dependent and independent functions of caspases. In: Los M, Walczak H, editors. Caspases—their role in cell death and cell survival Georgetown. Georgetown, Texas: Landes Bioscience; 2002. p.123.

[94] Alderson MR, Armitage RJ, Maraskovsky E, Tough TW, Roux E, Schooley K, et al. Fas transduces activation signals in normal human T lymphocytes. J Exp Med 1993;178:2231–5.

[95] Shinohara H, Yagita H, Ikawa Y, Oyaizu N. Fas drives cell cycle progression in glioma cells via extracellular signal-regulated kinase activation. Cancer Res 2000;60:1766–72.

[96] Chen L, Park S-M, Tumanov AV, Hau A, Sawada K, Feig C, et al. CD95 promotes tumour growth. Nature 2010;465:492–6.

[97] Hadji A, Ceppi P, Murmann AE, Brockway S, Pattanayak A, Bhinder B, et al. Death induced by CD95 or CD95 ligand elimination. Cell Rep 2014;7:208–22.

[98] Wisniewski P, Ellert-Miklaszewska A, Kwiatkowska A, Kaminska B. Cellular signalling. Cell Signal 2010;22:212–20.

[99] Lai YJ, Lin VTG, Zheng Y, Benveniste EN, Lin FT. The adaptor protein TRIP6 antagonizes fas-induced apoptosis but promotes its effect on cell migration. Mol Cell Biol 2010;30:5582–96.

[100] Tuettenberg J, Seiz M, Debatin K-M, Hollburg W, Staden von M, Thiemann M, et al. Int Immunopharmacol 2012;13:93–100.

[101] Bendszus M, Debus J, Wick W, Kobyakov G, Martens T, Heese O, et al. APG101_CD_002: a phase II, randomized, open-label, multi-center study of weekly APG101 plus reirradiation versus reirradiation in the treatment of patients with recurrent glioblastoma. J Clin Oncol 2012:30.

[102] Green DR, Ferguson TA. The role of Fas ligand in immune privilege. Nat Rev Mol Cell Biol 2001;2:917–24.

[103] OConnell J, O Sullivan GC, Collins JK, Shanahan F. The Fas counterattack: fas-mediated T cell killing by colon cancer cells expressing Fas ligand. J Exp Med 1996;184:1075–82.

[104] Didenko VV, Ngo HN, Minchew C, Baskin DS. Apoptosis of T lymphocytes invading glioblastomas multiforme: a possible tumor defense mechanism. J Neurosurg 2002;96:580–4.

[105] Jansen T, Tyler B, Mankowski JL, Recinos VR, Pradilla G, Legnani F, et al. FasL gene knock-down therapy enhances the antiglioma immune response. Neuro Oncol 2010;12:482–9.

[106] Kang S-M, Schneider DB, Lin Z, Hanahan D, Dichek DA, Stock PG, et al. Fas ligand expression in islets of Langerhans does not confer immune privilege and instead targets them for rapid destruction. Nat Med 1997;3:738–43.

[107] Choi C, Gillespie GY, Van Wagoner NJ, Benveniste EN. Fas engagement increases expression of interleukin-6 in human glioma cells. J Neurooncol 2002;56:13–9.

[108] Choi K, Benveniste EN, Choi C. Induction of intercellular adhesion molecule-1 by Fas ligation: proinflammatory roles of Fas in human astroglioma cells. Neurosci Lett 2003;352:21–4.

[109] Matsumoto N, Imamura R, Suda T. Caspase-8- and JNK-dependent AP-1 activation is required for Fas ligand-induced IL-8 production. FEBS J 2007;274:2376–84.

[110] Knight J, Hackett C, Soltys J, Mao-Draayer Y. Fas receptor modulates lineage commitment and stemness of mouse neural stem cells. Neurosci Med 2011:2.

[111] Matsuyama T, Hata R, Tagaya M, Yamamoto Y, Nakajima T, Furuyama J, et al. Fas antigen mRNA induction in postischemic murine brain. Brain Res 1994;657:342–6.

[112] Beier CP, Kölbl M, Beier D, Woertgen C, Bogdahn U, Brawanski A. CD95/Fas mediates cognitive improvement after traumatic brain injury. Cell Res 2007;17:732–4.

[113] Beer R, Franz G, Schöpf M, Reindl M, Zelger B, Schmutzhard E, et al. Expression of Fas and Fas ligand after experimental traumatic brain injury in the rat. J Cereb Blood Flow Metab 2000; 20:669–77.

[114] Bechmann I, Lossau S, Steiner B, Mor G, Gimsa U, Nitsch R. Reactive astrocytes upregulate Fas (CD95) and Fas ligand (CD95L) expression but do not undergo programmed cell death during the course of anterograde degeneration. Glia 2000;32:25–41.

Chapter 5

Role of Fundamental Pathways of Innate and Adaptive Immunity in Neural Differentiation: Focus on Toll-like Receptors, Complement System, and T-Cell-Related Signaling

Hélène Boudin[1] and Antoine Louveau[2]

[1]INSERM UMR913, IMAD, University of Nantes, Nantes, France; [2]Neuroscience Department, Center for Brain Immunology and Glia, University of Virginia, Charlottesville, VA, USA

5.1 MOLECULES FROM INNATE IMMUNITY

5.1.1 Toll-Like Receptors

The family of mammalian toll-like receptors (TLRs) includes at least 12 members that sense highly conserved structural motifs known as pathogen-associated molecular patterns (PAMPs), expressed by microbial pathogens, or danger-associated molecular patterns, which are endogenous molecules released from necrotic or dying cells. TLRs represent therefore central molecules in the rapid response to cellular stress and infection. TLRs are transmembrane proteins characterized by an ectodomain composed of leucine-rich repeats that are responsible for ligand recognition and a cytoplasmic domain homologous to the cytoplasmic region of the interleukin (IL)-1 receptor, known as the TIR domain, which is required for downstream signaling [1]. TLRs are classified into two subgroups depending on their cellular localization and respective PAMP ligands. One group is composed of TLR1 (CD281), TLR2 (CD282), TLR4 (CD284), TLR5, TLR6 (CD286), and TLR11, which are expressed on cell surfaces and recognize mainly microbial membrane components such as lipids, lipoproteins, and proteins; the other group is composed of TLR3 (CD283), TLR7, TLR8 (CD288), and TLR9 (CD289), which are expressed exclusively in intracellular vesicles such as the endoplasmic reticulum, endosomes, and lysosomes, where they recognize microbial nucleic acids. Signaling triggered by TLR activation results in a variety of cellular responses, including the production of a broad range of inflammatory

cytokines and effector cytokines involved in the modulation of the adaptive immune response [2]. In the immune system, TLRs are mainly expressed by antigen-presenting cells such as dendritic cells (DCs) and macrophages. TLRs are also expressed in the central nervous system (CNS), where their expression was initially believed to be limited to microglial cells for their role in inflammation in response to infection and injury [3,4]. However, several TLRs are also expressed in developing neurons and neuronal precursor cells, which has recently motivated several studies on the role of TLRs in neurodevelopmental processes. Whereas TLRs 2, 3, and 4 are expressed in neural stem cells (NSCs) and are involved in cell proliferation and specification, expression of TLRs 7 and 8 appears to be restricted to postmitotic neurons, in which they regulate neuronal polarization and early neuritogenesis (Table 5.1; Figure 5.1) [4–11].

Using immunohistofluorescence on adult mouse brain sections, both TLR2 and TLR4 have been found in doublecortin (DCX)-positive cells (DCX is a marker for neuroblasts) located in the neurogenic niches, the subventricular zone (SVZ) of the lateral ventricles, and the subgranular zone (SGZ) of the hippocampal dentate gyrus [5]. In addition, TLR2 and TLR4 mRNA and protein have been detected by polymerase chain reaction and immunofluorescence, respectively, in NSCs isolated from adult mouse brain [4]. However, TLR2 and TLR4 have distinct functions in NSC proliferation and differentiation. Whereas the lack of TLR2 affects neurogenesis as shown by a reduced percentage of newly formed BrdU$^+$/DCX$^+$ neuroblasts and BrdU$^+$/Tuj1$^+$neurons in TLR2-deficient mice [5], BrdU-positive proliferating

Neural Surface Antigens. http://dx.doi.org/10.1016/B978-0-12-800781-5.00005-0

TABLE 5.1 Molecules from the Innate and Adaptive Immunity in Neural Stem Cells (NSC) Functions and Neuronal Differentiation

		Function			
	Expression	NSC Proliferation	Cell-Fate Specification	Neuronal Differentiation	References
Molecules					
Innate Immunity					
TLR2	Embryonic NSC	Inhibits	No effect		[5,6]
	Adult NSC	No effect	Promotes neurogenesis		
TLR3	Embryonic NSC	Inhibits	No effect		[7–9]
	Adult NSC	No effect	Inhibits neurogenesis		
	Postmitotic neuron			Inhibits neurite outgrowth	
TLR4	Adult NSC	Inhibits	Inhibits neurogenesis		[5]
TLR7	Postmitotic neuron			Inhibits neurite outgrowth	[10]
TLR8	Postmitotic neuron			Inhibits neurite outgrowth	[11]
CR2	Adult NSC	Inhibits	Inhibits neurogenesis		[12]
C3aR	Adult NSC	Promotes	Promotes neurogenesis		[13–15]
	Postmitotic neuron			Promotes neurite outgrowth	
C5aR	Adult NSC but no ligand binding	No effect	No effect		[13,14]
Adaptive Immunity					
MHCI	Embryonic NSC, only after several days in culture				[16–23]
	Adult NSC	No effect	No effect		
	Postmitotic neuron			Membrane form: promotes neurite outgrowth	
				Soluble form: inhibits neurite outgrowth	
RAE-1	Embryonic NSC	Promotes			[24]
MHCII	Embryonic NSC (very low but upregulated by inflammation)				[16,25–27]
CD80/CD86	Embryonic and adult NSC (low but upregulated by inflammation)				[18,19,28,29]
CD3ζ	Postmitotic neuron			Inhibits neurite outgrowth	[30–32]

MHC, major histocompatibility complex; MHCI, MHC class I molecules; RAE-1, retinoic acid early induced transcript; MHCII, MHC class II molecules; TLR, toll-like receptor.

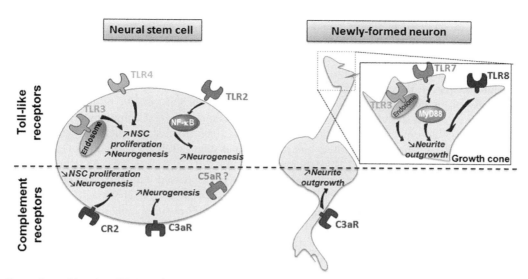

FIGURE 5.1 Expression and function of TLRs and complement receptors on neural stem cells (NSCs) and newly formed neurons. TLR and complement receptor subtypes can inhibit (↘) or stimulate (↗) NSC proliferation and neurogenesis. TLR, toll-like receptor.

cell numbers in the SVZ and SGZ were unaffected, suggesting that neither NSC proliferation nor survival is regulated by TLR2 [5]. These results were confirmed in cultures of NSCs derived from wild-type and TLR2-deficient adult mouse brain, and treatment of wild-type NSC cultures with a TLR2 agonist, the lipopeptide Pam3CysSK$_4$, significantly enhanced their neuronal differentiation through the NF-κB pathway [5]. These studies indicate that TLR2 supports neurogenesis, probably through a direct mechanism implicating NSCs expressing TLR2, and not other TLR2-expressing cells, such as microglia. Interestingly, the function of TLR2 expressed in embryonic NSCs differs from that described in adult NSCs. The ability of NSCs cultured from embryonic day 15 (E15) mouse brain to give rise to glial cells and neurons is not compromised by TLR2 deficiency compared with wild-type NSCs [6]. Basal cell proliferation is similarly unaffected by the lack of TLR2, but application of the TLR2 ligand Pam3CSK$_4$ (a synthetic bacterial lipopeptide) or FSL1 (a synthetic lipoprotein from *Mycoplasma salivarium*) to wild-type embryonic NSCs decreases cell proliferation, as fewer neurospheres are formed upon ligand treatment [6]. This apparent discrepancy with data obtained from adult NSCs probably results from differences in NSC-specific developmental transcriptional changes at embryonic and adult stages. In contrast to TLR2, which supports neuronal commitment without modifying cell proliferation of adult NSCs, TLR4 regulates both proliferation and differentiation of NSCs. Suppression of TLR4 expression in adult NSCs by TLR4-targeted siRNA or with NSCs derived from TLR4-deficient mice yields an increase in NSC proliferation along with an increase in neuronal differentiation, suggesting that TLR4 is involved in a negative control of both cell proliferation and neuronal differentiation [5]. TLR3 is highly expressed in embryonic mouse brain, during both the developmental period of cortical NSC expansion

(E10.5) and neurogenesis (E16.5). TLR3 immunoreactivity is detected throughout the neuroepithelium and is preferentially associated with Sox2-positive cells (multipotent NSCs) in the cortex, but is weakly expressed in Tuj1-positive cells (neurons) [7]. Embryonic NSC-containing neurosphere cultures derived from TLR3-deficient mice form greater numbers of neurospheres but display a similar glial and neuronal differentiation propensity compared to wild-type cultures. In vivo studies indicate an increased number of proliferating cells in the developing cortex of TLR3-deficient mice compared to wild-type mice, suggesting an inhibitory role of TLR3 in embryonic NSC proliferation [7]. Conversely, activation of TLR3 in NSC cultures by a specific TLR3 agonist, polyinosinic:polycytidylic acid (poly(I:C)), induces a decrease in the number of proliferating cells and in neurosphere formation [7]. TLR3 function in NSC proliferation and differentiation analyzed in the dentate gyrus of adult mice is different from that observed in embryonic NSCs [8]. In vivo analyses of TLR3-deficient mice show that cell proliferation, assessed by counting BrdU-labeled cells, is unaffected, whereas neurogenesis, assessed by counting cells colabeled for BrdU and NeuN, is increased in TLR3-deficient mice compared to wild-type mice [8].

Early neuronal morphological differentiation is also controlled by several TLRs, such as TLRs 3, 7, and 8, all of which negatively regulate neuronal process outgrowth (Table 5.1; Figure 5.1). In culture of postmitotic neurons dissociated from embryonic dorsal root ganglia, TLR3 proteins are concentrated at growth cones of nascent axons. Neurons treated with the TLR3 agonist poly(I:C) exhibit a marked growth cone collapse causing a reduced neurite outgrowth. This effect is independent of the canonical NF-κB pathway classically activated by TLR3 in immune cells [9].

TLR7 and TLR8 also show an inhibitory function on neurite outgrowth in cortical neuron cultures from embryonic mouse brain [10,11]. The finding that TLR7-knockout neurons show a greater axonal and dendritic outgrowth compared to control neurons suggests the existence of an endogenous ligand for TLR7 in the cell culture system. Natural ligands for TLR7 include single-stranded RNA (ssRNA), mRNA, and microRNA. Wild-type neurons treated with RNAse, which digests ssRNA, showed increased neurite length in a fashion similar to that seen in TLR7-deficient neurons, suggesting that ssRNA might be an endogenous ligand for neuronal TLR7 [10]. Exogenous agonist stimulation of TLR7 with CL075 and of TLR8 with resiquimod reduced neurite length of cultured cortical neurons [10,11]. However, resiquimod is also able to bind to TLR7 [33], which implies that resiquimod-induced effects on cortical neurons might also be partly attributable to TLR7 activation. The intracellular signaling underlying the regulation of neuritogenesis by TLR7 agonist stimulation involves Myd88, an adaptor protein interacting with several TLR subtypes, and the cytokine IL-6 [10]. Interestingly, TLR7-deficient mice show an increase in TLR8 transcript expression in vivo, in the embryonic mouse brain, and in cortical neuron cultures, suggesting the existence of redundant and compensatory mechanisms between TLR7 and TLR8 functions with regard to neurite outgrowth [10].

5.1.2 Complement System

The complement system provides a rapid response to infections by opsonizing foreign material, by attracting leukocytes to a site of inflammation, and via lysis of foreign cell membranes. In addition, the complement system plays a key role in the induction of a primary B cell response to antigens. Moreover, the complement system clears modified "self" cells, such as apoptotic cells and cellular debris. These effects are exerted through interaction of the activation products of complement factors C1, C3, and C5 with specific receptors on the responding cells [34,35]. The receptors can be divided into three categories: (1) those recognizing the anaphylatoxic polypeptides C3a, C5a, and C5a-desarg (C3aR and C5aR); (2) those binding the active C3 fragment, C3b, and its degradation products, iC3b and C3d (CR1–4); and (3) receptors for C1q and related collagenous lectins (C1qR). C3aR and C5aR (CD88) belong to the superfamily of G-protein-coupled receptors containing seven transmembrane segments. CR1 (CD35) and CR2 (CD21) share structural similarities with a group of proteins known as the regulators of complement activation family, which includes factor H, IL-2 receptor, and factor XIIIb. CR3 (CD11b) and CR4 (CD11c) are members of the integrin superfamily of adhesion proteins.

Activation of the complement system induces the formation of a membrane attack complex resulting in the elimination of pathogens and infected/apoptotic cells by lysis and phagocytosis mechanisms. In the brain, glial cells and neurons can produce complement molecules, which are implicated in the inflammatory response during injury and in neurodegenerative diseases, but also in developmental processes such as synaptic refinement [36]. In addition, the expression of several complement receptors in NSCs suggests a role for the complement system at early developmental stages (Table 5.1; Figure 5.1). To track complement receptor 2 (Cr2) gene expression in vivo, Cr2-cre transgenic mice were crossed with flox-stop enhanced green fluorescent protein (EGFP) to create a double transgenic mouse in which EGFP-positive cells express CR2 or descend from CR2-expressing precursors [12]. In this way, the authors showed, in adult brain, EGFP-positive cells colabeled with the multipotent NSC marker Sox2 in a proliferating area of the hippocampus. Expression of CR2 mRNA was further confirmed in freshly isolated neonatal NSCs [12]. The function of CR2 in NSC proliferation and differentiation was assessed using a CR2 agonist, complement C3d. Application of C3d to cultures of NSCs derived from postnatal mouse brain reduces the number of neurospheres without affecting cell survival, suggesting that CR2 activation specifically inhibits NSC proliferation. In vivo studies on CR2 function in basal neurogenesis assessed in CR2-deficient mice indicate that the lack of CR2 induces an increase in newly generated proliferating neuroblasts and mature neurons, identified by costaining of BrdU with DCX and NeuN, respectively. Conversely, intracerebral injection of C3d into the dentate gyrus of adult mice results in a decrease in the number of proliferating neuroblasts in the SGZ, suggesting that CR2 activation inhibits neurogenesis [12]. However, given the expression of complement receptors by other brain cells, such as microglia or oligodendrocyte progenitors, the effects produced by CR2 deletion and activation on neuronal differentiation might be indirectly mediated by glial cells.

Two other complement receptors, C3aR and C5aR, have been detected along the rostral migratory stream, a pathway through which neuroblasts exit the SVZ and tangentially migrate in the caudal-to-rostral direction to the olfactory bulb. In this pathway, both C3aR and C5aR are associated with nestin-positive cells in the SVZ, migrating neuroblasts (DCX-positive cells), and differentiated neurons (NeuN-positive cells) [13]. No colocalization of C3aR or C5aR is found with astrocytes, suggesting that both complement receptors are selectively associated with cells committed to a neuronal lineage. However, binding assays of C3a and C5a peptides to adult mouse brain-derived NSCs indicate that C3a, but not C5a, peptide binds significantly to NSCs, suggesting that C5aR might be improperly targeted to the cell surface [14]. Accordingly, the numbers of proliferating cells (BrdU-positive), newly formed migrating neuroblasts (BrdU/DCX-positive cells), and newly formed neurons (BrdU/NeuN-positive cells) are similar in C5aR-deficient mice and control mice, suggesting that C5aR is not involved in basal neurogenesis [13]. By contrast, mice deficient for the complement factor C3, or for C3aR, and wild-type mice treated with a C3aR antagonist all show a decrease in newly formed neuroblasts and neurons in the SVZ, SGZ, and olfactory bulb, suggesting that the C3–C3aR pathway supports adult neurogenesis [15]. Moreover, application of C3a to NSC cultures derived from adult mouse brain stimulates neuronal differentiation [14]. Interestingly, similar to the synergy described between C3a and the chemokine SDF-1α for the trafficking of bone marrow cells [37], C3a modulates NSC migration triggered by stromal cell-derived factor (SDF)-1α and inhibits SDF-1α-induced neuronal differentiation of NSCs [14]. Only one study reports a role for the complement system on early neuronal morphological differentiation. Neurons derived from NSCs treated with C3a exhibited longer processes than untreated cells, suggesting that C3aR activation promotes neuritogenesis [14]. At later developmental stages during synaptic refinement, C1q was shown to play a critical role in dendrite remodeling and synaptic pruning [38,39].

5.2 MOLECULES FROM ADAPTIVE IMMUNITY

5.2.1 Major Histocompatibility Complex Molecules

There are two major classes of major histocompatibility complex (MHC) molecules, both of which consist of an α and a β chain, but are from different sources. MHC class I molecules (MHCI) consist of one membrane-spanning α chain (heavy chain) produced by MHC genes and one β chain (light chain or β2-microglobulin) produced by the β2-microglobulin gene. MHC class II molecules (MHCII) consist of two membrane-spanning chains, α and β, of similar size and both produced by MHC genes [40]. There is a large diversity of MHCI and MHCII molecules that arises from the combination of multiple genes and multiple alleles, which makes this MHC protein family highly polygenic and polymorphic. MHCI and MHCII are both responsible for the antigenic presentation to T and natural killer (NK) cells, but are specialized to present different types of antigens, thereby eliciting different responses [41,42]. The MHCI molecules present endogenous antigens that originate from the cytoplasm. These antigens include not only self-proteins, but also foreign proteins produced within the cell, such as viral proteins. MHCII molecules present exogenously derived antigenic peptides. MHCI molecules are expressed by almost every cell in the body, whereas the MHCII complex is mostly restricted to professional antigen-presenting cells, such as DCs, macrophages, and microglia. The role of MHCII in the brain seems to be restricted to an antigen-presenting function and no study has reported a

nonimmune function in the differentiation or development of neural cells. Undetectable or very low levels of MHCII expression have been reported in NSCs from embryonic and newborn mouse brain [16,25] and embryonic human brain [26] (Table 5.1). Treatment with the proinflammatory molecules interferon-γ (IFN-γ) or tumor necrosis factor-α (TNF-α) increases MHCII mRNA and protein expression in NSCs [25–27], suggesting a potential function of NSCs as antigen-presenting cells in an inflammatory environment. Accordingly, intracerebral allotransplantation of NSCs in a mouse model of demyelination induced by infection with the JHM strain of mouse hepatitis virus indicated that these allogeneic NSCs are antigenic and induce T cell proliferation and expression of the chemokines CXCL9 and CXCL10 [27].

In contrast to MHCII, nonimmune functions of MHCI molecules have been demonstrated, in the developing and adult brain, in axonal refinement, synaptogenesis, and synaptic plasticity [43,44]. The expression of MHCI by NSCs has long been controversial and remains a subject of debate. The source of MHCI antibodies, the procedure used to isolate NSCs, and the culture conditions might result in discrepancies between studies. Freshly isolated NSCs from human fetuses or newborn mice show no or a low level of MHCI expression [16–18], whereas NSC cultures expanded under proliferating conditions show a notable increase in MHCI expression [18–20] (Table 5.1). In vivo, immunohistofluorescence analysis of MHCI distribution in brain sections of E9.5/E10.5 mouse embryos indicated that only a subset of nestin-positive NSCs located in the neuroepithelium and olfactory placode expressed MHCI, whereas the vast majority of cells committed to a neuronal lineage labeled with the Pax6 or Tuj1 markers also expressed MHCI, suggesting that increased levels of MHCI expression accompanied neuronal differentiation [21]. Despite the close relationship between MHCI expression and neuronal differentiation, MHCI molecules appear to not be involved in adult neurogenesis as mice deficient for the transporter associated with antigen processing 1 (TAP1), in which cell-surface expression of MHCI is drastically reduced [45] (Table 5.2), show similar NSC proliferation or neurogenesis in the dentate gyrus and in the SVZ of wild-type and TAP1-deficient mice [46]. However, an MHCI-related ligand for NKG2D receptors, retinoic acid early induced transcript (RAE-1), promotes NSC proliferation [24]. Considering the large diversity of MHCI family proteins, one could expect that different members might exert specific effects on NSC proliferation and neuronal differentiation. Moreover, the level of expression of MHCI in NSCs is strongly upregulated by the proinflammatory factors IFN-γ and TNF-α [16,20] and by Japanese encephalitis virus infection [28]. Whether the role of MHCI molecules in NSC proliferation and differentiation is dependent on the inflammatory environment remains to be studied. Although MHCI molecules probably

TABLE 5.2 Transgenic and Knockout Mouse Lines Used to Study MHCI Function in Neural Differentiation

Name	Background	Overexpression/Knockout	Affected Cell Type	Phenotype	References
NSE-D^b	C57Bl6	Overexpression	Neurons	Neuronal overexpression of H-2D^b haplotype	[47]
TAP1$^{-/-}$	C57Bl6	Knockout	All types of cells	Lack of MHCI surface expression	[45]
K^{b}D$^{b-/-}$	C57Bl6	Knockout	All types of cells	Lack of H-2K^b and H-2D^b expression	[48]

MHC, major histocompatibility complex; MHCI, MHC class I molecules.

do not play a major role in NSC proliferation or in the initial cell-fate decision, several studies point out the importance of MHCI in early neuronal morphological differentiation. Cultures of hippocampal neurons from transgenic NSE-D^b mice, which have elevated levels of neuronal MHCI [47] (Table 5.2), show an increased neurite outgrowth during the stage of neurite emergence and elongation. Conversely, MHCI-deficient neurons prepared from K^{b}D$^{b-/-}$ mice display reduced neuritogenesis [22,48] (Table 5.2). Interestingly, the soluble form of MHCI, shed from intact cell-surface MHCI, inhibits neurite and axonal outgrowth in an embryonic retinothalamic explant coculture model [23], and addition of recombinant MHCI molecules to embryonic dorsal root ganglia explant cultures also inhibits axon outgrowth [49]. Thus, the activity of MHCI molecules in the early stages of neuronal polarization is conformation-dependent and probably involves different sets of MHCI neuronal receptors, including PirB, Ly49, and other yet-uncharacterized receptors [50,51].

5.2.2 Costimulatory Molecules

Costimulatory factors have been widely studied in the immune system for their critical role in T cell activation. Among the known multiple costimulatory signals, one of the best described pathways involves the cytotoxic T lymphocyte antigen-4 (CTLA-4, also known as CD152). The CD28/CTLA-4-CD80/CD86 pathway critically controls the nature and duration of the T cell response [52,53]. After the initiation of T cell activation through the interaction with MHC/peptide complex, the binding of CD80/86 expressed on antigen-presenting cells to the costimulatory molecule CD28 or CTLA-4, both present on T cells, promotes or limits T cell response, respectively [52]. Expression of CD80/CD86 was reported to be very low under basal conditions in nestin-positive cells of the SVZ in rat and mouse brain and in neurospheres cultures [18,19,28,29] (Table 5.1). Upon inflammation induced by IFN-γ or TNF-α, cell-surface expression of CD80 and CD86 is markedly increased in rat and mouse, but not human, NSCs [18,28,29]. Similarly, in the course of CNS inflammatory diseases, such as

experimental autoimmune encephalomyelitis or Japanese encephalitis virus infection, CD80 and CD86 are upregulated in nestin-positive cells in SVZ of rat and mouse brain [28,29]. The impact of costimulatory molecule expression on NSC proliferation capacity or neural differentiation has not been investigated, but has been well characterized in the context of immune response. NSCs expressing CD80/CD86 are capable of providing functional costimulation to allogeneic T cells, resulting in increased T cell proliferation [28,29].

5.2.3 CD3ζ

CD3ζ (also called CD247) is a transmembrane signaling adaptor protein first described in T lymphocytes as a component of the CD3 complex, the signaling module of the T cell receptor (TCR). Later, it was also found to transduce signaling of the activating receptors NKp46/NKp30 and the low-affinity Fc receptor for IgG CD16 in NK cells [54]. CD3ζ forms homodimers in T cells and homo- or heterodimers with the γ chain of the receptor for IgE in NK cells [55]. In its cytoplasmic domain, CD3ζ comprises three immunoreceptor tyrosine-based activation motifs (ITAMs), which, upon tyrosine phosphorylation by a Src family protein tyrosine kinase, recruit ZAP-70/Syk tyrosine kinase family members that in turn trigger several downstream signaling cascades leading to T cell activation [56]. In addition, CD3ζ is also involved in T cell differentiation and cell-surface expression of the TCR [57–60]. CD3ζ represents a good example of so-called immune molecules that have been shown to have important, and often similar, functions in cell differentiation and activation in the nervous system. CD3ζ was found to be expressed in the CNS, mostly by neurons, and to play a critical role in neuronal differentiation and function throughout development (Table 5.1; Figure 5.2). During the neurogenic period in E15–E16 embryonic rat brain, CD3ζ immunoreactivity is absent in the proliferating ventricular zone harboring the NSC, but is selectively expressed by early postmitotic Tuj1-positive neurons in the cerebral cortex, hippocampus, and olfactory bulb [30]. Accordingly, in neurosphere-derived neural cell cultures, CD3ζ is virtually undetectable in nestin-positive

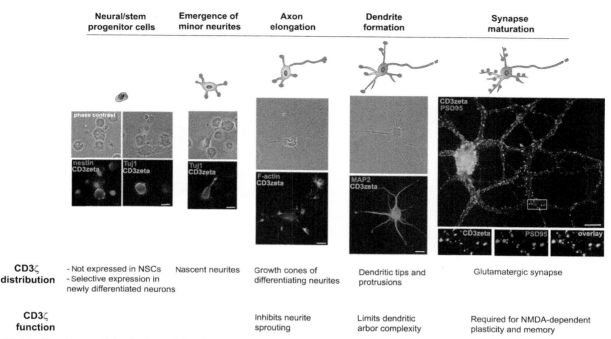

FIGURE 5.2   Developmental distribution and function of CD3ζ during neuronal differentiation. CD3ζ protein distribution is detected by double immunofluorescence for CD3ζ (green) and several developmentally relevant markers (red) in neurons derived from rat embryonic NSCs and in hippocampal neurons in culture. Nestin identifies NSCs; Tuj1, neurons; F actin labeled with TRITC–phalloidin, growth cones; MAP2, dendrites; and PSD95, glutamatergic synapses. Scale bars, 10 μm. NSC, neural stem cell. (For interpretation of the references to color in this figure legend, the reader is referred to the online version of this book.)

NSCs, but largely expressed in newly born neurons shortly after their differentiation and before morphological polarization (Figure 5.2). In contrast, other neural cell subtypes, namely astrocytes and oligodendrocytes, exhibit only minimal CD3ζ expression. The coincidence of CD3ζ expression with neuronal commitment suggests that CD3ζ is probably not involved in NSC proliferation or in the initial cell-fate decision, but could play a role in the maintenance of a neuronal phenotype shortly after NSC differentiation. The selective expression of CD3ζ in newly born neurons, as opposed to NSCs and glial cells, makes the potential utilization of CD3ζ as a selection marker an attractive option for sorting neuroblasts from a mixed population for developmental, pharmacological, and therapeutic applications. CD3ζ expression is maintained throughout neuronal development and plays specific roles at different critical stages of neuronal polarization and maturation. During the early steps of neuronal morphological differentiation leading to the emergence of minor neurites, CD3ζ proteins redistribute to nascent neurites and concentrate at the tips of growing neurites, within growth cones (Figure 5.2). Gain- and loss-of-function approaches indicate that overexpression of CD3ζ in postmitotic neurons represses neurite outgrowth, whereas neurons deficient for CD3ζ exhibit an enhanced neurite number [31]. The upstream and downstream signals involved in CD3ζ-mediated inhibition of neurite outgrowth at early developmental stages have been

partially identified and involve the receptor tyrosine kinase EphA4 and its ligand ephrinA1, a strong repulsion cue implicated in neurite outgrowth and guidance [61–63]. In young neurons, CD3ζ relays ephrinA1/EphA4 activation to the downstream effector molecules Vav2, a Rho guanine triphosphatase guanine-nucleotide exchange factor, and ZAP-70/Syk, a protein tyrosine kinase [31] (Figure 5.3). In older neurons, at the time of dendrite specification, CD3ζ becomes preferentially expressed in dendrites associated with discrete subdomains involved in dendrite elongation and branching, i.e., filopodia and growth cones [32] (Figure 5.2). As at earlier developmental stages, CD3ζ negatively regulates dendritic arbor complexity. This effect is dependent on the ITAMs of the intracellular region of CD3ζ [32]. In mature neurons, when synapses are formed, CD3ζ is concentrated at glutamatergic synapses (Figure 5.2), where it interacts with the N-Methyl-D-aspartate (NMDA) glutamate receptor subunit GluN2A [32,64]. By using CD3ζ-knockout mice, several studies have highlighted the role of CD3ζ as a regulator of glutamatergic synaptic plasticity [64–66]. By regulating the interaction of GluN2A with the key NMDA downstream signaling protein calcium/calmodulin-dependent protein kinase II (CaMKII), CD3ζ participates in the regulation of the synaptic recruitment of the α-amino-3-hydroxy-5-methyl-4-isoxazolepropionic acid (AMPA) receptor subunit GluA1 during long-term potentiation (Figure 5.3) [64].

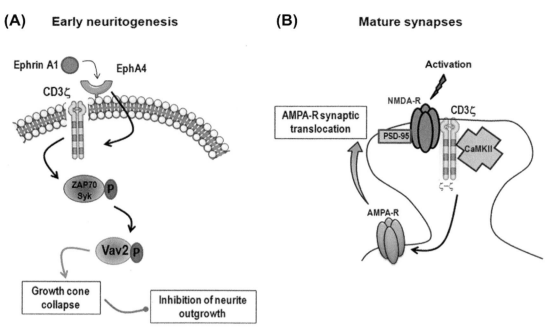

FIGURE 5.3   CD3ζ-mediated signaling pathways during neuronal development. (A) During early neuritogenesis, CD3ζ negatively regulates neurite outgrowth by relaying ephrinA1/EphA4 signaling. A downstream underlying mechanism may involve the activating Rho GTPase guanine-nucleotide exchange factor Vav2 and the protein tyrosine kinases of the ZAP-70/Syk family. (B) At glutamatergic synapses in mature neurons, CD3ζ stabilizes NMDA receptor–PSD95 protein complex and is required for NMDA-dependent activation of CAMKII and consequently for the synaptic translocation of AMPA receptors.

5.3 CONCLUSION

A range of surface molecules originally discovered in the context of their function in immune responses was subsequently found to play important roles in NSC proliferation, cell-fate decision, and differentiation. A better knowledge of the immune molecular repertoire expressed on the cell surface of NSCs and their neural derivatives is particularly important in two ways. First, differentiation of NSCs often results in variable and heterogeneous cultures of neurons, glia, and undifferentiated cells, making the characterization and purification of distinct cell types particularly challenging. The functional studies of NSCs and their derivatives described here will probably contribute to a refinement of specific cell surface markers, similar to what has been accomplished in studies of hematopoiesis. This achievement should benefit downstream applications that require purified or defined cell populations, such as in vitro assays, transplantation, and microarrays. Second, the expression of molecules from the innate and adaptive immunity on the cell surface of NSCs and their cell derivatives makes them potentially sensitive and reactive to an immune challenge. The mediators secreted during the immune response can directly modulate NSC proliferation and influence their neurogenic potential, which in turn might affect critical periods of CNS development and induce long-term functional changes in the adult. Many groups have reported that maternal viral or bacterial infection during pregnancy is a significant risk factor for several neuropsychiatric disorders with presumed developmental origins, including schizophrenia and autism [67]. In addition to the association based on epidemiological studies, animal models have been established to demonstrate linkage between maternal immune activation, NSC biology, and neuropsychiatric disorders. In a mouse model of maternal immune activation via injection of poly(I:C) into pregnant mice on gestation days 9.5–12.5, adult neurogenesis in the offspring is impaired, as measured in the dentate gyrus of the hippocampus and in the SVZ–olfactory bulb pathway [68,69]. Moreover, analogous experimental paradigms in a nonhuman primate model yield offspring with abnormal repetitive behaviors, communication, and social interactions resembling those observed in both autism and schizophrenia [70]. A better knowledge of the expression and function of immune-related molecules in NSCs and their neural derivatives might therefore have an important impact upon the understanding of neurodevelopmental psychiatric disorders and the design of novel therapeutic strategies.

REFERENCES

[1] Kawai T, Akira S. The role of pattern-recognition receptors in innate immunity: update on toll-like receptors. Nat Immunol 2010;11: 373–84.

[2] Kawai T, Akira S. Toll-like receptors and their crosstalk with other innate receptors in infection and immunity. Immunity 2011;34: 637–50.

[3] Kielian T. Toll-like receptors in central nervous system glial inflammation and homeostasis. J Neurosci Res 2006;83:711–30.

[4] Lehnardt S, Massillon L, Follett P, Jensen FE, Ratan R, Rosenberg PA, et al. Activation of innate immunity in the CNS triggers neurodegeneration through a toll-like receptor 4-dependent pathway. Proc Natl Acad Sci USA 2003;100:8514–9.

[5] Rolls A, Shechter R, London A, Ziv Y, Ronen A, Levy R, et al. Toll-like receptors modulate adult hippocampal neurogenesis. Nat Cell Biol 2007;9:1081–8.

[6] Okun E, Griffioen KJ, Son TG, Lee JH, Roberts NJ, Mughal MR, et al. TLR2 activation inhibits embryonic neural progenitor cell proliferation. J Neurochem 2010;114:462–74.

[7] Lathia JD, Okun E, Tang SC, Griffioen K, Cheng A, Mughal MR, et al. Toll-like receptor 3 is a negative regulator of embryonic neural progenitor cell proliferation. J Neurosci 2008;28:13978–84.

[8] Okun E, Griffioen K, Barak B, Roberts NJ, Castro K, Pita MA, et al. Toll-like receptor 3 inhibits memory retention and constrains adult hippocampal neurogenesis. Proc Natl Acad Sci USA 2010;107:15625–30.

[9] Cameron JS, Alexopoulou L, Sloane JA, DiBernardo AB, Ma Y, Kosaras B, et al. Toll-like receptor 3 is a potent negative regulator of axonal growth in mammals. J Neurosci 2007;27:13033–41.

[10] Liu HY, Hong YF, Huang CM, Chen CY, Huang TN, Hsueh YP. TLR7 negatively regulates dendrite outgrowth through the Myd88-c-Fos-IL-6 pathway. J Neurosci 2013;33:11479–93.

[11] Ma Y, Li J, Chiu I, Wang Y, Sloane JA, Lu J, et al. Toll-like receptor 8 functions as a negative regulator of neurite outgrowth and inducer of neuronal apoptosis. J Cell Biol 2006;175:209–15.

[12] Moriyama M, Fukuhara T, Britschgi M, He Y, Narasimhan R, Villeda S, et al. Complement receptor 2 is expressed in neural progenitor cells and regulates adult hippocampal neurogenesis. J Neurosci 2011;31:3981–9.

[13] Bogestal YR, Barnum SR, Smith PL, Mattisson V, Pekny M, Pekna M. Signaling through C5aR is not involved in basal neurogenesis. J Neurosci Res 2007;85:2892–7.

[14] Shinjyo N, Stahlberg A, Dragunow M, Pekny M, Pekna M. Complement-derived anaphylatoxin C3a regulates in vitro differentiation and migration of neural progenitor cells. Stem Cells 2009;27:2824–32.

[15] Rahpeymai Y, Hietala MA, Wilhelmsson U, Fotheringham A, Davies I, Nilsson AK, et al. Complement: a novel factor in basal and ischemia-induced neurogenesis. EMBO J 2006;25:1364–74.

[16] Chen Z, Phillips LK, Gould E, Campisi J, Lee SW, Ormerod BK, et al. MHC mismatch inhibits neurogenesis and neuron maturation in stem cell allografts. PLoS One 2011;6:e14787.

[17] Hori J, Ng TF, Shatos M, Klassen H, Streilein JW, Young MJ. Neural progenitor cells lack immunogenicity and resist destruction as allografts. Stem Cells 2003;21:405–16.

[18] Laguna Goya R, Busch R, Mathur R, Coles AJ, Barker RA. Human fetal neural precursor cells can up-regulate MHC class I and class II expression and elicit CD4 and CD8 T cell proliferation. Neurobiol Dis 2011;41:407–14.

[19] Sergent-Tanguy S, Veziers J, Bonnamain V, Boudin H, Neveu I, Naveilhan P. Cell surface antigens on rat neural progenitors and characterization of the CD3+/CD3− cell populations. Differentiation 2006;74:530–41.

[20] Yin L, Fu SL, Shi GY, Li Y, Jin JQ, Ma ZW, et al. Expression and regulation of major histocompatibility complex on neural stem cells and their lineages. Stem Cells Dev 2008;17:53–65.

[21] Chacon MA, Boulanger LM. MHC class I protein is expressed by neurons and neural progenitors in mid-gestation mouse brain. Mol Cell Neurosci 2013;52:117–27.

[22] Bilousova T, Dang H, Xu W, Gustafson S, Jin Y, Wickramasinghe L, et al. Major histocompatibility complex class I molecules modulate embryonic neuritogenesis and neuronal polarization. J Neuroimmunol 2012;247:1–8.

[23] Washburn LR, Zekzer D, Eitan S, Lu Y, Dang H, Middleton B, et al. A potential role for shed soluble major histocompatibility class I molecules as modulators of neurite outgrowth. PLoS One 2011;6:e18439.

[24] Popa N, Cedile O, Pollet-Villard X, Bagnis C, Durbec P, Boucraut J. RAE-1 is expressed in the adult subventricular zone and controls cell proliferation of neurospheres. Glia 2011;59:35–44.

[25] Mammolenti M, Gajavelli S, Tsoulfas P, Levy R. Absence of major histocompatibility complex class I on neural stem cells does not permit natural killer cell killing and prevents recognition by alloreactive cytotoxic T lymphocytes in vitro. Stem Cells 2004;22:1101–10.

[26] Ubiali F, Nava S, Nessi V, Frigerio S, Parati E, Bernasconi P, et al. Allorecognition of human neural stem cells by peripheral blood lymphocytes despite low expression of MHC molecules: role of TGF-β in modulating proliferation. Int Immunol 2007;19:1063–74.

[27] Weinger JG, Weist BM, Plaisted WC, Klaus SM, Walsh CM, Lane TE. MHC mismatch results in neural progenitor cell rejection following spinal cord transplantation in a model of viral-induced demyelination. Stem Cells 2012;30:2584–95.

[28] Das S, Ghosh D, Basu A. Japanese encephalitis virus induce immuno-competency in neural stem/progenitor cells. PLoS One 2009;4:e8134.

[29] Imitola J, Comabella M, Chandraker AK, Dangond F, Sayegh MH, Snyder EY, et al. Neural stem/progenitor cells express costimulatory molecules that are differentially regulated by inflammatory and apoptotic stimuli. Am J Pathol 2004;164:1615–25.

[30] Angibaud J, Baudouin SJ, Louveau A, Nerriere-Daguin V, Bonnamain V, Csaba Z, et al. Ectopic expression of the immune adaptor protein CD3ζ in neural stem/progenitor cells disrupts cell-fate specification. J Mol Neurosci 2012;46:431–41.

[31] Angibaud J, Louveau A, Baudouin SJ, Nerriere-Daguin V, Evain S, Bonnamain V, et al. The immune molecule CD3ζ and its downstream effectors ZAP-70/Syk mediate ephrin signaling in neurons to regulate early neuritogenesis. J Neurochem 2011;119:708–22.

[32] Baudouin SJ, Angibaud J, Loussouarn G, Bonnamain V, Matsuura A, Kinebuchi M, et al. The signaling adaptor protein CD3ζ is a negative regulator of dendrite development in young neurons. Mol Biol Cell 2008;19:2444–56.

[33] Jurk M, Heil F, Vollmer J, Schetter C, Krieg AM, Wagner H, et al. Human TLR7 or TLR8 independently confer responsiveness to the antiviral compound R-848. Nat Immunol 2002;3:499.

[34] Carroll MC. The complement system in regulation of adaptive immunity. Nat Immunol 2004;5:981–6.

[35] Zipfel PF, Skerka C. Complement regulators and inhibitory proteins. Nat Rev Immunol 2009;9:729–40.

[36] Stephan AH, Barres BA, Stevens B. The complement system: an unexpected role in synaptic pruning during development and disease. Annu Rev Neurosci 2012;35:369–89.

[37] Ratajczak MZ, Reca R, Wysoczynski M, Yan J, Ratajczak J. Modulation of the SDF_{-1}–$CXCR_4$ axis by the third complement component (C_3)—implications for trafficking of $CXCR_4^+$ stem cells. Exp Hematol 2006;34:986–95.

[38] Ma Y, Ramachandran A, Ford N, Parada I, Prince DA. Remodeling of dendrites and spines in the C1q knockout model of genetic epilepsy. Epilepsia 2013;54:1232–9.

[39] Stevens B, Allen NJ, Vazquez LE, Howell GR, Christopherson KS, Nouri N, et al. The classical complement cascade mediates CNS synapse elimination. Cell 2007;131:1164–78.

[40] Maenaka K, Jones EY. MHC superfamily structure and the immune system. Curr Opin Struct Biol 1999;9:745–53.

[41] Lanier LL. NK cell recognition. Annu Rev Immunol 2005;23:225–74.

[42] Natarajan K, Li H, Mariuzza RA, Margulies DH. MHC class I molecules, structure and function. Rev Immunogenet 1999;1:32–46.

[43] McAllister AK. Major histocompatibility complex I in brain development and schizophrenia. Biol Psychiatry 2014;75:262–8.

[44] Shatz CJ. MHC class I: an unexpected role in neuronal plasticity. Neuron 2009;64:40–5.

[45] Van Kaer L, Ashton-Rickardt PG, Ploegh HL, Tonegawa S. TAP1 mutant mice are deficient in antigen presentation, surface class I molecules, and CD4⁻8⁺ T cells. Cell 1992;71:1205–14.

[46] Laguna Goya R, Tyers P, Barker RA. Adult neurogenesis is unaffected by a functional knock-out of MHC class I in mice. Neuroreport 2010;21:349–53.

[47] Rall GF, Mucke L, Oldstone MB. Consequences of cytotoxic T lymphocyte interaction with major histocompatibility complex class I-expressing neurons in vivo. J Exp Med 1995;182:1201–12.

[48] Vugmeyster Y, Glas R, Perarnau B, Lemonnier FA, Eisen H, Ploegh H. Major histocompatibility complex (MHC) class I K^{b}D$^{b-/-}$ deficient mice possess functional CD8+ T cells and natural killer cells. Proc Natl Acad Sci USA 1998;95:12492–7.

[49] Wu ZP, Bilousova T, Escande-Beillard N, Dang H, Hsieh T, Tian J, et al. Major histocompatibility complex class I-mediated inhibition of neurite outgrowth from peripheral nerves. Immunol Lett 2011;135:118–23.

[50] Syken J, Grandpre T, Kanold PO, Shatz CJ. PirB restricts ocular-dominance plasticity in visual cortex. Science 2006;313:1795–800.

[51] Zohar O, Reiter Y, Bennink JR, Lev A, Cavallaro S, Paratore S, et al. Cutting edge: MHC class I-Ly49 interaction regulates neuronal function. J Immunol 2008;180:6447–51.

[52] Rudd CE, Taylor A, Schneider H. CD28 and CTLA-4 coreceptor expression and signal transduction. Immunol Rev 2009;229:12–26.

[53] Ueda H, Howson JM, Esposito L, Heward J, Snook H, Chamberlain G, et al. Association of the T-cell regulatory gene CTLA4 with susceptibility to autoimmune disease. Nature 2003;423:506–11.

[54] Lanier LL. On guard–activating NK cell receptors. Nat Immunol 2001;2:23–7.

[55] Baniyash M. TCR ζ-chain downregulation: curtailing an excessive inflammatory immune response. Nat Rev Immunol 2004;4:675–87.

[56] Wange RL, Samelson LE. Complex complexes: signaling at the TCR. Immunity 1996;5:197–205.

[57] D'Oro U, Munitic I, Chacko G, Karpova T, McNally J, Ashwell JD. Regulation of constitutive TCR internalization by the ζ-chain. J Immunol 2002;169:6269–78.

[58] Malissen M, Gillet A, Rocha B, Trucy J, Vivier E, Boyer C, et al. T cell development in mice lacking the CD3-zeta/eta gene. EMBO J 1993;12:4347–55.

[59] Shores EW, Huang K, Tran T, Lee E, Grinberg A, Love PE. Role of TCR zeta chain in T cell development and selection. Science 1994;266:1047–50.

[60] Sussman JJ, Bonifacino JS, Lippincott-Schwartz J, Weissman AM, Saito T, Klausner RD, et al. Failure to synthesize the T cell CD3ζ chain: structure and function of a partial T cell receptor complex. Cell 1988;52:85–95.

[61] Donoghue MJ, Merlie JP, Sanes JR. The Eph kinase ligand AL-1 is expressed by rostral muscles and inhibits outgrowth from caudal neurons. Mol Cell Neurosci 1996;8:185–98.

[62] Woo S, Rowan DJ, Gomez TM. Retinotopic mapping requires focal adhesion kinase-mediated regulation of growth cone adhesion. J Neurosci 2009;29:13981–91.

[63] Zhou X, Suh J, Cerretti DP, Zhou R, DiCicco-Bloom E. Ephrins stimulate neurite outgrowth during early cortical neurogenesis. J Neurosci Res 2001;66:1054–63.

[64] Louveau A, Angibaud J, Haspot F, Opazo MC, Thinard R, Thepenier V, et al. Impaired spatial memory in mice lacking CD3ζ is associated with altered NMDA and AMPA receptors signaling independent of T-cell deficiency. J Neurosci 2013;33:18672–85.

[65] Huh GS, Boulanger LM, Du H, Riquelme PA, Brotz TM, Shatz CJ. Functional requirement for class I MHC in CNS development and plasticity. Science 2000;290:2155–9.

[66] Xu HP, Chen H, Ding Q, Xie ZH, Chen L, Diao L, et al. The immune protein CD3ζ is required for normal development of neural circuits in the retina. Neuron 2010;65:503–15.

[67] Brown AS. Epidemiologic studies of exposure to prenatal infection and risk of schizophrenia and autism. Dev Neurobiol 2012;72:1272–6.

[68] Liu YH, Lai WS, Tsay HJ, Wang TW, Yu JY. Effects of maternal immune activation on adult neurogenesis in the subventricular zone–olfactory bulb pathway and olfactory discrimination. Schizophr Res 2013;151:1–11.

[69] Shi L, Smith SE, Malkova N, Tse D, Su Y, Patterson PH. Activation of the maternal immune system alters cerebellar development in the offspring. Brain Behav Immun 2009;23:116–23.

[70] Bauman MD, Iosif AM, Smith SE, Bregere C, Amaral DG, Patterson PH. Activation of the maternal immune system during pregnancy alters behavioral development of rhesus monkey offspring. Biol Psychiatry 2014;75:332–41.

Chapter 6

Neuropilins in Development and Disease of the Nervous System

Mathew Tata, Miguel Tillo and Christiana Ruhrberg

Department of Cell Biology, UCL Institute of Ophthalmology, London, UK

6.1 INTRODUCTION

Neuropilin 1 (NRP1; CD304) and neuropilin 2 (NRP2) are single-pass transmembrane proteins that regulate both cardiovascular and central nervous system (CNS) development [1]. Both neuropilins share 44% sequence homology at the amino acid level and have a similar domain structure comprised of a large N-terminal extracellular domain (835 amino acid residues [aa] for NRP1, 844 for NRP2), a short membrane-spanning domain (23 aa for NRP1, 25 for NRP2), and a small cytoplasmic domain (44 aa for NRP1, 42 for NRP2). The extracellular domain contains two complement-binding homology domains, termed a1 and a2, which are essential for binding to the SEMA domain present in axon guidance cues of the semaphorin family. Two coagulation factor V/VIII homology domains, termed b1 and b2, mediate binding to different classes and isoforms of the vascular endothelial growth factor (VEGF) family, and also have been implicated in cell adhesion. In addition, a meprin domain, termed c, separates the b2 and transmembrane domains and mediates dimerization with other receptors. The cytoplasmic tail contains a PDZ-domain binding motif; in the case of NRP1, this motif is responsible for binding synectin (GIPC1), which bridges NRP1 and a myosin 6-driven cell transport machinery for endocytic trafficking [1,2]. Due to the variety of structural domains present on both sides of the cell membrane, neuropilins can interact with several binding partners and thereby initiate a number of different signaling pathways to control neuronal and vascular cell behavior, as described below. Throughout this chapter, we have flagged some gaps that remain in our current knowledge of neuropilin function in development and disease, as they present opportunities for future research.

6.2 NEUROPILINS ASSOCIATE WITH PLEXINS TO MEDIATE SEMAPHORIN SIGNALING

The axons of both central and peripheral nervous system (PNS) neurons are guided by growth cones that sense guidance molecules [3]. These molecules are secreted as diffusible cues into the interstitial environment, tethered to the extracellular matrix (ECM) or bound to cell surfaces. One of the most studied groups of guidance cues is the class 3 semaphorins, which contain a 500-amino acid residue SEMA domain responsible for signaling activity [4,5]. Like other guidance molecule families, such as slits and ephrins, they regulate both vascular and nervous system development [1].

Seven secreted glycoproteins make up the class 3 semaphorin subfamily and are termed SEMA3A-3G. SEMA3 proteins require an interaction with transmembrane proteins of the plexin (PLXN) family to transduce signals, with the exception of SEMA3E, which binds a plexin (PLXND1) directly [6]. All other class 3 semaphorins bind to a neuropilin, which then forms a complex with an A-type plexin [6]. Neuropilins and plexins form *cis* complexes independently of class 3 semaphorins, but this interaction becomes stabilized by SEMA3 binding [7]. SEMA3 binding also induces PLXNA dimerization, ultimately leading to the formation of a coreceptor complex consisting of two neuropilin–plexin heterodimers that are linked by a dimeric semaphorin [7].

Plexins can activate several different signaling cascades involved in cytoskeletal remodeling to influence cell migration and axon guidance [8]. Thus, they regulate F-actin dynamics through RHO- and RAS-GTPases and impair microtubule assembly through inhibiting the interaction of the collapsing-response-mediator protein CRMP with tubulin heterodimers via the serine/threonine kinase GSK3β [9,10]. Both mechanisms result in the collapse and subsequent turning of axonal growth cones [10]. In addition, plexins signal through intracellular kinases such as proto-oncogene tyrosine-protein kinase Src (SRC), phosphoinositide 3-kinase (PI3K), and mitogen-activated protein kinase (MAPK) to modulate axon pathfinding [6]. Several tyrosine kinases, for example, proto-oncogene tyrosine-protein kinase Fes/Fps (FES) can phosphorylate plexins to diversify cellular responses to semaphorin signaling [11]. Neuropilin/plexin complexes also cooperate with L1-CAM (CD171) to modulate SEMA3-mediated

Neural Surface Antigens. http://dx.doi.org/10.1016/B978-0-12-800781-5.00006-2

growth cone collapse in certain neuron subtypes [12]. Specific examples of class 3 semaphorin roles in axon and cell migration will be discussed later in this chapter.

6.3 NEUROPILINS AND PLEXINS COOPERATE TO CONFER SPECIFICITY TO SEMAPHORIN SIGNALING

One of the crucial mechanisms required to modulate neuronal responses to semaphorins is the differential affinity of NRP1 and/or NRP2 for individual semaphorins, and the preferential interaction of the neuropilins with a coreceptor of the signal-transducing plexin family. Thus, several different plexin–neuropilin ternary signaling complex combinations act in axonal growth cones to modulate attraction and repulsion during axon pathfinding [6]. A variety of neuron subtypes in the developing CNS have been used to study the function of these combinations, some of which are described below. We will first discuss the role of neuropilin specificity for semaphorins in axon guidance, and then the involvement of plexin receptors in modulating signal transduction.

SEMA3C binds both neuropilins, such as in axons projecting from neurons in the dorsal root ganglia [13], but other class 3 semaphorins bind preferentially to NRP1 or NRP2 to enable axon guidance in the developing nervous system. Thus, NRP1 predominantly senses SEMA3A, while axons expressing NRP2 more typically respond to SEMA3F [14–18]. Interestingly, the SEMA3A/NRP1 and SEMA3F/NRP2 pathways cooperate to control axon pathfinding, for example, in the case of spinal motor neurons. Accordingly, the fasciculation of lateral motor column (LMC) nerve axons and their timing of limb invasion are regulated by SEMA3A/NRP1 signaling, whereas SEMA3F/NRP2 signaling guides a medial subset of LMC axons into the ventral compartment of the limb [19].

In addition to their preferential affinity for specific semaphorins, neuropilins preferentially recruit certain plexin coreceptors. PLXNA4 is a common coreceptor for NRP1 for axon sensing of SEMA3A, while PLXNA3 associates mainly with NRP2 to mediate SEMA3F signaling [20–22]. For example, axons from the visceromotor branch of the facial nerve (FVM) are organized by SEMA3A signaling through the NRP1/PLXNA4 complex [23,24], whereas axon pruning in the infrapyramidal tract is regulated by SEMA3F signaling through NRP2/PLXNA3 [16,17,22,25]. Both mechanisms may synergistically pattern other neuronal subtypes, as loss of either SEMA3A/NRP1/PLXNA4 or SEMA3F/NRP2/PLXNA3 causes fasciculation and projection defects of facial branchiomotor (FBM) axons and other cranial nerves [23,24]. Moreover, the combined loss of both plexins causes a significantly more severe phenotype, suggesting that SEMA3A and SEMA3F cooperate to regulate cranial nerve development [23,24].

Despite the typical preference of NRP1 for PLXNA4 and NRP2 for PLXNA3 [16,17], the analysis of trochlear axon patterning demonstrates that these combinations are not obligatory [20]. Here, axons are guided by SEMA3F/NRP2 signaling, but PLXNA4 can compensate for the loss of PLXNA3 [16,17,20]. Therefore, only the combined loss of both plexins recapitulates the pathfinding defects seen in *Sema3f−/−* mutants [20].

SEMA3E is the only class 3 semaphorin that does not require a neuropilin as a ligand-binding subunit, because it can bind directly to PLXND1 [26]. Thus, the axons of corticofugal and striatonigral neurons are repelled by SEMA3E via PLXND1 signaling [24]. However, the recruitment of NRP1 to the receptor complex can alter the axon response to SEMA3E (see below).

6.4 NEUROPILIN-MEDIATED REPULSION IN AXON GUIDANCE

Most of the examples for neuropilin-mediated axon guidance discussed above involve a repulsive response of growth cones to a semaphorin in the mammalian nervous system. Repulsive axon guidance is also mediated by neuropilins in aquatic species.

Zebrafish have two neuropilin 1 homologs, termed Nrp1a and Nrp1b. Nrp1a is expressed in zebrafish trunk motor axons to control their outgrowth from the spinal cord into the periphery [27]. Morpholino-mediated knockdown of Nrp1a led to displaced neuronal somata, multiple exit points and aberrant branching of motor axons. Surprisingly, the knockdown of either one of the Nrp1a ligands, Sema3a1 or Sema3a2, individually did not phenocopy these defects. Rather, the synergistic knock down of both ligands caused similar axon defects to those observed after Nrp1a morpholino injection [27]. Interestingly, the knockdown of Vegfa also functioned synergistically with targeting of Sema3a. However, the mechanistic reasons underlying this synergism between Vegfa and Sema3 in these neural responses has not yet been investigated for zebrafish motor neurons.

In both *Xenopus* and zebrafish, retinal ganglion cells (RGCs) project axons from each eye to the contralateral tectum in the midbrain, guided by attractive and repulsive cues in and around the midline, which is also known as the optic chiasm due to the cross shape of the ensuing optic nerve tracts. *Xenopus* retinal growth cones respond to Sema3a in vitro by initiating growth cone collapse, which is followed by branching of the treated axons; in vivo, Sema3a is expressed throughout the developing optic pathway and has varying roles in promoting the correct connectivity in the *Xenopus* retinotectal system [28]. Axon sensitivity of *Xenopus* RGCs to Sema3a can be regulated by the micro RNA miR-124 via its target CoREST, a cofactor for a transcriptional repressor that regulates Nrp1 expression to temporally restrict RGC responsiveness to Sema3a; in this fashion, miR-124 ensures correct targeting of RGC axons in the brain [29].

A role for NRP1 in RGC axon guidance has also been described for zebrafish. Similar to frogs zebrafish RGC axons project from each eye to their contralateral tectum by crossing at the optic chiasm. Morpholinos targeting Nrp1a increased the number of ipsilateral projections [30]. *Sema3d* and *Sema3e* are both expressed near the chiasm during RGC axon crossing, and their morpholino-mediated knockdown increased the number of ipsilateral RGC projections [30–32]. This study also suggested that multiple plexins may be required for correct axon RGC midline crossing, but it was not investigated whether RGC axon guidance by these semaphorins involves growth cone repulsion or attraction [30].

Sema3d also signals through Nrp1 to regulate axon guidance during the development of the habenular nucleus in the dorsal diencephalon, a model used to study left–right asymmetry in the zebrafish nervous system. Thus, *Nrp1a* expression is prominent in the habenular nucleus, and morpholino-mediated Nrp1a knockdown reduced the number of axons projecting from the habenular nucleus to the dorsal interpeduncular nucleus [33]. The morpholino-mediated Sema3d knockdown caused a similar innervation defect [33]. Low concentrations of the morpholinos targeting Sema3d and Nrp1a, which alone do not cause a phenotype, synergistically disrupted this innervation pattern, demonstrating that Nrp1a is the Sema3d receptor for habenular axon guidance [33].

6.5 NEUROPILINS CAN MEDIATE ATTRACTIVE RESPONSES BY AXONS TO SEMAPHORINS

Even though SEMA3 signaling through neuropilins is best known for roles in axon repulsion, some SEMA3 proteins elicit chemoattractive responses in certain contexts. The ability to stimulate either axon repulsion or attraction was initially demonstrated for different compartments of the same neuron; thus, SEMA3A functions as a chemoattractant for the apical dendrites of cortical pyramidal neurons, but a chemorepellent for their axons [34]. SEMA3F can also attract sprouting axons, such as those of cerebellar granule cells [35]. In both situations, increased intracellular cyclic guanosine monophosphate (GMP) levels are believed to convert repulsive into attractive responses [34]. Elevation of cyclic GMP is achieved through differential localization of soluble guanylate cyclase in the case of apical dendrites [34]. However, it is not yet known which neuropilin/plexin coreceptor complex is required for this pathway.

SEMA3B can also function as either a chemoattractant or chemorepellent in mice, as it repels commissural axons that extend between the paired olfactory bulbs but serves to attract crossing axons to the anterior commissure [36]. NRP2 mediates chemoattraction to SEMA3B through an interaction with the cell adhesion molecule NrCAM, rather than the recruitment of a plexin, which results in focal adhesion kinase (FAK) recruitment and signaling via SRC [36].

Even though SEMA3E is the only class 3 semaphorin that does not require a neuropilin to bind a plexin, the extracellular domains of NRP1 and PLXND1 are able to associate, and this interaction allows NRP1 to convert SEMA3E/PLXND1-mediated repulsion into attraction [24]. Specifically, SEMA3E attracts subiculo-mammilary axons expressing both PLXND1 and NRP1, which project from the hippocampus to the hypothalamus [24].

Alternative mechanisms for neuropilin-mediated chemoattractive responses to class 3 semaphorins have been identified in zebrafish. For example, Nrp2 can convert Nrp1-induced chemorepulsion into chemoattraction. Thus, Sema3d typically repels axons from the medial longitudinal fasciculus, which express Nrp1, but attracts axons derived from the telencephalon that express both Nrp1 and Nrp2 and are destined to form the anterior commissure that connects both hemispheres [37]. Sema3d binds Nrp1 directly, but it is not yet known whether it can also bind Nrp2 directly, or if it is recruited via Nrp1 to form a heterodimers for attraction [37]. It is also unknown if Nrp2-mediated attraction of commissural axons requires plexin recruitment for conversion of chemorepulsion to chemoattraction.

6.6 NEUROPILINS ORGANIZE AXON PROJECTIONS TO PROVIDE A SUBSTRATE FOR MIGRATING NEURONS

Semaphorin/neuropilin-mediated axon guidance can promote neuronal migration in situations where an axon scaffold guides migrating neuronal cell bodies. This was demonstrated for gonadotropin-releasing hormone (GnRH) neurons, which are born in the nasal placodes and then migrate along olfactory and vomeronasal axons into the brain to regulate reproductive functions through the hypothalamus [38]. Adult mice lacking NRP2 are infertile and possess fewer GnRH neurons in the hypothalamus [39,40]. Mice lacking semaphorin signaling through both neuropilins ($Nrp1^{Sema-/-}$ $Nrp2^{-/-}$) show an even more severe phenotype that manifests itself as an almost complete failure of GnRH neurons to enter the brain [39–41]. Instead, neurons accumulate outside the brain, where they appear trapped in a tangle of ectopic axons [39,41]. Similar phenotypes are seen in mice lacking SEMA3A, suggesting that it can signal through either NRP1 or NRP2 in nasal axons [42]. The plexin receptors that mediate neuropilin responses in nasal axons have not yet been identified.

6.7 SEMAPHORIN SIGNALING THROUGH NEUROPILINS IN NEURONS PROMOTES NEURONAL MIGRATION

Semaphorin signaling through neuropilins can also directly guide migrating neurons in the developing nervous system;

below, we provide examples of roles in both the tangential or radial migration of specific neuronal subtypes [43–46].

The progenitors of GABAergic cortical interneurons migrate tangentially from their birthplace in the ganglionic eminence (GE) around the striatum and into the cortex [44]. Both SEMA3A and SEMA3F are expressed in the developing striatum, and this region contains excess neurons in *Nrp2*-null embryos, suggesting a role for semaphorin-mediated repulsion in their guidance. Although the early embryonic lethality of *Nrp1*-null mutants prevents analysis at the relevant stages, a dominant negative form of NRP1 impairs the migration of cortical interneurons in a similar manner to NRP2 loss [44]. In addition, both neuropilins are involved in positioning cortical interneurons within the correct layer of the embryonic cortex [44]. Thus, SEMA3F guides NRP2-expressing interneurons into the intermediate zone, while SEMA3A guides NRP1-expressing interneurons into the subplate and cortical plate [44].

Semaphorin signaling through neuropilins also guides the tangential migration of lateral olfactory tract (LOT) neurons across the telencephalon, albeit in the opposite direction to cortical interneurons. Thus, LOT neurons migrate towards the GE before changing direction and spreading laterally along the interface between the neocortex and GE to guide LOT axons [45]. The mantle layer of the GE normally expresses SEMA3F, and many LOT neurons in *Sema3f−/−* mutants continue to migrate ectopically towards the GE without adopting the lateral migratory path, again suggesting the semaphorins serve as repulsive guidance cues [45]. It has not yet been identified if SEMA3A signaling through NRP1 is also involved in guiding LOT neurons.

NRP1 plays a key role in the "inside-out" organization of the mammalian neocortex, which relies on radial migration of cortical neurons along radial glia processes extending from the germinal ventricular zone to the superficial cortical layers. In contrast to its repulsive role for other types of migrating neurons, SEMA3A is a chemoattractant for migrating cortical neurons [46]. Accordingly, knockdown of NRP1 impairs the migration of cortical neurons to the outer layers of the neocortex, with PLXNA2 and PLXNA4 acting as coreceptors [46]. Cortical neurons also fail to extend neurites correctly in the *Nrp1*-null brain, raising the possibility that migration and projection rely on similar downstream signaling transduction pathways [34,46].

6.8 VEGF-A AS AN ALTERNATIVE NEUROPILIN LIGAND IN THE NERVOUS SYSTEM

The VEGF family of cysteine-knot growth factors contains several members that are expressed widely across the developing embryo and adult body of vertebrates. VEGF-A is the best-studied family member and promotes the process of angiogenesis, in which new blood vessels sprout from pre-existing ones to vascularize tissues such as the brain [47]. VEGF-A drives angiogenesis as a proliferation and chemotactic factor for endothelial cells (Figure 6.1(A)), in which it activates the tyrosine kinase receptor VEGFR2 (FLK, KDR) to elicit downstream signaling [48,49]. Defective VEGF-A impairs the vascularization of the nervous system and may thereby indirectly affect neuronal migration, survival, or axon guidance. During angiogenesis, NRP1 is an alternative VEGF-A receptor that also serves as a VEGFR2 coreceptor (Figure 6.1(A)). NRP1 is particularly important for CNS vascularization, for reasons that are not yet understood. In addition, a direct role for VEGF-A signaling through NRP1 has also been identified in a subset of CNS neurons (Figure 6.1(B) and (C)). Below, we will outline the mechanisms of VEGF-A binding to neuropilins and how VEGF-A/NRP1 signaling regulates neural development.

6.9 DIFFERENTIAL VEGF-A ISOFORM AFFINITY FOR THE TWO NEUROPILINS

VEGF-A exists in several different isoforms that arise through alternative splicing of the *VEGFA* gene, which comprises eight exons in the human genome [50]. This splicing results in the inclusion or exclusion of exons 6 and 7, which encode domains that enhance binding to heparin in vitro and heparan sulfate proteoglycans in vivo; they also enable cooperation with exon 8 for isoform binding to neuropilins [48,51,52]. The major human isoforms are VEGF189, VEGF165, and VEGF121, while their murine homologs are one aa shorter and termed VEGF188, VEGF164, and VEGF120. VEGF165 and VEGF145 contain the exon 7 and 6 sequence, respectively, which enables VEGF165 to bind NRP1 and VEGF145 to bind both NRP1 and NRP2 [53]. VEGF189 is also predicted to bind NRP1, given that it contains the exon 7 and 8 domains [54], but this has not yet been demonstrated in vivo. Even though VEGF121 contains the exon 8 domain, which can interact with NRP1, the exclusion of exon 6 or 7 sequences results in a significantly lower affinity of VEGF121 compared to VEGF164 for NRP1 in vitro [55,56]. Accordingly, VEGF121 is not believed to bind NRP1 with high specificity, although this idea has been contested in one in vitro study with cultured endothelial cells [57].

6.10 VEGF-A SIGNALING THROUGH NEUROPILIN 1

The interaction of VEGF-A with NRP1 was identified in cultured endothelial and tumor cells [58]. Different hypotheses exist for how NRP1 acts in endothelial cells: it was originally proposed to enhance VEGFR2 affinity for VEGF165 [59] and promote VEGFR2 clustering [60], but current research is focused on its role in VEGFR2 trafficking by endocytosis [61]. Both models agree that VEGF165

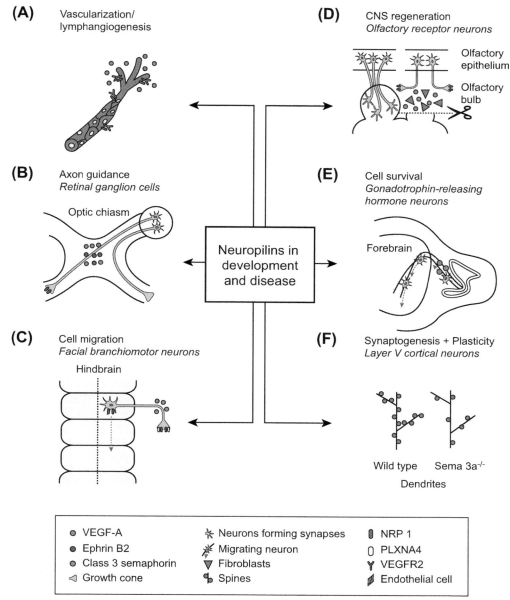

FIGURE 6.1 Neuropilins in development and disease of the nervous system. (A) Vascular endothelial growth factor (VEGF)-A binds to VEGFR2 (purple Y shape) and NRP1 (red rod) on endothelial cells in vessel sprouts for chemotactic guidance and to stimulate cell proliferation. VEGF-A (green dots) binding to NRP1 promotes the formation of a coreceptor complex with VEGFR2 via a VEGF-A bridge. (B) Retinal ganglion cells (RGCs) express NRP1 on their growth cones and are guided by VEGF-A across the optic chiasm to innervate the contralateral brain. RGCs that do not express NRP1, but the ephrin receptor EPHB1 on their growth cones are repelled by ephrin B2 (red dots) at the chiasm to project ipsilaterally. (C) The somata of facial branchiomotor neurons are guided by VEGF-A from rhombomere (r) 4 to r6, where they form the 7th cranial motor nuclei. Their axons also express NRP1 and PLXNA4 (white rods) on the growth cone and are repelled by SEMA3A (blue dots) when they extend from the hindbrain into the second branchial arch. (D) Olfactory receptor neurons (ORNs) project from the olfactory epithelium into the olfactory bulb, where they form synapses (left-hand side). Bulbectomy (right-hand side) increases the generation of new ORNs that express NRP1 on their growth cones, but fibroblasts (black triangles) derived from the scar tissue repel the growth cones by expressing SEMA3A (blue dots). (E) A subpopulation of newly born GnRH neurons expresses NRP1 and responds to VEGF-A along their migratory path to survive the journey from the vomeronasal organ through the nose into the forebrain. (F) During late vertebrate embryonic and early postnatal stages in vertebrates, synapses form between axons and neuronal dendrites. On dendritic trees (black lines), the postsynaptic site is located on membranous protrusions called dendritic spines (black dots). *Sema3a*$^{-/-}$ mice have reduced spine density in layer V cortical neurons. CNS, central nervous system. (For interpretation of the references to color in this figure legend, the reader is referred to the online version of this book.)

binding to NRP1 promotes complex formation between NRP1 and VEGFR2 via a VEGF165 bridge. This pathway appears to be particular important for arteriogenesis [62]. Other evidence suggests that NRP1 may also convey VEGF signals through its own cytoplasmic tail in endothelial cells, independently of VEGFR2. Accordingly, it was shown that the epidermal growth factor (EGF) stimulated the migration of cultured endothelial cells expressing a chimeric NRP1-EGF receptor [63]. Despite this finding, mice lacking the NRP1 cytoplasmic domain ($Nrp1^{cyto/cyto}$) do not have general defects in assembling blood vessel networks, with the exception of a subtle arteriovenous patterning defect in the developing retina [64] and impaired VEGFR2-dependent arteriogenesis due to abnormal VEGFR2 trafficking [62]. Whether the cytoplasmic tail of NRP1 might signal in neural development has so far only been explored in GnRH neurons; in these cells, VEGF164 signaling through NRP1 enhances survival independently of VEGFR2 through an unknown coreceptor, and the NRP1 cytoplasmic tail does not appear to play a role in this process [65].

6.11 NRP1 IN CNS ANGIOGENESIS

An extensive and properly patterned blood vessel network is crucial for maintaining tissue homeostasis and delivering physiological signals to both developing and adult neurons in the vertebrate CNS. Dividing neural precursor cells attract blood vessels by establishing chemoattractive VEGF-A gradients [47,48], which require the balanced expression of the matrix-binding and readily diffusible VEGF-A isoforms. Accordingly, mouse mutant embryos expressing only VEGF120 at the expense of other VEGF-A isoforms show impaired vessel sprouting and compensatory luminal growth in the developing brain and retina [48,49]. However, their severe vascular brain defects appear to be due to the abnormal extracellular distribution of VEGF-A, rather than a failed interaction with NRP1, because VEGF binding to NRP1 is dispensable for NRP1-mediated tissue vascularization in the brain [66]. In contrast, both VEGF-dependent and VEGF-independent NRP1 signaling may be involved in retinal angiogenesis [67,68]. VEGF-independent NRP1 signaling in angiogenesis may result from its ability to promote ECM signals provided by integrin ligands [67]. Mice lacking VEGF binding of NRP1 are yet to be studied for defects in neural development.

In contrast to NRP1, NRP2 is expressed mainly in veins and lymphatic vessels and can bind the lymphatic vascular growth factor VEGF-C; accordingly, genetic mouse mutants lacking NRP2 expression have fewer and smaller lymphatic capillaries [68–70]. However, it is not yet known whether NRP2 can act as a VEGF-A receptor in the nervous system, which lacks lymphatics. Thus, roles for VEGF-A signaling through NRP2 have not yet been described for either neurons or glia.

6.12 VEGF-A/NRP1 SIGNALING PROMOTES CONTRALATERAL AXON PROJECTION ACROSS THE OPTIC CHIASM

A role for VEGF-A signaling through NRP1 has not been identified for axon guidance in the PNS. However, NRP1 was found to regulate axon guidance in the CNS, where it promotes the sorting of RGC axons across the diencephalic commissure known as the optic chiasm [71]. Specifically, VEGF-A influences the choice of axons: whether to project across the optic chiasm as contralateral axons or instead to project ipsilaterally [71]. For this guidance event to proceed correctly, VEGF-A is expressed at the chiasm midline, and RGCs destined to project contralaterally express NRP1 (Figure 6.1(B)) [71]. In contrast, axons destined to project ipsilaterally express the EphB1 receptor and are repelled by ephrin B2 ligand at the optic chiasm (Figure 6.1(B)) [72,73]. The essential role for VEGF-A signaling though NRP1 in RGC axon guidance was demonstrated through the analysis of mice lacking NRP1 ($Nrp1^{-/-}$) or the NRP1-binding isoforms of VEGF-A ($Vegfa^{120/120}$); moreover, the NRP1-binding isoform VEGF164 stimulates extension and turning of cultured RGC axons, whereas the VEGF120 isoform with poor NRP1 binding does not [71]. On the other hand, class 3 semaphorins were not found to be expressed at the mouse chiasm during the period of axon crossing, and mice lacking semaphorin signaling through both NRP1 and NRP2 did not shown obvious crossing defects [71]. This contrasts with the role of semaphorins at the optic chiasm in zebrafish (see above) [30]. Whether VEGF also contributes to optic chiasm development in zebrafish has not yet been addressed. It is also not known whether possible differences in the neuropilin ligands used for RGC axons in the different species correlate with the fact that mammals, in contrast to zebrafish, naturally form both ipsilateral and contralateral projections to enable stereovision.

6.13 VEGF-A SIGNALS THROUGH NRP1 TO PROMOTE NEURONAL MIGRATION

In addition to semaphorins, VEGF also signals through NRP1 to guide neuronal migration. Thus, the migration of FBM neurons in the mouse hindbrain is regulated by VEGF164 signaling through NRP1 to guide them from rhombomere (r) 4 to r6 in the mouse hindbrain (Figure 6.1(C)) [74]. Whereas these neurons migrate in an organized stream and form paired nuclei on the pial side of the brain, FBM neurons in $Nrp1$-null mice migrate in disorganized streams and form misshaped and misplaced nuclei [74]. The finding that $Sema3a^{-/-}$ mice show normal FBM migration, while $Vegfa^{120/120}$ mice lacking the NRP1-binding isoform VEGF164 phenocopy the FBM defects seen in $Nrp1$-null mice, suggests that VEGF is the NRP1 ligand that controls FBM migration [74]. In contrast, SEMA3A, but not VEGF signaling through NRP1, is required for correct axon guidance of the facial axons (see above; Figure 6.1(C)) [23].

6.14 SEMAPHORIN SIGNALING THROUGH NRP1 IMPAIRS CNS, BUT NOT PNS REGENERATION

The vertebrate CNS has limited capacity for regeneration following injury or during aging. For example, the distal axonal segment degenerates if a nerve is severed and usually fails to regrow. The failure for axons to regrow stems partially from the repulsion of axons at injury sites (where they should be extending for repair) and additionally from secondary neuronal cell death. Semaphorin signaling via neuropilins has been implicated in both processes.

For example, SEMA3A/NRP1 signaling prevents olfactory receptor neuron (ORN) axons from reinnervating the olfactory bulb in the forebrain following transection of axon fibres [75]. During development, ORN axons express NRP1 in order to prevent premature entry into the telencephalic vesicle prior to olfactory bulb formation [76–78], whereby SEMA3A expression at the perimeter of the vesicle is spatiotemporally restricted to ensure timely axon ingrowth [79]. NRP1 is also expressed by axons extending from newly born ORNs in adult mice and ensures the correct target innervation of olfactory bulb glomeruli [77]. However, after injury, this persisting NRP1 expression prevents these axons from reaching the olfactory bulb, because glial scar-associated fibroblasts express SEMA3A (Figure 6.1(D)) [80,81]. The limited regrowth of axons across CNS scar tissue due to repulsive SEMA3A/NRP1 signaling also likely affects other regions of the CNS, such as the corticospinal and rubrospinal tracts. Thus, the axons of these descending motor pathways express both neuropilins as well as PLXNA1 following spinal cord injury, and fibroblasts invading the wound site express SEMA3 [82].

SEMA3A has also been implicated as a negative factor in neuronal survival after CNS injury. Thus, blocking SEMA3A or NRP1 function reduces dopamine-induced apoptosis after transection injury via the cytoplasmic tyrosine kinase proto-oncogene tyrosine-protein kinase FER (FER) [83,84]. Expression of both neuropilins and SEMA3A is also upregulated in neurons several days before their eventual death in a focal cerebral ischemia model [85]. These findings raise the possibility that targeting semaphorin signaling through neuropilins at neural lesions may improve CNS regeneration [85].

In contrast to CNS axons, peripheral nerves can regenerate over long distances, and this appears to correlate with low SEMA3A signaling through NRP1 in damaged peripheral nerves. For example, following transection or crush of the sciatic nerve, axons express NRP1, but SEMA3A levels are reduced compared to uninjured sciatic nerve until regeneration has taken place [86,87]. This suggests that regrowing sciatic nerve axons do not encounter lesion site-derived SEMA3A in the same manner as axons in the injured CNS. Surprisingly, NRP2 is required for efficient regeneration of injured sciatic nerves, although the mechanism is not presently understood [88].

Even though NRP1 has been implicated as an inhibitory semaphorin receptor that limits axon regeneration at CNS injury sites, NRP1 may also promote beneficial blood vessel growth around scar tissue to enable wound healing. In agreement, the VEGF receptors FLT, KDR, and NRP1 are all expressed in and around lesion areas in structures resembling blood vessels [89]. A similar observation was also made for sites of focal ischemia, where NRP1 expression is upregulated in vascular endothelium [90]. However, lesion-derived SEMA3A may compete with VEGF165 for NRP1 binding on blood vessels. Thus, hypoxic RGCs in the ischemic retina produce SEMA3A to repel sprouting vessels away from areas of tissue damage, and silencing neuronal SEMA3A expression improves vessel ingression into ischemic tissue to aid recovery of neuroretinal function [91]. The balanced expression of NRP1-binding VEGF and semaphorins may therefore be crucial to re-vascularizing sites of CNS injury.

6.15 VEGF SIGNALING THROUGH NRP1 PROMOTES SURVIVAL OF DEVELOPING NEURONS

In addition to mediating VEGF-driven axon guidance and neuronal migration, NRP1 promotes VEGF-mediated neuronal survival (Figure 6.1(E)) [92]. Thus, VEGF164 acts via NRP1 in immortalized GnRH neurons to activate survival signaling via PI3K and MAPK [92]. This finding was confirmed in vivo in embryos lacking neural NRP1 or NRP1-binding VEGF-A isoforms, as they lose GnRH neurons along their normal migratory path from the nasal cavity into the brain. During this process, NRP1 functions independently of VEGFR2, as mutants lacking VEGFR2 in the neural lineage have normal GnRH survival during their migration in the nose [92]. Importantly, NRP1 promotes cell survival cell-autonomously, rather than through blood vessel recruitment. Mice lacking NRP1 in CNS vasculature have a normal number of GnRH neurons, but mice lacking NRP1 in the neural lineage are affected [92]. VEGF-A signaling through VEGFR2 has also been implicated in neuroprotection of several different types of neurons [93]. It is not yet known whether NRP1 plays a role in these systems as a coreceptor for VEGFR2.

6.16 NEUROPILIN IN SYNAPTOGENESIS AND PLASTICITY

Once neurons have found their appropriate locations in the nervous system through cell body migration and axon guidance, they must make appropriate connections with each other to form a network. This is accomplished through various signaling pathways, including semaphorin signaling through neuropilins, which regulate the formation of synapses during development and modulate synaptic plasticity in the adult.

In vitro studies with primary mouse cortical neurons showed that treatment with the NRP1 ligand SEMA3A increases the density of dendritic spines, which are small protuberances where most synaptic contacts will form (Figure 6.1(F)) [94]. Moreover, SEMA3A promoted the colocalization of the postsynaptic marker PSD-95 with the presynaptic marker synapsin-1, indicative of increased synapse formation [94]. Because SEMA3A also acts as a chemoattractant for the dendrites of cortical apical neurons (see above), it appears to modulate neuronal connectivity at several levels. Interestingly, primary mouse hippocampal neurons respond to SEMA3A treatment differently, as they show decreased accumulation of PSD-95 and synaptophysin, indicating increased synapse elimination [95]. Together, these observations indicate that SEMA3A has distinct functions in different areas of the CNS during synaptogenesis.

Semaphorin signaling through neuropilins is also important for synaptic pruning. Thus, adult $Sema3f^{-/-}$ mice have an increase in the total number and distribution of dendritic spines at the level of the dentate gyrus and cortical layer V, and adult $Nrp2^{-/-}$ and $Plxna3^{-/-}$ mice replicate these phenotypes [96]. These results are supported by in vitro data, which show that SEMA3F treatment of dentate gyrus primary neurons reduces the number of PSD95-positive puncta, indicating synaptic elimination [96]. SEMA3E has also been shown to have a role in synaptic pruning, as the density of excitatory glutamatergic synapses on the primary dendrites of striatal neurons is increased in both $Sema3e^{-/-}$ and $Plxnd1^{-/-}$ mice [97]. However, it was not investigated if NRP1 may modulate these responses, as shown for axon guidance.

Semaphorin signaling through NRP1 also regulates synaptic plasticity in adults. Thus, rat models of epilepsy have shown increased levels of SEMA3A in the cortex and of SEMA3C and SEMA3F in the hippocampus [98,99]. Moreover, SEMA3F increases both the amplitude and frequency of mini excitatory postsynaptic currents (EPSCs) in hippocampal slices [100]. Interestingly, SEMA3A has the opposite effect, as adult mouse hippocampal slices showed a dose-dependent decrease in the number of EPSCs in response to SEMA3A treatment [95]. In vivo studies support a role for class 3 semaphorins in modulating synaptic plasticity. Thus, mice lacking SEMA3F are prone to seizures, while recordings from cortical layer V and dentate gyrus slices of $Nrp2^{-/-}$ mice revealed an increased frequency of mini EPSCs [96].

VEGF-A signaling has also been associated with synaptic development. One study showed that treatment of olfactory bulb cells with soluble VEGFR1, which sequesters VEGF-A, decreased the number and density of dendritic spines on newly born granule cell neurons [101]. Furthermore, this treatment impairs long-term potentiation (LTP) in the hippocampus [101], which indicates a role for VEGF-A in synaptic plasticity. Similar effects have also been shown in hippocampal cell cultures, in which VEGF-A

treatment enhances LTP [102]. Increased neuronal activity of mice, through seizures or environmental enrichment, also increases VEGF-A expression [103–105]. These studies suggest an important role for VEGF-A in modulating synaptic plasticity. Yet it remains to be established if these effects are a direct consequence of altered spine density or dendritic number, rather than a consequence of other, more general developmental defects. Moreover, it is not yet known if these effects reflect a direct effect of VEGF-A on neurons that express VEGF receptors such as neuropilin, or an indirect effect of blood vessels that are induced by VEGF.

6.17 OUTLOOK

Roles for semaphorin signaling through neuropilins have been identified in many different contexts for neural development, where it was found to regulate axon guidance, neuronal migration, and synaptic connectivity. More recently, VEGF-A isoforms were shown to also bind NRP1 to promote axon pathfinding, neuronal migration, and neuronal survival in some neuronal populations. Undoubtedly, further examples of such responses in different neural contexts will continue to emerge over the coming years to help explain the complexity of the nervous system. Most pressingly, we need to better understand neuropilin function in the adult CNS during normal physiology, pathological processes, and in regeneration. Initial evidence has suggested that blocking semaphorin signaling through neuropilins may facilitate repair after CNS injury, although the specific pathways involved are still poorly understood. The findings that semaphorins and VEGF-A both control synaptogenesis and plasticity in the developing brain raise the exciting possibility that they have opposing or synergistic functions, and that they may be suitable therapeutic targets to treat diseases associated with abnormal electrophysiological activity, such as epilepsy. However, on a cautionary note, neuropilin signaling may also be important for neovascularization around injury sites, and its therapeutic blockade may therefore elicit undesired effects on lesional vasculature.

REFERENCES

[1] Schwarz Q, Ruhrberg C. Neuropilin, you gotta let me know: should I stay or should I go? Cell Adhes Migr 2010;4(1):61–6.

[2] Raimondi C, Ruhrberg C. Neuropilin signalling in vessels, neurons and tumours. Semin Cell Dev Biol 2013;24(3):172–8.

[3] Kandel ER, Schwartz JH, Jessell TM. Principles of neural science, vol. 4. New York: McGraw-Hill; 2000.

[4] Nakamura F, Kalb RG, Strittmatter SM. Molecular basis of semaphorin-mediated axon guidance. J Neurobiol 2000;44(2): 219–29.

[5] Siebold C, Jones EY. Structural insights into semaphorins and their receptors. Semin Cell Dev Biol 2013;24(3):139–45.

[6] Zhou Y, Gunput RA, Pasterkamp RJ. Semaphorin signaling: progress made and promises ahead. Trends Biochem Sci 2008;33(4):161–70.

[7] Janssen BJ, Malinauskas T, Weir GA, Cader MZ, Siebold C, Jones EY, et al. Neuropilins lock secreted semaphorins onto plexins in a ternary signaling complex. Nat Struct Mol Biol 2012;19(12):1293–9.

[8] Sakurai A, Doci CL, Gutkind JS. Semaphorin signaling in angiogenesis, lymphangiogenesis and cancer. Cell Res 2012;22(1):23–32.

[9] Puschel AW. GTPases in semaphorin signaling. Adv Exp Med Biol 2007;600:12–23.

[10] Schmidt EF, Strittmatter SM. The CRMP family of proteins and their role in Sema3A signaling. Adv Exp Med Biol 2007;600:1–11.

[11] Franco M, Tamagnone L. Tyrosine phosphorylation in semaphorin signalling: shifting into overdrive. EMBO Rep 2008;9(9):865–71.

[12] Bechara A, Nawabi H, Moret F, Yaron A, Weaver E, Bozon M, et al. FAK-MAPK-dependent adhesion disassembly downstream of L1 contributes to semaphorin3A-induced collapse. EMBO J 2008;27(11):1549–62.

[13] Takahashi T, Nakamura F, Jin Z, Kalb RG, Strittmatter SM. Semaphorins A and E act as antagonists of neuropilin-1 and agonists of neuropilin-2 receptors. Nat Neurosci 1998;1(6):487–93.

[14] Taniguchi M, Yuasa S, Fujisawa H, Naruse I, Saga S, Mishina M, et al. Disruption of semaphorin III/D gene causes severe abnormality in peripheral nerve projection. Neuron 1997;19(3):519–30.

[15] Kitsukawa T, Shimizu M, Sanbo M, Hirata T, Taniguchi M, Bekku Y, et al. Neuropilin-semaphorin III/D-mediated chemorepulsive signals play a crucial role in peripheral nerve projection in mice. Neuron 1997;19(5):995–1005.

[16] Giger RJ, Cloutier JF, Sahay A, Prinjha RK, Levengood DV, Moore SE, et al. Neuropilin-2 is required in vivo for selective axon guidance responses to secreted semaphorins. Neuron 2000;25(1):29–41.

[17] Sahay A, Molliver ME, Ginty DD, Kolodkin AL. Semaphorin 3F is critical for development of limbic system circuitry and is required in neurons for selective CNS axon guidance events. J Neurosci 2003;23(17):6671–80.

[18] Chen H, Bagri A, Zupicich JA, Zou Y, Stoeckli E, Pleasure SJ, et al. Neuropilin-2 regulates the development of selective cranial and sensory nerves and hippocampal mossy fiber projections. Neuron 2000;25(1):43–56.

[19] Huber AB, Kania A, Tran TS, Gu C, De Marco Garcia N, Lieberam I, et al. Distinct roles for secreted semaphorin signaling in spinal motor axon guidance. Neuron 2005;48(6):949–64.

[20] Yaron A, Huang PH, Cheng HJ, Tessier-Lavigne M. Differential requirement for Plexin-A3 and -A4 in mediating responses of sensory and sympathetic neurons to distinct class 3 semaphorins. Neuron 2005;45(4):513–23.

[21] Suto F, Ito K, Uemura M, Shimizu M, Shinkawa Y, Sanbo M, et al. Plexin-a4 mediates axon-repulsive activities of both secreted and transmembrane semaphorins and plays roles in nerve fiber guidance. J Neurosci 2005;25(14):3628–37.

[22] Cheng HJ, Bagri A, Yaron A, Stein E, Pleasure SJ, Tessier-Lavigne M. Plexin-A3 mediates semaphorin signaling and regulates the development of hippocampal axonal projections. Neuron 2001;32(2):249–63.

[23] Schwarz Q, Waimey KE, Golding M, Takamatsu H, Kumanogoh A, Fujisawa H, et al. Plexin A3 and plexin A4 convey semaphorin signals during facial nerve development. Dev Biol 2008;324(1):1–9.

[24] Chauvet S, Cohen S, Yoshida Y, Fekrane L, Livet J, Gayet O, et al. Gating of Sema3E/PlexinD1 signaling by neuropilin-1 switches axonal repulsion to attraction during brain development. Neuron 2007;56(5):807–22.

[25] Bagri A, Cheng HJ, Yaron A, Pleasure SJ, Tessier-Lavigne M. Stereotyped pruning of long hippocampal axon branches triggered by retraction inducers of the semaphorin family. Cell 2003;113(3):285–99.

[26] Gu C, Yoshida Y, Livet J, Reimert DV, Mann F, Merte J, et al. Semaphorin 3E and plexin-D1 control vascular pattern independently of neuropilins. Science 2005;307(5707):265–8.

[27] Feldner J, Becker T, Goishi K, Schweitzer J, Lee P, Schachner M, et al. Neuropilin-1a is involved in trunk motor axon outgrowth in embryonic zebrafish. Dev Dyn 2005;234(3):535–49.

[28] Campbell DS, Regan AG, Lopez JS, Tannahill D, Harris WA, Holt CE. Semaphorin 3A elicits stage-dependent collapse, turning, and branching in Xenopus retinal growth cones. J Neurosci 2001;21(21):8538–47.

[29] Baudet ML, Zivraj KH, Abreu-Goodger C, Muldal A, Armisen J, Blenkiron C, et al. miR-124 acts through CoREST to control onset of Sema3A sensitivity in navigating retinal growth cones. Nat Neurosci 2012;15(1):29–38.

[30] Dell AL, Fried-Cassorla E, Xu H, Raper JA. cAMP-induced expression of neuropilin1 promotes retinal axon crossing in the zebrafish optic chiasm. J Neurosci 2013;33(27):11076–88.

[31] Sakai JA, Halloran MC. Semaphorin 3D guides laterality of retinal ganglion cell projections in zebrafish. Development 2006;133(6):1035–44.

[32] Liu Y, Berndt J, Su F, Tawarayama H, Shoji W, Kuwada JY, et al. Semaphorin3D guides retinal axons along the dorsoventral axis of the tectum. J Neurosci 2004;24(2):310–8.

[33] Kuan YS, Yu HH, Moens CB, Halpern ME. Neuropilin asymmetry mediates a left-right difference in habenular connectivity. Development 2007;134(5):857–65.

[34] Polleux F, Morrow T, Ghosh A. Semaphorin 3A is a chemoattractant for cortical apical dendrites. Nature 2000;404(6778):567–73.

[35] Ding S, Luo JH, Yuan XB. Semaphorin-3F attracts the growth cone of cerebellar granule cells through cGMP signaling pathway. Biochem Biophys Res Commun 2007;356(4):857–63.

[36] Falk J, Bechara A, Fiore R, Nawabi H, Zhou H, Hoyo-Becerra C, et al. Dual functional activity of semaphorin 3B is required for positioning the anterior commissure. Neuron 2005;48(1):63–75.

[37] Wolman MA, Liu Y, Tawarayama H, Shoji W, Halloran MC. Repulsion and attraction of axons by semaphorin3D are mediated by different neuropilins in vivo. J Neurosci 2004;24(39):8428–35.

[38] Cariboni A, Maggi R, Parnavelas J. From nose to fertility: the long migratory journey of gonadotropin-releasing hormone neurons. Trends Neurosci 2007;30(12):638–44.

[39] Cariboni A, Hickok J, Rakic S, Andrews W, Maggi R, Tischkau S, et al. Neuropilins and their ligands are important in the migration of gonadotropin-releasing hormone neurons. J Neurosci 2007;27(9):2387–95.

[40] Walz A, Rodriguez I, Mombaerts P. Aberrant sensory innervation of the olfactory bulb in neuropilin-2 mutant mice. J Neurosci 2002;22(10):4025–35.

[41] Cloutier JF, Sahay A, Chang EC, Tessier-Lavigne M, Dulac C, Kolodkin AL, Ginty DD. Differential requirements for semaphorin 3F and Slit-1 in axonal targeting, fasciculation, and segregation of olfactory sensory neuron projections. J Neurosci 2004;24(41):9087–96.

[42] Taniguchi M, Nagao H, Takahashi YK, Yamaguchi M, Mitsui S, Yagi T, et al. Distorted odor maps in the olfactory bulb of semaphorin 3A-deficient mice. J Neurosci 2003;23(4):1390–7.

[43] Marìn O, Rubenstein J. Cell migration in the forebrain. Annu Rev Neurosci 2003;26:441–83.

[44] Tamamaki N, Fujimori K, Nojyo Y, Kaneko T, Takauji R, et al. Evidence that Sema3A and Sema3F regulate the migration of GABAergic neurons in the developing neocortex. J Comp Neurol 2003;455(2):238–48.

[45] Ito K, Kawasaki T, Takashima S, Matsuda I, Aiba A, Hirata T. Semaphorin 3F confines ventral tangential migration of lateral olfactory tract neurons onto the telencephalon surface. J Neurosci 2008;28(17):4414–22.

[46] Chen G, Sima J, Jin M, Wang KY, Xue XJ, Zheng W, et al. Semaphorin-3A guides radial migration of cortical neurons during development. Nat Neurosci 2008;11(1):36–44.

[47] Risau W. Mechanisms of angiogenesis. Nature 1997;386(6626):671–4.

[48] Ruhrberg C, Gerhardt H, Golding M, Watson R, Ioannidou S, Fujisawa H, et al. Spatially restricted patterning cues provided by heparin-binding VEGF-A control blood vessel branching morphogenesis. Genes Dev 2002;16(20):2684–98.

[49] Gerhardt H, Golding M, Fruttiger M, Ruhrberg C, Lundkvist A, Abramsson A, et al. VEGF guides angiogenic sprouting utilizing endothelial tip cell filopodia. J Cell Biol 2003;161(6):1163–77.

[50] Tischer E, Mitchell R, Hartman T, Silva M, Gospodarowicz D, Fiddes JC, et al. The human gene for vascular endothelial growth factor. Multiple protein forms are encoded through alternative exon splicing. J Biol Chem 1991;266(18):11947–54.

[51] Park JE, Keller GA, Ferrara N. The vascular endothelial growth factor (VEGF) isoforms: differential deposition into the subepithelial extracellular matrix and bioactivity of extracellular matrix-bound VEGF. Mol Biol Cell 1993;4(12):1317–26.

[52] Houck KA, Leung DW, Rowland AM, Winer J, Ferrara N. Dual regulation of vascular endothelial growth factor bioavailability by genetic and proteolytic mechanisms. J Biol Chem 1992;267(36):26031–7.

[53] Gluzman-Poltorak Z, Cohen T, Herzog Y, Neufeld G. Neuropilin-2 is a receptor for the vascular endothelial growth factor (VEGF) forms VEGF-145 and VEGF-165 [corrected]. J Biol Chem 2000;275(24):18040–5.

[54] Vintonenko N, Pelaez-Garavito I, Buteau-Lozano H, Toullec A, Lidereau R, Perret GY, et al. Overexpression of VEGF189 in breast cancer cells induces apoptosis via NRP1 under stress conditions. Cell Adhes Migr 2011;5(4):332–43.

[55] Jia H, Bagherzadeh A, Hartzoulakis B, Jarvis A, Lohr M, Shaikh S, et al. Characterization of a bicyclic peptide neuropilin-1 (NP-1) antagonist (EG3287) reveals importance of vascular endothelial growth factor exon 8 for NP-1 binding and role of NP-1 in KDR signaling. J Biol Chem 2006;281(19):13493–502.

[56] Parker MW, Xu P, Li X, Vander Kooi CW. Structural basis for selective vascular endothelial growth factor-A (VEGF-A) binding to neuropilin-1. J Biol Chem 2012;287(14):11082–9.

[57] Pan Q, Chathery Y, Wu Y, Rathore N, Tong RK, Peale F, et al. Neuropilin-1 binds to VEGF121 and regulates endothelial cell migration and sprouting. J Biol Chem 2007;282(33):24049–56.

[58] Soker S, Takashima S, Miao HQ, Neufeld G, Klagsbrun M. Neuropilin-1 is expressed by endothelial and tumor cells as an isoform-specific receptor for vascular endothelial growth factor. Cell 1998;92(6):735–45.

[59] Whitaker GB, Limberg BJ, Rosenbaum JS. Vascular endothelial growth factor receptor-2 and neuropilin-1 form a receptor complex that is responsible for the differential signaling potency of VEGF(165) and VEGF(121). J Biol Chem 2001;276(27):25520–31.

[60] Soker S, Miao HQ, Nomi M, Takashima S, Klagsbrun M. VEGF165 mediates formation of complexes containing VEGFR-2 and neuropilin-1 that enhance VEGF165-receptor binding. J Cell Biochem 2002;85(2):357–68.

[61] Salikhova A, Wang L, Lanahan AA, Liu M, Simons M, Leenders WP, et al. Vascular endothelial growth factor and semaphorin induce neuropilin-1 endocytosis via separate pathways. Circ Res 2008;103(6):e71–9.

[62] Lanahan A, Zhang X, Fantin A, Zhuang Z, Rivera-Molina F, Speichinger K, et al. The neuropilin 1 cytoplasmic domain is required for VEGF-a-dependent arteriogenesis. Dev Cell 2013;25(2):156–68.

[63] Wang L, Dutta SK, Kojima T, Xu X, Khosravi-Far R, Ekker SC, et al. Neuropilin-1 modulates p53/caspases axis to promote endothelial cell survival. PLoS One 2007;2(11):e11–61.

[64] Fantin A, Schwarz Q, Davidson K, Normando EM, Denti L, Ruhrberg C. The cytoplasmic domain of neuropilin 1 is dispensable for angiogenesis, but promotes the spatial separation of retinal arteries and veins. Development 2011;138(19):4185–91.

[65] Cariboni A, Davidson K, Rakic S, Maggi R, Parnavelas JG, Ruhrberg C. Defective gonadotropin-releasing hormone neuron migration in mice lacking SEMA3A signalling through NRP1 and NRP2: implications for the aetiology of hypogonadotropic hypogonadism. Hum Mol Genet 2011;20(2):336–44.

[66] Fantin A, Herzog B, Mahmoud M, Yamaji M, Plein A, Denti L, et al. Neuropilin 1 (NRP1) hypomorphism combined with defective VEGF-A binding reveals novel roles for NRP1 in developmental and pathological angiogenesis. Development 2014;141(3):556–62.

[67] Raimondi C, Fantin A, Lampropoulou A, Denti L, Chikh A, Ruhrberg C. Imatinib inhibits VEGF-independent angiogenesis by targeting neuropilin 1-dependent ABL1 activation in endothelial cells. J Exp Med 2014;211(6):1167–83.

[68] Yuan L, Moyon D, Pardanaud L, Breant C, Karkkainen MJ, Alitalo K, et al. Abnormal lymphatic vessel development in neuropilin 2 mutant mice. Development 2002;129(20):4797–806.

[69] Herzog Y, Kalcheim C, Kahane N, Reshef R, Neufeld G. Differential expression of neuropilin-1 and neuropilin-2 in arteries and veins. Mech Dev 2001;109(1):115–9.

[70] Xu Y, Yuan L, Mak J, Pardanaud L, Caunt M, Kasman I, et al. Neuropilin-2 mediates VEGF-C-induced lymphatic sprouting together with VEGFR3. J Cell Biol 2010;188(1):115–30.

[71] Erskine L, Reijntjes S, Pratt T, Denti L, Schwarz Q, Vieira JM, et al. VEGF signaling through neuropilin 1 guides commissural axon crossing at the optic chiasm. Neuron 2011;70(5):951–65.

[72] Nakagawa S, Brennan C, Johnson KG, Shewan D, Harris WA, Holt CE, et al. Ephrin-B regulates the Ipsilateral routing of retinal axons at the optic chiasm. Neuron 2000;25(3):599–610.

[73] Williams SE, Mann F, Erskine L, Sakurai T, Wei S, Rossi DJ, et al. Ephrin-B2 and EphB1 mediate retinal axon divergence at the optic chiasm. Neuron 2003;39(6):919–35.

[74] Schwarz Q, Gu C, Fujisawa H, Sabelko K, Gertsenstein M, Nagy A, et al. Vascular endothelial growth factor controls neuronal migration and cooperates with Sema3A to pattern distinct compartments of the facial nerve. Genes Dev 2004;18(22):2822–34.

[75] Pasterkamp RJ, Verhaagen J. Semaphorins in axon regeneration: developmental guidance molecules gone wrong? Philos Trans R Soc Lond B Biol Sci 2006;361(1473):1499–511.

[76] Gong Q, Shipley MT. Evidence that pioneer olfactory axons regulate telencephalon cell cycle kinetics to induce the formation of the olfactory bulb. Neuron 1995;14(1):91–101.

[77] Gong Q, Shipley MT. Expression of extracellular matrix molecules and cell surface molecules in the olfactory nerve pathway during early development. J Comp Neurol 1996;366(1):1–4.

[78] Kobayashi H, Koppel AM, Luo Y, Raper JA. A role for collapsin-1 in olfactory and cranial sensory axon guidance. J Neurosci 1997; 17(21):8339–52.

[79] Schwarting GA, Kostek C, Ahmad N, Dibble C, Pays L, Puschel AW. Semaphorin 3A is required for guidance of olfactory axons in mice. J Neurosci 2000;20(20):7691–7.

[80] Doucette JR, Kiernan JA, Flumerfelt BA. The re-innervation of olfactory glomeruli following transection of primary olfactory axons in the central or peripheral nervous system. J Anat 1983;137(Pt 1):1–19.

[81] Pasterkamp RJ, De Winter F, Holtmaat AJ, Verhaagen J. Evidence for a role of the chemorepellent semaphorin III and its receptor neuropilin-1 in the regeneration of primary olfactory axons. J Neurosci 1998;18(23):9962–76.

[82] De Winter F, Oudega M, Lankhorst AJ, Hamers FP, Blits B, Ruitenberg MJ, et al. Injury-induced class 3 semaphorin expression in the rat spinal cord. Exp Neurol 2002;175(1):61–75.

[83] Shirvan A, Ziv I, Fleminger G, Shina R, He Z, Brudo I, et al. Semaphorins as mediators of neuronal apoptosis. J Neurochem 1999;73(3):961–71.

[84] Jiang SX, Whitehead S, Aylsworth A, Slinn J, Zurakowski B, Chan K, et al. Neuropilin 1 directly interacts with Fer kinase to mediate semaphorin 3A-induced death of cortical neurons. J Biol Chem 2010;285(13):9908–18.

[85] Fujita H, Zhang B, Sato K, Tanaka J, Sakanaka M. Expressions of neuropilin-1, neuropilin-2 and semaphorin 3A mRNA in the rat brain after middle cerebral artery occlusion. Brain Res 2001;914(1–2):1–14.

[86] Gavazzi I, Stonehouse J, Sandvig A, Reza JN, Appiah-Kubi LS, Keynes R, et al. Peripheral, but not central, axotomy induces neuropilin-1 mRNA expression in adult large diameter primary sensory neurons. J Comp Neurol 2000;423(3):492–9.

[87] Pasterkamp RJ, Giger RJ, Verhaagen J. Regulation of semaphorin III/collapsin-1 gene expression during peripheral nerve regeneration. Exp Neurol 1998;153(2):313–27.

[88] Bannerman P, Ara J, Hahn A, Hong L, McCauley E, Friesen K, et al. Peripheral nerve regeneration is delayed in neuropilin 2-deficient mice. J Neurosci Res 2008;86(14):3163–9.

[89] Skold M, Cullheim S, Hammarberg H, Piehl F, Suneson A, Lake S, et al. Induction of VEGF and VEGF receptors in the spinal cord after mechanical spinal injury and prostaglandin administration. Eur J Neurosci 2000;12(10):3675–86.

[90] Zhang ZG, Tsang W, Zhang L, Powers C, Chopp M. Up-regulation of neuropilin-1 in neovasculature after focal cerebral ischemia in the adult rat. J Cereb Blood Flow Metab 2001;21(5):541–9.

[91] Joyal JS, Sitaras N, Binet F, Rivera JC, Stahl A, Zaniolo K, et al. Ischemic neurons prevent vascular regeneration of neural tissue by secreting semaphorin 3A. Blood 2011;117(22):6024–35.

[92] Cariboni A, Davidson K, Dozio E, Memi F, Schwarz Q, Stossi F, et al. VEGF signalling controls GnRH neuron survival via NRP1 independently of KDR and blood vessels. Development 2011;138(17):3723–33.

[93] Mackenzie F, Ruhrberg C. Diverse roles for VEGF-A in the nervous system. Development 2012;139(8):1371–80.

[94] Morita A, Yamashita N, Sasaki Y, Uchida Y, Nakajima O, Nakamura F, et al. Regulation of dendritic branching and spine maturation by semaphorin3A-Fyn signaling. J Neurosci 2006;26(11):2971–80.

[95] Bouzioukh F, Daoudal G, Falk J, Debanne D, Rougon G, Castellani V, et al. Semaphorin3A regulates synaptic function of differentiated hippocampal neurons. Eur J Neurosci 2006;23(9):2247–54.

[96] Tran TS, Rubio ME, Clem RL, Johnson D, Case L, Tessier-Lavigne M, et al. Secreted semaphorins control spine distribution and morphogenesis in the postnatal CNS. Nature 2009;462(7276):1065–9.

[97] Ding JB, Oh WJ, Sabatini BL, Gu C. Semaphorin 3E-Plexin-D1 signaling controls pathway-specific synapse formation in the striatum. Nat Neurosci 2012;15(2):215–23.

[98] Barnes G, Puranam RS, Luo Y, McNamara JO. Temporal specific patterns of semaphorin gene expression in rat brain after kainic acid-induced status epilepticus. Hippocampus 2003;13(1):1–20.

[99] Holtmaat AJ, Gorter JA, De Wit J, Tolner EA, Spijker S, Giger RJ, et al. Transient downregulation of Sema3A mRNA in a rat model for temporal lobe epilepsy. A novel molecular event potentially contributing to mossy fiber sprouting. Exp Neurol 2003;182(1):142–50.

[100] Sahay A, Kim CH, Sepkuty JP, Cho E, Huganir RL, Ginty DD, et al. Secreted semaphorins modulate synaptic transmission in the adult hippocampus. J Neurosci 2005;25(14):3613–20.

[101] Licht T, Eavri R, Goshen I, Shlomai Y, Mizrahi A, Keshet E. VEGF is required for dendritogenesis of newly born olfactory bulb interneurons. Development 2010;137(2):261–71.

[102] Kim BW, Choi M, Kim YS, Park H, Lee HR, Yun CO, et al. Vascular endothelial growth factor (VEGF) signaling regulates hippocampal neurons by elevation of intracellular calcium and activation of calcium/calmodulin protein kinase II and mammalian target of rapamycin. Cell Signal 2008;20(4):714–25.

[103] Cao L, Jiao X, Zuzga DS, Liu Y, Fong DM, Young D, et al. VEGF links hippocampal activity with neurogenesis, learning and memory. Nat Genet 2004;36(8):827–35.

[104] Huang YF, Yang CH, Huang CC, Tai MH, Hsu KS. Pharmacological and genetic accumulation of hypoxia-inducible factor-1alpha enhances excitatory synaptic transmission in hippocampal neurons through the production of vascular endothelial growth factor. J Neurosci 2010;30(17):6080–93.

[105] McCloskey DP, Croll SD, Scharfman HE. Depression of synaptic transmission by vascular endothelial growth factor in adult rat hippocampus and evidence for increased efficacy after chronic seizures. J Neurosci 2005;25(39):8889–97.

Chapter 7

Growth and Neurotrophic Factor Receptors in Neural Differentiation and Phenotype Specification

Talita Glaser, Ágatha Oliveira, Laura Sardà-Arroyo and Henning Ulrich
Departamento de Bioquímica, Instituto de Química, Universidade de São Paulo, S.P., Brazil

7.1 INTRODUCTION

During development, growth factors known to regulate the proliferation of nonneuronal cells can also promote the growth of neural stem and precursor cells (NPCs) [1]. Some examples are insulin-like growth factor (IGF) [2,3], fibroblast growth factors (FGFs) [4], the epidermal growth factor (EGF) family, platelet-derived growth factor (PDGF), and neurotrophins [5]. Furthermore, growth and neurotrophic factors including glial cell line-derived neurotrophic factor (GDNF), nerve growth factor (NGF), brain-derived neurotrophic factor (BDNF), and neurotrophins-3 and -4/5 (NT-3 and NT-4/5) participate in differentiation induction and phenotype determination of neural cells [6–8]. These growth factors function by binding and activating specific receptor tyrosine kinases, which then trigger a cascade of events that specify programs of gene transcription and epigenetic and particular cellular responses [6,8]. Tables 7.1 and 7.2 give an overview of the effects of growth and neurotrophic factors on proliferation, differentiation, and stem cell fates, which are further described in the following. Moreover, in this chapter we focus on describing the actions of growth factors and their intracellular signaling pathways in the processes of neurogenesis and neuromaturation during embryo development and neural stem cell differentiation (Figure 7.1 and 7.2).

7.2 NEUROTROPHINS

The discovery of the neurotrophin family (Figure 7.3) and their tyrosine kinase receptors by Rita Levi-Montalcini encouraged studies to identify a genetic system that contributes to the development and maintenance of the vertebrate nervous system, regulating neuronal survival, axonal growth and guidance, synaptic plasticity, and long-term potentiation events [9–12]. This family includes BDNF, NT-3, and NT-4/5 [13–16]. Science has made tremendous progress in the identification and characterization of the intracellular signaling pathways triggered by neurotrophic factors. These signaling mechanisms support the elucidation of diverse cellular responses including proliferation, differentiation, and programmed cell death. Neurotrophic factor receptor stimulation usually activates cellular tyrosine kinases propagating signal transduction within cells [17–23], resulting in subsequent changes in gene expression [8,24–28]. NGF, BDNF, and NT-3 activate respectively the tropomyosin-related kinase receptors A, B, and C (TrkA, TrkB, and TrkC) [29–41]. A second neurotrophin, NT-4/5, also binds to the TrkB receptor [8]. The most described adaptor proteins for Trk receptors are the Src homology 2 domain-containing protein (SHC) and the FGF receptor substrate 2 (FRS2), initiating downstream phosphorylation cascades by activating the RAS/RAF/extracellular signal-regulated kinase (ERK) mitogen-activated protein kinase and the phosphoinositide 3-kinase (PI3K)/AKT pathways, which are widely attributed to classical Trk-promoted neuronal survival and differentiation. The tyrosine phosphorylation recruits phospholipase C-γ1 (PLC-γ1) protein, leading to the modulation of synaptic plasticity produced by the Trk system [12].

Neurotrophins can also bind to the p75 panneurotrophin receptor (p75NTR or CD271), a member of the tumor necrosis factor receptor family. This multiligand receptor is composed of an extracellular domain with four cysteine-rich regions that optimizes neurotrophin binding, a transmembrane portion, and an intracellular domain, the flexibility of which is responsible for the activation of intracellular signaling, because these receptors do not exhibit an enzymatic function induced by ligand binding, unlike Trk receptors [42].

Neural Surface Antigens. http://dx.doi.org/10.1016/B978-0-12-800781-5.00007-4

TABLE 7.1 Effects of Neurotrophic and Growth Factors on Neural Differentiation

Model	Factors	Receptor	Differentiation Fate/Effect	References
SVZ-derived NPC	BDNF	TrkB	Neurite outgrowth	[83]
NSC	BDNF	p75	Neuroblast	[84]
NPC	BDNF	p75	Neuronal differentiation	[86]
In vivo rat hippocampal NSC	IGF-1	IGF-1R	Neuronal differentiation	[190]
NSC	IGF-1	IGF-1R	Oligodendrocytes	[191]
Human ESC	FGF-2	All	Maintenance of pluripotency	[150]
Mouse ESC	FGF-2	FGFR-1/3	Promotion of self-renewal	[151]
ESC	FGF-2	–	NSC	[152]
Mouse NSC	FGF-4	–	NPC	[153]
Human NSC	FGF-2/8 and 20	–	NPC	[153]
NPC	FGF-2 + EGF short-term removal	–	Neuron fate and neurite outgrowth	[124,125]
NPC	Low FGF-2 concentration	–	Neuronal differentiation	[126]
NPC	High FGF-2 concentration	–	Astrocyte	[126]
Telencephalic progenitors	FGF-8	–	Olfactory bulb neurons	[194]
Human ESC + mouse stromal cells	FGF-20	FGFR-1	Dopaminergic neurons	[195]
–	NT-3	TrkC and p75NTR	Neuronal differentiation	[95]
Cortical NPC	NT-3	TrkC	Neuronal differentiation	[96,199]
Cerebellar granule precursor cells	NT-3	TrkC	Increased neurite fasciculation	[7]
Striatal precursors	NGF + FGF-2	–	Proliferation	[68]
Different kinds of stem cell	NGF + BDNF	–	Neuronal and astroglial differentiation	[70,71]
Neural precursors	NGF	TrkA	Glutamatergic and sensory neuron differentiation, neurite outgrowth	[73,74,198]
NSC from subgranular zone	EGF	–	Proliferation	[78,193]
SVZ-derived NPC	EGF	–	Premyelinating and myelinating oligodendrocytes	[78,110]
Fetal telencephalon	EGF	–	Ventrolateral and radial migration	[78,111,113]
Enteric neural crest	GDNF	–	Migration, proliferation, differentiation, and survival to form enteric nervous system	[162,169]
Ventral mesencephalic neurons	GDNF	–	Differentiation and survival of TH-positive cells	[171,196,197]
Ventral mesencephalic neurons	GDNF	–	GABAergic neurons	[172]
NPC	GDNF	–	Survival and differentiation into TH-positive cells	[170]

TABLE 7.1 Effects of Neurotrophic and Growth Factors on Neural Differentiation—cont'd

Model	Factors	Receptor	Differentiation Fate/Effect	References
Multipotent astrocytic stem cells	Prolonged exposure to GDNF in vitro	–	Increased migration and neuronal differentiation	[173]
NPC	GDNF	–	Migration	[174]
Dentate gyrus-derived NPC	GDNF + retinoic acid	–	Astrocytes	[183]
ESC	GDNF + retinoic acids + other factors	–	Motor neurons	[177,178]

TABLE 7.2 Signaling Pathways Involved in Receptor Activation

Factors	Receptor	Signaling Pathways	References
NGF	TrkA	PI3K/AKT, PLC-γ/DAG/PKC	[64–66]
NGF	TrkA	RAS/RAF/ERK	[77]
NGF/proNGF	p75NTR	JNK/NF-κB	[61–63]
BDNF	TrkB	RAS/RAF/ERK, PI3K/AKT, PLC-γ/DAG/PKC	[80]
NT-3	TrkC	RAS/RAF/ERK, PLC-γ/DAG/PKC	[84]
NT-3	TrkC in Schwann cells	Cdc42–GEF/Dbl	[94]
EGF	EGFR	RAS/RAF/ERK, PI3K/AKT/mTOR, cAMP/PKA/CREB	[78,107,112]
FGF	FGFR	RAS/RAF/ERK, PI3K/AKT, PLC-γ/DAG/PKC	[116]
GDNF	GFR/RET complex	RAS/RAF/ERK, PI3K/AKT, PLC-γ/DAG/PKC	[141,142]
GDNF	GDNF/NCAM complex	RAS/RAF/ERK	[147,148]
GDNF	GFRα1	RAS/RAF/ERK, PI3K/AKT	[140,175]
IGF-1	IGF-1R	RAS/RAF/ERK, PI3K/AKT	[189]

Docked PLC-γ1 is activated through Trk-mediated phosphorylation and then hydrolyzes phosphatidylinositol 4,5-bisphosphate to generate inositol triphosphate (IP3) and diacylglycerol (DAG). IP3 formation results in Ca^{2+} release from cytoplasmic stores, whereas DAG stimulates activation of classical isoforms of protein kinase C (PKC). These signaling molecules activate intracellular enzymes including Ca^{2+}–calmodulin-dependent protein kinases and other Ca^{2+}–calmodulin-regulated targets. Moreover, PLC-γ1 stimulates PKC-δ, which is required for NGF-promoted activation of mitogen-activated protein kinase (MAPK) kinase (MEK) 1 and ERK1/2 [43], with subsequent expression and/or activity control of ion channels and transcription factors [44–47].

Evidence has shown that Trk may also modulate cancer growth such as neuroblastoma and medulloblastoma, whereas a rearranged Trk oncogene is often observed in nonneuronal neoplasms such as colon and papillary thyroid cancers [48]. However, further chapters of this book discuss this theme.

Between all existing neurotrophins, just four are expressed in mammals; these are NGF, BDNF, NT-3, and NT-4 [46]. We focus here on the effects of NGF, BDNF, and NT-3 in neural development and differentiation.

7.3 NERVE GROWTH FACTOR

In 1950, Levi-Montalcini and Hamburger observed neuron growth following mouse sarcoma transplantation into the chick embryo peripheral nervous system. By tissue culture techniques, they identified this tumor factor and named it nerve growth factor, the first identified member of the neurotrophin family (Figure 7.3) [49–51]. NGF exerts its functions on restricted target populations regulating axon guidance, synaptic function, and neuronal differentiation [11] and promotes survival of specific populations of sensory, sympathetic, and central nervous system (CNS) neurons. It is also involved in the maintenance of basal forebrain cholinergic neurons, which project to the hippocampus and are important for memory processes [52–55].

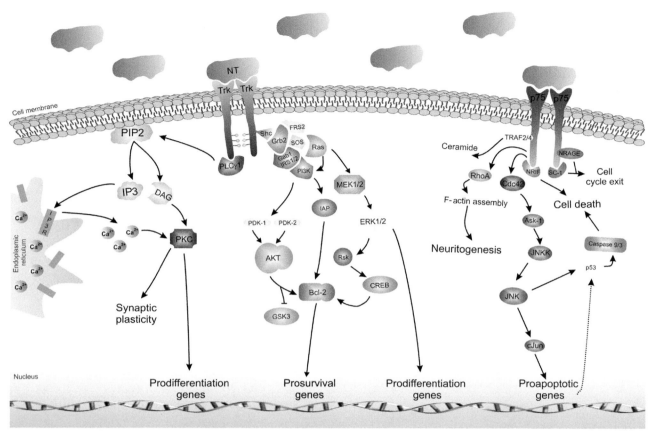

FIGURE 7.1 Neurotrophin signaling. This figure depicts the interactions of neurotrophins (NGF, BDNF, NT-3) with Trk and p75NTR receptors and major intracellular signaling pathways activated through each receptor. The p75NTR receptor triggers two major signaling pathways. One is RhoA activation resulting in F-actin assembly, promoting neuritogenesis and growth cone motility. The Jun kinase pathway controls the activities of several genes, with some of them promoting neuronal apoptosis. Proapoptosis actions of p75NTR appear to require the presence of sortilin, which functions as a coreceptor for neurotrophins. Sortilin is not depicted in this figure. Each Trk receptor controls three major signaling pathways. Activation of RAS results in activation of the MEK/ERK signaling cascade, which promotes neuronal differentiation including neurite outgrowth. Activation of PI3K through RAS or Gab1 promotes survival and growth of neurons and neuroblasts. Activation of PLC-γ1 results in activation of Ca^{2+}- and protein kinase C-regulated pathways that promote synaptic plasticity and differentiation of specific neuron subtypes. Each of these neurotrophin-activated signaling pathways also regulates gene transcription. Additional signaling proteins for p75NTR and Trk receptor pathways are not depicted for reasons of simplicity; however, these are described in more detail in the text.

Therefore, this neurotrophic factor is an important target for treating memory diseases such as Alzheimer disease.

NGF is initially synthesized as a large precursor protein called proNGF and is released in an activity-dependent manner as a precursor form (proNGF) into the extracellular space together with convertases and proteases, which cleave proNGF into its mature form (mNGF) and degrade free unbound mNGF [56]. This proteolytic cleavage has important roles during development and in the adult organism [57,58].

Mature NGF exerts its cellular effects preferentially through the activation of two different receptors, TrkA and p75NTR [11,59]. NGF triggers apoptosis in mistargeting conditions (binding to p75NTR) during neural development [60] and after nervous injury, inflammation, or stress conditions through p75NTR activation, increasing JNK and consequent NF-κB levels and inducing proapoptotic protein

synthesis [61–63]. ProNGF is more effective than mNGF at binding p75NTR and inducing this apoptotic process [62]. In addition to these proapoptotic events promoted by NGF–p75NTR interaction, NGF binds specifically to TrkA, and this activation usually promotes neuroprotective actions such as cell survival and differentiation.

Like TrkC, TrkA autophosphorylates at specific phosphotyrosine residues as soon as this neurotrophin binds to the activation site. The activated subunits turn into recruitment docking sites for signaling effectors, such as PI3K and PLC-γ and adaptor proteins SHC and FRS2 [64–66]. The main intracellular docking domains are Tyr490 and Tyr790 [59]. Consequently, these interactions trigger intracellular signal transduction cascades such as PLC-γ and others that will be further discussed. After prolonged activation, the receptor–ligand complex can be internalized into vesicles

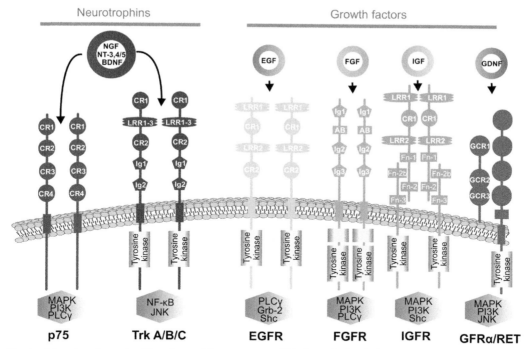

FIGURE 7.2 Structural domains of neurotrophin and growth factor receptors. This figure depicts the interactions of neurotrophins (NGF, BDNF, NT-3) with Trk and p75NTR receptors and major intracellular signaling pathways activated through each receptor and the interactions of growth factors (EGF, FGF, IGF, and GDNF) with their respective receptors (EGFR, FGFR, IGFR, GFRα/RET). Cysteine-rich domain (CR), leucine-rich-repeat domain (LRR), immunoglobulin-like domain (Ig), acid box (AB), fibronectin type III domain (Fn), and globular cysteine-rich domain (GCR). Additional structural domains are not depicted for reasons of simplicity; however, these are described in more detail in the text.

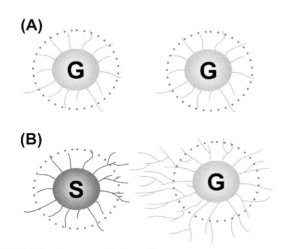

FIGURE 7.3 Nerve growth factor discovery. In 1950, Levi-Montalcini and Hamburger observed a huge innervation growth caused by mouse sarcoma 180 transplanted on the chick embryo peripheral nervous system. By tissue culture techniques they identified the chemical tumor factor as nerve growth factor, the first identified member of the neurotrophin family. (A) When a chick embryo sensory ganglion is cultivated in the presence of another ganglion there are many fibroblasts but few nerve fibers. (B) When a chick embryo sensory ganglion is cultivated in the presence of mouse sarcoma 180 there are many nerve fibers and few fibroblasts, showing the "halo" effect.

and kept activated for as long as the ligand remains associated with the receptor [67].

Some actions of NGF in cell differentiation were described by Cattaneo and Mckay in 1990, who showed that NGF together with FGF-2 stimulates proliferation of striatal precursors [68]. Combined treatment with NGF/BDNF promotes neural and astroglial, but not oligodendrocyte, differentiation [69–72]. Furthermore, NGF promotes not only glutamatergic and sensory neuron differentiation but also neurite outgrowth from different neural precursors in a concentration-dependent manner [16,73,74]. A possible pathway resulting in differentiation through NGF activation is the downregulation of ATF5 [75] and upregulation of tissue inhibitor of metalloproteinases-2 expression [76].

NGF also activates the RAS-mediated induction of the MAPK pathway. This pathway is initiated through recruitment and activation of SHC, which leads to RAS activation through GRB-2 and SOS-1. The MAPK cascade includes RAF, MEK, and ERK. The downstream effectors of the RAS pathway include activation of c-Fos and Jun to form AP-1, activating genes through this transcription factor, leading to cell fate choice, synaptic plasticity, and also neoplastic transformation causing cancer [77].

EGR (early growth-response protein) and CREB (cAMP-response element-binding protein) are transcription factors involved in NGF responses. The EGR family of transcription factors, as well as the MEK/ERK pathway, contributes to NGF-induced neurite formation, and the CREB family is involved in NGF-induced survival of sympathetic neurons. For instance, NGF treatment of PC12 cells induces continuous ERK activity, whereas EGF treatment provides a transient activation [78]. NGF-signaling mechanisms may be further studied to understand how modulation of NGF-induced responses can be exploited for the treatment of neurodegenerative diseases.

7.4 BRAIN-DERIVED NEUROTROPHIC FACTOR

Barde, Edgar, and Thoenen, in 1982, motivated by the fact that no molecule had yet been isolated from a tissue directly involved in nervous system development, purified "a neuronal survival factor" from pig brain, which later was named brain-derived neurotrophic factor [79].

BDNF is a neurotrophin that acts in neural development as well as in adult neurogenesis. BDNF is synthesized as a precursor isoform (proBDNF) that is proteolytically cleaved to produce its mature form (mBDNF). Like NGF, BDNF activates two classes of receptors: the selective TrkB and the aforementioned multiligand p75NTR [80]. mBDNF induces TrkB receptor dimerization and autophosphorylation of tyrosine residues, which induces various intracellular signaling pathways, such as the previously described RAS/MAPK, PI3K/AKT, and PLC-γ1 pathways [81].

Secretion of BDNF and expression of its receptors by embryonic stem cells occur mainly during embryogenesis phases. The lack of this neurotrophin during development of mouse embryos results in a loss of neurons in the sensory ganglia and affects especially the developing thalamus, the substantia nigra, and the cerebellum, yielding poor motor coordination and body-balanced mice [82].

It is well known that TrkB activation by BDNF induces intracellular signaling pathways related to effects that are beneficial for the nervous system, by inducing long-term potentiation in the hippocampus. This physiological enhancement in signal transmission promotes proliferation, survival, and neuronal differentiation of neural stem cells (NSCs) and NPCs [1]. Furthermore, BDNF binding to TrkB is suggested to induce dendritic outgrowth of subventricular zone (SVZ)-derived NPCs in vitro [83]. However, BDNF per se is incapable of promoting proliferation and differentiation of NPCs derived from rat embryo telencephalon [78].

The primary function of p75NTR activation is the induction of apoptosis in some stages of development, though this receptor is expressed in a neurogenic area of the brain by mitotically active immature cells of the hippocampal SVZ, indicating a role in proliferation. Thus BDNF increases neuroblast generation and determines neuronal differentiation fate through activation of p75NTR during development of dorsal root ganglia [84–86]. However, the mechanisms underlying the involvement of p75NTR in neuroblast generation by BDNF are not well understood and seem unlikely to affect NSC and NPC survival, but BDNF-promoted p75NTR activation through promotion of cell cycle progression and differentiation may have a role in final fate determination [84].

By exploring the neurogenic effects of BDNF in the CNS and combining it with cell therapy, novel strategies have been developed for treatment of neurological diseases. For example, mesenchymal stem cells (MSC) are known to secrete BDNF. In view of this, transplantation of these cells should increase BDNF levels and subsequently endogenous neurogenesis in diseases that causes neurodegeneration, such as Parkinson and Huntington diseases, stroke, and multiple sclerosis [87].

BDNF, as well as supplementation of the growth factors IGF-1 and EGF, facilitates the development of human embryonic stem cell (ESC) line derivation [88]. This is important, because of the obvious concerns with regard to the therapeutic use of ESCs obtained from fertilized embryos [40]. Accordingly, Lameu and coworkers showed that BDNF reverses the delay of neural differentiation of NSCs caused by inhibition of nitric oxide (NO) production together with upregulated p75NTR expression. As a possible mechanism, the lack of NO induces BDNF gene overexpression in these cells, probably in an attempt to compensate for deficient NO signaling and to maintain the progress of neural differentiation [89].

In addition to promoting cell differentiation, BDNF has neuroprotective properties that may sustain transplanted cells in cell therapy, because human MSCs (hMSCs) cultured in supernatant derived from ischemic brain extracts increased production of BDNF, NGF, VEGF, and hepatocyte growth factor. The capacity of these adult stem cells to increase expression of growth and trophic factors may be the key to the benefit provided by transplanted hMSCs in the ischemic brain [90].

7.5 NEUROTROPHIN-3

NT-3-induced dimerization of TrkC receptors results in receptor autophosphorylation and rapid generation of phosphorylated docking sites for adaptor cytoplasmic proteins as proteins containing phosphotyrosine-binding and/or Src homology 2 domains.

There are inactive splicing isoforms of TrkC presenting short cytoplasmic motifs without a tyrosine kinase domain that compete with the productive TrkC isoform for NT-3 binding, leading to a decreased bioavailability of NT-3 and subsequent TrkC activation [46]. Furthermore, a truncated isoform of TrkC associates with the postsynaptic density-95/discs large/zona occludens-1 domain of the scaffolding protein tamalin in the presence of NT-3, forming a complex with

ADP-ribosylation factor and guanine nucleotide-exchange factor (ARF–GEF) [91]. Therefore NT-3-mediated signaling through truncated TrkC results in activation of ARF6, which activates RAS-related C3 botulinum toxin substrate (Rac) and leads to membrane disturbance. Differential splicing of TrkC mRNA also results in expression of a TrkC isoform with an amino acid insert within the tyrosine kinase domain. This insert may modify the substrate specificity of this tyrosine kinase, inhibiting its activation for several substrates and interfering with its ability to promote neuronal differentiation [46].

Moreover, neurotrophins activate GTPases, such as cell division control protein 42 homolog (Cdc42) and Rac, through not yet elucidated mechanisms and pathways. Interestingly, neurotrophin effects on axonal growth and growth cone guidance depend on activity regulation of these G proteins [92]. In Schwann cells, stimulation of TrkC results in phosphorylation and activation of the Cdc42–GEF/Double's Big Sister (Dbl) signaling cascade [93] and activates other GEFs directly in neurons. Subsequent GEF–Son of Sevenless complex activity is regulated through phosphorylation by c-Abl and other tyrosine kinases [94].

NT-3 and BDNF stimulation of truncated TrkC receptors promotes differentiation into neurons only in the presence of p75NTR [95]. However, the p75NTR biological response to NT-3 versus BDNF can be quite different: NT-3 enhances neuronal differentiation of the cortical NPCs, whereas BDNF promotes their survival [96]. Similarly, cerebellar granule precursor cells express both TrkB and TrkC. In these cells BDNF promotes survival and enhances axonal elongation, whereas NT-3 increases neurite fasciculation [7,12].

7.6 EPIDERMAL GROWTH FACTOR

Cohen isolated and characterized EGF while trying to understand what made newborn mouse eyes open and teeth erupt precociously when treated with male mouse salivary gland extracts. Later he discovered that the EGF receptor is a ligand-activated tyrosine kinase protein using membranes from a human epidermoid carcinoma cell, a tumor that expresses high levels of EGF receptor (EGFR) [97,98]. Nowadays, there is plenty of information available on the structure of the EGFR [99], its relationship to oncogene products [100,101], and mitogenic responses induced by this receptor [102]. Recent investigations focus on second-messenger pathways that mediate biological responses to EGF.

The EGFR is composed of a single polypeptide chain of 1186 amino acid residues containing an N-linked oligosaccharide. A hydrophobic sequence separates an extracellular ligand-binding domain from a cytoplasmic domain that encodes an EGF-regulated tyrosine kinase. The organizational motif of the EGFR is similar to that of other growth factor receptors (PDGF, insulin, IGF-1, colony-stimulating factor 1, and FGF) and its tyrosine kinase activity has a central role in the regulation of cell proliferation. EGF and EGF-like ligands bind with high affinity to the extracellular domain containing 10 or 11 N-linked oligosaccharide chains [103].

Usually this tyrosine kinase activity leads to activation of some intracellular signaling pathways such as the PI3K/AKT/mammalian target of rapamycin (mTOR) and ERK1/2 pathways [78]. Furthermore, activation of adenylate cyclase and inhibition of cAMP-specific phosphodiesterase activities may be induced by EGF, resulting in intracellular cAMP accumulation and protein kinase A (PKA) activation and consequent CREB stimulation [78]. Cells may migrate, proliferate, or differentiate depending on the pathway induced by EGFR activation. Distinct activation patterns in different cell types lead to a variety of signaling pathways, which in turn promote specific cell responses in specific cellular contexts as described below.

The EGFR has important functions in embryo development [104], including in the nervous system, by induction of proliferation and migration of neural stem cells during neural development (mouse embryonic day 10) and stimulating proliferation of neuroepithelial cells and NSCs from the SVZ (mouse embryonic day 14.5) [78,105,106]. The EGFR acts also during adult neurogenesis by promoting proliferation of NSCs of the SVZ through the cAMP/PKA/CREB pathway [107] and modulation of Sox2 and Pax6 (self-renewal markers) expression in NSCs and NPCs. The EGFR contributes to the maintenance of multipotency without disturbing the neurogenic potency, suggesting that EGFR might not promote differentiation [78,107–109]. Contradicting the hypothesis that EGFR functions are limited to NSC and NPC self-renewal, experiments in vivo showed that exogenous infusion of EGF into the SVZ induced NPC differentiation into pre-myelinating and myelinating oligodendrocytes after primary precursor proliferation and migration induction [78,110]. In line with this observation, migration during establishment of the developing nervous system occurs by EGF-stimulated ERK1/2 and PI3K/AKT pathways and focal adhesion kinase (FAK) activation [78,107,111–113]. This growth factor promotes ventrolateral migration of fetal telencephalon NPCs in the lateral cortical stream and radial migration in the direction of the cortical plate [78,111,113]. Furthermore, transit-amplifying precursor cells that later on differentiate into neuroblast cells are sensitive to stimulation by EGF and express high levels of EGFR [114].

7.7 FIBROBLAST GROWTH FACTOR

FGF is a polypeptide involved in several physiological processes, including paracrine signaling that modulates neural development and adult neurogenesis. It contains 23 isoforms (FGF-15 is not expressed in human tissue) and activates five different tyrosine-kinase receptors (FGFRs) that are expressed in distinct patterns accordingly to the tissue type [115]. Receptor binding can induce the activation of several pathways including PLC-γ, PI3K, and ERK, which are regulated by the negative feedback of downstream transcriptional targets of these pathways [116]. Moreover, secreted FGF bioavailability can be modulated by its binding to extracellular heparin sulfate proteoglycans that prevents its degradation. When partially degraded, this complex releases FGF–glycosaminoglycan, an active FGFR ligand [117].

In the adult mammalian brain, FGF-2 is found in astrocyte cells of the SVZ and subgranular zone of the hippocampal dentate gyrus. However, proliferating NPCs express FGFR-1, -2, and -4, indicating that neurogenesis may be promoted by astrocytes releasing FGF-2 [118]. Functions of FGF in embryonic development are well established, and expression patterns of FGFR depend on the embryo stage. FGF-2 deficiency produces mice with a reduced number of neurons in the deep cortical layers. Moreover, mutations of the FGF-20 gene have been related to a predisposition for Parkinson disease development [115].

The importance of in vitro differentiation of stem cells has been recognized, keeping in mind possible therapeutic applications of cultured stem cells in neurological disorders. Thus, the modulation of stem cell fate during differentiation is currently a focus of study. FGF-2 is usually added to culture cell medium to maintain human ESC pluripotency by induction of multiple genes. Further, inhibition of the FGF signaling pathway causes human ESC differentiation [119]. It is noteworthy that mouse ESCs remain undifferentiated without FGF-2 supplementation in vitro, and leukemia-inhibitory factor and activin are enough to maintain the stem cell characteristics of pluripotency and self-renewal [120]. This can be promoted in rats in vivo by coactivating FGFR-1 and -3 along with higher concentrations of FGF-2 [121]. Beyond that, FGF-2 per se improves the commitment of ESCs to generate proliferative NSCs that can differentiate into all neural cell types [122]. Mouse stem cells can be induced to differentiation by FGF-4 as well as FGF-2, and human by FGF-2, -8, and -20 [123]. Once the neural fate of stem cells is established, proliferation can be induced by FGF-2 together with high concentrations of EGF. Short-term removal of these growth factors promotes neurogenesis and neurite outgrowth [124,125]. Moreover, inhibition of FGFR-1 activity reduces cell proliferation promoted by FGF-2 [118], and restriction of FGF-2 in the culture medium induces NSCs to differentiate into neurons, whereas at higher concentrations both neurons and astrocytes are obtained [126].

The presence of distinct FGF isoforms promotes differentiation into specialized neural phenotypes. For instance, FGF-8 induces telencephalic progenitors to differentiate into olfactory bulb neurons, whereas controlled FGF-2 treatment can induce cholinergic phenotype formation with motor-neuron characteristics, and FGF-20 has the ability to generate dopaminergic neurons from human ESCs when they are cocultured with mouse stromal cells [78].

7.8 GLIAL CELL LINE-DERIVED NEUROTROPHIC FACTOR

GDNF is a member of the neurotrophic factor family that was purified and characterized in 1993 from the supernatant of the B49 glial cell line based on its effects on the survival of midbrain dopaminergic neurons [127,128]. More recently, an increase in peripheral and central nervous system neurons was also related to GDNF-induced effects [129]. GDNF as well as other GDNF family ligands (GFLs) including neurturin, artemin, and persephin are part of the structurally related transforming growth factor-β superfamily. All GFLs possess enriched cysteine dimerization domains and are functional as homodimers [129].

Different from its suggestive name, GDNF is mainly expressed by neurons in some brain structures such as septum, striatum, and thalamus [130,131]. However, GDNF mRNA transcription has been detected in astrocyte and microglia cultures [132]. Several studies revealed that in pathological conditions such as inflammation, there is an increase in GDNF expression by astrocytes and microglia [133,134]. The GFLs have some signaling characteristics in common. They need the activation of two receptor types to exert their intracellular response: 1). the tyrosine kinase receptor RET, a single-pass transmembrane protein with an extracellular domain that transmits the signal to the cytoplasm; 2). the glycosylphosphatidylinositol-anchored coreceptor GFRα1 (GDNF family receptor α1), necessary for GDNF binding, containing an intracellular domain important for transmission of the intracellular signaling pathway [135,136]. RET, originally characterized by its functions as a proto-oncogene [137], is the most common receptor for all the GFLs, including GDNF [131,133,134.137–139].

As an agreed molecular mechanism of activation, canonical GDNF signaling is initialized by GDNF binding to GFRα1, forming a homodimer [140], usually occurring at lipid rafts [129], followed by recruitment of RET receptors to these structures. Then, the GFRα1–GDNF complex interacts with RET, which dimerizes and turns into the active tyrosine-phosphorylated form. In this state, various signaling pathways have been elucidated, including PI3K/AKT signaling for promotion of neural survival and transmission, activation of PLC-γ1 as a regulator of neural transmission, and RAS/MAPK for cell survival and neurogenesis [141,142].

RET and GFRα1 are differentially expressed in some regions of the CNS, as GFRα1 is highly expressed when RET expression is almost undetectable [129,142]. There are two explanations for this situation. First of all, it has been described that GFRα1 can act in a non-cell-autonomous fashion, activating RET receptor *in trans* [143–145]. On one hand, GFRα1 could act from the surface of other cells or be in a soluble form. On the other hand, GFRα1 could signal through another kind of receptor in a fashion independent of RET. This hypothesis led to the discovery of neural cell adhesion molecule (NCAM; CD56) as an alternative signal receptor for GFRα1 [146]. This pathway involving GFRα1 and NCAM is called RET-independent signalization. Unlike RET, NCAM, specifically isoform p140^{NCAM}, can bind GDNF by itself, but in an inefficient way. In this situation GDNF interacts with two GRFα1 receptor molecules to form a complex capable of interaction with NCAM [146]. This interaction produces a signal similar to that produced by homophilic NCAM–NCAM interactions. NCAM is associated in its intracellular domain with Fyn (a Src family member) [147], which becomes activated and recruits FAK, stimulating the MAPK pathway [148]. This GDNF–GFRα1–NCAM pathway participates in migration, specification of neural morphology, and synaptogenesis [146,149–153].

GDNF is involved in physiological functions of various tissues including proliferation, migration, and differentiation of neural cells, as well as morphogenesis of the kidney; development of the enteric system; maintenance of sensory, parasympathetic, and enteric neurons; and maintenance and neuroprotection of dopaminergic and catecholaminergic populations [127,154–159]. In other words, GDNF participates in proliferation, migration, and differentiation of neural cells [129,160–163]. GDNF is essential for the formation of the enteric nervous system by regulating migration, proliferation, differentiation, and survival of enteric neural crest via RET–GFRα1 signaling [160,162–169]. In vitro, GDNF acts on NPCs, promoting survival and differentiation into tyrosine hydroxylase (TH)-positive cells [170,171], and also on a subset of ventral mesencephalic neurons. It is able to generate, in addition to dopaminergic neurons, small-sized GABAergic cells [172].

In vitro studies of migration and differentiation of multipotent astrocytic stem cells show that GDNF added to the culture medium increases migration in prolonged exposure and most efficiently promotes neuronal differentiation compared to other neurotrophins [173]. Accordingly, some findings showed that GDNF acts as an attractant compound for neural precursors [174]. This neurotrophin at 50 ng/mL concentration promotes the differentiation and tangential migration of GABAergic neuronal precursors from the medial ganglionic eminence in a mechanism dependent on GFRα1 and independent of RET and NCAM. This differentiation is promoted by activation of the MAPK pathway, by phosphorylation of the ERK1/2 kinases, and by moderate phosphorylation of AKT after 1 h of treatment [140,175].

In vitro protocols use a combination of GDNF and other factors after stimulus with retinoic acid and Sonic hedgehog for induction of neural differentiation of ESCs into motor neurons [176,177]. It has been shown that infusion of GDNF together with other neurotrophic factors contributes to the incorporation of predifferentiated motor neuron precursors into neural circuits [178–181]. In agreement with these findings, Garcia-Bennett and colleagues promoted, via delivery of mimetic growth factors through a mesoporous silica vehicle, specific differentiation of ESCs into motor neurons. Infusion of mimetic particles as well as of growth factor into the damaged area could be an alternative to the treatment of motor neurodegenerative disorders such as amyotrophic lateral sclerosis [182].

Depending on the context, GDNF may also favor glial differentiation. Boku and colleagues [183] showed that a 50 ng/mL GDNF treatment promotes an increase in the ratio of astrocyte-like glial fibrillary acidic protein-positive cells over Tuj1-positive cells after differentiation induction by *all-trans* retinoic acid. This astrogliogenic effect from neural precursors of the dentate gyrus is promoted by increasing the phosphorylation of STAT3, affecting neither proliferation nor apoptosis rates [183].

7.9 INSULIN-LIKE GROWTH FACTOR

The discovery of IGF goes back to 1963 when Froesch and coworkers observed that there was a serum factor similar to insulin, but its induced effects were not suppressed even in the presence of anti-insulin antibodies. Subsequent studies showed growth-promoting effects by IGF and the characterization of isoforms I and II [184]. The insulin-like growth factor family signaling system embraces two polypeptides (IGF-1 and IGF-2) and type 1 and 2 IGF receptors (IGF-1R, -2R). Both ligands can bind to IGF-1R with different affinities, triggering diverse biological responses [185].

IGF-1R is a class II tyrosine kinase receptor that, differing from TrkA, B, and C, is a heterotetramer composed of two extracellular α subunits containing the binding site and two β subunits that penetrate the membrane and include the tyrosine kinase domain [186]. This system also has IGF-binding proteins responsible for controlling the amount of free IGF-1 and -2 in the circulatory system, therefore modulating their binding to the receptors [185].

In the brain, IGF-1 and IGF-1R are expressed in similar areas, indicating an autocrine and/or paracrine action on this tissue. IGF-1 available in the CNS can be produced in peripheral tissues, being captured by the blood–brain barrier via specific receptors and reaching the hypothalamus and hippocampus. Evidence for a low synthesis of these peptides in the brain exists, although is not well defined [185].

IGF-1 signaling is present throughout development. The growth factor increases the number of embryos becoming blastocysts in normal and critical situations. For instance, IGF-1-mediated responses are directed against different types of stressors that could lead to a decreased survival ratio, such as toxicity, oxidative stress, and exposure to tumor necrosis factor-α. Although IGF-1 gene-knockout mice are viable, they show a retarded development of the CNS with loss of mass in several brain areas, especially involving the white matter [187]. Moreover, IGF is also related to cancer development, because cerebrospinal fluid samples from patients with glioblastoma multiforme show elevated IGF-2 levels and IGF-2-dependently stimulated stem cell proliferation [188].

In vivo studies endorse the importance of this growth factor in the cell fate of stem cells [189]. Hippocampal NSCs of rats submitted to IGF-1 infusion proliferate and differentiate in neurons [190]. Rat hippocampal NSCs can be stimulated in vitro to differentiate into oligodendrocytes in the presence of IGF-1 [71,191].

Furthermore, IGF-1 has antiapoptotic effects; this factor promotes cell survival of cortical neurons cultured on serum-free medium, in other words, medium that lacks essential nutrients for cell survival [192].

7.10 CONCLUSION

In summary, various intracellular pathways can be triggered depending on the expression patterns of growth and neurotrophic factors and their respective receptors in NSCs and NPCs. These signals, via subsequent modulation of transcription factors and gene expression, thereby determine stem cell fate, governing proliferation versus differentiation (see Table 7.1 for some examples).

Thus, as an oversimplified synopsis, whereas NT-3, BDNF, FGF, and GDNF mostly have implications for the progress of neurogenesis and neuronal phenotype determination, proliferation is favored in the presence of EGF and FGF. However, removal or combination or even different concentrations of these factors may lead to diverse effects by modulating NSC and NPC fate and neuronal and astroglial fate specification. Overall, the specific effects mediated by the discussed growth and neurotrophic receptors very much depend on the cellular context.

REFERENCES

[1] Cho T, et al. Long-term potentiation promotes proliferation/survival and neuronal differentiation of neural stem/progenitor cells. PLoS One 2013;8(10):e76860.

[2] LeRoith D, Roberts Jr CT. Insulin-like growth factors. Ann NY Acad Sci 1993;692:1–9.

[3] LeRoith D, Roberts Jr CT. Insulin-like growth factors and their receptors in normal physiology and pathological states. J Pediatr Endocrinol 1993;6(3–4):251–5.

[4] Reuss B, von Bohlen, Halbach O. Fibroblast growth factors and their receptors in the central nervous system. Cell Tissue Res 2003;313(2):139–57.

[5] Hsu YC, Lee DC, Chiu IM. Neural stem cells, neural progenitors, and neurotrophic factors. Cell Transplant 2007;16(2):133–50.

[6] Heldin CH. Dimerization of cell surface receptors in signal transduction. Cell 1995;80(2):213–23.

[7] Segal RA. Selectivity in neurotrophin signaling: theme and variations. Annu Rev Neurosci 2003;26:299–330.

[8] Segal RA, Greenberg ME. Intracellular signaling pathways activated by neurotrophic factors. Annu Rev Neurosci 1996;19:463–89.

[9] Huang EJ, Reichardt LF. Neurotrophins: roles in neuronal development and function. Annu Rev Neurosci 2001;24:677–736.

[10] Poo MM. Neurotrophins as synaptic modulators. Nat Rev Neurosci 2001;2(1):24–32.

[11] Huang EJ, Reichardt LF. Trk receptors: roles in neuronal signal transduction. Annu Rev Biochem 2003;72:609–42.

[12] Benito-Gutierrez E, Garcia-Fernandez J, Comella JX. Origin and evolution of the Trk family of neurotrophic receptors. Mol Cell Neurosci 2006;31(2):179–92.

[13] Snider WD, Johnson Jr EM. Neurotrophic molecules. Ann Neurol 1989;26(4):489–506.

[14] Eide FF, Lowenstein DH, Reichardt LF. Neurotrophins and their receptors–current concepts and implications for neurologic disease. Exp Neurol 1993;121(2):200–14.

[15] Korsching S. The neurotrophic factor concept: a reexamination. J Neurosci 1993;13(7):2739–48.

[16] Zhang L, et al. NGF and NT-3 have differing effects on the growth of dorsal root axons in developing mammalian spinal cord. J Neurosci 1994;14(9):5187–201.

[17] Barbacid M, et al. The trk family of tyrosine protein kinase receptors. Biochim Biophys Acta 1991;1072(2–3):115–27.

[18] Bothwell M. Tissue localization of nerve growth factor and nerve growth factor receptors. Curr Top Microbiol Immunol 1991;165: 55–70.

[19] Chao MV. Neurotrophin receptors: a window into neuronal differentiation. Neuron 1992;9(4):583–93.

[20] Chao MV. Growth factor signaling: where is the specificity? Cell 1992;68(6):995–7.

[21] Chao MV, Battleman DS, Benedetti M. Receptors for nerve growth factor. Int Rev Cytol 1992;137B:169–80.

[22] Barbacid M. Nerve growth factor: a tale of two receptors. Oncogene 1993;8(8):2033–42.

[23] Raffioni S, Bradshaw RA, Buxser SE. The receptors for nerve growth factor and other neurotrophins. Annu Rev Biochem 1993;62:823–50.

[24] Cohen-Cory S, et al. Depolarizing influences increase low-affinity NGF receptor gene expression in cultured Purkinje neurons. Exp Neurol 1993;119(2):165–75.

[25] Greenberg ME, Hermanowski AL, Ziff EB. Effect of protein synthesis inhibitors on growth factor activation of c-fos, c-myc, and actin gene transcription. Mol Cell Biol 1986;6(4):1050–7.

[26] Greenberg ME, Ziff EB, Greene LA. Stimulation of neuronal acetylcholine receptors induces rapid gene transcription. Science 1986;234(4772):80–3.

[27] Pincus DW, DiCicco-Bloom EM, Black IB. Vasoactive intestinal peptide regulates mitosis, differentiation and survival of cultured sympathetic neuroblasts. Nature 1990;343(6258):564–7.

[28] Schwartz JP. Neurotransmitters as neurotrophic factors: a new set of functions. Int Rev Neurobiol 1992;34:1–23.

[29] Berkemeier LR, et al. Neurotrophin-5: a novel neurotrophic factor that activates trk and trkB. Neuron 1991;7(5):857–66.

[30] Cordon-Cardo C, et al. The trk tyrosine protein kinase mediates the mitogenic properties of nerve growth factor and neurotrophin-3. Cell 1991;66(1):173–83.

[31] Glass DJ, et al. TrkB mediates BDNF/NT-3-dependent survival and proliferation in fibroblasts lacking the low affinity NGF receptor. Cell 1991;66(2):405–13.

[32] Hempstead BL, et al. High-affinity NGF binding requires coexpression of the trk proto-oncogene and the low-affinity NGF receptor. Nature 1991;350(6320):678–83.

[33] Kaplan DR, et al. The trk proto-oncogene product: a signal transducing receptor for nerve growth factor. Science 1991;252(5005):554–8.

[34] Kaplan DR, Martin-Zanca D, Parada LF. Tyrosine phosphorylation and tyrosine kinase activity of the trk proto-oncogene product induced by NGF. Nature 1991;350(6314):158–60.

[35] Klein R, et al. The trk proto-oncogene encodes a receptor for nerve growth factor. Cell 1991;65(1):189–97.

[36] Klein R, et al. The trkB tyrosine protein kinase is a receptor for brain-derived neurotrophic factor and neurotrophin-3. Cell 1991;66(2):395–403.

[37] Lamballe F, Klein R, Barbacid M. trkC, a new member of the trk family of tyrosine protein kinases, is a receptor for neurotrophin-3. Cell 1991;66(5):967–79.

[38] Lamballe F, Klein R, Barbacid M. The trk family of oncogenes and neurotrophin receptors. Princess Takamatsu Symp 1991;22:153–70.

[39] Nebreda AR, et al. Induction by NGF of meiotic maturation of Xenopus oocytes expressing the trk proto-oncogene product. Science 1991;252(5005):558–61.

[40] Soppet D, et al. The neurotrophic factors brain-derived neurotrophic factor and neurotrophin-3 are ligands for the trkB tyrosine kinase receptor. Cell 1991;65(5):895–903.

[41] Squinto SP, et al. trkB encodes a functional receptor for brain-derived neurotrophic factor and neurotrophin-3 but not nerve growth factor. Cell 1991;65(5):885–93.

[42] Chen Y, et al. Multiple roles of the p75 neurotrophin receptor in the nervous system. J Int Med Res 2009;37(2):281–8.

[43] Corbit KC, Foster DA, Rosner MR. Protein kinase Cdelta mediates neurogenic but not mitogenic activation of mitogen-activated protein kinase in neuronal cells. Mol Cell Biol 1999;19(6):4209–18.

[44] Klein M, Hempstead BL, Teng KK. Activation of STAT5-dependent transcription by the neurotrophin receptor Trk. J Neurobiol 2005;63(2):159–71.

[45] Minichiello L, et al. Mechanism of TrkB-mediated hippocampal long-term potentiation. Neuron 2002;36(1):121–37.

[46] Reichardt LF. Neurotrophin-regulated signalling pathways. Philos Trans R Soc Lond B Biol Sci 2006;361(1473):1545–64.

[47] Toledo-Aral JJ, et al. A single pulse of nerve growth factor triggers long-term neuronal excitability through sodium channel gene induction. Neuron 1995;14(3):607–11.

[48] Nakagawara A. Trk receptor tyrosine kinases: a bridge between cancer and neural development. Cancer Lett 2001;169(2):107–14.

[49] Levi-Montalcini R, Angeletti PU. Nerve growth factor. Physiol Rev 1968;48(3):534–69.

[50] Levi-Montalcini R, et al. In vitro effects of the nerve growth factor on the fine structure of the sensory nerve cells. Brain Res 1968;8(2):347–62.

[51] Thoenen H, Barde YA. Physiology of nerve growth factor. Physiol Rev 1980;60(4):1284–335.

[52] Chen KS, et al. Disruption of a single allele of the nerve growth factor gene results in atrophy of basal forebrain cholinergic neurons and memory deficits. J Neurosci 1997;17(19):7288–96.

[53] Cuello AC. Trophic factor therapy in the adult CNS: remodelling of injured basalo-cortical neurons. Prog Brain Res 1994;100:213–21.

[54] Ebendal T. Function and evolution in the NGF family and its receptors. J Neurosci Res 1992;32(4):461–70.

[55] Hefti F, Knusel B, Lapchak PA. Protective effects of nerve growth factor and brain-derived neurotrophic factor on basal forebrain cholinergic neurons in adult rats with partial fimbrial transections. Prog Brain Res 1993;98:257–63.

[56] Bruno MA, Cuello AC. Activity-dependent release of precursor nerve growth factor, conversion to mature nerve growth factor, and its degradation by a protease cascade. Proc Natl Acad Sci USA 2006;103(17):6735–40.

[57] Fahnestock M, Yu G, Coughlin MD. ProNGF: a neurotrophic or an apoptotic molecule? Prog Brain Res 2004;146:101–10.

[58] Fahnestock M, et al. The nerve growth factor precursor proNGF exhibits neurotrophic activity but is less active than mature nerve growth factor. J Neurochem 2004;89(3):581–92.

[59] Chao MV, Rajagopal R, Lee FS. Neurotrophin signalling in health and disease. Clin Sci (Lond) 2006;110(2):167–73.

[60] Majdan M, Miller FD. Neuronal life and death decisions functional antagonism between the Trk and p75 neurotrophin receptors. Int J Dev Neurosci 1999;17(3):153–61.

[61] Dobrowsky RT, Carter BD. p75 neurotrophin receptor signaling: mechanisms for neurotrophic modulation of cell stress? J Neurosci Res 2000;61(3):237–43.

[62] Lee R, et al. Regulation of cell survival by secreted proneurotrophins. Science 2001;294(5548):1945–8.

[63] Roux PP, Barker PA. Neurotrophin signaling through the p75 neurotrophin receptor. Prog Neurobiol 2002;67(3):203–33.

[64] Kaplan DR, Miller FD. Neurotrophin signal transduction in the nervous system. Curr Opin Neurobiol 2000;10(3):381–91.

[65] Lemmon MA, et al. Independent binding of peptide ligands to the SH2 and SH3 domains of Grb2. J Biol Chem 1994;269(50):31653–8.

[66] Lemmon MA, Schlessinger J. Regulation of signal transduction and signal diversity by receptor oligomerization. Trends Biochem Sci 1994;19(11):459–63.

[67] Bergeron JJ, et al. Endosomes, receptor tyrosine kinase internalization and signal transduction. Biosci Rep 1995;15(6):411–8.

[68] Cattaneo E, McKay R. Proliferation and differentiation of neuronal stem cells regulated by nerve growth factor. Nature 1990;347(6295):762–5.

[69] Levenberg S, et al. Neurotrophin-induced differentiation of human embryonic stem cells on three-dimensional polymeric scaffolds. Tissue Eng 2005;11(3–4):506–12.

[70] Nakajima K, et al. Essential role of NKCC1 in NGF-induced neurite outgrowth. Biochem Biophys Res Commun 2007;359(3):604–10.

[71] Choi KC, et al. Effect of single growth factor and growth factor combinations on differentiation of neural stem cells. J Korean Neurosurg Soc 2008;44(6):375–81.

[72] Lachyankar MB, et al. Embryonic precursor cells that express Trk receptors: induction of different cell fates by NGF, BDNF, NT-3, and CNTF. Exp Neurol 1997;144(2):350–60.

[73] Ernsberger U. Role of neurotrophin signalling in the differentiation of neurons from dorsal root ganglia and sympathetic ganglia. Cell Tissue Res 2009;336(3):349–84.

[74] Singh RP, et al. Retentive multipotency of adult dorsal root ganglia stem cells. Cell Transpl 2009;18(1):55–68.

[75] Angelastro JM, et al. Regulated expression of ATF5 is required for the progression of neural progenitor cells to neurons. J Neurosci 2003;23(11):4590–600.

[76] Jaworski DM, Perez-Martinez L. Tissue inhibitor of metalloproteinase-2 (TIMP-2) expression is regulated by multiple neural differentiation signals. J Neurochem 2006;98(1):234–47.

[77] Hess J, Angel P, Schorpp-Kistner M. AP-1 subunits: quarrel and harmony among siblings. J Cell Sci 2004;117(Pt 25):5965–73.

[78] Oliveira SL, et al. Functions of neurotrophins and growth factors in neurogenesis and brain repair. Cytom A 2013;83(1):76–89.

[79] Barde YA, Edgar D, Thoenen H. Purification of a new neurotrophic factor from mammalian brain. EMBO J 1982;1(5):549–53.

[80] Lu B, Pang PT, Woo NH. The yin and yang of neurotrophin action. Nat Rev Neurosci 2005;6(8):603–14.

[81] Binder DK, Scharfman HE. Brain-derived neurotrophic factor. Growth Factors 2004;22(3):123–31.

[82] Bartkowska K, Turlejski K, Djavadian RL. Neurotrophins and their receptors in early development of the mammalian nervous system. Acta Neurobiol Exp (Wars) 2010;70(4):454–67.

[83] Gascon E, et al. Sequential activation of p75 and TrkB is involved in dendritic development of subventricular zone-derived neuronal progenitors in vitro. Eur J Neurosci 2005;21(1):69–80.

[84] Young KM, et al. p75 neurotrophin receptor expression defines a population of BDNF-responsive neurogenic precursor cells. J Neurosci 2007;27(19):5146–55.

[85] Pruginin-Bluger M, Shelton DL, Kalcheim C. A paracrine effect for neuron-derived BDNF in development of dorsal root ganglia: stimulation of Schwann cell myelin protein expression by glial cells. Mech Dev 1997;61(1–2):99–111.

[86] Hosomi S, et al. The p75 receptor is required for BDNF-induced differentiation of neural precursor cells. Biochem Biophys Res Commun 2003;301(4):1011–5.

[87] Paul G, Anisimov SV. The secretome of mesenchymal stem cells: potential implications for neuroregeneration. Biochimie 2013; 95(12):2246–56.

[88] Fan Y, et al. Improved efficiency of microsurgical enucleated tri-pronuclear zygotes development and embryonic stem cell derivation by supplementing epidermal growth factor, brain-derived neuro-trophic factor, and insulin-like growth factor-1. Stem Cells Dev 2014;23(6):563–75.

[89] Lameu C, et al. Interactions between the NO-citrulline cycle and brain-derived neurotrophic factor in differentiation of neural stem cells. J Biol Chem 2012;287(35):29690–701.

[90] Chen X, et al. Ischemic rat brain extracts induce human marrow stromal cell growth factor production. Neuropathology 2002; 22(4):275–9.

[91] Esteban PF, et al. A kinase-deficient TrkC receptor isoform activates Arf6-Rac1 signaling through the scaffold protein tamalin. J Cell Biol 2006;173(2):291–9.

[92] Yuan XB, et al. Signalling and crosstalk of Rho GTPases in mediating axon guidance. Nat Cell Biol 2003;5(1):38–45.

[93] Blanke S, Jackle H. Novel guanine nucleotide exchange factor GEFmeso of Drosophila melanogaster interacts with Ral and Rho GTPase Cdc42. FASEB J 2006;20(6):683–91.

[94] Sini P, et al. Abl-dependent tyrosine phosphorylation of Sos-1 mediates growth-factor-induced Rac activation. Nat Cell Biol 2004; 6(3):268–74.

[95] Hapner SJ, et al. Neural differentiation promoted by truncated trkC receptors in collaboration with p75(NTR). Dev Biol 1998;201(1):90–100.

[96] Ghosh A, Greenberg ME. Distinct roles for bFGF and NT-3 in the regulation of cortical neurogenesis. Neuron 1995;15(1):89–103.

[97] Cohen S. Origins of growth factors: NGF and EGF. Ann NY Acad Sci 2004;1038:98–102.

[98] Cohen S. Origins of growth factors: NGF and EGF. J Biol Chem 2008;283(49):33793–7.

[99] Needham SR, et al. Structure-function relationships and supramolecular organization of the EGFR (epidermal growth factor receptor) on the cell surface. Biochem Soc Trans 2014;42(1):114–9.

[100] Burgess AW, et al. EGF receptor family: twisting targets for improved cancer therapies. Growth Factors 2014;32(2):74–81.

[101] Tomas A, Futter CE, Eden ER. EGF receptor trafficking: consequences for signaling and cancer. Trends Cell Biol 2014;24(1):26–34.

[102] Aasrum M, et al. The involvement of the docking protein Gab1 in mitogenic signalling induced by EGF and HGF in rat hepatocytes. Biochim Biophys Acta 2013;1833(12):3286–94.

[103] Carpenter G, Cohen S. Epidermal growth factor. J Biol Chem 1990;265(14):7709–12.

[104] Reinchisi G, et al. Sonic Hedgehog modulates EGFR dependent proliferation of neural stem cells during late mouse embryogenesis through EGFR transactivation. Front Cell Neurosci 2013;7:166.

[105] Threadgill DW, et al. Targeted disruption of mouse EGF receptor: effect of genetic background on mutant phenotype. Science 1995;269(5221):230–4.

[106] Kornblum HI, et al. Prenatal ontogeny of the epidermal growth factor receptor and its ligand, transforming growth factor alpha, in the rat brain. J Comp Neurol 1997;380(2):243–61.

[107] Ciccolini F, et al. Prospective isolation of late development multipotent precursors whose migration is promoted by EGFR. Dev Biol 2005;284(1):112–25.

[108] Hu Q, et al. The EGF receptor-sox2-EGF receptor feedback loop positively regulates the self-renewal of neural precursor cells. Stem Cells 2010;28(2):279–86.

[109] Jia H, et al. Pax6 regulates the epidermal growth factor-responsive neural stem cells of the subventricular zone. Neuroreport 2011; 22(9):448–52.

[110] Gonzalez-Perez O, et al. Epidermal growth factor induces the progeny of subventricular zone type B cells to migrate and differentiate into oligodendrocytes. Stem Cells 2009;27(8):2032–43.

[111] Caric D, et al. EGFRs mediate chemotactic migration in the developing telencephalon. Development 2001;128(21):4203–16.

[112] Jiang Q, et al. EGF-induced cell migration is mediated by ERK and PI3K/AKT pathways in cultured human lens epithelial cells. J Ocul Pharmacol Ther 2006;22(2):93–102.

[113] Burrows RC, et al. Response diversity and the timing of progenitor cell maturation are regulated by developmental changes in EGFR expression in the cortex. Neuron 1997;19(2):251–67.

[114] Cesetti T, et al. Analysis of stem cell lineage progression in the neonatal subventricular zone identifies EGFR+/NG2− cells as transit-amplifying precursors. Stem Cells 2009;27(6):1443–54.

[115] Itoh N, Ornitz DM. Fibroblast growth factors: from molecular evolution to roles in development, metabolism and disease. J Biochem 2011;149(2):121–30.

[116] Lanner F, Rossant J. The role of FGF/Erk signaling in pluripotent cells. Development 2010;137(20):3351–60.

[117] Ornitz DM, Itoh N. Fibroblast growth factors. Genome Biol 2001;2(3):REVIEWS3005.

[118] Mudo G, et al. The FGF-2/FGFRs neurotrophic system promotes neurogenesis in the adult brain. J Neural Transm 2009;116(8): 995–1005.

[119] Vallier L, Alexander M, Pedersen RA. Activin/Nodal and FGF pathways cooperate to maintain pluripotency of human embryonic stem cells. J Cell Sci 2005;118(Pt 19):4495–509.

[120] Daheron L, et al. LIF/STAT3 signaling fails to maintain self-renewal of human embryonic stem cells. Stem Cells 2004;22(5):770–8.

[121] Maric D, et al. Self-renewing and differentiating properties of cortical neural stem cells are selectively regulated by basic fibroblast growth factor (FGF) signaling via specific FGF receptors. J Neurosci 2007;27(8):1836–52.

[122] Guillemot F, Zimmer C. From cradle to grave: the multiple roles of fibroblast growth factors in neural development. Neuron 2011; 71(4):574–88.

[123] Kosaka N, et al. FGF-4 regulates neural progenitor cell proliferation and neuronal differentiation. FASEB J 2006;20(9):1484–5.

[124] Schwindt TT, et al. Short-term withdrawal of mitogens prior to plating increases neuronal differentiation of human neural precursor cells. PLoS One 2009;4(2):e4642.

[125] Schwindt TT, et al. Effects of FGF-2 and EGF removal on the differentiation of mouse neural precursor cells. Acad Bras Cienc 2009;81(3):443–52.

[126] Gage FH, et al. Survival and differentiation of adult neuronal progenitor cells transplanted to the adult brain. Proc Natl Acad Sci USA 1995;92(25):11879–83.

[127] Tomac A, et al. Protection and repair of the nigrostriatal dopaminergic system by GDNF in vivo. Nature 1995;373(6512):335–9.

[128] Lin LF, et al. GDNF: a glial cell line-derived neurotrophic factor for midbrain dopaminergic neurons. Science 1993;260(5111):1130–2.

[129] Airaksinen MS, Saarma M. The GDNF family: signalling, biological functions and therapeutic value. Nat Rev Neurosci 2002;3(5):383–94.

[130] Pochon NA, et al. Neuronal GDNF expression in the adult rat nervous system identified by in situ hybridization. Eur J Neurosci 1997;9(3):463–71.

[131] Trupp M, et al. Complementary and overlapping expression of glial cell line-derived neurotrophic factor (GDNF), c-ret proto-oncogene, and GDNF receptor-alpha indicates multiple mechanisms of trophic actions in the adult rat CNS. J Neurosci 1997;17(10):3554–67.

[132] Saavedra A, Baltazar G, Duarte EP. Driving GDNF expression: the green and the red traffic lights. Prog Neurobiol 2008;86(3):186–215.

[133] Bresjanac M, Antauer G. Reactive astrocytes of the quinolinic acid-lesioned rat striatum express GFRα1 as well as GDNF in vivo. Exp Neurol 2000;164(1):53–9.

[134] Hughes PE, et al. Activity and injury-dependent expression of inducible transcription factors, growth factors and apoptosis-related genes within the central nervous system. Prog Neurobiol 1999;57(4): 421–50.

[135] Treanor JJ, et al. Characterization of a multicomponent receptor for GDNF. Nature 1996;382(6586):80–3.

[136] Jing S, et al. GDNF-induced activation of the ret protein tyrosine kinase is mediated by GDNFR-α a novel receptor for GDNF. Cell 1996;85(7):1113–24.

[137] Takahashi M. The GDNF/RET signaling pathway and human diseases. Cytokine Growth Factor Rev 2001;12(4):361–73.

[138] Durbec P, et al. GDNF signalling through the Ret receptor tyrosine kinase. Nature 1996;381(6585):789–93.

[139] Trupp M, et al. Functional receptor for GDNF encoded by the c-ret proto-oncogene. Nature 1996;381(6585):785–9.

[140] Pozas E, Ibanez CF. GDNF and GFRα1 promote differentiation and tangential migration of cortical GABAergic neurons. Neuron 2005;45(5):701–13.

[141] Sariola H, Saarma M. Novel functions and signalling pathways for GDNF. J Cell Sci 2003;116(Pt 19):3855–62.

[142] Santoro M, et al. Minireview: RET: normal and abnormal functions. Endocrinology 2004;145(12):5448–51.

[143] Ledda F, Paratcha G, Ibanez CF. Target-derived GFRα1 as an attractive guidance signal for developing sensory and sympathetic axons via activation of Cdk5. Neuron 2002;36(3):387–401.

[144] Paratcha G, et al. Released GFRα1 potentiates downstream signaling, neuronal survival, and differentiation via a novel mechanism of recruitment of c-Ret to lipid rafts. Neuron 2001;29(1):171–84.

[145] Worley DS, et al. Developmental regulation of GDNF response and receptor expression in the enteric nervous system. Development 2000;127(20):4383–93.

[146] Paratcha G, Ledda F, Ibanez CF. The neural cell adhesion molecule NCAM is an alternative signaling receptor for GDNF family ligands. Cell 2003;113(7):867–79.

[147] Beggs HE, et al. NCAM140 interacts with the focal adhesion kinase p125(fak) and the SRC-related tyrosine kinase p59(fyn). J Biol Chem 1997;272(13):8310–9.

[148] Paratcha G, Ledda F. GDNF and GFRα a versatile molecular complex for developing neurons. Trends Neurosci 2008;31(8): 384–91.

[149] Iwase T, et al. Glial cell line-derived neurotrophic factor-induced signaling in Schwann cells. J Neurochem 2005;94(6):1488–99.

[150] Ledda F, et al. GDNF and GFRα1 promote formation of neuronal synapses by ligand-induced cell adhesion. Nat Neurosci 2007;10(3):293–300.

[151] Nielsen J, et al. Role of glial cell line-derived neurotrophic factor (GDNF)-neural cell adhesion molecule (NCAM) interactions in induction of neurite outgrowth and identification of a binding site for NCAM in the heel region of GDNF. J Neurosci 2009;29(36):11360–76.

[152] Niethammer P, et al. Cosignaling of NCAM via lipid rafts and the FGF receptor is required for neuritogenesis. J Cell Biol 2002; 157(3):521–32.

[153] Sjostrand D, et al. Disruption of the GDNF binding site in NCAM dissociates ligand binding and homophilic cell adhesion. J Biol Chem 2007;282(17):12734–40.

[154] Arenas E, et al. GDNF prevents degeneration and promotes the phenotype of brain noradrenergic neurons in vivo. Neuron 1995;15(6):1465–73.

[155] Gash DM, et al. Functional recovery in parkinsonian monkeys treated with GDNF. Nature 1996;380(6571):252–5.

[156] Gerlach TH, Zile MH. Effect of retinoic acid and apo-RBP on serum retinol concentration in acute renal failure. FASEB J 1991;5(1): 86–92.

[157] Kordower JH, et al. Neurodegeneration prevented by lentiviral vector delivery of GDNF in primate models of Parkinson's disease. Science 2000;290(5492):767–73.

[158] Pascual A, et al. Absolute requirement of GDNF for adult catecholaminergic neuron survival. Nat Neurosci 2008;11(7):755–61.

[159] Ungerstedt U. 6-Hydroxy-dopamine induced degeneration of central monoamine neurons. Eur J Pharmacol 1968;5(1):107–10.

[160] Young HM, et al. GDNF is a chemoattractant for enteric neural cells. Dev Biol 2001;229(2):503–16.

[161] Gianino S, et al. GDNF availability determines enteric neuron number by controlling precursor proliferation. Development 2003; 130(10):2187–98.

[162] Natarajan D, et al. Requirement of signalling by receptor tyrosine kinase RET for the directed migration of enteric nervous system progenitor cells during mammalian embryogenesis. Development 2002;129(22):5151–60.

[163] Taraviras S, et al. Signalling by the RET receptor tyrosine kinase and its role in the development of the mammalian enteric nervous system. Development 1999;126(12):2785–97.

[164] Heuckeroth RO, et al. Neurturin and GDNF promote proliferation and survival of enteric neuron and glial progenitors in vitro. Dev Biol 1998;200(1):116–29.

[165] Iwashita T, et al. Hirschsprung disease is linked to defects in neural crest stem cell function. Science 2003;301(5635):972–6.

[166] Moore SW. The contribution of associated congenital anomalies in understanding Hirschsprung's disease. Pediatr Surg Int 2006; 22(4):305–15.

[167] Pichel JG, et al. Defects in enteric innervation and kidney development in mice lacking GDNF. Nature 1996;382(6586):73–6.

[168] Sanchez MP, et al. Renal agenesis and the absence of enteric neurons in mice lacking GDNF. Nature 1996;382(6586):70–3.

[169] Uesaka T, Nagashimada M, Enomoto H. GDNF signaling levels control migration and neuronal differentiation of enteric ganglion precursors. J Neurosci 2013;33(41):16372–82.

[170] Sun ZH, et al. GDNF augments survival and differentiation of TH-positive neurons in neural progenitor cells. Cell Biol Int 2004; 28(4):323–5.

[171] Widmer HR, et al. Glial cell line-derived neurotrophic factor stimulates the morphological differentiation of cultured ventral mesencephalic calbindin- and calretinin-expressing neurons. Exp Neurol 2000;164(1):71–81.

[172] Schaller B, et al. Effect of GDNF on differentiation of cultured ventral mesencephalic dopaminergic and non-dopaminergic calretinin-expressing neurons. Brain Res 2005;1036(1–2):163–72.

[173] Douglas-Escobar M, et al. Neurotrophin-induced migration and neuronal differentiation of multipotent astrocytic stem cells in vitro. PLoS One 2012;7(12):e51706.

[174] Cornejo M, et al. Effect of NRG1, GDNF, EGF and NGF in the migration of a Schwann cell precursor line. Neurochem Res 2010; 35(10):1643–51.

[175] Perrinjaquet M, et al. MET signaling in GABAergic neuronal precursors of the medial ganglionic eminence restricts GDNF activity in cells that express GFRα1 and a new transmembrane receptor partner. J Cell Sci 2011;124(Pt 16):2797–805.

[176] Li XJ, et al. Specification of motoneurons from human embryonic stem cells. Nat Biotechnol 2005;23(2):215–21.

[177] Marchetto MC, Winner B, Gage FH. Pluripotent stem cells in neurodegenerative and neurodevelopmental diseases. Hum Mol Genet 2010;19(R1):R71–6.

[178] Li XJ, et al. Directed differentiation of ventral spinal progenitors and motor neurons from human embryonic stem cells by small molecules. Stem Cells 2008;26(4):886–93.

[179] Rakowicz WP, et al. Glial cell line-derived neurotrophic factor promotes the survival of early postnatal spinal motor neurons in the lateral and medial motor columns in slice culture. J Neurosci 2002;22(10):3953–62.

[180] Wichterle H, et al. Directed differentiation of embryonic stem cells into motor neurons. Cell 2002;110(3):385–97.

[181] Zurn AD, et al. Combined effects of GDNF, BDNF, and CNTF on motoneuron differentiation in vitro. J Neurosci Res 1996;44(2):133–41.

[182] Garcia-Bennett AE, et al. Delivery of differentiation factors by mesoporous silica particles assists advanced differentiation of transplanted murine embryonic stem cells. Stem Cells Transl Med 2013;2(11):906–15.

[183] Boku S, et al. Tricyclic antidepressant amitriptyline indirectly increases the proliferation of adult dentate gyrus-derived neural precursors: an involvement of astrocytes. PLoS One 2013;8(11):e79371.

[184] Rinderknecht E, Humbel RE. Polypeptides with nonsuppressible insulin-like and cell-growth promoting activities in human serum: isolation, chemical characterization, and some biological properties of forms I and II. Proc Natl Acad Sci USA 1976;73(7):2365–9.

[185] Werner H, Leroith D. Insulin and insulin-like growth factor receptors in the brain: physiological and pathological aspects. Eur Neuropsychopharmacol 2014;1036(1–2):163–72.

[186] Yarden Y, Ullrich A. Growth factor receptor tyrosine kinases. Annu Rev Biochem 1988;57:443–78.

[187] O'Kusky J, Ye P. Neurodevelopmental effects of insulin-like growth factor signaling. Front Neuroendocrinol 2012;33(3):230–51.

[188] Lehtinen MK, et al. The cerebrospinal fluid provides a proliferative niche for neural progenitor cells. Neuron 2011;69(5):893–905.

[189] Annenkov A. The insulin-like growth factor (IGF) receptor type 1 (IGF1R) as an essential component of the signalling network regulating neurogenesis. Mol Neurobiol 2009;40(3):195–215.

[190] Aberg MA, et al. Peripheral infusion of IGF-I selectively induces neurogenesis in the adult rat hippocampus. J Neurosci 2000;20(8):2896–903.

[191] Hsieh J, et al. IGF-I instructs multipotent adult neural progenitor cells to become oligodendrocytes. J Cell Biol 2004;164(1):111–22.

[192] Yamada M, et al. Differences in survival-promoting effects and intracellular signaling properties of BDNF and IGF-1 in cultured cerebral cortical neurons. J Neurochem 2001;78(5):940–51.

[193] Tropepe V, et al. Distinct neural stem cells proliferate in response to EGF and FGF in the developing mouse telencephalon. Dev Biol 1999;208(1):166–88.

[194] Eiraku M, et al. Self-organized formation of polarized cortical tissues from ESCs and its active manipulation by extrinsic signals. Cell Stem Cell 2008;3(5):519–32.

[195] Correia AS, et al. Fibroblast growth factor-20 increases the yield of midbrain dopaminergic neurons derived from human embryonic stem cells. Front Neuroanat 2007;1:4.

[196] Beck KD, et al. Mesencephalic dopaminergic neurons protected by GDNF from axotomy-induced degeneration in the adult brain. Nature 1995;373(6512):339–41.

[197] Hou JG, Lin LF, Mytilineou C. Glial cell line-derived neurotrophic factor exerts neurotrophic effects on dopaminergic neurons in vitro and promotes their survival and regrowth after damage by 1-methyl-4-phenylpyridinium. J Neurochem 1996;66(1):74–82.

[198] Zhang L, Jiang H, Hu Z. Concentration-dependent effect of nerve growth factor on cell fate determination of neural progenitors. Stem Cells Dev 2011;20(10):1723–31.

[199] Ghosh A, Greenberg ME. Calcium signaling in neurons: molecular mechanisms and cellular consequences. Science 1995;268(5208):239–47.

Chapter 8

Glycolipid Antigens in Neural Stem Cells

Yutaka Itokazu[1,2] and Robert K. Yu[1,2]

[1]*Department of Neuroscience and Regenerative Medicine, Medical College of Georgia, Georgia Regents University, Augusta, GA, USA;* [2]*Charlie Norwood VA Medical Center, Augusta, GA, USA*

ABBREVIATIONS

CD Cluster of differentiation
Cer Ceramide
CNS Central nervous system
CST Cerebroside sulfotransferase
EGF Epidermal growth factor
ERK Extracellular signal-regulated protein kinase
FGF Fibroblast growth factor
FUT Fucosyltransferase
GalCer Galactosylceramide
GalNAcT *N*-acetylgalactosaminyltransferase
GalT Galactosyltransferase
Ganglioside The nomenclature for gangliosides and their components is based on that of Svennerholm and the IUPAC–IUBMB Joint Commission on Biochemical Nomenclature [1,2]
GlcCer Glucosylceramide
GlcT Glucosyltransferase
GRP Glial restricted precursor
GSL Glycosphingolipid
LacCer Lactosylceramide
mAb Monoclonal antibody
MAPK Mitogen-activated protein kinase
NEC Neuroepithelial cell
NG2 Nerve/glial antigen 2
NRP Neuronal restricted progenitor
NSC Neural stem cell
OPC Oligodendrocyte precursor cell
PSA–NCAM Polysialic acid–neural cell adhesion molecule
RGC Radial glial cell
SGZ Subgranular zone
SSEA Stage-specific embryonic antigen
ST Sialyltransferase
SVZ Subventricular zone
VZ Ventricular zone

8.1 INTRODUCTION

8.1.1 Glycosphingolipids and Development

During neural development, dramatic and consistent changes in the composition of glycoconjugates, including glycolipids, glycoproteins, and proteoglycans, occur [3–5].

Glycosphingolipids (GSLs) contain one or more monosaccharide residues bound by a glycosidic linkage to a hydrophobic moiety, ceramide. It is known that changes in the expression of glycolipids, including gangliosides that contain one or more sialic acid residues, in the nervous system correlate with neurodevelopmental events [6,7]. For example, in fertilized eggs, the *globo*-series of glycolipids (e.g., stage-specific embryonic antigen, SSEA-3/SSEA-4) is robustly expressed. As cell division proceeds, the *lacto*-series GSLs (e.g., lactoneotetraosylceramide) are expressed at embryonic day (E) 1.5, followed by the *ganglio*-series GSLs in the developing brain (from E7). The lipid portion of GSLs, including gangliosides, is ceramide, which is synthesized primarily in the endoplasmic reticulum (ER) from a sphingosine base and a fatty acid residue. Ceramide is transferred to the Golgi apparatus and glucosylated to glucosylceramide (GlcCer) and catalyzed by glucosyltransferase (GlcT) on the cytosolic leaflet of the *cis*-Golgi. The expression of GlcCer is abundant in early embryonic brain and decreases from E16 on [3]. The synthesized GlcCer is transported to the *trans*-Golgi by the phosphoinositol 4-phosphate adaptor protein 2 (FAPP2) and translocated to the luminal side of the Golgi vesicle [8,9]. In the other direction, FAPP2 also mediates GlcCer transportation from the *cis*-Golgi back to the ER [10]. Then, GlcCer is flip-flopped into the lumen of the ER, and GlcCer ultimately reaches the lumen of the Golgi by vesicular transport [11]. At the luminal side of the Golgi, GlcCer is converted to lactosylceramide (LacCer). The sequential action catalyzed by sialyltransferases and galactosyltransferase can elongate the carbohydrate moieties to gangliosides. Another pathway from ceramide is catalyzed by galactosyltransferase III (GalT-III) in the ER to form galactosylceramide (GalCer), sulfatide, and GM4 [12]. Each step is catalyzed by a unique, specifically controlled glycosyltransferase (Figure 8.1).

8.1.2 Gangliosides

Gangliosides are sialic acid-containing GSLs expressed primarily, but not exclusively, on the outer leaflet of the

Neural Surface Antigens. http://dx.doi.org/10.1016/B978-0-12-800781-5.00008-6

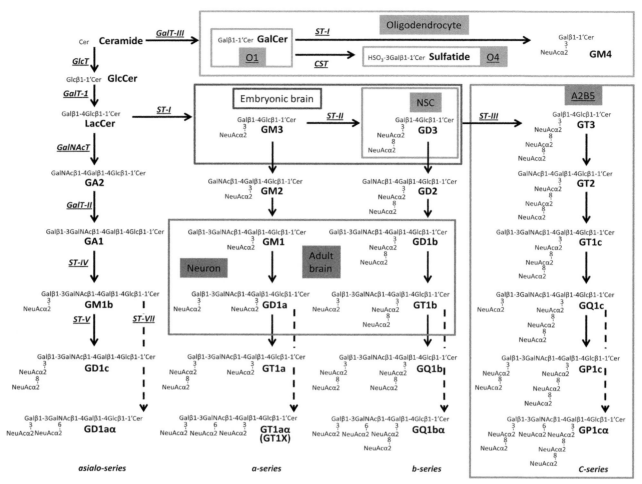

FIGURE 8.1 Structures and biosynthetic pathways of glycosphingolipids (GSLs). Cer, ceramide; CST, cerebroside sulfotransferase (*Gal3st1*, sulfatide synthase); GalNAcT, *N*-acetylgalactosaminyltransferase I (*B4galnt1*, GA2/GM2/GD2/GT2 synthase); GalT-I, galactosyltransferase I (*B4galt6*, lactosylceramide synthase); GalT-II, galactosyltransferase II (*B3galt4*, GA1/GM1/GD1b/GT1c synthase); GalT-III, galactosyltransferase III (*Ugt8a*, galactosylceramide synthase); GlcT, glucosyltransferase (Ugcg, glucosylceramide synthase); ST-I, sialyltransferase I (*St3gal5*, GM3 synthase); ST-II, sialyltransferase II (*St8Sia1*, GD3 synthase); ST-III, sialyltransferase III (*St8Sia3*, GT3 synthase); ST-IV, sialyltransferase IV (*St3gal2*, GM1b/GD1a/GT1b/GQ1c synthase); ST-V, sialyltransferase V (*St8sia5*, GD1c/GT1a/GQ1b/GP1c synthase); ST-VII, sialyltransferase VII (*St6galnac6*, GD1aα/GT1aα/GQ1bα/GP1cα synthase). Official symbols of genes are represented in *italics*. GM3 and GD3 are abundant in embryonic brain (blue) and NSCs express GD3 (light blue). The c-series gangliosides are A2B5 antigens (green), and astrocytes express GM3 (green). GM1, GD1a, GD1b, and GT1b are the most abundant ganglioside species in adult mammalian brain (red). Oligodendrocyte markers O1 and O4 are GalCer and sulfatide, respectively (orange). (For interpretation of the references to color in this figure legend, the reader is referred to the online version of this book.)

plasma membrane of cells in all vertebrates. In early embryonic rodent brains, the pattern of ganglioside expression is characterized by the expression of a large amount of simple gangliosides, such as GM3 and GD3. In later developmental stages, more complex gangliosides prevail, particularly GM1, GD1a, GD1b, and GT1b, which account for more than 80% of the total gangliosides [3,5]. Thus, the expression of neural gangliosides changes dramatically during cellular differentiation and brain development. Correlations between ganglioside expression in the nervous system and neurodevelopmental events are summarized schematically in Figure 8.2. This unique expression pattern of specific gangliosides can be used for specific cell lineage markers and may reflect the functional

roles they play at specific developmental stages. Abundant evidence supports the notion that GSLs, including gangliosides, serve regulatory roles in cellular events, including proliferation and neural differentiation, as exemplified by neuritogenesis, axonogenesis, and synaptogenesis [3,7,13–17]. The functional importance of gangliosides has been evaluated using specific enzyme gene-knockout (KO) mice. With the advent of contemporary molecular genetics and biology, several lines of genetically modified mice have been established in which the expression of gangliosides and other GSLs has been altered or depleted, and this has greatly facilitated elucidating their biological functions. For example, GM2/GD2 synthase (GalNAcT) is one of the key enzymes needed for the synthesis

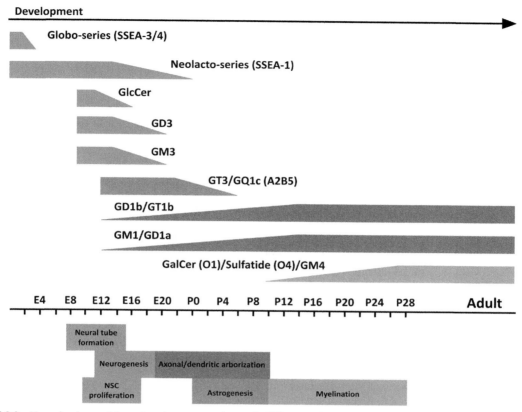

FIGURE 8.2   Neurodevelopmental events and concurrent changes in GSL expression. "E" denotes embryonic day, and "P," postnatal day.

of the major "brain-type" gangliosides, including GM1, GD1a, GD1b, and GT1b. Mice lacking this enzyme do not express GalNAc-containing gangliosides. As a result they are developmentally abnormal and appear to have neurological problems such as axonal degeneration; sensory, motor, and behavioral deficits; and other neurological dysfunctions [18–23]. During brain development, gangliosides are assumed to modulate ceramide (Cer)-induced apoptosis and to maintain cellular survival and differentiation [13]. GM3 synthase (sialyltransferase I, ST-I) is a critical enzyme for the synthesis of all complex gangliosides. Mutation of GM3 synthase is associated with human autosomal recessive infantile-onset symptomatic epilepsy syndrome [24], and an alteration in the GalNAcT gene was reported in patients with hereditary spastic paraplegias [25]. These studies clearly demonstrated that deletion of complex gangliosides can be associated with human diseases. A lack of b- and c-series gangliosides results in clear and subtle developmental and behavioral deficits in mice, which exhibited sudden death from audiogenic seizures [26]. Both GalNAcT- and ST-I-deficient mice, which lack all gangliosides, die soon after weaning at 3 weeks of age [27]. Taken together, these observations clearly indicate that GSLs, including gangliosides, have important biological functions in the developing nervous system. In

this chapter, we introduce these GSLs expressed during neural development (Figure 8.3).

8.2 STAGE-SPECIFIC EMBRYONIC ANTIGEN-1 (CD15)

8.2.1 SSEA-1 in Early Embryogenesis

Although the SSEA-1 carbohydrate antigen is carried not only by glycolipids but also by glycoproteins, its functional aspect is important to discuss in this chapter. After fertilization, the fertilized egg undergoes cleavage to two-, four-, and eight-cell stages. From the eight- to the 32-cell stage, the spherical cells undergo changes in morphology to a cubic shape. The cells bind tightly to each other, forming compact spheres, and this stage is called the compaction stage [28]. At this stage, cell surface glycoconjugate markers, SSEA-1/Lewis X/CD15 (Galβ1-4(Fucα1-3)GlcNAcβ-) and others start to emerge. SSEA-1 was established as the antigen of a monoclonal antibody (mAb) generated by immunization of mice with F9 embryonic carcinoma cells [29]. Because incomplete antigens (haptens) of Lewis X have been reported to inhibit the cell compaction process in mouse embryos [29,30], it is believed that SSEA-1 may play an important role in early embryogenesis.

FIGURE 8.3 A model for neural cell lineages derived from mouse neural stem cells (NSCs). The known glycoconjugate markers are underlined. NRP, neuronal restricted precursor; GRP, glial restricted precursor; OPC, oligodendrocyte precursor cell.

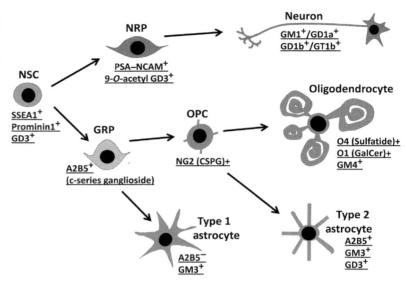

Other stage-specific antigens, such as SSEA-3 and SSEA-4, are also expressed at the early stages of mouse embryogenesis. SSEA-1 is carried not only by GSLs, but also glycoproteins. However, SSEA-3 and SSEA-4 are mostly carried by GSLs. The expression of SSEA-3 usually peaks at the four- to eight-cell stages, whereas SSEA-4 expression peaks at the morula and early blastocyst stages with some overlap with that of SSEA-3 [31]. It is well known that SSEA-1 is expressed in mouse pluripotent stem cells, such as embryonic stem (ES) and induced pluripotent stem cells. The expression patterns of these stage-specific antigens are different in human and mouse. SSEA-1 is not expressed in human ES cells. Instead, human ES cells express SSEA-3, SSEA-4, and keratin sulfate antigens (TRA)-1-60, TRA-181, GCTM2, and GCT343 [32,33]. Analysis of mice deficient in SSEA-1 (fucosyltransferase 9 (FUT9)-deficient mice) revealed increased anxiety-like behavior, but no distinguishable morphological phenotypes in brain development [34,35]. Although mice deficient in SSEA-3 and SSEA-4 expression (α1,4-galactosyltransferase-deficient mice) were resistant to Shiga-like toxins (also known as verotoxin), they showed no apparent abnormality in development [36]. These studies suggest that the functions of SSEAs may be compensated for by other carbohydrate molecules or are not essential for neural development.

8.2.2 SSEA-1 in Neural Stem Cells

SSEA-1 is expressed on neuroepithelial cells (NECs) at early stages of neural development and in the subventricular zone (SVZ) and subgranular zone in adult brain [37–39]. This suggests a functional role for SSEA-1 in sustaining stem and progenitor cell growth. Glycolipids as well as glycoproteins are known to carry the SSEA-1 epitope in neural stem cells (NSCs) [40]. SSEA-1 can bind and regulate

fibroblast growth factor 2 (FGF-2), which is known as a mitogen that maintains the stemness of NSCs [41,42]. In addition, the SSEA-1 epitope is also associated with chondroitin sulfate proteoglycan [43], β1-integrin, glycolipids [40], lysosome-associated membrane protein 1 [44], the extracellular matrix protein tenascin-C [45], phosphacan [45,46], and Wnt1 [37]. Strong SSEA-1 expression can be observed during embryonic development of NSCs in neurogenic regions, such as the hippocampal primordium and the embryonic cerebral cortex; its expression remains clearly visible until E19 [38,39]. SSEA-1$^+$ cells typically have bipolar morphology, radial orientation, and glial processes, and they resemble a subtype of radial glial cells (RGCs) (E12–E14) [39,47].

8.2.3 SSEA-1 and Integrin

NSCs can be isolated as neurospheres in serum-free floating culture with epidermal growth factor (EGF) and/or FGF-2. Treatment of NSCs with an anti-SSEA-1 antibody, AK97, drastically repressed the migration of the cells on a fibronectin-coated dish [40]. Because the fibronectin receptors on the cell surface are integrin heterodimers, this suggests that adhesion of NSCs to the fibronectin substratum is dependent on integrin molecules. It has been shown that the peptide Arg-Gly-Asp (RGD) is capable of inhibiting the binding activity of α5β1 and αvβ1-integrins (CD49e/CD29, CD51/CD29) among the fibronectin receptors [48]. The RGD peptide suppressed migration of NSCs in a dose-dependent manner, which strongly suggests that NSC migration is dependent on the interaction between fibronectin and integrin. Immunoprecipitation experiments revealed that β1-integrin, a common component of fibronectin receptors, can be considered as one of the SSEA-1-carrying proteins. These results indicate that the SSEA-1 epitope is involved in the regulation of NSC migration via the carbohydrate chains of the β1-integrin molecule [40,49].

8.2.4 SSEA-1 and Notch Signaling

FUT9 is a key enzyme for the synthesis of the SSEA-1 carbohydrate epitope. Knockdown of FUT9 expression by short interfering RNA in NSCs reduced the number of and the ability of the cells to form neurospheres. Meanwhile, no significant difference could be detected by TUNEL assay (an indicator of programmed cell death accompanied by DNA fragmentation) nor activation of caspase-3 (a critical executioner of apoptosis or programmed cell death signaling) in FUT9-knockdown NSCs. Hence, these knockdown analyses revealed that the expression of SSEA-1 promotes the proliferation of NSCs without affecting the cell death pathways [50]. FUT9-knockdown NSCs had significantly lower levels of Musashi-1 and slightly higher levels of cell differentiation makers. Musashi-1 plays a crucial role in maintaining the undifferentiated state of NSCs via activation of the Notch signaling pathway [51,52]. SSEA-1 may regulate proliferation of NSCs via modulation of the expression of Musashi-1. Thus, the SSEA-1 carbohydrate epitope not only serves as an important marker of stem cells, but also plays important roles in cell proliferation, differentiation, migration, and adhesion in NSCs [50].

8.3 SSEA-4

The *globo*-series GSLs, including SSEA-3 (Galβ1-3GalNAcβ1-3Galα1-4Galβ1-4Glcβ1-1′Cer) and SSEA-4 (NeuAcα2-3Galβ1-3GalNAcβ1-3Galα1-4Galβ1-4Glcβ1-1′Cer), are expressed during embryogenesis [31]. SSEA-3 and SSEA-4, but not SSEA-1, are expressed human ES cells and human teratocarcinoma cells [53–55]. It has been reported that SSEA-4 is expressed on human NSCs from spinal cord and forebrain [56,57]. SSEA-4 is expressed in the proliferative areas (ventricular zone (VZ) and SVZ) and choroid plexus of human embryonic brain. At human developmental stage 7–9 weeks, some population of the dissociated forebrain cells expresses SSEA-4. Prominin-1 (CD133), a unique NSC marker, is expressed on 22.2% of the cells in dissociated forebrain at 7 weeks of development. At the same age, 8.2 and 4.9% of these cells express SSEA-4 and SSEA-1, respectively. According to SSEA expression, there are at least two major populations of Prominin-1⁺ cells in human NSCs. Usually, in Prominin⁺ cells, SSEA-4⁺ cells do not express SSEA-1. In the same way, the majority of SSEA-1⁺ cells do not express SSEA-4. Prominin-1⁺/SSEA-4⁺ cells and Prominin-1⁺/SSEA-1⁺ cells are able to produce 8 and 3.5 times more neurospheres, respectively, than Prominin-1⁺/SSEA-4⁻/SSEA-1⁻ cells. Prominin-1⁺/SSEA-4⁺ cells are confirmed to differentiate into neurons and glia in vitro [56]. These studies suggest that human

NSCs in embryo forebrain and spinal cord include SSEA-4⁺ cells. The functional role of SSEA-4 in NSCs, however, remains to be investigated.

8.4 GD3 GANGLIOSIDE

8.4.1 GD3 as a Specific Marker for NSCs

The b-series ganglioside GD3 (CD60a) is expressed in the neural tube during early development and can be detected using a GD3-specific mAb, R24 [58]. Upon closer examination, GD3 was found to be robustly expressed on NECs in the neural tube and in RGCs of the VZ in embryos and the SVZ of postnatal and adult rodents [59–63]. GD3⁺ cells also express SSEA-1 in the SVZ of mouse brain [63]. GD3 is the predominant ganglioside species in cultured mouse neurospheres, accounting for more than 80% of the total gangliosides. In neurospheres, in addition to GD3, another b-series ganglioside, GD2, has also been shown to be highly expressed in human NSCs [64]. NSCs are costained with GD3 and SSEA-1, and GD3-positive cells also express other NSC markers, such as Sox2, nestin, and Musashi-1. After cell differentiation, GD3 expression is decreased in cultured NSCs. For this reason, GD3 has been proposed to be a useful biomarker for mouse NSCs. Most living cells (70–80%) were positive for GD3 in E14 striata and in postnatal day (P) 2 and P10 SVZs. In contrast, fewer cells (30%) were positive for GD3 in P30 and adult SVZs. A similar strong expression of GD3 is observed in NSCs isolated from embryonic, postnatal, and adult brains. Most GD3⁺ cells isolated by FACS were also positive for nestin and SSEA-1, but negative for differentiated cell markers. GD3⁺ cells can generate neurospheres more efficiently compared to GD3⁻ cells. The GD3⁺ cells are capable of differentiating into neuronal and glial cells. Those results clearly suggest that GD3 is a useful cell-surface marker for isolating living NSCs, especially from adult brain tissues, in which NSCs are less abundant [63].

8.4.2 GD3-Synthase-KO NSCs

To clarify whether GD3 modulates the self-renewal capacity or cell fate determination of NSCs, NSCs from GD3 synthase (GD3S)-KO mice were analyzed. Surprisingly, no significant difference in the proliferation rate or expression of lineage-associated markers was found between GD3S⁺/⁺ and GD3S-KO NSCs that were cultured in the presence of FGF-2 but in the absence of EGF [65]. GD3S-KO NSCs that were cultured with EGF, however, showed suppressed proliferation [66]. The expression of nestin and EGF receptor (EGFR) was also strongly downregulated and the mitogen-activated protein kinase (MAPK)/extracellular signal-regulated protein kinase (ERK) pathway signaling was impaired in GD3S-KO NSCs. Additionally, EGFR degradation and the reduction of p-EGFR and p-ERK1/2 were correlated in the GD3S-KO NSCs after EGF stimulation. Subsequently a decrease in the MAPK/ERK proliferation pathway was found in GD3S-KO NSCs. Furthermore, more membrane EGFRs were expressed on the cell surface of GD3S$^{+/+}$ NSCs (53%) than in GD3S-KO NSCs (22%). Likewise, EGFR and GD3 were found to be colocalized in NSCs and they interacted in the lipid raft region of the cell surface as well as in intracellular vesicles. Interestingly, EGFR was found to exist on non-lipid-raft fractions in the GD3S-KO NSCs. Those findings provide convincing evidence that EGFR and GD3 are colocalized in lipid raft microdomains and support the notion that GD3 is essential for the maintenance of the self-renewal capability in NSCs by recruiting EGFR to the microdomains to sustain the EGF-induced downstream signaling [66].

8.4.3 GD3 and Endocytosis

Endocytosis is a basic cellular process that is used by cells to internalize a variety of molecules. Cells are estimated to internalize, via endocytosis, about half their plasma membrane per hour [67]. GSLs are recognized to undergo endocytosis. Once internalized, GSLs can be (1) recycled to the plasma membrane; (2) sorted to the Golgi complex, in which they can be reglycosylated; or (3) degraded in the lysosome. It was found that the number of NSCs with internalized EGF was significantly reduced in the absence of GD3. In the GD3S-KO NSCs, EGFR was increased in lysosomes and decreased in recycling endosomes and in early endosomes. A large amount of EGFR in GD3S-KO NSCs underwent the endosomal–lysosomal degradative pathway, whereas a greater portion of EGFR was subject to the recycling pathway in GD3S$^{+/+}$ NSCs than in GD3S$^{-/-}$ NSCs. Hence, the interaction of GD3 and EGFR in NSCs is responsible for sustaining EGFR surface expression and downstream signaling to maintain the self-renewal capability of the cells [66].

8.5 9-O-ACETYL GD3

Ganglioside 9-O-acetyl GD3 (CD60b) was detected in neuroblasts during neural development using the JONES antibody [68,69]. 9-O-acetyl GD3 is expressed in the SVZ and along the rostral migration stream in both embryonic and adult brains [70]. Most of the migrating neuroblasts expressing 9-O-acetyl GD3 also express polysialic acid–neural cell adhesion molecule (Polysialic acid–neural cell adhesion molecule) [71]. A more recent study casts some doubt, however, on the importance of 9-O-acetyl GD3 in these studies. GD3S-KO mice, in which GD3 and its downstream products, including 9-O-acetyl GD3, are missing, appear grossly normal in development [72]. Addition of the mAb JONES to the culture medium blocked the neuronal migration in

cerebellum explant culture of not only wild-type mice but also GD3S-KO mice. Thin-layer chromatography-immunostaining analysis detected no glycolipids with the JONES antibody in the lipid extract of GD3S-KO mice. However, Western blot analysis revealed at least three JONES-positive protein bands, one of which was β1-integrin. This raises the intriguing question whether the 9-O-acetyl sialic acid residue is conjugated with a protein and functions in a manner similar to that of 9-O-acetyl GD3.

8.6 GM1 GANGLIOSIDE

During neuronal differentiation, the concentration of GD3, which is the predominant ganglioside in NSCs, is rapidly decreased. Concomitantly, the levels of GM1, GD1a, GD1b, and GT1b continuously increase in young animals, reaching a plateau during adulthood [3,73,74]. This pattern change follows closely the upregulation of N-acetylgalactosaminyltransferase (GalNAcT, GM2 synthase) expression [3]. The dramatic changes in the expression profile of gangliosides during neuronal cell differentiation clearly reflect the biological needs at particular stages during brain development (Figure 8.2). During neuronal development, GM1-expressing cells are considered neuronal progenitor cells and neurons [75–77].

8.6.1 GM1 and Its Epigenetic Regulation

Our laboratory demonstrated for the first time that histone acetylation in chromatin of the 5′ region of the mouse GalNAcT gene can regulate mRNA expression, and this epigenetic modification is highly correlative to the stage-specific alterations in GalNAcT mRNA levels in mouse brain during development [74,78]. It is known that GM1 enhances neurite outgrowth in primary neuronal cultures. It has been shown that GM1 is also present in nuclear membranes and that upregulation of GM1 in the nuclear membrane accompanies the process of neurite outgrowth. This observation prompted Ledeen et al. to propose that elevated GM1 has a modulatory effect on Ca^{2+} homeostasis in the nucleus, which is mediated by a tight association of GM1 with the Na^{+}/Ca^{2+} exchanger [79]. Interestingly, NSCs cultured with a supplement of GM1 for 7 days also exhibited a significantly enhanced neurogenic effect [78]. It is possible that in this enhanced neurogenic process, exogenous GM1 induced NSCs to transcribe more GalNAcT mRNA with a higher level of acetylated histone H4 (AcH4) on the GalNacT promoter region, to which more transcription factors are recruited. On the other hand, the ST-II gene did not show significant changes in mRNA expression or AcH4 binding in response to the added GM1. This result might represent a potential mechanism accounting for the correlations between the ganglioside pattern shift and the epigenetic modifications of ganglioside synthase expression over neuronal differentiation and neural development. Therefore, GM1 might generate a positive feedback loop for NSCs to promote neuronal differentiation and to produce more GM1 and "brain-type" gangliosides, such as GD1a, GD1b, and GT1b.

8.6.2 Integrin–Focal Adhesion Kinase Signaling

Another critical role of GM1 in neuronal differentiation is modulating the integrin–focal adhesion kinase (FAK) signaling pathway. GM1 is suggested to modulate the integrin–FAK signaling pathway at the leading edge of migrating cells, and disruption of the microdomain structures failed to activate downstream signals [80,81]. Neurite outgrowth is a key process in neuronal differentiation, and GM1 has been proposed to regulate integrin signaling on neurite outgrowth. GM1, β1-integrin, and FAK are colocalized in the membrane microdomains, and the GM1–integrin complex modulates FAK activities and their downstream signaling during neurite outgrowth [82]. It is known that laminin-1, an extracellular matrix molecule, also binds integrins and promotes neurite outgrowth [83]. This laminin-1-mediated neurite outgrowth requires nerve growth factor-dependent TrkA signaling [84]. TrkA and β1-integrin have been found to accumulate in GM1-enriched membrane microdomains with laminin-1, and the accumulated focal clustering on GM1-enriched domains could activate signaling pathways to promote neurite outgrowth [85]. These studies strongly suggest that GM1 ganglioside and membrane microdomains are essential for activating the integrin signaling pathways and neurite outgrowth.

8.7 THE C-SERIES GANGLIOSIDES

The first GSL antigen expressed in cells of the glial lineage is the A2B5 antigen (Figure 8.1). A2B5 is a mAb originally developed using chicken embryonic retina cells as the immunogen [86]. The antigens recognized by mAb A2B5 have been established as the c-series gangliosides, including GQ1c, GT1c, and GT3 [87,88], and the antibody recognizes the Neu5Acα2-8-Neu5Acα2-8Neu5Acα-or α2,8-trisialosyl structure [89]. These c-series gangliosides are abundant in fish brains and in mammalian embryonic, but not adult, brains [90–94]. During development, the expression of c-series gangliosides is diminished in favor of the a- and b-series gangliosides, and the rate-limiting enzyme appears to be ST-III [91]. Glial restricted precursors (GRPs) have been recognized by the expression of the A2B5 epitope [95]. It is uncertain, however, whether GRPs exist in vivo. A2B5 antigens in these progenitors have been identified as GT3 and *O*-acetyl GT3 [96]. In the developing mouse brain, it has been reported that mAb A2B5 also reacts with four glycoproteins in addition to c-series gangliosides [89].

8.8 GALCER AND SULFATIDE

As oligodendrocyte development proceeds, unique GSLs appear on the oligodendrocyte plasma membrane and myelin. The two most common GSLs are the O4 (sulfatide; HSO_3-3Galβ1-1′Cer) and O1 antigens (GalCer; Galβ1-1′Cer), which also have been utilized as specific markers to define immature and mature myelinating oligodendrocytes, respectively [97]. The O1 and O4 antigens play important roles as modulators of oligodendrocyte development, and they also function as major components of the myelin sheath to facilitate nerve conduction. A series of studies has clearly shown that KO mice deficient in GalT-III or sulfatide synthase (cerebroside sulfotransferase, CST) present severe neurological deficits, such as tremor, progressive ataxia, and reduction in nerve conduction velocity [98–100]. In these KO mice, morphologically normal-appearing compact myelin is preserved, but the paranodal loops are absent from the axon and paranodal junctions are abnormal [100]. The number of oligodendrocytes is increased in sulfatide-KO mice, indicating that sulfatide is a crucial molecule for the negative regulation of terminal differentiation of oligodendrocytes [101]. GalCer expression factor-1, a rat homolog of hepatocyte growth factor-regulated tyrosine kinase substrate, has been cloned as an inducer of O1 antigen expression [102]. Overexpression of this molecule causes suppression of cell proliferation, causing the cells to undergo dramatic morphological changes to become fibroblast-like in appearance [103]. Although GalCer expression factor-1 may regulate the expression of O1 and O4 antigens during glial development, its function in NSCs and glial precursor cells remains to be investigated.

8.9 SIALOSYL GALACTOSYLCERAMIDE

The *gala*-series glycolipids, including GalCer (O1), sulfatide (O4), and GM4 (NeuAcα2-3Galβ1-1′Cer), are enriched in myelin of the CNS (central nervous system), but not of the peripheral nervous system. The occurrence of GM4 is also species-dependent, being most abundant in myelin of humans (20.3% of total myelin ganglioside) and avian species such as chicken (31.8%), but less so in rodents (5%) and other vertebrates [104,105]. In addition, GM4 is reported to be expressed in astrocytes of chicken cerebellum [106]. The enzyme activity of GM4 synthase has been demonstrated in mouse brain [107]. Recently, ST-I has been revealed as a common enzyme generating GM3 and GM4 from LacCer and GalCer, respectively [108–110].

With respect to the divergence of *gala*- and *lacto*-series GSLs, it is interesting to observe that the biosynthesis of monoglycosylceramide expression constitutes an important point of control of ganglioside biosynthesis during CNS development. It has been reported that the expression of GlcCer and that of GalCer are complementary to each other

[111]. GlcCer is the only monoglycoceramide in embryonic brain, and the enzyme responsible for its synthesis, GlcT, is continuously expressed during the whole life span of animals. On the other hand, GalCer appears postnatally, and the enzyme responsible for its biosynthesis, GalT-III, is drastically increased after birth, particularly during the period of active myelination [3]. This result suggests that the expression of GalCer after birth is the result of the increased GalT-III activity. It was reported that an increase in the *gala*-series glycolipids, including GM4, was found in mouse whole-embryo culture treated with a GlcCer synthesis inhibitor, *N*-butyldeoxygalactonojirimycin [112]. Therefore, the dominant enzyme activity of GalT-III, i.e., the robust expression of GalCer, is a key step to produce GM4. The physiological roles of GM4 in CNS development need to be studied.

8.10 PHOSPHATIDYLGLUCOSIDE

A phosphoglycerolipid, phosphatidylglucoside (PtdGlc), is expressed in astrocytes and radial glia in rat embryonic brain [113]. PtdGlc is localized in the plasma membrane microdomains [114]. A PtdGlc mAb, DIM21, has been developed that labels RGCs at E12.5–E14.5, astrocytes in late embryonic to early postnatal stages, and RGC-like cells in the adult SVZ [115–117]. In an in vitro study, the association of EGFRs and PtdGlc-enriched membrane microdomains was confirmed in NSCs and were found to control NSC-to-astrocyte differentiation through EGF signaling [115]. PtdGlc-positive cells can generate neurospheres and produce neurons and glia in vitro [118].

8.11 CONCLUSIONS AND PROSPECTIVE STUDIES

NSCs are characterized by their capacity for self-renewal and their ability to differentiate into neurons and glia. Remarkably, they can be isolated not only from embryonic brains, but from adult CNS tissues as well [119]. The ability to manipulate NSC fate determination in vitro has greatly facilitated an understanding of the properties and regulatory mechanisms of NSCs in the developing nervous system and adult brain that would have been difficult to decipher in vivo. Glycolipids are predominantly expressed on the cell surface. Because of their structural diversity, they have been used effectively as cell surface biomarkers for identification and isolation of specific cell types. During neural development and neuronal/glial cell differentiation, glycolipids frequently undergo dramatic qualitative and quantitative changes that correlate with cellular changes and events. There is an urgent need to answer the question of whether these changes are merely consequences of differentiation or, in fact, represent essential mediators of critical biological processes such as cell–cell recognition, adhesion,

migration, and cell proliferation. Recent evidence has shed light on their roles in modulating signaling pathways during self-renewal, cellular differentiation, and reprogramming. For example, we found that cell surface SSEA-1 modulates the stemness, fate determination, and migration of those cells [40,50]. The emerging evidence points to a functional role of glycoconjugate antigens in regulating growth factor activities. Thus, GM3 can modulate EGFR function by inhibiting its internalization and subsequent tyrosine kinase activity [120]. Most recently, we showed that GD3 is tightly associated with EGFR to modulate NSC proliferation [66]. Additionally, it is well known that the epithelial–mesenchymal transition (EMT) participates in and is a critical step in the developmental process. GSLs play an important role in EMT; changes in cell surface glycolipid expression by inhibition of GlcCer synthesis convert epithelial cells to a fibroblastic morphology [121]. This study indicates that cell surface glycoconjugates may control cell fate to effect *trans*-differentiation of one cell type into another. Clearly this represents a fruitful area of future research.

The other critical area for future investigation is the basis of induced pluripotent stem cell (iPSC) generation. Cell surface glycoconjugates again occupy an important area for study. For example, in human fibroblasts, less than 1% of the cells express SSEA-3 (3Galβ1-3GalNAcβ1-3Galα1-4Galβ1-4Glcβ1-1'Cer) and SSEA-3+ dermal fibroblast-enhanced iPSC generation, whereas no iPSC's could be generated from the SSEA-3− cell population [122,123]. SSEA-3+ fibroblast and bone marrow stromal cells host a multipotent stem cell population that can generate the three germ layers without Yamanaka factors, such as Oct3/4, Sox2, c-Myc, and Klf4 [124]. This clearly shows that SSEA-3 plays a crucial role during reprogramming of fibroblasts to stem cells to maintain cell stemness. Although SSEA-1, SSEA-3, SSEA-4, and GD3 are all expressed in stem cells, mice deficient in one of these molecules show only subtle phenotypic abnormalities compared with wild-type animals. Clearly, the biological function of one glycoconjugate can be substituted by another, albeit with less efficiency. The "biological redundancy" phenomenon governing cellular events needs to be further defined. Future studies in this regard should contribute greatly to regenerative and reparative biology.

Another future field of study is the epigenetic regulation of glycogenes. Epigenetic regulation such as chromatin remodeling, DNA methylation, histone modification, and noncoding RNA modulates developmental genes in the CNS [125–127]. However, the epigenetic mechanisms of glycoconjugates in the CNS have not been clearly elucidated. Recently, we have reported that histone acetylation of GalNAcT and ST-II genes is highly correlated to their mRNA expression levels during development [74]. Treatment of NSCs with a histone deacetylase inhibitor could change the ganglioside expression pattern. Further, we demonstrated that GM1 addition promoted neuronal differentiation with epigenetic activation of the GalNAcT gene [78]. Epigenetic regulation should be proven important for controlling the expression pattern of gangliosides during cellular differentiation and brain development. Ganglioside expression profiles and glycogene expression patterns are associated not only with CNS development but also with pathogenic mechanisms of CNS diseases such as Alzheimer disease [128–130], Parkinson disease [131], Huntington disease [132], amyotrophic lateral sclerosis [133,134], and multiple sclerosis [135,136]. Epigenetic studies of cell surface GSL expression provide clues as to the pathogenic mechanisms, which should be useful in providing novel strategies for disease treatment and neural repair.

ACKNOWLEDGMENTS

This work has been supported by grants from the NIH and VA Merit Awards to RKY. The authors also gratefully acknowledge the many collaborators who contributed to the studies cited in this review. We also thank Dr Rhea Markowitz, Georgia Regents University, for her editorial assistance.

REFERENCES

[1] The nomenclature of lipids. Recommendations (1976) IUPAC-IUB commission on biochemical nomenclature. Lipids 1977;12:455–68.

[2] Svennerholm L. Chromatographic separation of human brain gangliosides. J Neurochem 1963;10:613–23.

[3] Ngamukote S, Yanagisawa M, Ariga T, Ando S, Yu RK. Developmental changes of glycosphingolipids and expression of glycogenes in mouse brains. J Neurochem 2007;103:2327–41.

[4] Yanagisawa M, Yu RK. The expression and functions of glycoconjugates in neural stem cells. Glycobiology 2007;17:57R–74R.

[5] Yu RK, Macala LJ, Taki T, Weinfield HM, Yu FS. Developmental changes in ganglioside composition and synthesis in embryonic rat brain. J Neurochem 1988;50:1825–9.

[6] Xu YH, Barnes S, Sun Y, Grabowski GA. Multi-system disorders of glycosphingolipid and ganglioside metabolism. J Lipid Res 2010;51:1643–75.

[7] Yu RK, Nakatani Y, Yanagisawa M. The role of glycosphingolipid metabolism in the developing brain. J Lipid Res 2009;50(Suppl.):S440–5.

[8] D'Angelo G, Polishchuk E, Di Tullio G, Santoro M, Di Campli A, Godi A, et al. Glycosphingolipid synthesis requires FAPP2 transfer of glucosylceramide. Nature 2007;449:62–7.

[9] D'Angelo G, Uemura T, Chuang CC, Polishchuk E, Santoro M, Ohvo-Rekila H, et al. Vesicular and non-vesicular transport feed distinct glycosylation pathways in the Golgi. Nature 2013;501:116–20.

[10] Halter D, Neumann S, van Dijk SM, Wolthoorn J, de Maziere AM, Vieira OV, et al. Pre- and post-Golgi translocation of glucosylceramide in glycosphingolipid synthesis. J Cell Biol 2007;179:101–15.

[11] Chalat M, Menon I, Turan Z, Menon AK. Reconstitution of glucosylceramide flip-flop across endoplasmic reticulum: implications for mechanism of glycosphingolipid biosynthesis. J Biol Chem 2012;287:15523–32.

[12] Yu RK, Tsai YT, Ariga T. Functional roles of gangliosides in neurodevelopment: an overview of recent advances. Neurochem Res 2012;37:1230–44.

[13] Bieberich E, MacKinnon S, Silva J, Yu RK. Regulation of apoptosis during neuronal differentiation by ceramide and b-series complex gangliosides. J Biol Chem 2001;276:44396–404.

[14] Fang Y, Wu G, Xie X, Lu ZH, Ledeen RW. Endogenous GM1 ganglioside of the plasma membrane promotes neuritogenesis by two mechanisms. Neurochem Res 2000;25:931–40.

[15] Wu G, Fang Y, Lu ZH, Ledeen RW. Induction of axon-like and dendrite-like processes in neuroblastoma cells. J Neurocytol 1998;27:1–14.

[16] Wu G, Lu ZH, Xie X, Li L, Ledeen RW. Mutant NG108-15 cells (NG-CR72) deficient in GM1 synthase respond aberrantly to axonogenic stimuli and are vulnerable to calcium-induced apoptosis: they are rescued with LIGA-20. J Neurochem 2001;76:690–702.

[17] Yu RK, Bieberich E, Xia T, Zeng G. Regulation of ganglioside biosynthesis in the nervous system. J Lipid Res 2004;45:783–93.

[18] Furukawa K, Aixinjueluo W, Kasama T, Ohkawa Y, Yoshihara M, Ohmi Y, et al. Disruption of GM2/GD2 synthase gene resulted in overt expression of 9-O-acetyl GD3 irrespective of Tis21. J Neurochem 2008;105:1057–66.

[19] Sheikh KA, Sun J, Liu Y, Kawai H, Crawford TO, Proia RL, et al. Mice lacking complex gangliosides develop Wallerian degeneration and myelination defects. Proc Natl Acad Sci USA 1999;96:7532–7.

[20] Sugiura Y, Furukawa K, Tajima O, Mii S, Honda T. Sensory nerve-dominant nerve degeneration and remodeling in the mutant mice lacking complex gangliosides. Neuroscience 2005;135:1167–78.

[21] Susuki K, Baba H, Tohyama K, Kanai K, Kuwabara S, Hirata K, et al. Gangliosides contribute to stability of paranodal junctions and ion channel clusters in myelinated nerve fibers. Glia 2007;55:746–57.

[22] Takamiya K, Yamamoto A, Furukawa K, Yamashiro S, Shin M, Okada M, et al. Mice with disrupted GM2/GD2 synthase gene lack complex gangliosides but exhibit only subtle defects in their nervous system. Proc Natl Acad Sci USA 1996;93:10662–7.

[23] Wu G, Lu ZH, Kulkarni N, Amin R, Ledeen RW. Mice lacking major brain gangliosides develop parkinsonism. Neurochem Res 2011;36:1706–14.

[24] Simpson MA, Cross H, Proukakis C, Priestman DA, Neville DC, Reinkensmeier G, et al. Infantile-onset symptomatic epilepsy syndrome caused by a homozygous loss-of-function mutation of GM3 synthase. Nat Genet 2004;36:1225–9.

[25] Boukhris A, Schule R, Loureiro JL, Lourenco CM, Mundwiller E, Gonzalez MA, et al. Alteration of ganglioside biosynthesis responsible for complex hereditary spastic paraplegia. Am J Hum Genet 2013;93:118–23.

[26] Kawai H, Allende ML, Wada R, Kono M, Sango K, Deng C, et al. Mice expressing only monosialoganglioside GM3 exhibit lethal audiogenic seizures. J Biol Chem 2001;276:6885–8.

[27] Yamashita T, Wada R, Sasaki T, Deng C, Bierfreund U, Sandhoff K, et al. A vital role for glycosphingolipid synthesis during development and differentiation. Proc Natl Acad Sci USA 1999;96:9142–7.

[28] Purves D, Lichtman JW. Geometrical differences among homologous neurons in mammals. Science 1985;228:298–302.

[29] Solter D, Knowles BB. Monoclonal antibody defining a stage-specific mouse embryonic antigen (SSEA-1). Proc Natl Acad Sci USA 1978;75:5565–9.

[30] Fenderson BA, Zehavi U, Hakomori S. A multivalent lacto-*N*-fucopentaose III-lysyllysine conjugate decompacts preimplantation mouse embryos, while the free oligosaccharide is ineffective. J Exp Med 1984;160:1591–6.

[31] Fenderson BA, Eddy EM, Hakomori S. Glycoconjugate expression during embryogenesis and its biological significance. Bioessays 1990;12:173–9.

[32] Adewumi O, Aflatoonian B, Ahrlund-Richter L, Amit M, Andrews PW, Beighton G, et al. Characterization of human embryonic stem cell lines by the International Stem Cell Initiative. Nat Biotechnol 2007;25:803–16.

[33] Muramatsu T, Muramatsu H. Carbohydrate antigens expressed on stem cells and early embryonic cells. Glycoconj J 2004;21:41–5.

[34] Kudo T, Fujii T, Ikegami S, Inokuchi K, Takayama Y, Ikehara Y, et al. Mice lacking α1,3-fucosyltransferase IX demonstrate disappearance of Lewis x structure in brain and increased anxiety-like behaviors. Glycobiology 2007;17:1–9.

[35] Kudo T, Ikehara Y, Togayachi A, Kaneko M, Hiraga T, Sasaki K, et al. Expression cloning and characterization of a novel murine α1, 3-fucosyltransferase, mFuc-TIX, that synthesizes the Lewis x (CD15) epitope in brain and kidney. J Biol Chem 1998;273:26729–38.

[36] Okuda T, Tokuda N, Numata S, Ito M, Ohta M, Kawamura K, et al. Targeted disruption of Gb_3/CD_{77} synthase gene resulted in the complete deletion of globo-series glycosphingolipids and loss of sensitivity to verotoxins. J Biol Chem 2006;281:10230–5.

[37] Capela A, Temple S. LeX is expressed by principle progenitor cells in the embryonic nervous system, is secreted into their environment and binds Wnt-1. Dev Biol 2006;291:300–13.

[38] Hennen E, Czopka T, Faissner A. Structurally distinct LewisX glycans distinguish subpopulations of neural stem/progenitor cells. J Biol Chem 2011;286:16321–31.

[39] Mai JK, Andressen C, Ashwell KW. Demarcation of prosencephalic regions by CD15-positive radial glia. Eur J Neurosci 1998;10:746–51.

[40] Yanagisawa M, Taga T, Nakamura K, Ariga T, Yu RK. Characterization of glycoconjugate antigens in mouse embryonic neural precursor cells. J Neurochem 2005;95:1311–20.

[41] Dvorak P, Hampl A, Jirmanova L, Pacholikova J, Kusakabe M. Embryoglycan ectodomains regulate biological activity of FGF-2 to embryonic stem cells. J Cell Sci 1998;111(Pt 19):2945–52.

[42] Jirmanova L, Pacholikova J, Krejci P, Hampl A, Dvorak P. O-linked carbohydrates are required for FGF-2-mediated proliferation of mouse embryonic cells. Int J Dev Biol 1999;43:555–62.

[43] Kabos P, Matundan H, Zandian M, Bertolotto C, Robinson ML, Davy BE, et al. Neural precursors express multiple chondroitin sulfate proteoglycans, including the lectican family. Biochem Biophys Res Commun 2004;318:955–63.

[44] Yagi H, Yanagisawa M, Kato K, Yu RK. Lysosome-associated membrane protein 1 is a major SSEA-1-carrier protein in mouse neural stem cells. Glycobiology 2010;20:976–81.

[45] Hanjan SN, Kearney JF, Cooper MD. A monoclonal antibody (MMA) that identifies a differentiation antigen on human myelo-monocytic cells. Clin Immunol Immunopathol 1982;23:172–88.

[46] Tole S, Kaprielian Z, Ou SK, Patterson PH. FORSE-1: a positionally regulated epitope in the developing rat central nervous system. J Neurosci 1995;15:957–69.

[47] Mo Z, Moore AR, Filipovic R, Ogawa Y, Kazuhiro I, Antic SD, et al. Human cortical neurons originate from radial glia and neuron-restricted progenitors. J Neurosci 2007;27:4132–45.

[48] Pierschbacher MD, Ruoslahti E. Cell attachment activity of fibronectin can be duplicated by small synthetic fragments of the molecule. Nature 1984;309:30–3.

[49] von Holst A, Sirko S, Faissner A. The unique 473HD-Chondroitin-sulfate epitope is expressed by radial glia and involved in neural precursor cell proliferation. J Neurosci 2006;26:4082–94.

[50] Yagi H, Saito T, Yanagisawa M, Yu RK, Kato K. Lewis X-carrying *N*-glycans regulate the proliferation of mouse embryonic neural stem cells via the Notch signaling pathway. J Biol Chem 2012;287:24356–64.

[51] Imai T, Tokunaga A, Yoshida T, Hashimoto M, Mikoshiba K, Weinmaster G, et al. The neural RNA-binding protein Musashi1 translationally regulates mammalian *numb* gene expression by interacting with its mRNA. Mol Cell Biol 2001;21:3888–900.

[52] Okano H, Kawahara H, Toriya M, Nakao K, Shibata S, Imai T. Function of RNA-binding protein Musashi-1 in stem cells. Exp Cell Res 2005;306:349–56.

[53] Henderson JK, Draper JS, Baillie HS, Fishel S, Thomson JA, Moore H, et al. Preimplantation human embryos and embryonic stem cells show comparable expression of stage-specific embryonic antigens. Stem Cells 2002;20:329–37.

[54] Kannagi R, Cochran NA, Ishigami F, Hakomori S, Andrews PW, Knowles BB, et al. Stage-specific embryonic antigens (SSEA-3 and -4) are epitopes of a unique globo-series ganglioside isolated from human teratocarcinoma cells. EMBO J 1983;2:2355–61.

[55] Solter D, Knowles BB. Developmental stage-specific antigens during mouse embryogenesis. Curr Top Dev Biol 1979;13(Pt 1):139–65.

[56] Barraud P, Stott S, Mollgard K, Parmar M, Bjorklund A. In vitro characterization of a human neural progenitor cell coexpressing SSEA4 and CD133. J Neurosci Res 2007;85:250–9.

[57] Piao JH, Odeberg J, Samuelsson EB, Kjaeldgaard A, Falci S, Seiger A, et al. Cellular composition of long-term human spinal cord- and forebrain-derived neurosphere cultures. J Neurosci Res 2006;84:471–82.

[58] Rosner H, al-Aqtum M, Rahmann H. Gangliosides and neuronal differentiation. Neurochem Int 1992;20:339–51.

[59] Bannerman PG, Oliver TM, Xu Z, Shieh A, Pleasure DE. Effects of FGF-1 and FGF-2 on GD3 immunoreactive spinal neuroepithelial cells. J Neurosci Res 1996;45:549–57.

[60] Cammer W, Zhang H. Ganglioside GD3 in radial glia and astrocytes in situ in brains of young and adult mice. J Neurosci Res 1996;46:18–23.

[61] Cammer W, Zhang H. Carbonic anhydrase II in microglia in forebrains of neonatal rats. J Neuroimmunol 1996;67:131–6.

[62] Goldman JE, Hirano M, Yu RK, Seyfried TN. GD3 ganglioside is a glycolipid characteristic of immature neuroectodermal cells. J Neuroimmunol 1984;7:179–92.

[63] Nakatani Y, Yanagisawa M, Suzuki Y, Yu RK. Characterization of GD3 ganglioside as a novel biomarker of mouse neural stem cells. Glycobiology 2010;20:78–86.

[64] Yanagisawa M, Yoshimura S, Yu RK. Expression of GD2 and GD3 gangliosides in human embryonic neural stem cells. ASN Neuro 2011;3.

[65] Yu RK, Yanagisawa M. Glycosignaling in neural stem cells: involvement of glycoconjugates in signal transduction modulating the neural stem cell fate. J Neurochem 2007;103(Suppl. 1):39–46.

[66] Wang J, Yu RK. Interaction of ganglioside GD3 with an EGF receptor sustains the self-renewal ability of mouse neural stem cells in vitro. Proc Natl Acad Sci USA 2013;110:19137–42.

[67] Steinman RM, Mellman IS, Muller WA, Cohn ZA. Endocytosis and the recycling of plasma membrane. J Cell Biol 1983;96:1–27.

[68] Blum AS, Barnstable CJ. O-acetylation of a cell-surface carbohydrate creates discrete molecular patterns during neural development. Proc Natl Acad Sci USA 1987;84:8716–20.

[69] Mendez-Otero R, Schlosshauer B, Barnstable CJ, Constantine-Paton M. A developmentally regulated antigen associated with neural cell and process migration. J Neurosci 1988;8:564–79.

[70] Mendez-Otero R, Cavalcante LA. Expression of 9-O-acetylated gangliosides is correlated with tangential cell migration in the rat brain. Neurosci Lett 1996;204:97–100.

[71] Miyakoshi LM, Todeschini AR, Mendez-Otero R, Hedin-Pereira C. Role of the 9-O-acetyl GD3 in subventricular zone neuroblast migration. Mol Cell Neurosci 2012;49:240–9.

[72] Yang CR, Liour SS, Dasgupta S, Yu RK. Inhibition of neuronal migration by JONES antibody is independent of 9-O-acetyl GD3 in GD3-synthase knockout mice. J Neurosci Res 2007;85:1381–90.

[73] Hirschberg K, Zisling R, van Echten-Deckert G, Futerman AH. Ganglioside synthesis during the development of neuronal polarity. Major changes occur during axonogenesis and axon elongation, but not during dendrite growth or synaptogenesis. J Biol Chem 1996;271:14876–82.

[74] Suzuki Y, Yanagisawa M, Ariga T, Yu RK. Histone acetylation-mediated glycosyltransferase gene regulation in mouse brain during development. J Neurochem 2011;116:874–80.

[75] Androutsellis-Theotokis A, Walbridge S, Park DM, Lonser RR, McKay RD. Cholera toxin regulates a signaling pathway critical for the expansion of neural stem cell cultures from the fetal and adult rodent brains. PLoS One 2010;5:e10841.

[76] Liour SS, Dinkins MB, Su CY, Yu RK. Spatiotemporal expression of GM1 in murine medial pallial neural progenitor cells. J Comp Neurol 2005;491:330–8.

[77] Maric D, Maric I, Chang YH, Barker JL. Prospective cell sorting of embryonic rat neural stem cells and neuronal and glial progenitors reveals selective effects of basic fibroblast growth factor and epidermal growth factor on self-renewal and differentiation. J Neurosci 2003;23:240–51.

[78] Tsai YT, Yu RK. Epigenetic activation of mouse ganglioside synthase genes: implications for neurogenesis. J Neurochem 2014;128:101–10.

[79] Ledeen RW, Wu G, Lu ZH, Kozireski-Chuback D, Fang Y. The role of GM1 and other gangliosides in neuronal differentiation. Overview and new finding. Ann NY Acad Sci 1998;845:161–75.

[80] Itokazu Y, Pagano RE, Schroeder AS, O'Grady SM, Limper AH, Marks DL. Reduced GM$_1$ ganglioside in CFTR-deficient human airway cells results in decreased β$_1$-integrin signaling and delayed wound repair. Am J Physiol Cell Physiol 2014;306(9):C819–30.

[81] Palazzo AF, Eng CH, Schlaepfer DD, Marcantonio EE, Gundersen GG. Localized stabilization of microtubules by integrin- and FAK-facilitated Rho signaling. Science 2004;303:836–9.

[82] Wu G, Lu ZH, Obukhov AG, Nowycky MC, Ledeen RW. Induction of calcium influx through TRPC5 channels by cross-linking of GM1 ganglioside associated with α5β1 integrin initiates neurite outgrowth. J Neurosci 2007;27:7447–58.

[83] Ivins JK, Yurchenco PD, Lander AD. Regulation of neurite outgrowth by integrin activation. J Neurosci 2000;20:6551–60.

[84] Tucker BA, Rahimtula M, Mearow KM. Integrin activation and neurotrophin signaling cooperate to enhance neurite outgrowth in sensory neurons. J Comp Neurol 2005;486:267–80.

[85] Ichikawa N, Iwabuchi K, Kurihara H, Ishii K, Kobayashi T, Sasaki T, et al. Binding of laminin-1 to monosialoganglioside GM1 in lipid rafts is crucial for neurite outgrowth. J Cell Sci 2009;122:289–99.

[86] Eisenbarth GS, Walsh FS, Nirenberg M. Monoclonal antibody to a plasma membrane antigen of neurons. Proc Natl Acad Sci USA 1979;76:4913–7.

[87] Kasai N, Yu RK. The monoclonal antibody A2B5 is specific to ganglioside G$_{Q1c}$. Brain Res 1983;277:155–8.

[88] Saito M, Kitamura H, Sugiyama K. The specificity of monoclonal antibody A2B5 to c-series gangliosides. J Neurochem 2001;78:64–74.

[89] Inoko E, Nishiura Y, Tanaka H, Takahashi T, Furukawa K, Kitajima K, et al. Developmental stage-dependent expression of an α2,8-trisialic acid unit on glycoproteins in mouse brain. Glycobiology 2010;20:916–28.

[90] Ando S, Yu RK. Isolation and characterization of two isomers of brain tetrasialogangliosides. J Biol Chem 1979;254:12224–9.

[91] Freischutz B, Saito M, Rahmann H, Yu RK. Activities of five different sialyltransferases in fish and rat brains. J Neurochem 1994;62:1965–73.

[92] Freischutz B, Saito M, Rahmann H, Yu RK. Characterization of sialyltransferase-IV activity and its involvement in the c-pathway of brain ganglioside metabolism. J Neurochem 1995;64:385–93.

[93] Rosner H, Greis C, Henke-Fahle S. Developmental expression in embryonic rat and chicken brain of a polysialoganglioside-antigen reacting with the monoclonal antibody Q 211. Brain Res 1988;470:161–71.

[94] Yu RK, Ando S. Structures of some new complex gangliosides of fish brain. Adv Exp Med Biol 1980;125:33–45.

[95] Rao MS, Subbarao V. Effect of dexamethasone on ciprofibrate-induced cell proliferation and peroxisome proliferation. Fundam Appl Toxicol 1997;35:78–83.

[96] Farrer RG, Quarles RH. GT3 and its O-acetylated derivative are the principal A2B5-reactive gangliosides in cultured O2A lineage cells and are down-regulated along with O-acetyl GD3 during differentiation to oligodendrocytes. J Neurosci Res 1999;57:371–80.

[97] Zhang SC. Defining glial cells during CNS development. Nat Rev Neurosci 2001;2:840–3.

[98] Bosio A, Binczek E, Stoffel W. Functional breakdown of the lipid bilayer of the myelin membrane in central and peripheral nervous system by disrupted galactocerebroside synthesis. Proc Natl Acad Sci USA 1996;93:13280–5.

[99] Coetzee T, Fujita N, Dupree J, Shi R, Blight A, Suzuki K, et al. Myelination in the absence of galactocerebroside and sulfatide: normal structure with abnormal function and regional instability. Cell 1996;86:209–19.

[100] Honke K, Hirahara Y, Dupree J, Suzuki K, Popko B, Fukushima K, et al. Paranodal junction formation and spermatogenesis require sulfoglycolipids. Proc Natl Acad Sci USA 2002;99:4227–32.

[101] Hirahara Y, Bansal R, Honke K, Ikenaka K, Wada Y. Sulfatide is a negative regulator of oligodendrocyte differentiation: development in sulfatide-null mice. Glia 2004;45:269–77.

[102] Ogura K, Kohno K, Tai T. Molecular cloning of a rat brain cDNA, with homology to a tyrosine kinase substrate, that induces galactosylceramide expression in COS-7 cells. J Neurochem 1998;71:1827–36.

[103] Ogura K, Tai T. Molecular cloning and characterization of galactosylceramide expression factor-1 (GEF-1). Neurochem Res 2002;27:779–84.

[104] Cochran Jr FB, Yu RK, Ledeen RW. Myelin gangliosides in vertebrates. J Neurochem 1982;39:773–9.

[105] Ueno K, Ando S, Yu RK. Gangliosides of human, cat, and rabbit spinal cords and cord myelin. J Lipid Res 1978;19:863–71.

[106] Ozawa H, Kotani M, Kawashima I, Numata M, Ogawa T, Terashima T, et al. Generation of a monoclonal antibody specific for ganglioside GM4: evidence for GM4 expression on astrocytes in chicken cerebellum. J Biochem 1993;114:5–8.

[107] Yu RK, Lee SH. In vitro biosynthesis of sialosylgalactosylceramide (G7) by mouse brain microsomes. J Biol Chem 1976;251:198–203.

[108] Berselli P, Zava S, Sottocornola E, Milani S, Berra B, Colombo I. Human GM_3 synthase: a new mRNA variant encodes an NH_2-terminal extended form of the protein. Biochim Biophys Acta 2006;1759:348–58.

[109] Chisada S, Yoshimura Y, Sakaguchi K, Uemura S, Go S, Ikeda K, et al. Zebrafish and mouse $\alpha 2,3$-sialyltransferases responsible for synthesizing GM4 ganglioside. J Biol Chem 2009;284:30534–46.

[110] Uemura S, Go S, Shishido F, Inokuchi J. Expression machinery of GM_4: the excess amounts of GM_3/GM_4S synthase (ST_3GAL_5) are necessary for GM_4 synthesis in mammalian cells. Glycoconj J 2014;31:101–8.

[111] Dasgupta S, Everhart MB, Bhat NR, Hogan EL. Neutral monoglycosylceramides in rat brain: occurrence, molecular expression and developmental variation. Dev Neurosci 1997;19:152–61.

[112] Brigande JV, Platt FM, Seyfried TN. Inhibition of glycosphingolipid biosynthesis does not impair growth or morphogenesis of the postimplantation mouse embryo. J Neurochem 1998;70:871–82.

[113] Nagatsuka Y, Kasama T, Ohashi Y, Uzawa J, Ono Y, Shimizu K, et al. A new phosphoglycerolipid, 'phosphatidylglucose', found in human cord red cells by multi-reactive monoclonal anti-i cold agglutinin, mAb GL-1/GL-2. FEBS Lett 2001;497:141–7.

[114] Nagatsuka Y, Hara-Yokoyama M, Kasama T, Takekoshi M, Maeda F, Ihara S, et al. Carbohydrate-dependent signaling from the phosphatidylglucoside-based microdomain induces granulocytic differentiation of HL60 cells. Proc Natl Acad Sci USA 2003;100:7454–9.

[115] Kinoshita MO, Furuya S, Ito S, Shinoda Y, Yamazaki Y, Greimel P, et al. Lipid rafts enriched in phosphatidylglucoside direct astroglial differentiation by regulating tyrosine kinase activity of epidermal growth factor receptors. Biochem J 2009;419:565–75.

[116] Kinoshita MO, Shinoda Y, Sakai K, Hashikawa T, Watanabe M, Machida T, et al. Selective upregulation of 3-phosphoglycerate dehydrogenase (Phgdh) expression in adult subventricular zone neurogenic niche. Neurosci Lett 2009;453:21–6.

[117] Nagatsuka Y, Horibata Y, Yamazaki Y, Kinoshita M, Shinoda Y, Hashikawa T, et al. Phosphatidylglucoside exists as a single molecular species with saturated fatty acyl chains in developing astroglial membranes. Biochemistry 2006;45:8742–50.

[118] Kaneko J, Kinoshita MO, Machida T, Shinoda Y, Nagatsuka Y, Hirabayashi Y. Phosphatidylglucoside: a novel marker for adult neural stem cells. J Neurochem 2011;116:840–4.

[119] Reynolds BA, Weiss S. Generation of neurons and astrocytes from isolated cells of the adult mammalian central nervous system. Science 1992;255:1707–10.

[120] Handa K, Hakomori SI. Carbohydrate to carbohydrate interaction in development process and cancer progression. Glycoconj J 2012;29:627–37.

[121] Guan F, Handa K, Hakomori SI. Specific glycosphingolipids mediate epithelial-to-mesenchymal transition of human and mouse epithelial cell lines. Proc Natl Acad Sci USA 2009;106:7461–6.

[122] Reijo Pera RA, DeJonge C, Bossert N, Yao M, Hwa Yang JY, Asadi NB, et al. Gene expression profiles of human inner cell mass cells and embryonic stem cells. Differentiation 2009;78:18–23.

[123] Wakao S, Kitada M, Kuroda Y, Shigemoto T, Matsuse D, Akashi H, et al. Multilineage-differentiating stress-enduring (Muse) cells are a primary source of induced pluripotent stem cells in human fibroblasts. Proc Natl Acad Sci USA 2011;108:9875–80.

[124] Kuroda Y, Kitada M, Wakao S, Nishikawa K, Tanimura Y, Makinoshima H, et al. Unique multipotent cells in adult human mesenchymal cell populations. Proc Natl Acad Sci USA 2010;107:8639–43.

[125] Hirabayashi Y, Gotoh Y. Epigenetic control of neural precursor cell fate during development. Nat Rev Neurosci 2010;11:377–88.

[126] Jobe EM, McQuate AL, Zhao X. Crosstalk among epigenetic pathways regulates neurogenesis. Front Neurosci 2012;6:59.

[127] Takizawa T, Nakashima K, Namihira M, Ochiai W, Uemura A, Yanagisawa M, et al. DNA methylation is a critical cell-intrinsic determinant of astrocyte differentiation in the fetal brain. Dev Cell 2001;1:749–58.

[128] Ariga T, Itokazu Y, McDonald MP, Hirabayashi Y, Ando S, Yu RK. Brain gangliosides of a transgenic mouse model of Alzheimer's disease with deficiency in GD3-synthase: expression of elevated levels of a cholinergic-specific ganglioside, $GT1a\alpha$. ASN Neuro 2013;5:141–8.

[129] Ariga T, Wakade C, Yu RK. The pathological roles of ganglioside metabolism in Alzheimer's disease: effects of gangliosides on neurogenesis. Int J Alzheimers Dis 2011;2011:193618.

[130] Itokazu Y, Yu RK. Amyloid β-peptide 1-42 modulates the proliferation of mouse neural stem cells: upregulation of fucosyltransferase IX and notch signaling. Mol Neurobiol 2014 (in press).

[131] Wu G, Lu ZH, Kulkarni N, Ledeen RW. Deficiency of ganglioside GM1 correlates with Parkinson's disease in mice and humans. J Neurosci Res 2012;90:1997–2008.

[132] Maglione V, Marchi P, Di Pardo A, Lingrell S, Horkey M, Tidmarsh E, et al. Impaired ganglioside metabolism in Huntington's disease and neuroprotective role of GM1. J Neurosci 2010;30:4072–80.

[133] Rapport MM. Implications of altered brain ganglioside profiles in amyotrophic lateral sclerosis (ALS). Acta Neurobiol Exp (Wars) 1990;50:505–13.

[134] Rapport MM, Donnenfeld H, Brunner W, Hungund B, Bartfeld H. Ganglioside patterns in amyotrophic lateral sclerosis brain regions. Ann Neurol 1985;18:60–7.

[135] Yu RK, Ledeen RW, Eng LF. Ganglioside abnormalities in multiple sclerosis. J Neurochem 1974;23:169–74.

[136] Yu RK, Ueno K, Glaser GH, Tourtellotte WW. Lipid and protein alterations of spinal cord and cord myelin of multiple sclerosis. J Neurochem 1982;39:464–77.

Chapter 9

NG2 (Cspg4): Cell Surface Proteoglycan on Oligodendrocyte Progenitor Cells in the Developing and Mature Nervous System

Akiko Nishiyama, Aaron Lee and Christopher B. Brunquell

Department of Physiology and Neurobiology, University of Connecticut, Storrs, CT, USA

9.1 INTRODUCTION

NG2 was first discovered and described by Bill Stallcup and colleagues in the early 1980s during their search for developmentally important neural cell surface antigens. It was initially characterized from a rat neural cell line called B49 [1], which had properties that were neither typically neuronal nor glial, as defined by the ability of the cells to generate action potentials and conduct Na+ and K+, and their reactivity with antibodies to previously characterized neuron- and glial-specific antigens [2]. With an interest in studying cell surface antigens involved in the development of neurons and glia, Stallcup and colleagues took an immunological approach to identify molecules expressed on these cells with intermediate pseudo-glial properties. Rabbits were immunized with B49 cells, and the resulting antiserum was absorbed with the typical neuronal cell lines that could generate action potentials. The absorbed antiserum called NG2 recognized pseudo-glial and pseudo-neuronal cells but did not react with any cell lines that were unambiguously neuronal or glial [3]. Subsequently, the then relatively new hybridoma technology [4] was used to generate monoclonal antibodies against B49 cells, and those with reactivity to the NG2 antigen were isolated. The monoclonal antibodies to NG2 allowed further purification and characterization of the NG2 antigen and the investigation of its cellular localization in the developing and mature rat brain.

9.2 THE STRUCTURE OF NG2

The antigen on B49 cells recognized by the anti-NG2 antibodies is a large cell surface proteoglycan of 400–800 kDa, which is converted into a discrete core glycoprotein of 300 kDa by treatment with chondroitinase ABC but not with heparitinase [5], suggesting that the NG2 core protein carries primarily chondroitin sulfate and not heparan sulfate chains. In addition to the O-linked glycosaminoglycan chains, NG2 also carries N-linked carbohydrates. Further biochemical characterization followed by elucidation of the primary structure of rat NG2 revealed a type 1 integral membrane core protein of 2326 amino acids with a large extracellular domain and 76 amino acids in the short cytoplasmic tail [6] (Figure 9.1(A)). It is encoded by the Cspg4 gene. The extracellular portion of the core protein can be largely divided into three domains. The first amino-terminal globular domain (D1 domain) and the third juxtamembrane domain (D3 domain) are separated by an extended nonglobular second domain (D2 domain), which contains a single chondroitin sulfate attachment site. Interestingly, the amount of glycosaminoglycan modification is cell-type-specific and undergoes an age-dependent change. NG2 from cell lines such as the rat B49 and mouse Oli-neu cells is much more heavily modified with glycosaminoglycan chains than NG2 extracted from nervous system tissues (see below). Currently it is unclear what regulates the extent of chondroitin sulfate chain modification on NG2. The D1 domain contains two laminin-G-like domains, whose function remains unknown (Figure 9.1(A) and (B)) [7, 8]. In addition, 15 repeats called chondroitin sulfate proteoglycan (CSPG) repeats have been identified and found to span a large part of the ectodomain from the end of the laminin-G-like domain to the juxtamembrane region. They appear to be distantly related to cadherin repeats and are predicted to have a β-fold structure (Figure 9.1(B)) [8]. The juxtamembrane extracellular domain D3 contains sites for proteolytic cleavage, and cleavage may be regulated by intracellular signals (Figure 9.1(A)) [9].

The short cytoplasmic domain lacks any known signal-transducing motifs but contains two threonine residues that are phosphorylated by different intracellular signaling molecules (see below) and ends with the sequence, QYWV, which constitutes a motif used to recruit intracellular proteins that have a PDZ (PSD-95, DISC-large, ZO-1) domain.

Neural Surface Antigens. http://dx.doi.org/10.1016/B978-0-12-800781-5.00009-8

9.2.1 NG2 in Different Species

Independent of the Stallcup lab, Ralph Reisfeld and colleagues, who had been using monoclonal antibodies to study antigens that affect proliferation and migration of malignant melanoma cells, identified a high-molecular-weight CSPG with biochemical properties similar to those of rat NG2 on a subset of highly malignant human melanoma cells [10]. The 9.2.27 and 155.8 monoclonal antibodies to this antigen, now known as the melanoma-associated chondroitin sulfate proteoglycan (MCSP), were shown to inhibit spreading and anchorage-independent growth of human melanoma cells, suggesting a role for MCSP in the regulation of tumor growth [11]. Subsequently, the cDNA encoding MCSP (CSPG4 transcript) was cloned and mapped to chromosome 15 [12], and the deduced amino sequence revealed it to be the human ortholog of NG2.

More than a decade later, a third group led by Jacky Trotter generated rat monoclonal antibodies against the Oli-neu mouse oligodendroglial cell line, derived by transforming oligodendrocyte-enriched cultures from embryonic mouse cerebellum with a retrovirus encoding v-ErbB2 [13] (the neu oncogene). One of the monoclonal antibodies, clone AN2 1E6, was described as a single-cell clone of an "interesting" hybridoma line [14]. Although the authors do not describe in what ways this particular clone was interesting, they had been generally interested in studying cell surface proteins involved in neuron–glial interaction, and Oli-neu cells had been shown to recognize and ensheathe demyelinated axons [13]. The AN2 monoclonal antibody was then used to purify a glycoprotein of 330 kDa from P9–10 mouse brains, in which the antigen was most abundantly expressed [14]. Further characterization of the AN2 antigen from Oli-neu cells revealed that in addition to N-linked carbohydrates, it contained chondroitin sulfate glycosaminoglycans, similar to rat NG2. The final confirmation that the AN2 antigen was the mouse homolog of NG2 was obtained by proteomic analysis. Matrix-assisted laser desorption/ionization time-of-flight (MALDI-TOF) mass spectrometry was performed on AN2 antigen purified from P9–10 mouse brains [15]. Thirty-one of the tryptic peptide masses matched the predicted masses of putative tryptic fragments of rat NG2 [6], and peptide sequences obtained from Edman degradation of two of the tryptic fragments of the mouse AN2 antigen were found to be identical to the rat NG2 sequence.

The mammalian Cspg4 gene consists of 10 exons spanning 35 kb and has a large first intron of 18 kb that separates the first and second exons, which encode the signal peptide and the beginning of the mature NG2 core protein, respectively (Figure 9.1(C), top). The exon/intron organization and

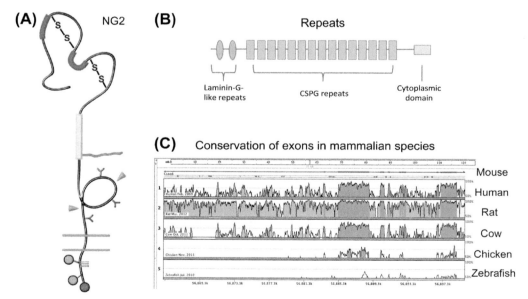

FIGURE 9.1 Structure of NG2. (A) A cartoon depicting functional domains and features of NG2. The large extracellular domain consists of an N-terminal globular domain with disulfide bonds (S–S), a central extended domain (yellow) to which a single chondroitin sulfate chain (red) is attached, and a juxtamembrane domain, which contains proteolytic cleavage sites (arrowheads). In the cytoplasmic domain (below the green double lines depicting the plasma membrane), there are two threonine residues that are phosphorylated (blue circles) and the C-terminal PDZ binding motif (green circle). *(Copied with permission from Ref. [7].)* (B) Structural motifs predicted by the amino acid sequence of NG2. Ovals, laminin-G-like repeats at the N-terminus; blue rectangles, putative CSPG repeats, remotely resembling cadherin repeats; green box, cytoplasmic domain. *(Adapted from Ref. [8].)* (C) Comparison of Cspg4 sequences across vertebrate species. The mouse exon/intron structure is shown on top, followed by an alignment of human, rat, cow, chicken, and zebrafish sequences. Blue indicates regions of a high degree of homology. Sequences are highly conserved among mammals, but diverge in lower vertebrates. Snapshot of alignment performed using the Vista software. (For interpretation of the references to color in this figure legend, the reader is referred to the online version of this book.)

the amino acid sequence are highly conserved among mammalian species but diverge in other lower vertebrate species (Figure 9.1(C)). The above-mentioned CSPG repeats are seen not only among vertebrate species but also in the Kon-tiki protein (CG10275 gene product) in *Drosophila melanogaster* and C48E7.6 in *Caenorhabditis elegans*, as well as in the sea urchin *Lytechinus variegatus* in the form of ECM3 protein [8] (www.uniprot.org).

9.3 EXPRESSION OF NG2 IN THE NERVOUS SYSTEM

The expression of NG2 is not confined to the nervous system. NG2 is detected on proliferative tissue-specific progenitor cells of various organs, including muscle, cartilage, bone, skin, adipose tissue, and vascular mural cells [7]. NG2 is not present in multipotent stem cells, and its expression is downregulated as the proliferative progenitor cells undergo terminal differentiation. The expression of NG2 is highly dynamic and is often upregulated as the cells become activated in response to injury or under neoplastic transformation.

9.3.1 NG2 in the Central Nervous System

In the central nervous system (CNS), NG2 is expressed on oligodendrocyte progenitor cells (OPCs) and vascular mural cells (Figure 9.2(A)–(C)). In the vasculature, NG2 is expressed by pericytes but not endothelial cells. In the CNS parenchyma, NG2 is not present on neural stem cells or the cells in the germinal zone [17] (Figure 9.2(C)). NG2 becomes detectable once the cell exits the germinal zone and becomes committed to the glial lineage, as revealed by glia-associated transcription factors such as Sox10. NG2 continues to be expressed robustly on OPCs that proliferate and migrate during late embryonic to early postnatal stages to uniformly occupy the entire CNS parenchyma by the end of the first postnatal week. The developmental origin of these

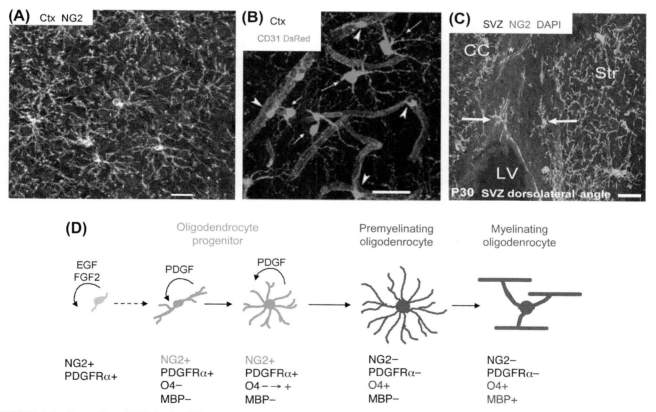

FIGURE 9.2 Expression of NG2 in the CNS. (A)–(C) Localization of NG2 in P30 mouse neocortex. (A) Coronal section immunolabeled for NG2 showing typical NG2 cells. (B) Coronal section from an NG2-DsRed transgenic mouse (The Jackson Laboratory stock #008241) labeled for the endothelial marker CD31 in green *(reproduced from Ref. [16])*, showing NG2 transcriptional activity (DsRed+) in vascular mural cells (arrowheads) as well as in NG2 glial cells (arrows). (C) NG2 immunolabeling in dorsolateral SVZ showing a paucity of NG2 cells (arrows) in the SVZ compared with their abundance in the adjacent corpus callosum (CC) and striatum (Str). Asterisk indicates NG2 expression on a capillary. *(Reproduced from Ref. [17].)* Scale bars, (A) and (B) 25 μm, (C) 50 μm. (D) A scheme showing the stage of oligodendrocyte lineage when NG2 is expressed *(reproduced with modification from Ref. [18]).* Curved arrows indicate self-renewal mediated by growth factors. (For interpretation of the references to color in this figure legend, the reader is referred to the online version of this book.)

cells is reviewed in more detail elsewhere [19]. Beyond this stage and throughout adulthood, NG2-expressing cells are evenly distributed throughout the CNS parenchyma in both gray and white matter with the exception of the neurogenic niches of the subventricular zone where they are less abundant (see below). As the NG2 cells begin to undergo terminal differentiation into mature oligodendrocytes, they lose the expression of NG2 and acquire the expression of oligodendrocyte antigens and myelin genes (Figure 9.2(D)). NG2 is also expressed on proliferating OPCs in culture but is lost as they differentiate into oligodendrocytes.

There has been an intense debate regarding what the NG2-expressing cells are in the developing and mature CNS parenchyma, particularly in the adult, in which the bulk of myelin has already been formed by oligodendrocytes. A number of Cre–loxP-mediated genetic fate mapping studies have shown that the predominant fate of NG2 cells in the postnatal white matter is the oligodendrocyte (reviewed in Ref. [20]). However, approximately 50% of NG2-expressing cells in the gray matter remain as NG2-expressing cells without differentiating into oligodendrocytes, which raises the question of whether all NG2 cells are OPCs although all NG2 cells in gray matter appear similar in other respects. Furthermore, some NG2-expressing cells in certain gray matter regions of embryonic and early postnatal forebrain generate protoplasmic astrocytes, although again the astrogliogenic and oligodendrogliogenic NG2-expressing cells are indistinguishable morphologically and antigenically and are found side by side with each other. These observations and the finding that NG2-expressing cells receive synaptic input from neurons and respond to vesicularly released neurotransmitters from presynaptic neuronal terminals [21] have challenged the idea that the sole function of NG2-expressing cells is to generate oligodendrocytes. Thus, the field has not yet reached a consensus on the name of NG2-expressing cells, and the terms NG2 cells, OPCs, and polydendrocytes are used interchangeably to refer to NG2-expressing glial cells in the CNS [19]. Here we use the term NG2 cells to refer to NG2-expressing glial cells in the CNS.

9.3.2 NG2 in the Peripheral Nervous System

In the peripheral nervous system, there had been some confusion over the cell type that expresses NG2. The use of NG2-EYFP transgenic mice has allowed unambiguous localization of NG2 to Schwann cell precursor cells and nonmyelinating Schwann cells but not to myelinating Schwann cells [15]. The degree of glycosaminoglycan modification seems to decrease with age in the sciatic nerve, where the proteoglycan form is detected more readily during the first postnatal week than in the nerves from the adult, which contain predominantly the core glycoprotein form. NG2 from the mouse brain appears to be predominantly in the core glycoprotein form rather than the proteoglycan form through all developmental stages [14].

NG2 from glial cell lines has been shown to undergo proteolytic cleavage in the juxtamembrane region to generate soluble forms [9]. In the normal sciatic nerve and after nerve crush, NG2 core glycoprotein appeared as a single 330 kDa form, and no other proteolytically cleaved forms were detected in the extracts. This is in contrast to another report suggesting that NG2 is shed from perineurial fibroblasts [22]. Further studies are needed to identify the mechanism of cleavage and the role of the cleaved products in cellular behavior.

9.4 THE ROLE OF NG2 IN CELL ATTACHMENT AND MIGRATION

The structure of NG2 as a type 1 integral membrane glycoprotein makes it a good candidate molecule for sensing extracellular signals and translating them to intracellular signaling events to elicit a cellular response. Collagen VI (C6) was the first extracellular protein shown to interact with NG2 [23]. On rat glioma cell lines, C6 is colocalized with NG2 in a punctate manner. On another rat glioma cell line that does not express NG2, C6 exhibits a more extracellular fibrillary pattern, but when NG2 is artificially expressed, the distribution of C6 shifts to the punctate cell surface pattern that coincides with that of NG2 [24]. Studies using deletion mutants of rat NG2 lacking various extracellular regions have revealed that NG2 interacts with both collagen V (C5) and C6 through its central nonglobular domain [25,26]. NG2-expressing melanoma cells show greater adhesion to C6 substrate compared with parent cells that were not transfected with NG2 [27]. This study also showed by coimmunoprecipitation that NG2 interacts with CD44 and $\alpha 4\beta 1$-integrin, and a separate study showed that NG2 collaborates with $\alpha 4\beta 1$-integrin to modulate melanoma cell spreading on fibronectin [28]. Studies using cells that lack $\beta 1$-integrin, which is known to bind C6, indicate that adhesion of NG2-expressing cells to C5 and C6 occurs independent of $\beta 1$-integrin and that NG2-expressing cells lacking $\beta 1$-integrin spread and reorganize their actin cytoskeleton only in response to C6 substrate and not to C5, even though the latter is capable of binding to the same domain of NG2 [29].

Although earlier work had suggested that C6 does not exist in the nervous system, a more recent study using hippocampal neurons revealed that amyloid-beta (Aβ) 42 peptide increased the expression of C6, and the presence of C6 could reduce Aβ42 toxicity by creating large Aβ aggregates [30]. In fibroblasts, soluble C6 has been shown to prevent apoptosis by suppressing the induction of the apoptotic protein Bax under serum-starved conditions [31]. In another study on myelination in the peripheral nerve, loss of C6 was shown to cause hypermyelination by increasing phosphorylation of intracellular signaling molecules [32]. It remains to be determined whether NG2 plays any role in these C6-mediated phenomena.

The role of NG2 on cellular function has been extensively studied in the vascular system, in which NG2 is expressed on smooth muscle cells and pericytes, which are known to influence the behavior of endothelial cells. Endothelial cells adhere to and spread on NG2-coated surfaces, and NG2 stimulates the migration of endothelial cells and promotes corneal angiogenesis [33]. To identify molecules that interact with NG2 and mediate these effects, NG2-interacting proteins were isolated from endothelial cell extracts by affinity purification on CL-4B Sepharose beads coupled to the extracellular domain of NG2. One protein of 30 kDa that bound to NG2 was identified by MALDI-TOF as galectin-3, which was then shown to directly interact with NG2 [33] via N-linked carbohydrates in the D3 extracellular domain of NG2 [34]. Furthermore, $\alpha 3\beta 1$-(CD49c/CD29) but not $\alpha 6\beta 1$-integrin (CD49e/CD29) expressed on endothelial cells is found in a complex with both galectin-3 and NG2, and endothelial migration on NG2 could be blocked by antibodies to galectin-3 or $\alpha 3\beta 1$-integrin [33]. In the CNS, galectin-3 has been implicated in oligodendrocyte differentiation [35], and galectin-1 was shown to promote axonal regeneration after spinal cord injury [36]. Thus, it would be interesting to explore whether NG2 is involved in mediating these effects.

9.5 THE ROLE OF NG2 IN AXON–NG2 CELL INTERACTIONS

The relationship between growing axons and NG2-expressing cells has been a subject of considerable debate. A series of studies has suggested that NG2 is repulsive to growing axons. However, this poses a challenge in interpreting the more recent observations that NG2 cells form synapses with axon terminals and respond to vesicularly released neurotransmitters [21]. The following are some observations related to the role of NG2 in growing axons.

9.5.1 NG2 and Developing Axons

A series of studies by Levine and colleagues showed that growing neurites avoid membrane stripes prepared from NG2 cells [37], and different domains of the NG2 core protein induce axonal growth cone collapse [38]. In contrast, work from the Trotter lab revealed that cerebellar granule neurons adhere to and extend neurites equally well on NG2-coated and NG2-free surfaces [14]. When added to laminin, NG2 does not reduce neurite growth from dorsal root ganglion explants [15]. We have shown that neurites from neonatal rat hippocampal neurons avoid mature oligodendrocytes, as previously shown [39], but grow toward and form stable contacts with NG2 cells [40]. Furthermore, the ability of axons to grow on NG2 cells is not altered by changing the levels of NG2 on the surface. In vivo, axons that traverse the perinatal rat corpus callosum are extensively

contacted by NG2 cell processes [40], again suggesting that NG2 expressed in the context of other secreted and surface proteins on NG2 glial cells is not repulsive to growing axons. The inconsistent observations reported in the literature could in part arise from the nature of the in vitro assay, in which the effects of NG2 are compared with those of a very potent neurite-promoting protein such as laminin.

In the developing rodent somatosensory cortex, thalamocortical axons carrying information from each whisker project somatotopically to a specific barrel, and these inputs are segregated by the surrounding barrel septa [41]. If NG2 were to play a role in negatively directing growing thalamocortical axons to their correct target into their corresponding barrels, one would predict NG2 expression to be greater in the septa than in the barrel hollows. However, contrary to the expectation, the distribution of NG2 was uniform across the barrel hollows [42], which also reinforces the notion that NG2 does not play a major inhibitory role in axon growth during development.

9.5.2 NG2 and Axons in the Injured CNS

Spinal cord injury and other types of traumatic injury to the CNS are accompanied by successive reactions of microglia, NG2 cells, and astrocytes, leading to deposition of a potent growth-inhibitory extracellular matrix [43]. The lectican family of CSPGs, the prototype of which is aggrecan, strongly inhibits axonal growth mainly through the chondroitin sulfate chains. Removal of chondroitin sulfate with chondroitinase ABC significantly improves regeneration and functional recovery [44]. In addition to these matrix proteoglycans, NG2 is a major CSPG that appears in the injured spinal cord [45]. Local application of anti-NG2 antibodies to the injured rat spinal cord led to increased regeneration of ascending sensory axons [46]. Injection of anti-NG2 antibodies also facilitated the conduction of action potentials [47], although the mechanism by which this occurs remains unclear. By contrast, no difference was seen in the extent of axonal regeneration in wild-type and NG2-knockout mice after lesioning of the corticospinal tract or the dorsal column, and no enhancement of axonal regeneration was observed in the NG2-knockout mice [48,49]. Others also observed that despite dense accumulation of NG2 in the injured spinal cord, regenerating axons were often found aligned with dense NG2 immunoreactivity, as if they were growing along NG2-positive cellular elements [50,51].

To further evaluate how severed axons behave in response to contacting NG2 cells, we transplanted green fluorescent protein (GFP)-expressing NG2 cells from perinatal mice into completely severed corpus callosum of wild-type adult mice (Figure 9.3(A) and (B)) and examined the spatial relationship between the severed axons labeled with biotinylated dextran amine (BDA) and donor GFP+ NG2 cells.

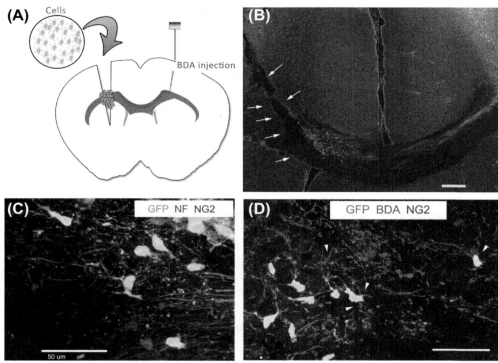

FIGURE 9.3 Spatial relationship between severed adult callosal axons and NG2 cells. (A) Schematic illustration of the model. The corpus callosum was completely transected, and GFP+ NG2 cells were transplanted into the lesion on the same day. Twenty-four hours before the animal was sacrificed, BDA was injected directly contralateral to the lesion. (B) An example of a lesion. Montage of images from a coronal section of a control animal that was lesioned but did not receive transplanted cells, 1 day postlesioning (1 dpl). White arrows indicate the site of lesion. BDA injected into the contralateral hemisphere (right) at the time of lesioning and detected with Cy3-conjugated streptavidin (red) reveals labeled axons terminating proximal to the lesion, confirming that a complete transection had been made. Scale bar, 100 μm. (C) An 8-dpl lesion showing NG2 expression (blue) on GFP+ grafted cells and endogenous NG2 cells (GFP−). Axons that are labeled with the smi-31 antibody to phosphorylated neurofilaments (NF; red) appear to navigate through an area rich in NG2. (D) The relationship between grafted GFP+ cells, NG2 cells (blue), and BDA-labeled axons (red) at 17 dpl. Most of the transplanted cells (GFP+, green) expressed NG2 (blue). BDA-labeled severed axons (red) were intermingled with transplanted (GFP+) and endogenous (GFP−) NG2 cells (arrowheads). Scale bars in (C) and (D), 50 μm. (For interpretation of the references to color in this figure legend, the reader is referred to the online version of this book.)

In successful grafts in which donor GFP+ cells survived and were found in close proximity to BDA+ axons, severed axons were seen to grow toward NG2 cells in the graft (Figure 9.3(C) and (D)), and their terminals seemed to land on NG2 cells (arrowheads in Figure 9.3(D)). The severed axons were seen intertwined with both grafted and existing NG2 cells, suggesting that severed, regrowing axons are not repelled by NG2 expressed on NG2 cells. The close apposition of axon terminals during regeneration and development suggests that there is a neuronal receptor for NG2 on axons that is capable of transmitting signal from NG2 cells to the axon, but the nature of such signal transduction in axons has not been elucidated.

9.6 THE ROLE OF NG2 IN CELL PROLIFERATION

NG2 has been shown to bind with high affinity two growth factors, platelet-derived growth factor-AA (PDGF-AA) and fibroblast growth factor-2 (FGF2) [52]. These are two key growth factors that regulate self-renewal and differentiation of oligodendrocyte lineage cells (Figure 9.2(D)) [53,54]. Unlike the interaction of heparan sulfate proteoglycans with growth factors, which occurs primarily through their glycosaminoglycan chains, the ability of NG2 to interact with these growth factors does not depend on the presence of chondroitin sulfate chains. Interestingly, the high-affinity binding of these growth factors by NG2 is highly specific, and NG2 does not bind PDGF-BB or other growth factors, including epidermal growth factor and VEGF (vascular endothelial growth factor). This was rather surprising, as NG2 is also expressed by vascular pericytes that express PDGFRβ.

PDGF-AA (PDGF-1) shares 56.8% amino acid identity with PDGF-BB (PDGF-2) in the mature peptide region and belongs to the same subclass of the PDGF/VEGF family of growth factors [55]. PDGF-AA is the most well characterized mitogen that is required for the developmental expansion of oligodendrocyte lineage cells, and it promotes their proliferation as well as migration through

PDGFRα [56,57]. On rat NG2 cells in culture, NG2 is found in a molecular complex that includes PDGFRα, and blocking NG2 with a rabbit anti-NG2 antibody reduced the number of these cells, suggesting a role for NG2 in PDGFRα-mediated proliferation and/or survival [58]. Similarly, blocking NG2 on aortic smooth muscle cells, which express PDGFRα, reduced their proliferation and migration in response to PDGF-AA but not to PDGF-BB [59]. Similarly, smooth muscle cells isolated from NG2-knockout mice exhibited significantly reduced proliferation and migration in response to PDGF-AA but not to PDGF-BB [60]. Furthermore, NG2 cells from NG2-knockout mice could not be maintained in the progenitor state even in the presence of PDGF-AA or PDGF-AA plus FGF2 and differentiated prematurely into O4+ late-stage progenitor/preoligodendrocytes [61]. These observations indicate a specific role for NG2 in mediating cellular effects triggered by PDGF-AA but not PDGF-BB. In NG2-knockout mice, proliferation of NG2 cells in the cerebellar white matter is reduced during the first postnatal week, resulting in a reduced number of oligodendrocytes, but their numbers are restored by the third postnatal week and no overt phenotype is observed [62]. After acute chemically induced demyelination in the spinal cord, NG2 cell proliferation is also compromised in NG2-knockout mice, and the repair outcome reflects compromised proliferation of the three types of NG2-expressing cells that appear in the lesion: NG2 cells, pericytes, and macrophages [63]. The mechanism by which loss of NG2 reduces cell proliferation in these in vivo models remains unclear.

Consistent with the greater propensity of cells expressing NG2 to grow tumors when transplanted into host mice [27], NG2 is detected in a variety of human tumors. Analysis of human glioma tissues revealed the presence of NG2+ tumor cells in 7/7 oligodendroglioma, 3/3 in pilocytic astrocytoma, and 1/6 glioblastoma multiforme cases [64]. Another study revealed a higher frequency of NG2 expression in high-grade brain tumors compared with low-grade tumors and upregulation of NG2 expression on cultured glioma cells by FGF2 [65].

Outside the CNS, NG2 interacts with FGFR1 and FGFR3 and promotes cell proliferation in response to FGF2. NG2-knockout mice exhibited defective corneal neovascularization upon stimulation with FGF2 [66]. Furthermore, pericyte-specific knockout of NG2 reduced pericyte proliferation and coverage of endothelial cells by pericytes and basal lamina in microvessels that supply intracranial melanoma, leading to defective formation of tight junctions and hence leaky vessels [67]. This study also showed that NG2 is required for the activation of β1-integrins and focal adhesion kinase not only on pericytes but also in endothelial cells. Thus, NG2 seems to be able to activate β1-integrin both in *cis* and in *trans* and play an essential role in maintaining the vascular integrity in the CNS.

9.7 THE ROLE OF NG2 IN INTRACELLULAR SIGNALING

The studies described above suggest that NG2 serves as a coreceptor for other integral membrane signaling molecules such as integrins and receptor tyrosine kinases. How might the short cytoplasmic tail of NG2 interact with the intracellular signaling machinery?

9.7.1 Activation of Rho Family GTPases

The C-terminus of the cytoplasmic tail of NG2 contains the consensus QYWV sequence used for interacting with intracellular proteins that contain a PDZ domain. By the yeast two-hybrid assay, the multi-PDZ-binding domain protein 1 (Mupp1) was identified from mouse embryos as a protein that interacts with the PDZ-binding domain of NG2 [68]. The Trotter lab has identified additional PDZ proteins that interact with the cytoplasmic domain of NG2, including glutamate receptor-interacting protein and the adaptor protein Syntenin-1, also known as Syndecan-binding protein 1 [69,70], but the exact function of these interactions remains unknown. Several studies have shown that NG2 activates members of the Rho family of GTPases, which play an important role in cell spreading, process extension, and migration. Engagement or clustering of NG2 on melanoma cells leads to phosphorylation of breast cancer anti-estrogen resistance protein 1 (also known as CRK-associated substrate or p130[cas]), which is mediated by activation of the GTPase Cdc42 [71]. A new study suggests a role for NG2 in activating another Rho GTPase family member, RhoA, through Mupp1 or Syx1 [72], which appears to be important for establishing cell polarity in response to FGF2 or in mediating contact inhibition. Protein kinase Cα (PKCα), which causes membrane protrusions and redistribution of NG2, phosphorylates the cytoplasmic domain of NG2 on threonine 2256 (T2256) [73]. In cells transfected with the T2256E mutant of NG2, which mimics phosphorylation at this site, NG2 becomes colocalized with β1-integrin on lamellipodia at the leading edges [7], and the cells become more motile even without stimulation by PKCα.

9.7.2 Extracellular Signal-Regulated Kinase Signaling

Another threonine residue in the cytoplasmic domain of NG2, T2314, was identified as a potential site of docking with and phosphorylation by extracellular signal-regulated kinase (ERK), and this was experimentally confirmed by in vitro phosphorylation [74]. Glioma cells transfected with the phosphomimetic mutant T2314E exhibited a higher level of basal proliferation, similar to cells transfected with constitutively active MEK-DD. Intriguingly, cells transfected with the T2256E mutant of NG2 exhibited reduced proliferation, suggesting that phosphorylation of the two threonine residues in the cytoplasmic domain by ERK and PKCα has opposing effects on cell motility and migration.

9.7.3 Akt Signaling

A novel role for NG2 in mediating Akt-dependent cell survival and chemoresistance was described in a study using human glioma cell lines expressing various levels of NG2 [75]. Knockdown of NG2 in glioma cell lines by small interfering RNA dramatically increased apoptosis in response to tumor necrosis factor α and other cytotoxic compounds. Interestingly, this effect of NG2 knockdown could be reversed by blocking the function of the β1 subunit of α3β1-integrin, which forms a complex with NG2. Thus, the antiapoptotic effect appears to involve an α3β1-integrin signal. In this model, treatment of NG2-expressing glioma cells with the irreversible Akt inhibitor wortmannin significantly increased apoptosis, and knockdown of NG2 reduced the level of phosphorylated Akt. These observations suggest that NG2 confers cell survival and chemoresistance by activating the Akt pathway. Akt has been shown to play a major role in NG2 cell proliferation [76] and oligodendrocyte differentiation [77,78], and it remains to be shown whether NG2 is involved in mediating the upstream signaling leading to its activation.

9.8 CONCLUDING REMARKS

In the CNS, NG2 has served as a valuable tool with which to study and genetically manipulate NG2-expressing glial cells, also known as NG2 cells, OPCs, or polydendrocytes. The ability to quantitatively and consistently identify these cells at the light-microscopy level has revealed a fourth glial cell population with unique biochemical, morphological, and electrophysiological properties, whose presence had been largely overlooked until the end of the twentieth century. Many of the key studies on the function of NG2 have been conducted on nonneural cells but they have revealed that NG2 is not merely a useful marker but rather plays a critical role in fine-tuning the cellular responses to growth factors and the extracellular matrix. It not only influences the cell that expresses NG2 but also affects the cell that interacts with NG2 in *trans*. Thus NG2 could provide bidirectional signaling important for development and tumor growth. The findings obtained on the function of NG2 in the vasculature could be used to direct future studies toward understanding the role of NG2 in the nervous system.

REFERENCES

[1] Schubert D, Heinemann S, Carlisle W, Tarikas H, Kimes B, Patrick J, et al. Clonal cell lines from the rat central nervous system. Nature 1974;249:224–7.

[2] Stallcup WB, Cohn M. Correlation of surface antigens and cell type in cloned cell lines from the rat central nervous system. Exp Cell Res 1976;98:285–97.

[3] Wilson SS, Baetge EE, Stallcup WB. Antisera specific for cell lines with mixed neuronal and glial properties. Dev Biol 1981;83:146–53.

[4] Kohler G, Milstein C. Continuous cultures of fused cells secreting antibody of predefined specificity. Nature 1975;256:495–7.

[5] Stallcup WB, Beasley L, Levine J. Cell-surface molecules that characterize different stages in the development of cerebellar interneurons. Cold Spring Harb Symp Quant Biol 1983;XLVIII:761–74.

[6] Nishiyama A, Dahlin KJ, Prince JT, Johnstone SR, Stallcup WB. The primary structure of NG2, a novel membrane-spanning proteoglycan. J Cell Biol 1991;114:359–71.

[7] Stallcup WB, Huang FJ. A role for the NG2 proteoglycan in glioma progression. Cell Adh Migr 2008;2:192–201.

[8] Staub E, Hinzmann B, Rosenthal A. A novel repeat in the melanoma-associated chondroitin sulfate proteoglycan defines a new protein family. FEBS Lett 2002;527:114–8.

[9] Nishiyama A, Lin XH, Stallcup WB. Generation of truncated forms of the NG2 proteoglycan by cell surface proteolysis. Mol Biol Cell 1995;6:1819–32.

[10] Bumol TF, Reisfeld RA. Unique glycoprotein-proteoglycan complex defined by monoclonal antibody on human melanoma cells. Proc Natl Acad Sci USA 1982;79:1245–9.

[11] Harper JR, Bumol TF, Reisfeld RA. Characterization of monoclonal antibody 155.8 and partial characterization of its proteoglycan antigen on human melanoma cells. J Immunol 1984;132:2096–104.

[12] Pluschke G, Vanek M, Evans A, Dittmar T, Schmid P, Itin P, et al. Molecular cloning of a human melanoma-associated chondroitin sulfate proteoglycan. Proc Natl Acad Sci USA 1996;93:9710–5.

[13] Jung M, Kramer E, Grzenkowski M, Tang K, Blakemore W, Aguzzi A, et al. Lines of murine oligodendroglial precursor cells immortalized by an activated *neu* tyrosine kinase show distinct degrees of interaction with axons *in vitro* and *in vivo*. Eur J Neurosci 1995;7:1245–65.

[14] Niehaus A, Stegmuller J, Diers-Fenger M, Trotter J. Cell-surface glycoprotein of oligodendrocyte progenitors involved in migration. J Neurosci 1999;19:4948–61.

[15] Schneider S, Bosse F, D'Urso D, Muller H, Sereda MW, Nave K, et al. The AN2 protein is a novel marker for the Schwann cell lineage expressed by immature and nonmyelinating Schwann cells. J Neurosci 2001;21:920–33.

[16] Zhu X, Bergles DE, Nishiyama A. NG2 cells generate both oligodendrocytes and gray matter astrocytes. Development 2008;135:145–57.

[17] Komitova M, Zhu X, Serwanski DR, Nishiyama A. NG2 cells are distinct from neurogenic cells in the postnatal mouse subventricular zone. J Comp Neurol 2009;512:702–16.

[18] Nishiyama A. Polydendrocytes: NG2 cells with many roles in development and repair of the CNS. Neuroscientist 2007;13:62–76.

[19] Hill RA, Nishiyama A. NG2 cells (polydendrocytes): listeners to the neural network with diverse properties. Glia 2014;62:1195–2010.

[20] Richardson WD, Young KM, Tripathi RB, McKenzie I. NG2-glia as multipotent neural stem cells: fact or fantasy? Neuron 2011;70:661–73.

[21] Bergles DE, Roberts JD, Somogyi P, Jahr CE. Glutamatergic synapses on oligodendrocyte precursor cells in the hippocampus. Nature 2000;405:187–91.

[22] Martin S, Levine AK, Chen ZJ, Ughrin Y, Levine JM. Deposition of the NG2 proteoglycan at nodes of Ranvier in the peripheral nervous system. J Neurosci 2001;21:8119–28.

[23] Stallcup WB, Dahlin K, Healy P. Interaction of the NG2 chondroitin sulfate proteoglycan with type VI collagen. J Cell Biol 1990;111:3177–88.

[24] Nishiyama A, Stallcup WB. Expression of NG2 proteoglycan causes retention of type VI collagen on the cell surface. Mol Biol Cell 1993;4:1097–108.

[25] Tillet E, Ruggiero F, Nishiyama A, Stallcup WB. The membrane-spanning proteoglycan NG2 binds to collagens V and VI through the central nonglobular domain of its core protein. JBC 1997;272:10769–76.

[26] Burg MA, Nishiyama A, Stallcup WB. A central segment of the NG2 proteoglycan is critical for the ability of glioma cells to bind and migrate toward type VI collagen. Exp Cell Res 1997;235:254–64.

[27] Burg MA, Grako KA, Stallcup WB. Expression of the NG2 proteoglycan enhances the growth and metastatic properties of melanoma cells. J Cell Physiol 1998;177:299–312.

[28] Iida J, Meijne AML, Spiro RC, Roos E, Furcht LT, McCarthy JB. Spreading and focal contact formation of human melanoma cells in response to the stimulation of both melanoma-associated proteoglycan (NG2) and alpha 4 beta 1 integrin. Cancer Res 1995;55:2177–85.

[29] Tillet E, Gential B, Garrone R, Stallcup WB. NG2 proteoglycan mediates beta1 integrin-independent cell adhesion and spreading on collagen VI. J Cell Biochem 2002;86:726–36.

[30] Cheng JS, Dubal DB, Kim DH, Legleiter J, Cheng IH, Yu GQ, et al. Collagen VI protects neurons against Abeta toxicity. Nat Neurosci 2009;12:119–21.

[31] Ruhl M, Sahin E, Johannsen M, Somasundaram R, Manski D, Riecken EO, et al. Soluble collagen VI drives serum-starved fibroblasts through S phase and prevents apoptosis via down-regulation of Bax. J Biol Chem 1999;274:34361–8.

[32] Chen P, Cescon M, Megighian A, Bonaldo P. Collagen VI regulates peripheral nerve myelination and function. FASEB J 2014;28:1145–56.

[33] Fukushi J, Makagiansar IT, Stallcup WB. NG2 proteoglycan promotes endothelial cell motility and angiogenesis via engagement of galectin-3 and alpha3beta1 integrin. Mol Biol Cell 2004;15:3580–90.

[34] Wen Y, Makagiansar IT, Fukushi J, Liu FT, Fukuda MN, Stallcup WB. Molecular basis of interaction between ng2 proteoglycan and galectin-3. J Cell Biochem 2006;98:115–27.

[35] Pasquini LA, Millet V, Hoyos HC, Giannoni JP, Croci DO, Marder M, et al. Galectin-3 drives oligodendrocyte differentiation to control myelin integrity and function. Cell Death Differ 2011;18:1746–56.

[36] Quinta HR, Pasquini JM, Rabinovich GA, Pasquini LA. Glycan-dependent binding of galectin-1 to neuropilin-1 promotes axonal regeneration after spinal cord injury. Cell Death Differ 2014;21:941–55.

[37] Chen ZJ, Ughrin Y, Levine JM. Inhibition of axon growth by oligodendrocyte precursor cells. Mol Cell Neurosci 2002;20:125–39.

[38] Ughrin YM, Chen ZJ, Levine JM. Multiple regions of the NG2 proteoglycan inhibit neurite growth and induce growth cone collapse. J Neurosci 2003;23:175–86.

[39] Schwab ME, Caroni P. Oligodendrocytes and CNS myelin are non-permissive substrates for neurite growth and fibroblast spreading in vitro. J Neurosci 1988;8:2381–93.

[40] Yang Z, Suzuki R, Daniels SB, Brunquell CB, Sala CJ, Nishiyama A. NG2 glial cells provide a favorable substrate for growing axons. J Neurosci 2006;26:3829–39.

[41] Woolsey TA, Van der Loos H. The structural organization of layer IV in the somatosensory region (SI) of mouse cerebral cortex. The description of a cortical field composed of discrete cytoarchitectonic units. Brain Res 1970;17:205–42.

[42] Hill RA, Natsume R, Sakimura K, Nishiyama A. NG2 cells are uniformly distributed and NG2 is not required for barrel formation in the somatosensory cortex. Mol Cell Neurosci 2011;46:689–98.

[43] Yiu G, He Z. Glial inhibition of CNS axon regeneration. Nat Rev Neurosci 2006;7:617–27.

[44] Zhao RR, Fawcett JW. Combination treatment with chondroitinase ABC in spinal cord injury—breaking the barrier. Neurosci Bull 2013;29:477–83.

[45] Jones LL, Yamaguchi Y, Stallcup WB, Tuszynski MH. NG2 is a major chondroitin sulfate proteoglycan produced after spinal cord injury and is expressed by macrophages and oligodendrocyte progenitors. J Neurosci 2002;22:2792–803.

[46] Tan AM, Colletti M, Rorai AT, Skene JH, Levine JM. Antibodies against the NG2 proteoglycan promote the regeneration of sensory axons within the dorsal columns of the spinal cord. J Neurosci 2006;26:4729–39.

[47] Petrosyan HA, Hunanyan AS, Alessi V, Schnell L, Levine J, Arvanian VL. Neutralization of inhibitory molecule NG2 improves synaptic transmission, retrograde transport, and locomotor function after spinal cord injury in adult rats. J Neurosci 2013;33:4032–43.

[48] de Castro Jr R, Tajrishi R, Claros J, Stallcup WB. Differential responses of spinal axons to transection: influence of the NG2 proteoglycan. Exp Neurol 2005;192:299–309.

[49] Hossain-Ibrahim MK, Rezajooi K, Stallcup WB, Lieberman AR, Anderson PN. Analysis of axonal regeneration in the central and peripheral nervous systems of the NG2-deficient mouse. BMC Neurosci 2007;8:80.

[50] Jones LL, Sajed D, Tuszynski MH. Axonal regeneration through regions of chondroitin sulfate proteoglycan deposition after spinal cord injury: a balance of permissiveness and inhibition. J Neurosci 2003;23:9276–88.

[51] McTigue DM, Tripathi R, Wei P. NG2 colocalizes with axons and is expressed by a mixed cell population in spinal cord lesions. J Neuropathol Exp Neurol 2006;65:406–20.

[52] Goretzki L, Burg MA, Grako KA, Stallcup WB. High-affinity binding of basic fibroblast growth factor and platelet-derived growth factor-AA to the core protein of the NG2 proteoglycan. JBC 1999;274:16831–7.

[53] Bögler O, Wren D, Barnett SC, Land H, Noble M. Cooperation between two growth factors promotes extended self-renewal and inhibits differentiation of oligodendrocyte-type-2 astrocyte (O-2A) progenitor cells. Proc Natl Acad Sci USA 1990;87:6368–72.

[54] Pfeiffer SE, Warrington AE, Bansal R. The oligodendrocyte and its many cellular processes. Trends Cell Biol 1993;3:191–7.

[55] Andrae J, Gallini R, Betsholtz C. Role of platelet-derived growth factors in physiology and medicine. Genes Dev 2008;22:1276–312.

[56] Richardson WD, Pringle N, Mosley MJ, Westermark B, Dubois-Dalcq M. A role for platelet-derived growth factor in normal gliogenesis in the central nervous system. Cell 1988;53:309–19.

[57] Calver AR, Hall AC, Yu W-P, Walsh FS, Heath JK, Betsholtz C, et al. Oligodendrocyte population dynamics and the role of PDGF in vivo. Neuron 1998;20:869–82.

[58] Nishiyama A, Lin XH, Giese N, Heldin CH, Stallcup WB. Interaction between NG2 proteoglycan and PDGF alpha-receptor on O2A progenitor cells is required for optimal response to PDGF. J Neurosci Res 1996;43:315–30.

[59] Grako K, Stallcup WB. Participation of the NG2 proteoglycan in rat aortic smooth muscle cell responses to platelet-derived growth factor. Exp Cell Res 1995;22:231–40.

[60] Grako K, Ochiya T, Barritt D, Nishiyama A, Stallcup WB. PDGF α-receptor is unresponsive to PDGF-AA in aortic smooth muscle cells from the NG2 knockout mouse. J Cell Sci 1999;112:905–15.

[61] Stallcup WB. The NG2 proteoglycan: past insights and future prospects. J Neurocytol 2002;31:423–35.

[62] Kucharova K, Stallcup WB. The NG2 proteoglycan promotes oligodendrocyte progenitor proliferation and developmental myelination. Neuroscience 2010;166:185–94.

[63] Kucharova K, Chang Y, Boor A, Yong VW, Stallcup WB. Reduced inflammation accompanies diminished myelin damage and repair in the NG2 null mouse spinal cord. J Neuroinflammation 2011;8:158.

[64] Shoshan Y, Nishiyama A, Chang A, Mork S, Barnett GH, Cowell JK, et al. Expression of oligodendrocyte progenitor cell antigens by gliomas: implications for the histogenesis of brain tumors. Proc Natl Acad Sci USA 1999;96:10361–6.

[65] Chekenya M, Rooprai HK, Davies D, Levine JM, Butt AM, Pilkington GJ. The NG2 chondroitin sulfate proteoglycan: role in malignant progression of human brain tumours. Int J Dev Neurosci 1999;17:421–35.

[66] Cattaruzza S, Ozerdem U, Denzel M, Ranscht B, Bulian P, Cavallaro U, et al. Multivalent proteoglycan modulation of FGF mitogenic responses in perivascular cells. Angiogenesis 2013;16:309–27.

[67] You WK, Yotsumoto F, Sakimura K, Adams RH, Stallcup WB. NG2 proteoglycan promotes tumor vascularization via integrin-dependent effects on pericyte function. Angiogenesis 2014;17:61–76.

[68] Barritt DS, Pearn MT, Zisch AH, Lee SS, Javier RT, Pasquale EB, et al. The multi-PDZ domain protein MUPP1 is a cytoplasmic ligand for the membrane-spanning proteoglycan NG2. J Cell Biochem 2000;79:213–24.

[69] Stegmuller J, Werner H, Nave KA, Trotter J. The proteoglycan NG2 is complexed with alpha-amino-3-hydroxy-5-methyl-4-isoxazolepropionic acid (AMPA) receptors by the PDZ glutamate receptor interaction protein (GRIP) in glial progenitor cells. Implications for glial-neuronal signaling. J Biol Chem 2003;278:3590–8.

[70] Chatterjee N, Stegmuller J, Schatzle P, Karram K, Koroll M, Werner HB, et al. Interaction of syntenin-1 and the NG2 proteoglycan in migratory oligodendrocyte precursor cells. J Biol Chem 2008;283:8310–7.

[71] Eisenmann KM, McCarthy JB, Simpson MA, Keely PJ, Guan JL, Tachibana K, et al. Melanoma chondroitin sulphate proteoglycan regulates cell spreading through Cdc42, Ack-1 and p130cas. Nat Cell Biol 1999;1:507–13.

[72] Biname F, Sakry D, Dimou L, Jolivel V, Trotter J. NG2 regulates directional migration of oligodendrocyte precursor cells via Rho GTPases and polarity complex proteins. J Neurosci 2013;33:10858–74.

[73] Makagiansar IT, Williams S, Dahlin-Huppe K, Fukushi J, Mustelin T, Stallcup WB. Phosphorylation of NG2 proteoglycan by protein kinase C-alpha regulates polarized membrane distribution and cell motility. J Biol Chem 2004;279:55262–70.

[74] Makagiansar IT, Williams S, Mustelin T, Stallcup WB. Differential phosphorylation of NG2 proteoglycan by ERK and PKCalpha helps balance cell proliferation and migration. J Cell Biol 2007;178:155–65.

[75] Chekenya M, Krakstad C, Svendsen A, Netland IA, Staalesen V, Tysnes BB, et al. The progenitor cell marker NG2/MPG promotes chemoresistance by activation of integrin-dependent PI3K/Akt signaling. Oncogene 2008;27:5182–94.

[76] Hill RA, Patel KD, Medved J, Reiss AM, Nishiyama A. NG2 cells in white matter but not gray matter proliferate in response to PDGF. J Neurosci 2013;33:14558–66.

[77] Tyler WA, Gangoli N, Gokina P, Kim HA, Covey M, Levison SW, et al. Activation of the mammalian target of rapamycin (mTOR) is essential for oligodendrocyte differentiation. J Neurosci 2009;29:6367–78.

[78] Narayanan SP, Flores AI, Wang F, Macklin WB. Akt signals through the mammalian target of rapamycin pathway to regulate CNS myelination. J Neurosci 2009;29:6860–70.

Chapter 10

Comprehensive Overview of CD133 Biology in Neural Tissues across Species

József Jászai[1], Denis Corbeil[2] and Christine A. Fargeas[2]

[1]*Institute of Anatomy, Medizinische Fakultät der Technischen Universität Dresden, Dresden, Germany;* [2]*Tissue Engineering Laboratories (BIOTEC), Medizinische Fakultät der Technischen Universität Dresden, Dresden, Germany*

10.1 INTRODUCTION

Prominin-1 (hereafter CD133) was initially identified as a novel antigen (13A4) of the apical domain of polarized epithelial cells present in the mouse embryonic neuroepithelium, adult ependymal layer, and proximal tubules of the kidney and characterized by its specific subcellular location in plasma membrane protrusions [1,2]. Its name was chosen from the Latin word *"prominere"* to reflect this unique feature. At the same time, its human homolog was independently characterized in hematopoietic stem and progenitor cells by its AC133 epitope [3,4] and rapidly gained interest as a useful antigenic marker for the isolation of human stem (and cancer stem) cells in various organs. Thus, CD133+ progenitor cells were successfully isolated from human fetal and postmortem brains [5,6]. Later neural progenitors from postnatal mouse cerebellum and adult hippocampus were isolated similarly [7,8].

10.1.1 CD133

The structure of the CD133 glycoprotein, which comprises up to 865 amino acids, is depicted (Figure 10.1) [1,4]. It shares this unique membrane topology, as well as structural and biochemical features, with the other prominin family members [9]. Two related prominin genes (*prom1* and *prom2*) are present in mammals, whereas in nonmammalian vertebrates a third relative (*prom3*) exists [10–12]. In zebrafish, *prom2* is absent and *prom1*, unlike *prom3*, is duplicated, presenting two co-orthologs (a and b) [10,12].

Several CD133 splice variants (named s1–s12) affecting its protein sequence have been described both in rodent and in human [13–18], and additional ones in other species [11,19]. The inclusion/exclusion of facultative exons mainly alters the extracellular (EC) 1 and intracellular (IC) 3 domains (Figure 10.1). In the N-terminal domain (EC1) the inclusion or not of a 9 amino acid stretch (encoded by a 27-bp exon) distinguishes CD133.s1 and s2 variants. It is of note that both splice variants in human can confer AC133 epitope positivity (see below) [4,20]. Probably reflecting the functional importance of the cytoplasmic C-terminal domain (IC3), 10 different CD133 C-termini (referred to as types A to J) can be generated [19,21]. Their expression is tissue-specific and may be developmentally regulated [16,18,20]. For instance, mouse myelin sheath contains a CD133 variant with a lower molecular weight (100 kDa vs. 115 kDa (kidney)), which may correspond to CD133.s3, which harbors a shorter C-terminus (type B) and is expressed in cultured oligodendrocytes. On the other hand, cultured astrocytes express mainly s1 variant (type A) [18]. Recently, CD133.s6 (type C), which lacks the portion of the C-terminal domain after the membrane-proximal tyrosine phosphorylation consensus site ($\underline{R}_{821}MDSEDV\underline{Y}_{828}$) was detected in the adult mouse subventricular zone (SVZ) [8]. The various C-terminal domains display diverse potential PSD95/Dlg/ZO-1 (PDZ)-binding motifs, suggesting that various interactions of CD133 with cytoplasmic PDZ-domain-containing protein-interacting partners may occur [17]. Moreover, one tyrosine residue (Y_{828}) is conserved in all types of alternative C-termini [17]. Interestingly, Y_{828} is, with tyrosine 852 (Y_{852}), one of the two targets of Src and Fyn kinases in human medulloblastoma cells [22]. The phosphorylation of Y_{828} was shown to regulate the interaction of CD133 with the 85-kDa regulatory subunit of phosphoinositide 3-kinase (PI3K) and the subsequent activation of the PI3K/Akt pathway, which brings to light a possible role for CD133 in cancer stem cell self-renewal (Figure 10.1) [23]. The IC1 domain has been shown to mediate the interaction of CD133, possibly via lysine residue 129 (K_{129}), with histone deacetylase 6 (HDAC6), which modulates the acetylation of tubulin (Figure 10.1) [24]. Such interaction is proposed to negatively regulate CD133 trafficking down to the endosomal compartment (Figure 10.2). CD133–HDAC6 complexes would favor the stabilization of β-catenin, a central molecule of the canonical Wnt signaling pathway, leading to the activation of β-catenin signaling targets [24].

Neural Surface Antigens. http://dx.doi.org/10.1016/B978-0-12-800781-5.00010-4

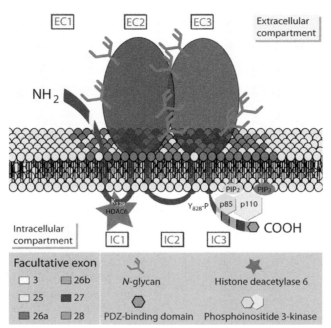

FIGURE 10.1 Membrane organization of CD133 and its potential interacting partners. CD133 contains five transmembrane segments separating two small intracellular (IC1 and IC2) domains and two large extracellular (EC2 and EC3) loops that contain all eight potential N-glycosylation sites (green forks). The N-terminal domain (EC1) is located in the EC compartment, whereas the C-terminal domain (IC3) is inside the cell. Within the membrane, CD133 is incorporated into cholesterol-based membrane microdomains (lipid rafts; red). Various facultative exons have been described. For instance, the CD133 splice variant s2 contains an additional exon encoding 9 amino acids (yellow area) within the EC1 domain that are missing in variant s1. Likewise, the C-terminal domain is subjected to alternative splicing that involves exons 25–28. The numbering of the exons begins with the exon bearing the initial start codon. Distinct classes of PDZ-binding motif are found at the C-terminal domain of certain CD133 splice variants, suggesting its interaction with an unidentified PDZ-domain-containing protein-interacting partner. Two cytoplasmic signal transducer enzymes have been described to interact with CD133. First, the histone deacetylase 6 (HDAC6) binds to the IC1 domain, involving lysine residue 129 (K_{129}) to negatively regulate the trafficking of CD133 down the endosomal–lysosomal pathway for degradation. Second, the 85-kDa regulatory subunit (p85) of PI3K interacts with the phosphorylated tyrosine residue 828 (Y_{828}-P) of CD133 resulting in the activation of the PI3K/Akt pathway. PI3K catalyzes the conversion of phosphoinositide 4,5-biphosphate (PIP_2) to phosphoinositide 3,4,5-trisphosphate (PIP_3) at the inner leaflet of the plasma membrane, and the latter lipid becomes an anchor and activator of pleckstrin homology domain-containing enzymes such as Akt, which in turns phosphorylates numerous kinases and transcription factors. PDZ, PSD95/Dlg/ZO-1; PI3K, phosphoinositide 3-kinase. (For interpretation of the references to color in this figure legend, the reader is referred to the online version of this book.)

In a given species, distinct and overlapping expression patterns are observed among prominin paralogs [9,25–28]. For instance, both *prom1* and *prom2* genes are expressed in human and mouse kidney as well as in the digestive tract, with the notable exception of the esophagus (only prominin-2; [9]), whereas the hematopoietic compartment and the retina express solely CD133 [9,29,30]. In tissues expressing two distinct prominin molecules, the knockdown of one of them may be compensated for by an upregulation of another one, as illustrated in the dentate gyrus of CD133-null mice [8].

CD133 and the other prominin relatives are selectively concentrated in plasma membrane protrusions irrespective of the cell type. For instance, murine and human CD133 are targeted to the apical membrane of polarized epithelial cells, such as neural progenitors and kidney proximal cells, and selectively retained in microvilli and primary cilium [1,31–33]. Remarkably, no targeting/retention motif appears in the amino acid sequences of prominin family members,

and CD133 maintains its specific subcellular localization through a direct interaction with membrane cholesterol and its association with specific membrane microdomains (lipid rafts) [34–37]. An interaction of CD133 with membrane microdomain-associated lipids other than cholesterol has also been suggested. Fantini et al. have indeed proposed the presence of two potential ganglioside-binding sites in the CD133 structure [38]. This would be in line with the observation that CD133 and the glycolipid GM_1 are colocalized in microvilli of epithelial cells [39]. The membrane topology of CD133, with its two large glycosylated EC2 and EC3 domains, and the conical shape of GM_1 may be reflected by their preference for curved membranes [40]. For an overview of GM_1 biology in neural cells please refer to Chapter 8 by Itokazu and Yu.

As yet, the physiological function of CD133 remains elusive. Nonetheless, CD133-knockout mice suffer from a defect in the morphogenesis and/or maintenance of the photoreceptor neuron, notably the outer segment

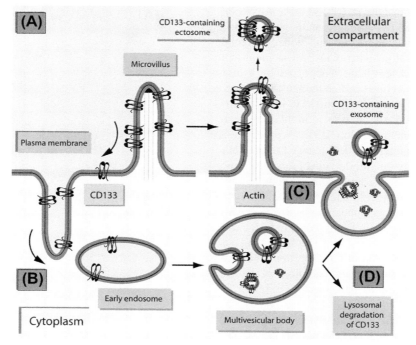

FIGURE 10.2 Distinct molecular mechanisms regulate the expression of CD133 at the plasma membrane. At the plasmalemma of epithelial and non-epithelial cells, CD133 is selectively concentrated in membrane protrusions such as microvilli and primary cilium (not depicted), and a cholesterol-based membrane microdomain (red) is involved in its specific retention. The surface expression of CD133 protein can be modulated by three distinct molecular mechanisms that are not mutually exclusive. First, CD133 could be released into the extracellular compartment by the budding of small membrane vesicles (referred to as ectosomes) from membrane protrusions (A). Second, CD133 could be internalized and transported to the early endosomal compartment (B). Afterward, CD133 is delivered to late endosomal multivesicular bodies, which involves the inward invagination and budding of the limiting membrane of the endosome toward its own lumen. Multivesicular bodies could either fuse with the plasma membrane, and hence release internal vesicles (exosomes) (C), or lead to the proteolytic degradation of CD133 by a heterotypic fusion with lysosomes (D). Biochemically, the release of CD133-containing ectosomes and exosomes might involve a specific membrane microdomain. (For interpretation of the references to color in this figure legend, the reader is referred to the online version of this book.)

(a specialized cilium), which leads to blindness. This suggests a structural role for CD133 in membrane architecture (Section 10.7) [29]. Within the plasma membrane the specific interaction of CD133 with the surrounding protein and lipid environments may modify the general organization and/or composition of the plasma membrane microdomains, particularly in the protrusions. Similarly, the potential interaction of CD133 with cytoplasmic partners such as PDZ-domain-containing proteins, PI3K, and HDAC6 might also regulate the biogenesis and maintenance of microvilli and primary cilium, respectively. For instance, the activation of HDAC6 through phosphorylation by Aurora A leads to tubulin deacetylation and, hence, disassembly of the axoneme, the central microtubule complex within the cilium [41]. The association of PDZ-domain-containing proteins with lipid rafts is also interesting in this context [42]. A complex interplay between these adaptor proteins, CD133, and lipid raft components may be essential for the generation and maintenance of apical plasma membrane protrusions and other membrane projections involved in various cellular processes including migration of CD133+ cancer cells [2,43].

10.1.2 Detecting Mammalian (Mouse and Human) CD133

The molecular cloning of murine CD133 resulted from the immunological screening of a kidney cDNA expression library using the rat 13A4 monoclonal antibody (mAb) [1]. This first antibody was generated by injecting a rat with telencephala of 12-day-old mouse embryos, consisting mostly of neuroepithelial cells. MAb 13A4 was identified by its labeling of the apical domain of neuroepithelial cells [1]. This well-characterized and commercially available antibody does not show the apparent stem and progenitor cell-restricted pattern of some antibodies directed against human CD133 (e.g., AC133, see below). Nonetheless, it has proved a useful tool to characterize CD133 expression in murine tissues as well as to isolate neural stem cells [1,44–46]. Its epitope, although not fully mapped, is located in the second half of the CD133 EC3 domain (Figure 10.1) [47]. Technically, 13A4 mAb can be applied to flow cytometry, immunofluorescence, immunocyto/histochemistry, immunogold electron microscopy, and immunoblotting [1,8,16,33].

The general interest in human CD133 as a novel cell surface antigen expressed by various somatic stem cells is reflected in a variety of commercial and noncommercial antibodies [48]. Historically, human CD133 was identified as the novel antigenic determinant recognized by a new antibody generated by injecting purified human CD34$^+$ progenitors into mice. The first antibody, mAb AC133, specifically labeled in flow cytometry analysis retinoblastoma cell lines and a subset of hematopoietic CD34^{bright} progenitors from fetal liver and bone marrow [3]. Human CD133 cDNA sequence was obtained by polymerase chain reaction using the information gained from the sequencing of peptide fragments derived from immunoprecipitated AC133 antigen [4]. The AC133 (CD133/1) epitope appears to be dependent on its three-dimensional conformation and seems to be located in the EC3 domain (Figure 10.1) [32,49]. The AC133 mAb is frequently used for the isolation of CD133$^+$ stem and progenitor cells from various healthy and pathological tissues, including solid tumors (Sections 10.1.3 and 10.6) [50–52].

The utilization of alternative CD133 antibodies has brought controversial data, particularly concerning the exclusive CD133 detection on human stem cells [25,32,53]. Technically, its AC133 epitope (or others) may be masked under native conditions and an appropriate antigen retrieval technique should be applied [54] (reviewed in Refs [55,56]). For instance, the interaction of CD133 with certain lipids and/or unidentified protein partners might prevent its detection. When human tissue samples are available, an immunoblot should be performed using a mAb directed against a linear epitope (e.g., mAb's 80B258 and C24B9) or an antiserum [25,48,57]. A negative staining resulting from the expression of an alternative CD133 splice variant, as during brain development, might be contradicted by the use of antibodies raised against different domains of CD133 [18]. Otherwise, immunodetection methods could be complemented by in situ hybridization to detect CD133 transcripts in a given cell type and/or tissue [19,28]. This issue is particularly relevant in view of the specific subcellular localization of CD133. Evaluating CD133 expression exclusively by means of antigen detection might lead to false negative scoring of certain cell types considering that it is associated with distant projections of the plasmalemma. Glial fibrillary acidic protein (GFAP)$^+$ adult neural stem cells, in which CD133 strictly localizes at the unique cilium lining the ventricular surface, are one example (Figure 10.3(A)) [58,59]. A similar situation may prevail for oligodendrocytes and potential CD133$^+$ Müller cell subpopulation in adult brain and retina, respectively [18,19]. Finally, given that CD133 could be internalized into the endosomal compartment (Section 10.2.1), its detection in any cell type under physiological or pathological conditions, for instance in glioblastoma, may require a permeabilization step prior to the immunolabeling procedure [60–62]. This is especially true when the cells of interest ought to be

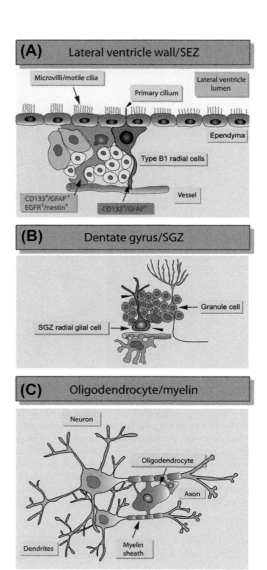

FIGURE 10.3 CD133 in the adult mammalian brain. Both major telencephalic neurogenic niches, the subependymal zone (SEZ) of the lateral ventricle wall (A) and the subgranular zone (SGZ) of the dentate gyrus of the hippocampal formation (B), are characterized by the presence of CD133$^+$ neural stem cells. Yet, the subcellular distribution of CD133 appears to be markedly distinct in both compartments. In the SEZ, CD133 is confined to the cerebrospinal fluid-contacting cilium ((A); arrowhead) of the quiescent type B radial glial cells (CD133$^+$/GFAP$^+$; red), which are also morphologically and phenotypically distinguished from the activated cells (no contacting cilium, EGFR$^+$/nestin$^+$; green). In the SGZ, CD133 is distributed over the soma and membrane process of the radial glial cells ((B); arrowhead). Owing to the unrestricted expression of CD133 by multiple nonneurogenic cell types, including multiciliated/microvillar ependymal cells (A) and some subgranular and hilar astrocytes (B) directly within or in the vicinity of neurogenic zones, a simultaneous detection of CD133, GFAP, and other antigenic markers is required for the prospective identification of adult neural stem cells. Myelinating oligodendrocytes are additional nonneurogenic CD133-expressing cells in the central nervous system (C). EGFR, epithelial growth factor receptor; GFAP, glial fibrillary acidic protein. (For interpretation of the references to color in this figure legend, the reader is referred to the online version of this book.)

analyzed by flow cytometry. Immunolabeling solely of the cell surface might lead as well to a false CD133-negative scoring [60].

10.1.3 CD133: A Unique Cell Surface Antigen for Cell-Based Therapy and Tissue Engineering

CD133 is one of the most extensively studied cell surface antigens of somatic stem cell populations. Beyond the neural and hematopoietic systems, CD133 has emerged as a marker of somatic stem cells originating from the prostate, kidney, skin, muscles, and eye [63–67].

In humans, the isolation of stem cells relies essentially on the accessibility of either of its distinct epitopes AC133 (CD133/1) and AC141/293C3 (CD133/2), which might label only a minute subpopulation of CD133+ cells, i.e., those harboring stem cell properties. In fact, human CD133 protein per se, just like its murine counterpart, is abundantly expressed in terminally differentiated cells found in numerous tissues, particularly in glandular organs such as pancreas, liver, and salivary glands [25,54,68]. Within these glands, CD133 is highly expressed in ductal epithelia, which are proposed to host cells with dedifferentiation capacities [69]. Thus, CD133 might highlight facultative stem cells acting during regeneration. The increased CD133 expression during epimorphic regeneration of the axolotl spinal neural tube is consistent with such a role (Section 10.5). Whether mammalian CD133+ terminally differentiated cells possess the capacity to dedifferentiate and acquire multipotency is still an open question [46,70]. The medical value of CD133 as a cell surface antigen is rapidly rising. It is already used for the isolation and transplantation of CD133+ hematopoietic stem cells [71] or serves as a potential therapeutic target of cancer stem cells [72]. Other promising therapeutic applications of CD133+ cells include muscular dystrophies [73] or regenerative cardiac therapy strategies [74].

The study of CD133 itself in the context of tissue formation and maintenance has already revealed fundamental phenomena regarding stem and cancer stem cell properties and membrane organization and dynamics (Sections 10.2 and 10.7). Such basic cell biological knowledge is important to establish potential protocols for stem cell-based and gene replacement therapies.

10.2 CELL BIOLOGY OF CD133 PROTEIN IN THE NERVOUS SYSTEM

The differential expression of CD133 splice variants is highly controlled and seems to be tissue-specific, developmentally regulated, and/or conditioned by microenvironmental clues. Six alternative promoters of CD133 have been described. For more details concerning their activities the readers are invited to consult the following reports [75–79]. Beyond its regulation at a transcriptional level, and possibly by microRNAs [80–82], the expression level of CD133 glycoprotein is also tightly controlled in neural stem cells.

10.2.1 CD133+ Membrane Particles

An active remodeling of the apical surface of neural progenitors (i.e., neuroepithelial cells and radial glial cells) that are in contact with the ventricular lumen occurs during the development of the mammalian neocortex (see Chapter 1). The apical compartment of neural progenitors is characterized by the coexistence of two major plasma membrane subdomains [83]. As in the case of absorptive epithelial cells, planar, nonprotruding areas separate numerous microvilli, and both subdomains contain a specific set of membrane proteins and lipids [34,39]. A primary cilium also emerges from the apical surface. This structure acts as a sensory organelle and ciliary function has been linked to two key signaling pathways (i.e., Wnt and sonic hedgehog (Shh)) known to regulate proliferation versus differentiation of progenitors [84]. A third pleomorphic and large protuberant structure found at the apical domain is the midbody, a cytoplasmic bridge connecting the nascent daughter cells that is formed transiently at late stages of cell division [85]. The central portion of the midbody, the so-called midbody ring, is normally inherited by one of the daughter cells after abscission, the terminal step of cytokinesis. However, neural progenitors can discharge it into the EC space during neurogenesis [33]. Remarkably, CD133 is selectively concentrated in these three apical protruding structures, a property observed even when neuroepithelial cells have lost functional tight junctions, suggesting that other fence mechanisms are operational to maintain the protrusion-specific localization of CD133 [31]. A direct interaction of CD133 with a specific cholesterol-based membrane microdomain seems to be implicated in its retention (Figure 10.2) [36].

An important observation made by Kosodo et al. was that the switch from proliferative to neurogenic divisions of neural progenitors is concomitant with a reduction in the size of their apical plasma membrane [86]. In fact, the abundance of microvilli and the length of the primary cilium decrease with the onset of neurogenesis [33,87]. Although a direct contribution of CD133 in favor of proliferation versus differentiation of neural progenitors remains to be determined [88], the occurrence of membrane particles (MPs) containing CD133 within the ventricular fluid and their increase during the process of neurogenesis argues for its implication. Indeed, several lines of evidence suggest a link between the release of CD133+ MPs and cell differentiation. In neural progenitors, the release of CD133+ MPs may contribute to the morphological remodeling and composition of their apical compartment [87]. Actually, two major subpopulations of

CD133+ MPs are found in the murine embryonic cerebrospinal fluid. Large MPs (>500 nm in diameter) arise from the cleavage and release of the midbody ring and show a ring-like CD133+ immunostaining, whereas small ones (<100 nm) bud from either the tip of the microvilli and primary cilium or the remnant of the midbody. Because they are directly released from the plasma membrane, both types of EC particles are referred to as CD133-containing ectosomes (Figure 10.2(A)) [33,87]. They can be recovered from cerebrospinal fluids upon ultracentrifugation or immunoisolated via CD133. The molecular mechanism underlying the release of the midbody ring into the EC space involves the same players as midbody abscission [89]. The release of small MPs might involve a change in the organization of membrane microdomains particularly at the tip of the apical protrusions (Figure 10.2(A)). Consistent with this model, the biochemical properties of CD133 within these particles (i.e., detergent solubility and specific interaction with membrane cholesterol) are identical to those of CD133 in the donor membranes, the microvilli [35].

The discharge of CD133+ MPs is not unique to neural progenitors. Other types of stem (and cancer stem) cells release them as well. Others and we could show the secretion of small CD133+ MPs by hematopoietic stem cells and neoplastic cells such as neuroblastoma and malignant melanoma cells [90,91]. In these three examples, CD133+ MPs do not bud directly from the plasma membrane. Instead, CD133 is endocytosed to the early endosomal compartment en route to multivesicular bodies (Figure 10.2(B)). These structures could then fuse either with the plasma membrane, and release their internal CD133+ MPs into the EC fluid as exosomes (Figure 10.2(C)), or with lysosomes by a heterotypic fusion, thus providing another mechanism (i.e., proteolytic degradation) accounting for the reduction in CD133 protein (Figure 10.2(D)) [24,60,92]. The IC trafficking of CD133 may be regulated by ubiquitination, syntenin-1, and HDAC6 [24,93]. Irrespective of the molecular mechanism supporting CD133 clearance, it is concomitant with cell differentiation.

The function of CD133+ MPs is unknown, but the presence of microRNAs in CD133+ exosomes derived from melanoma cells and probably those found in embryonic cerebrospinal fluid suggests that they might play a role in IC communication and/or cancer progression [60,94]. Indeed, a study has shown that the embryonic cerebrospinal fluid serves as a medium for the distribution of nanovesicles and the coordinated transfer of regulators of embryonic neural stem cell amplification during corticogenesis in rodents and humans [95]. Clinically, the analysis of CD133+ MPs in cerebrospinal fluids for diagnostic purposes is under investigation in view of their differential levels in relation with neural diseases (e.g., glioma, temporal lobe epilepsy) [58,96]. By extrapolation to other pathologies (e.g., kidney and prostate cancers), the expression levels of CD133+ MPs in other bodily fluids, notably urine and seminal fluids, may have a prognostic value [97].

10.2.2 Symmetric versus Asymmetric Distribution of CD133

The release of CD133+ MPs is not the sole mechanism contributing to the reduction and/or loss of CD133 at the plasma membrane of neural progenitors. Similar to the cell division observed in *Drosophila* embryonic neuroblasts [98], mammalian neural progenitors can distribute, symmetrically or asymmetrically, certain constituents during mitosis, notably apical proteins, including CD133. Specifically, the switch of neural progenitors from proliferative to neurogenic divisions is not associated with a clear rotation of the cleavage plane from parallel to perpendicular relative to their apical–basal axis as observed in *Drosophila* neuroblasts. Rather, a subtle change in the distribution of the apical compartment, which represents a minute fraction of the total plasma membrane (see above), leads from a symmetric to an asymmetric inheritance by the daughter cells [86]. As a net result, one daughter cell will be depleted of CD133 and will adopt a neuronal fate, whereas the CD133+ cell will maintain its stem cell properties. The mechanisms underlying neural stem cell division and daughter cell fate specification have been reviewed elsewhere [99,100].

Again, the symmetric/asymmetric distribution of CD133 during cell division is not limited to neural progenitors, but is also observed in hematopoietic stem cells, glioma stem cells, and melanoma and lung cancer cells, highlighting a phenomenon common to stem (cancer stem) cells [53,101–104]. Microenvironmental factors and specific cell-cycle regulators might be the underlying players involved in the growth and division characteristics of CD133+ stem and cancer stem cells (Section 10.6) [103,105]. Given the association of CD133 with a particular membrane microdomain, we hypothesize that the disposal of a specific "stem-cell-characteristic membrane microdomain" that contains key determinants allowing the maintenance of their stem/progenitor properties by means of (1) MP release, (2) asymmetric distribution, and (3) degradation may contribute to cell differentiation [53,87]. It is recognized that cholesterol-dependent membrane microdomains (lipid rafts) are actively involved in signal transduction. In the near future, the complete proteome of CD133-containing membrane microdomains might provide new insights into the cell differentiation process.

10.3 COMPARTMENTALIZATION OF CD133 IN MAMMALIAN NEURAL TISSUES

Since its discovery by Weigmann et al. in 1997 [1], CD133 and the biological tools detecting it (e.g., antibodies, cDNA probes) have contributed significantly to the prospective identification and isolation of mammalian cells with neurogenic capabilities, from both the embryonic and the postnatal nervous system [5,7,8,32,44,45,106–114]. The characterization of CD133 expression throughout brain development appears crucial to both deciphering cell biological details of neural progenitors and establishing potential stem-cell-based treatment for neurodegenerative diseases. In the following sections, the compartmentalization of CD133 expression in the developing and adult brain is summarized to provide a comprehensive and comparative view of germinative zones of the central nervous system (CNS) in mammals and nonmammalian vertebrates.

10.3.1 Developing Mammalian Brain

In prenatal murine brain, the expression of CD133 begins before the completion of neurulation. CD133+ cells are distributed in a nonhomogeneous fashion at the neural plate stage [28]. In fruit flies and anamniotic aquatic vertebrates, such patchy distribution might be indicative of a lateral inhibition mechanism characterizing certain steps of neurogenesis that is less established in mammals [115,116]. After the closure of the neural tube, CD133+ cells that become abundant along the ventricular progenitor zones of the murine and human brain and the spinal neural tube are colabeled with proliferation markers. The level of CD133 expression is not homogeneous along the dorsoventral axis of the developing CNS, as cells with ventral coordinates, including the midline floor, are more intensely labeled. The functional significance of this phenomenon is not established yet, but may be related to the distinct spatial and temporal pattern of neurogenesis along the dorsalventral axis. CD133 expression correlates with germinative zones irrespective of the morphological and phenotypic transitions (i.e., from early neuroepithelial to later radial glial) of the neural progenitors observed during ontogenesis [1,32,87,110,113]. Basal progenitors found in the SVZ (in mouse and human) and/or outer SVZ (only in human) express no markers of the apical (luminal) domain of ventricular zone progenitors, including CD133 [117,118] (reviewed in Ref. [119]).

CD133 expression drastically regresses in the newborn mammalian brain coincident with the consumption of germinative matrix [28]. It is then mainly confined to the late-developing parts of the postnatal brain with active neurogenesis, such as the dentate gyrus and olfactory bulb [28]. Although, the ependyma does not seem to be the major source of novel neuronal cells under physiological conditions, these multiciliated epithelial cells derived from radial glia inherit CD133. Both the transcriptional downregulation and the diminished protein concentration at apical membranes indicate a potential causal link between

CD133 and the maintenance of the proliferation status of neural progenitors. As mentioned above, an increased shedding of CD133+ MPs could promote cell differentiation resulting in a net decrease in CD133 in their apical compartment [83]. In addition to the neuronal progenitors, CD133 is associated with cell populations distributed in various extraventricular, parenchymal locations. The topography and marker analysis revealed that part of them could belong to glial lineages and others could represent terminally differentiated cells. By the second postnatal week, CD133 is upregulated and CD133+ cells can be seen as a subpopulation of Olig-2+ glial cells [18,28]. Whether these cells represent a quiescent pool of oligodendrocyte progenitors derived from the ventricular zone or correspond to multipotent neural progenitors as reported in the subcortical white matter of the human brain is as yet not evidenced [120]. Extraventricular populations of CD133+ cells are also observed in the cerebellum and correspond to neuronal cells inasmuch as the CD133 reactivity is detected in the (inner) granule cell layer [28]. However, the germinative external granule layer generating these excitatory interneurons does not express CD133 [7,28]. On the other hand, a population of multipotent progenitor cells, which can be prospectively isolated based on the presence of CD133 and lack of markers associated with neuronal and glial lineages, was observed in the cerebellum [7]. These progenitors can form self-renewing neurospheres and are able to differentiate into astrocytes, oligodendrocytes, and nongranule neurons in vitro, and each of these lineages can be generated after transplantation. Such a plurality of expression is not unique to CD133 as it is also observed among established markers of progenitor cells. For instance, Sox2 and musashi-1 are detected in terminally differentiated neuroepithelium-derived retinal cells [121]. Moreover, multipotential CD133− clonogenic cells have been reported, albeit in neural stem cell lines in vitro [122]. Therefore a careful evaluation and a combination of antigens in either positive or negative selection are essential to identify stem and progenitor cells.

10.3.2 Adult Mammalian Brain

In adult mammals, neurogenesis persists into adulthood or is reinitiated, adding new neuronal cells well beyond the weaning age. The generation of new neuroblasts requires specialized microenvironments that maintain and support the self-renewal capacity of adult stem cells [123]. These neurogenic niches are spatially restricted. The best-characterized compartments are the subependymal zone (SEZ) of the lateral ventricle and the subgranular zone (SGZ) of the dentate gyrus (Figure 10.3(A,B)). Therein, the mitotically active radial cells represent a reservoir of stem cells throughout the life span [124–126]. Whereas the cells generated in the former niche migrate a significant distance to supply the olfactory bulb through a restricted cell migratory pathway called the rostral migratory stream, those produced in the latter compartment remain within the dentate gyrus. The identity of adult neural stem cells in the postnatal forebrain as to the expression of stem cell markers and location in the SVZ has been debated [45,127]. Interestingly, CD133 appears to define sets of adult GFAP-expressing radial neural stem cells in both compartments, as elegantly demonstrated by a molecular genetic tracing method allowing for the discrimination of neural stem cells, astrocytes, and ependymal cells [112,113]. Displaced from the ventricular zone by the epithelial ependymal cells, the stem cells might have retained their fundamental embryonic/fetal neuroepithelial (morphologic and molecular) properties [125,127]. The ependymal cells, in contrast, do not retain a self-renewing ability, although they can be stimulated to proliferate and act as progenitors [128]. CD133 expression is apically restricted in the multiciliated ependymal lining and is apparently absent from the subependymal stem cell compartment of the lateral ventricle [108,127]. This seeming contradiction regarding the assignment of CD133 to stem and progenitor cells was resolved by showing its concentration in the primary cilium of the radial type B1 cells. This slender protrusion is incorporated between ependymal cells and contacts the cerebrospinal fluid (Figure 10.3(A)) [109,112,127]. SEZ progenitors appear in two—quiescent and activated—states that differ morphologically and phenotypically (Figure 10.3(A)) [114].

In the SGZ of the dentate gyrus, CD133 is distributed all along the soma and radial processes of radial astrocyte-like cells in both the murine and the human brain (Figure 10.3(B)), being in sharp contrast with its apically restricted confinement in the SEZ [8,112]. A further, less conventional, neurogenic compartment was described to express CD133: the ependymal lining of the central canal of the adult murine spinal cord. It is composed of at least three morphologically distinct cell types, among them radial ependymal cells harboring a long basal process and preferentially located in the midline. Irrespective of their morphology, they are characterized by the expression of stem and progenitor cell markers (e.g., nestin, vimentin, Msi1, Sox2, Sox3, Sox9), and upon injury they all contribute to the production of scar-forming astroglial cells and some myelinating oligodendrocytes [129,130].

Finally, we could demonstrate by means of a combination of 13A4 mAb (recognizing all identified CD133 splice variants) and an antiserum specifically detecting CD133.s1 (but not s3) that an isoform switch occurs during brain development from embryonic day 10 to adulthood, isoform s3 becoming the major variant in the adult mammalian brain at the expense of s1. CD133.s3 is a constituent of myelin sheaths in both the central and the peripheral nervous system owing to its expression in oligodendrocytes and Schwann cells (Figure 10.3(C)). Consistently, its

expression is extremely reduced or absent in brain extracts of myelin-deficient mice [18].

10.4 CD133+ NEURAL STEM AND PROGENITOR CELLS ACROSS SPECIES

Molecular comparisons of neurogenic stem cells between mammalian and nonmammalian systems are important to reveal common traits and decipher the specific features allowing for the impressive regenerative capacity of non-mammalian vertebrates (Section 10.5). Until now, the prospective identification and isolation of nonmammalian neural progenitors have been largely hampered by the lack of appropriate cell surface antigens. As a consequence, nonmammalian neural progenitors are only partly characterized. The identification of CD133 (prominin-1) orthologs in nonmammalian vertebrates and the information gained in the murine system have enabled the mapping of cellular compartmentalization of CD133 in relation to proliferative cell zones in the CNS in these model organisms. Here, we summarize the essentials concerning the widespread neurogenic zones of nonmammalian vertebrates, pointing out common and distinguishing features among mammals, aquatic vertebrate species, and birds.

In contrast to mammals, cold-blooded (poikilothermic) nonmammalian aquatic vertebrates and, to certain extent, embryonic chicks have an intrinsic ability for complete regeneration, being able to restore complex anatomical structures and, amazingly, even portions of their CNS [131]. This feature, peculiar to anamniote vertebrates, is seemingly related to their continuous growth, implying that beyond homeostatic tissue replacement, newly generated cells are regularly added to the existing ones, leading to a net growth. The CNS of adult nonmammalian vertebrates contains several neurogenic foci distributed along the entire extent of the cerebral ventricular zone [132]. The proliferating progenitors and putative stem cells found therein were shown to share some key phenotypic and morphologic characteristics with mammalian fetal neural progenitors having a radial glial morphology. These cells are phenotypically heterogeneous and a significant population of them is quiescent, depending on their position along the neuraxis, whereas the frequency of proliferating cells is uneven along the ventricular surface [133–135]. Expression of progenitor markers is not limited to actively proliferating cells. Thus, quiescent cells (in the case of zebrafish) could also express musashi-1, an evolutionarily conserved stem cell antigen. In addition, radial ependymoglial cells of essentially nongerminative zones of the ventricular surface can be stimulated to reenter the cell cycle and act as multipotential progenitors [136]. The zebrafish, despite its simple brain morphology, expresses two co-orthologs (prominin-1a and -1b) of the single mammalian CD133, which adds a further level of complexity to the compartmentalization of CD133. Thus, both coparalog genes are expressed in a complementary pattern along the rostrocaudal axis of the ventricle with only a minimal degree of overlap, indicating a potential subfunctionalization of these proteins. Remarkably, their global expression recapitulates the distribution of musashi-1 in proliferative zones, with the exception of the cerebellum [28]. In addition, prominin-1a+ cells were found in several extraventricular areas. It is not yet resolved if these cells belong to the numerous populations of quiescent parenchymal Olig-2+ oligodendroglial progenitors found in the zebrafish brain [137].

In the ambystomatid salamanders (a group of tailed amphibians) including axolotl (*Ambystoma mexicanum*), the brain is, owing to a secondary simplification, more primitive than is expected based on their phylogenetic position as tetrapods [138]. Thus, the postmitotic neuronal cells are kept in an embryonic position, forming a stratum griseum around the circumference of the ventricle similar to the mantle zone of the embryonic mammalian neural tube. Consequently, the gray matter is not segregated into nuclei or laminae that would be comparable to those of amniote vertebrates. A persistent proliferative activity was described in numerous brain regions of postembryonic axolotl [139], and interestingly, almost all the cells that express the proliferating cell nuclear antigen also coexpressed CD133. In addition, a significant amount of extraventricular CD133+ cells was observed in the axolotl brain [28].

In the developing chicken brain (*Gallus gallus*), CD133+ cells are distributed along the ventricular zone of the embryonic telencephalon in a way similar to that in developing mammals (Figure 10.4(A)). In addition to the evolutionarily conserved ventricular zone expression, significant CD133+ cell populations were observed in both the pallial and the subpallial parenchyma as well as in the mesencephalic and pontine tegmentum, reminiscent of the distribution of double CD133+/Olig-2+ cells observed in postnatal mice [28].

The analysis of distinct nonmammalian vertebrate species has revealed a markedly conserved confinement of CD133+ cells in the ventricular proliferative zone reminiscent of the distribution seen in mammals. CD133 represents apparently a molecular determinant common to all

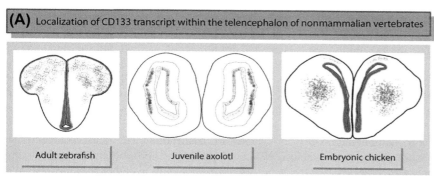

(A) Localization of CD133 transcript within the telencephalon of nonmammalian vertebrates

Adult zebrafish | Juvenile axolotl | Embryonic chicken

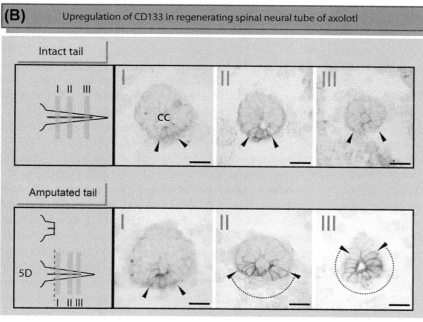

(B) Upregulation of CD133 in regenerating spinal neural tube of axolotl

Intact tail

I II III | I | II | III

CC

Amputated tail

5D | I | II | III

FIGURE 10.4 Distribution of CD133+ cells within the telencephalon of nonmammalian vertebrates and in the regenerating spinal neural tube of the axolotl. Cold-blooded, anamniote vertebrates with constitutive postembryonic neurogenesis, such as zebrafish and axolotl, display and/or retain a distribution of CD133 (prominin-1a in the case of zebrafish) in neurogenic ventricular zones similar to that of amniote vertebrates (e.g., chick), around the circumference of their telencephalic brain vesicles, as revealed by in situ hybridization (blue). In addition, a number of CD133+ cells are found in various parenchymal, extraventricular locations (A). Intact (top) and 5-day postamputation regenerated (5D, bottom) tails of larval axolotl were probed for CD133 expression. The position of the amputation (dash line) and section (I–III) planes along the rostrocaudal axis of the spinal neural tube are indicated. An expansion of CD133+ cells (blue, arrowheads) toward the lateral and dorsal coordinates is observed in the regenerating neural tube, especially in the caudal portions (II and III; dotted lines), compared to intact samples (B). CC, central canal. Scale bars, 50 μm. *(Panel (B) is adapted from our publication released in PLoS One in 2013 [28].)* (For interpretation of the references to color in this figure legend, the reader is referred to the online version of this book.)

proliferating and latent/quiescent neurogenic stem cells conserved during evolution. Moreover, CD133+ cell populations detected in distinct extraventricular locations within the CNS are suggestive of further, nonneurogenic neural functions of this molecule (Section 10.7). Altogether, CD133 as a cell surface antigen might serve as a useful tool for prospective isolation and functional comparison of neural progenitors from distinct vertebrate species irrespective of their phylogenetic position.

10.5 CD133+ CELLS AND REGENERATION

Deciphering the mechanisms underlying neural regeneration in nonmammalian vertebrates might lead to novel strategies in the mammalian neural stem cell field and hence be of medical interest. The stimulation of endogenous neurogenesis upon disease or trauma may be a simple and an effective approach for cell-based replacement.

One characteristic model of epimorphic regeneration (i.e., complete regeneration of complex anatomical structures) is the spinal cord regeneration of tailed amphibians upon surgical amputation [140,141]. The molecular mechanisms underlying this process are being unfolded [142]. Nonetheless, uncertainties remain as to the contribution and/or rearrangement of cellular components involved in tissue regeneration. It is still a matter of debate whether a preexisting pool of stem cells or dedifferentiating cells is the primary source of regeneration. The upregulation of CD133 expression and its expansion from a limited number of cells of the ventral midline floor toward more dorsal coordinates in postamputation regenerating spinal neural tube of larval axolotl highlight the fact that, irrespective of the fine details of morphology, the expression of evolutionarily conserved antigens is a general phylotypic property of progenitor cells (Figure 10.4(B)) [28].

Mechanistically, the ependymoglial cells of the regenerating spinal neural tube migrate and rearrange upon injury to form a sealed structure, the terminal vesicle, over the cut edge of the central canal. The proliferation of ependymal cells gives rise to an elongating ependymal tube, resembling the developing vertebrate neural tube, from which the new spinal cord is then generated [143]. The ependymal cells can give rise to all the cell types of a new spinal cord [144]. Interestingly, the quiescent radial–ependymal cells of the axolotl do not express nestin, in contrast to homeostatic neural progenitors of the zebrafish [135,145]. Whether ependymal cells of the axolotl spinal neural tube are stem cells *per definitionem* is not evidenced as yet. Nevertheless, they have to undergo an activation process upon injury to act as neural progenitors. This activation is reflected by the increased expression of CD133, similar to that of nestin.

The expansion of the initially floor-plate-restricted expression of CD133 toward more dorsal coordinates is reminiscent of what has been observed for Shh after injury [142,146]. These findings argue for the status of the floor plate and the expanded Shh-signal-emitting cells in coordinating the repair of spinal cord tissue, beyond the established function of the midline floor in elaborating positional identity along the dorsoventral axis of the spinal cord in vertebrates [147]. Expansion of the CD133 expression domain upon injury suggests further similarities between mammalian and nonmammalian vertebrate neurogenic progenitors, in addition to their glial features. Whether CD133 expression could be influenced by Shh is an intriguing question, as data obtained from various systems, including telencephalic development and brain tumor models, do hint at the control of CD133 expression by Shh and its downstream effectors [148,149]. It might not be a coincidence that both CD133 and players in the Shh signaling pathway share ciliary localization [150,151].

10.6 CD133 AND NEURAL DISEASES

Evidence for brain cancer stem cells first came from Dirks et al. by studying human glioblastoma samples [50,51]. They demonstrated, by isolating cells with tumorigenic potential based on their CD133 expression, that the concept of cancer stem cells introduced in 1997 by Bonnet and Dick with regard to acute myeloid leukemia [152] may be applicable to some solid tumors. They showed that these cells generated orthotopic tumor xenografts in immunodeficient mice with a higher efficiency than CD133− cells. Since then, CD133+ cells have been isolated from various types of pediatric and adult brain neoplasms, including glioma, glioblastoma, anaplastic oligoastrocytomas, medulloblastoma, neuroblastoma, and atypical teratoid/rhabdoid tumors [153–159]. In Chapter 17, Mahendram et al. further develop the identification and characterization of brain tumor-initiating cells. It is difficult to conclude whether the discrepancies that have appeared as to the utility of CD133 in identifying tumorigenic cells result from technical issues related to its detection (Section 10.1.2) or the use of different models and the cellular heterogeneity within tumors. They might also reflect the complex regulation of CD133 biology. Its expression may be modulated under low oxygen levels, the general consensus being that hypoxia increases CD133 expression and favors the expansion of CD133+ cells [79,155,158]; for more details, see our review [48]. Interestingly, self-renewing CD133− cell types could generate CD133+ and CD133− progeny, supporting the concept of hierarchical lineages with an uncertain position for CD133+ cells [160].

Studies conducted in several human tumor cell lines concluded that CD133 regulates the proliferation and colony-forming ability of cancer cells [161]. Its role in brain tumorigenesis has been therefore addressed by knockdown techniques in neurospheres derived from neuroblastoma and glioblastoma and it was shown that by interfering with CD133 expression one could drive neuroblastoma cells toward differentiation [62,162]. Irrespective of its precise function, further support for an essential role of CD133 in carcinogenesis is given by a study in glioma that also illustrates the difficulties that could arise in its utilization as a therapy target [62].

In 2011, Lathia and co-workers showed at the single-cell level, by monitoring CD133 expression, that glioma stem cells can adopt different division modes depending on the cellular environment and that symmetric division will generate cellular heterogeneity [102]. Interestingly, it was reported that an inhibitor of the c-MYC proto-oncogene suppresses not only the expression of MYCN (v-myc avian myelocytomatosis viral oncogene neuroblastoma derived homolog), a key player in self-renewal of normal neural stem and precursor cells [163,164], but also CD133 expression and symmetric cell division, and it induces asymmetric cell division in a neuroblastoma cell line [165]. The balance between symmetric and asymmetric cell division of glioma and other brain cancer cells monitored by CD133 (Section 10.2.2) could be modulated by therapeutic agents to promote the differentiation process, i.e., the release and/or degradation of CD133 might be an alternative way to neutralize cancer growth.

In a little more than 10 years, while the cancer stem cell concept evolved [166], CD133 gained prominence despite the debate about its relevance in various cancer contexts. One should keep in mind that alternative splicing, for instance, might affect the detection of both the glycoprotein and the mRNA. Moreover it may affect the still elusive function of CD133 and explain contradictory results obtained in the various systems. The crosstalk of CD133 with other signaling pathways and/or surface receptors needs to be further dissected. Yet, there is ever-growing evidence that CD133 is associated with tumor progression and recurrence and poor prognosis. It is therefore considered a major target for cancer eradication. Effective therapies should take into account its complex cell biology to

- immunotarget cell surface CD133;
- downregulate its expression (e.g., RNA interference);
- promote its clearance (e.g., release of MPs) and/or asymmetric cell division.

The relevance of CD133 in targeting brain neoplasms also relates to its expression in endothelial cells in the context of tumor vascularization, as it was demonstrated in a recombinant mouse model that CD133+ endothelium sustained the growth of proneural glioma [167].

10.7 CD133 AND PHOTORECEPTOR NEURON MORPHOGENESIS

CD133 is also of great interest in the eye and vision research fields. Retinal dystrophies account for millions of blindness cases worldwide, and deciphering the molecular mechanisms causing visual impairment and establishing regenerative therapies will have a significant socioeconomic impact. Shortly after its identification, Maw et al. demonstrated that a mutation in the *PROM1* gene caused retinal degeneration

[47]. Since then, numerous recessive and dominant mutations affecting the open reading frame of CD133 have been described (reviewed in Refs [168,169]), all causing various forms of retinal phenotypes, including macular degeneration or retinitis pigmentosa. Remarkably, genetically modified mice carrying either a null allele or a dominant negative form of human CD133 mimic the human eye diseases [29,170]. CD133 is thereby a potential candidate for gene replacement therapy. The effect of CD133 defects on retina relates to its specific subcellular localization in neuroepithelial cell-derived photoreceptor cells. In vertebrates, CD133 is concentrated in newly synthesized plasma membrane evaginations located at the base of the rod outer segment. These are the precursor structures of photoreceptor discs. In cones, CD133 is distributed in the outer rims of open disc lamellae (Figure 10.5(A)) [47,169,171]. Remarkably, CD133 (prom) also plays a crucial role in the

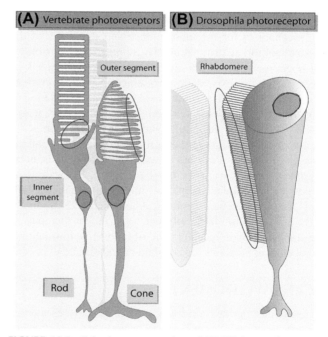

FIGURE 10.5 Selective concentration of CD133 in membrane outgrowths of vertebrate and fly photoreceptor neurons. Two photoreceptor cell types are found in vertebrates, rods, and cones (A). In mammals (e.g., mouse, human) and in tailless amphibians (e.g., *Xenopus laevis*), CD133 is concentrated at the outer rims of newly synthesized plasma membrane evaginations growing from the connecting cilium at the base of the rod outer segment. They are the precursor structures of photoreceptor discs. In the cones, CD133 is distributed in the outer rims of open disc lamellae. In fruit flies, photoreceptors harbor an elaborated apical surface, named the rhabdomere, which is composed of about 50,000 densely packed microvilli (B). The rhabdomere is the functional equivalent of the outer segment of vertebrate photoreceptors. The *Drosophila* CD133 ortholog (prom) is highly concentrated at the microvillar tip. Red circles delineate strong CD133 expression domains. Note that this protein is also sparsely present in the mature discs of the rod outer segment and at the base of microvilli on the rhabdomere (not depicted). (For interpretation of the references to color in this figure legend, the reader is referred to the online version of this book.)

morphogenesis and maintenance of fruit fly photoreceptors despite an entirely different cell organization. CD133 is indeed concentrated in plasma membrane protrusions, i.e., microvilli, within the rhabdomere domains of fly photoreceptors (Figure 10.5(B)) [172,173]. Moreover, expression of a mutated form of human CD133 in fly rhabdomeric photoreceptor cells could phenocopy the morphogenic disruption observed in mammalian systems. It is therefore tempting to speculate that the cell biological function of CD133 in fly photoreceptor disc membranes is comparable to that in human and thus provides an attractive system for studying its role [174]. Indeed, in both photoreceptor types, CD133 in concert with other protein partners (e.g., protocadherin 21 in the case of vertebrates and eyes shut/spacemaker in flies) is implicated in the architecture of functional plasma membrane subdomains by a mechanism that is conserved throughout evolution [170,172].

Together, the experimental and clinical observations reported until now support the concept that CD133, in association with other proteins and lipids, provides a scaffolding mechanism for the maintenance and dynamics of the plasma membrane protrusions. Thus, numerous physiological phenomena involving either an active (e.g., outer segment morphogenesis, cellular interaction, cell migration) or a passive (signal transduction, absorption) remodeling of plasma membrane protrusions might require CD133, which could explain its wide cellular and tissue distribution.

In conclusion, CD133 is ever more prominent as a unique antigen of somatic stem and diverse cancer stem cells, notably those derived from the nervous system, but also as an important structural element in terminally differentiated cells such as photoreceptor neurons. Altogether, CD133 might serve as a potential target of gene-replacement therapies in the context of photoreceptors and of innovative translational therapeutic approaches well beyond neural diseases.

ACKNOWLEDGMENTS

We apologize to investigators whose work could not be cited owing to space constraints. J.J. was supported by intramural funds from the Medical Faculty of Technische Universität Dresden and the Deutsche Forschungsgemeinschaft (DFG) (MeDDrive38 grants and CO298/5-1, respectively). D.C. was supported by the DFG (SFB 655 B3; TRR83 TP6; CO298/5-1) and Sächsisches Staatsministerium für Wissenschaft und Kunst.

REFERENCES

[1] Weigmann A, Corbeil D, Hellwig A, Huttner WB. Prominin, a novel microvilli-specific polytopic membrane protein of the apical surface of epithelial cells, is targeted to plasmalemmal protrusions of non-epithelial cells. Proc Natl Acad Sci USA 1997;94:12425–30.

[2] Corbeil D, Röper K, Fargeas CA, Joester A, Huttner WB. Prominin: a story of cholesterol, plasma membrane protrusions and human pathology. Traffic 2001;2:82–91.

[3] Yin AH, Miraglia S, Zanjani ED, Almeida-Porada G, Ogawa M, Leary AG, et al. AC133, a novel marker for human hematopoietic stem and progenitor cells. Blood 1997;90:5002–12.

[4] Miraglia S, Godfrey W, Yin AH, Atkins K, Warnke R, Holden JT, et al. A novel five-transmembrane hematopoietic stem cell antigen: isolation, characterization, and molecular cloning. Blood 1997;90:5013–21.

[5] Uchida N, Buck DW, He D, Reitsma MJ, Masek M, Phan TV, et al. Direct isolation of human central nervous system stem cells. Proc Natl Acad Sci USA 2000;97:14720–5.

[6] Schwartz PH, Bryant PJ, Fuja TJ, Su H, O'Dowd DK, Klassen H. Isolation and characterization of neural progenitor cells from post-mortem human cortex. J Neurosci Res 2003;74:838–51.

[7] Lee A, Kessler JD, Read TA, Kaiser C, Corbeil D, Huttner WB, et al. Isolation of neural stem cells from the postnatal cerebellum. Nat Neurosci 2005;8:723–9.

[8] Walker TL, Wierick A, Sykes AM, Waldau B, Corbeil D, Carmeliet P, et al. Prominin-1 allows prospective isolation of neural stem cells from the adult murine hippocampus. J Neurosci 2013;33:3010–24.

[9] Fargeas CA, Florek M, Huttner WB, Corbeil D. Characterization of prominin-2, a new member of the prominin family of pentaspan membrane glycoproteins. J Biol Chem 2003;278:8586–96.

[10] McGrail M, Batz L, Noack K, Pandey S, Huang Y, Gu X, et al. Expression of the zebrafish CD133/prominin1 genes in cellular proliferation zones in the embryonic central nervous system and sensory organs. Dev Dyn 2010;239:1849–57.

[11] Han Z, Papermaster DS. Identification of three prominin homologs and characterization of their messenger RNA expression in *Xenopus laevis* tissues. Mol Vis 2011;17:1381–96.

[12] Fargeas CA. Prominin-2 and other relatives of CD133. Adv Exp Med Biol 2013;777:25–40.

[13] Miraglia S, Godfrey W, Buck D. A response to AC133 hematopoietic stem cell antigen: human homologue of mouse kidney prominin or distinct member of a novel protein family? Blood 1998;91:4390–1.

[14] Corbeil D, Fargeas CA, Huttner WB. Rat prominin, like its mouse and human orthologues, is a pentaspan membrane glycoprotein. Biochem Biophys Res Commun 2001;285:939–44.

[15] Fargeas CA, Corbeil D, Huttner WB. AC133 antigen, CD133, prominin-1, prominin-2, etc.: prominin family gene products in need of a rational nomenclature. Stem Cells 2003;21:506–8.

[16] Fargeas CA, Joester A, Missol-Kolka E, Hellwig A, Huttner WB, Corbeil D. Identification of novel prominin-1/CD133 splice variants with alternative c-termini and their expression in epididymis and testis. J Cell Sci 2004;117:4301–11.

[17] Fargeas CA, Huttner WB, Corbeil D. Nomenclature of prominin-1 (CD133) splice variants – an update. Tissue Antigens 2007;69:602–6.

[18] Corbeil D, Joester A, Fargeas CA, Jászai J, Garwood J, Hellwig A, et al. Expression of distinct splice variants of the stem cell marker prominin-1 (CD133) in glial cells. Glia 2009;57:860–74.

[19] Jászai J, Fargeas CA, Graupner S, Tanaka EM, Brand M, Huttner WB, et al. Distinct and conserved prominin-1/CD133-positive retinal cell populations identified across species. PLoS One 2011;6:e17590.

[20] Yu Y, Flint A, Dvorin EL, Bischoff J. AC133-2, a novel isoform of human AC133 stem cell antigen. J Biol Chem 2002;277:20711–6.

[21] Corbeil D, Karbanová J, Fargeas CA, Jászai J. Prominin-1 (CD133): molecular and cellular features across species. Adv Exp Med Biol 2013;777:3–24.

[22] Boivin D, Labbe D, Fontaine N, Lamy S, Beaulieu E, Gingras D, et al. The stem cell marker CD133 (prominin-1) is phosphorylated on cytoplasmic tyrosine-828 and tyrosine-852 by Src and Fyn tyrosine kinases. Biochemistry 2009;48:3998–4007.

[23] Wei Y, Jiang Y, Zou F, Liu Y, Wang S, Xu N, et al. Activation of PI3K/Akt pathway by CD133-p85 interaction promotes tumorigenic capacity of glioma stem cells. Proc Natl Acad Sci USA 2013;110:6829–34.

[24] Mak AB, Nixon AM, Kittanakom S, Stewart JM, Chen GI, Curak J, et al. Regulation of CD133 by HDAC6 promotes β-catenin signaling to suppress cancer cell differentiation. Cell Rep 2012;2:951–63.

[25] Florek M, Haase M, Marzesco AM, Freund D, Ehninger G, Huttner WB, et al. Prominin-1/CD133, a neural and hematopoietic stem cell marker, is expressed in adult human differentiated cells and certain types of kidney cancer. Cell Tissue Res 2005;319:15–26.

[26] Jászai J, Janich P, Farkas LM, Fargeas CA, Huttner WB, Corbeil D. Differential expression of prominin-1 (CD133) and prominin-2 in major cephalic exocrine glands of adult mice. Histochem Cell Biol 2007;128:409–19.

[27] Jászai J, Farkas LM, Fargeas CA, Janich P, Haase M, Huttner WB, et al. Prominin-2 is a novel marker of distal tubules and collecting ducts of the human and murine kidney. Histochem Cell Biol 2010;133:527–39.

[28] Jászai J, Graupner S, Tanaka EM, Funk RH, Huttner WB, Brand M, et al. Spatial distribution of prominin-1 (CD133)-positive cells within germinative zones of the vertebrate brain. PLoS One 2013; 8:e63457.

[29] Zacchigna S, Oh H, Wilsch-Bräuninger M, Missol-Kolka E, Jászai J, Jansen S, et al. Loss of the cholesterol-binding protein prominin-1/CD133 causes disk dysmorphogenesis and photoreceptor degeneration. J Neurosci 2009;29:2297–308.

[30] Arndt K, Grinenko T, Mende N, Reichert D, Portz M, Ripich T, et al. CD133 is a modifier of hematopoietic progenitor frequencies but is dispensable for the maintenance of mouse hematopoietic stem cells. Proc Natl Acad Sci USA 2013;110:5582–7.

[31] Corbeil D, Röper K, Hannah MJ, Hellwig A, Huttner WB. Selective localization of the polytopic membrane protein prominin in microvilli of epithelial cells – a combination of apical sorting and retention in plasma membrane protrusions. J Cell Sci 1999;112: 1023–33.

[32] Corbeil D, Röper K, Hellwig A, Tavian M, Miraglia S, Watt SM, et al. The human AC133 hematopoietic stem cell antigen is also expressed in epithelial cells and targeted to plasma membrane protrusions. J Biol Chem 2000;275:5512–20.

[33] Dubreuil V, Marzesco AM, Corbeil D, Huttner WB, Wilsch-Bräuninger M. Midbody and primary cilium of neural progenitors release extracellular membrane particles enriched in the stem cell marker prominin-1. J Cell Biol 2007;176:483–95.

[34] Röper K, Corbeil D, Huttner WB. Retention of prominin in microvilli reveals distinct cholesterol-based lipid micro-domains in the apical plasma membrane. Nat Cell Biol 2000;2:582–92.

[35] Marzesco AM, Wilsch-Bräuninger M, Dubreuil V, Janich P, Langenfeld K, Thiele C, et al. Release of extracellular membrane vesicles from microvilli of epithelial cells is enhanced by depleting membrane cholesterol. FEBS Lett 2009;583:897–902.

[36] Corbeil D, Marzesco AM, Fargeas CA, Huttner WB. Prominin-1: a distinct cholesterol-binding membrane protein and the organisation of the apical plasma membrane of epithelial cells. Subcell Biochem 2010;51:399–423.

[37] Florek M, Bauer N, Janich P, Wilsch-Braeuninger M, Fargeas CA, Marzesco AM, et al. Prominin-2 is a cholesterol-binding protein associated with apical and basolateral plasmalemmal protrusions in polarized epithelial cells and released into urine. Cell Tissue Res 2007;328:31–47.

[38] Taïeb N, Maresca M, Guo XJ, Garmy N, Fantini J, Yahi N. The first extracellular domain of the tumour stem cell marker CD133 contains an antigenic ganglioside-binding motif. Cancer Lett 2009;278:164–73.

[39] Janich P, Corbeil D. GM1 and GM3 gangliosides highlight distinct lipid microdomains within the apical domain of epithelial cells. FEBS Lett 2007;581:1783–7.

[40] Iglic A, Hagerstrand H, Veranic P, Plemenitas A, Kralj-Iglic V. Curvature-induced accumulation of anisotropic membrane components and raft formation in cylindrical membrane protrusions. J Theor Biol 2006;240:368–73.

[41] Pugacheva EN, Jablonski SA, Hartman TR, Henske EP, Golemis EA. HEF1-dependent Aurora A activation induces disassembly of the primary cilium. Cell 2007;129:1351–63.

[42] Bruckner K, Pablo Labrador J, Scheiffele P, Herb A, Seeburg PH, Klein R. EphrinB ligands recruit GRIP family PDZ adaptor proteins into raft membrane microdomains. Neuron 1999;22:511–24.

[43] Bauer N, Fonseca AV, Florek M, Freund D, Jászai J, Bornhäuser M, et al. New insights into the cell biology of hematopoietic progenitors by studying prominin-1 (CD133). Cells Tissues Organs 2008;188:127–38.

[44] Kania G, Corbeil D, Fuchs J, Tarasov KV, Blyszczuk P, Huttner WB, et al. Somatic stem cell marker prominin-1/CD133 is expressed in embryonic stem cell-derived progenitors. Stem Cells 2005;23:791–804.

[45] Coskun V, Wu H, Blanchi B, Tsao S, Kim K, Zhao J, et al. CD133+ neural stem cells in the ependyma of mammalian postnatal forebrain. Proc Natl Acad Sci USA 2008;105:1026–31.

[46] Corbeil D, Fargeas CA, Jászai J. CD133 might be a pan marker of epithelial cells with dedifferentiation capacity. Proc Natl Acad Sci USA 2014;111:E1451–2.

[47] Maw MA, Corbeil D, Koch J, Hellwig A, Wilson-Wheeler JC, Bridges RJ, et al. A frameshift mutation in prominin (mouse)-like 1 causes human retinal degeneration. Hum Mol Genet 2000;9:27–34.

[48] Grosse-Gehling P, Fargeas CA, Dittfeld C, Garbe Y, Alison MR, Corbeil D, et al. CD133 as a biomarker for putative cancer stem cells in solid tumours: limitations, problems and challenges. J Pathol 2013;229:355–78.

[49] Kemper K, Sprick MR, de Bree M, Scopelliti A, Vermeulen L, Hoek M, et al. The AC133 epitope, but not the CD133 protein, is lost upon cancer stem cell differentiation. Cancer Res 2010;70:719–29.

[50] Singh SK, Clarke ID, Terasaki M, Bonn VE, Hawkins C, Squire J, et al. Identification of a cancer stem cell in human brain tumors. Cancer Res 2003;63:5821–8.

[51] Singh SK, Hawkins C, Clarke ID, Squire JA, Bayani J, Hide T, et al. Identification of human brain tumour initiating cells. Nature 2004;432:396–401.

[52] O'Brien CA, Pollett A, Gallinger S, Dick JE. A human colon cancer cell capable of initiating tumour growth in immunodeficient mice. Nature 2007;445:106–10.

[53] Fargeas CA, Fonseca AV, Huttner WB, Corbeil D. Prominin-1 (CD133): from progenitor cells to human diseases. Future Lipidol 2006;1:213–25.

[54] Immervoll H, Hoem D, Sakariassen PO, Steffensen OJ, Molven A. Expression of the "stem cell marker" CD133 in pancreas and pancreatic ductal adenocarcinomas. BMC Cancer 2008;8:48.

[55] Bidlingmaier S, Zhu X, Liu B. The utility and limitations of glycosylated human CD133 epitopes in defining cancer stem cells. J Mol Med (Berl) 2008;86:1025–32.

[56] Fargeas CA, Karbanová J, Jászai J, Corbeil D. CD133 and membrane microdomains: old facets for future hypotheses. World J Gastroenterol 2011;17:4149–52.

[57] Missol-Kolka E, Karbanová J, Janich P, Haase M, Fargeas CA, Huttner WB, et al. Prominin-1 (CD133) is not restricted to stem cells located in the basal compartment of murine and human prostate. Prostate 2011;71:254–67.

[58] Huttner HB, Janich P, Kohrmann M, Jászai J, Siebzehnrubl F, Blumcke I, et al. The stem cell marker prominin-1/CD133 on membrane particles in human cerebrospinal fluid offers novel approaches for studying central nervous system disease. Stem Cells 2008;26:698–705.

[59] Fischer J, Beckervordersandforth R, Tripathi P, Steiner-Mezzadri A, Ninkovic J, Gotz M. Prospective isolation of adult neural stem cells from the mouse subependymal zone. Nat Protoc 2011;6:1981–9.

[60] Bauer N, Wilsch-Bräuninger M, Karbanová J, Fonseca AV, Strauss D, Freund D, et al. Haematopoietic stem cell differentiation promotes the release of prominin-1/CD133-containing membrane vesicles—a role of the endocytic–exocytic pathway. EMBO Mol Med 2011;3:398–409.

[61] Campos B, Zeng L, Daotrong PH, Eckstein V, Unterberg A, Mairbaurl H, et al. Expression and regulation of AC133 and CD133 in glioblastoma. Glia 2011;59:1974–86.

[62] Brescia P, Ortensi B, Fornasari L, Levi D, Broggi G, Pelicci G. CD133 is essential for glioblastoma stem cell maintenance. Stem Cells 2013;31:857–69.

[63] Richardson GD, Robson CN, Lang SH, Neal DE, Maitland NJ, Collins AT. CD133, a novel marker for human prostatic epithelial stem cells. J Cell Sci 2004;117:3539–45.

[64] Alessandri G, Pagano S, Bez A, Benetti A, Pozzi S, Iannolo G, et al. Isolation and culture of human muscle-derived stem cells able to differentiate into myogenic and neurogenic cell lineages. Lancet 2004;364:1872–83.

[65] Bussolati B, Bruno S, Grange C, Buttiglieri S, Deregibus MC, Cantino D, et al. Isolation of renal progenitor cells from adult human kidney. Am J Pathol 2005;166:545–55.

[66] Charruyer A, Strachan LR, Yue L, Toth AS, Cecchini G, Mancianti ML, et al. CD133 is a marker for long-term repopulating murine epidermal stem cells. J Invest Dermatol 2012;132:2522–33.

[67] Carter DA, Balasubramaniam B, Dick AD. Functional analysis of retinal microglia and their effects on progenitors. Methods Mol Biol 2013;935:271–83.

[68] Lardon J, Corbeil D, Huttner WB, Ling Z, Bouwens L. Stem cell marker prominin-1/AC133 is expressed in duct cells of the adult human pancreas. Pancreas 2008;36:e1–6.

[69] Yanger K, Stanger BZ. Facultative stem cells in liver and pancreas: fact and fancy. Dev Dyn 2011;240:521–9.

[70] Kusaba T, Lalli M, Kramann R, Kobayashi A, Humphreys BD. Differentiated kidney epithelial cells repair injured proximal tubule. Proc Natl Acad Sci USA 2014;111:1527–32.

[71] Bornhäuser M, Eger L, Oelschlaegel U, Auffermann-Gretzinger S, Kiani A, Schetelig J, et al. Rapid reconstitution of dendritic cells after allogeneic transplantation of CD133+ selected hematopoietic stem cells. Leukemia 2005;19:161–5.

[72] Lorico A, Mercapide J, Rappa G. Prominin-1 (CD133) and metastatic melanoma: current knowledge and therapeutic perspectives. Adv Exp Med Biol 2013;777:197–211.

[73] Meregalli M, Farini A, Belicchi M, Torrente Y. CD133+ cells for the treatment of degenerative diseases: update and perspectives. Adv Exp Med Biol 2013;777:229–43.

[74] Donndorf P, Steinhoff G. CD133-positive cells for cardiac stem cell therapy: current status and outlook. Adv Exp Med Biol 2013;777:215–27.

[75] Shmelkov SV, Jun L, St Clair R, McGarrigle D, Derderian CA, Usenko JK, et al. Alternative promoters regulate transcription of the gene that encodes stem cell surface protein AC133. Blood 2004;103:2055–61.

[76] Tabu K, Sasai K, Kimura T, Wang L, Aoyanagi E, Kohsaka S, et al. Promoter hypomethylation regulates CD133 expression in human gliomas. Cell Res 2008;18:1037–46.

[77] Tabu K, Kimura T, Sasai K, Wang L, Bizen N, Nishihara H, et al. Analysis of an alternative human CD133 promoter reveals the implication of Ras/ERK pathway in tumor stem-like hallmarks. Mol Cancer 2010;9:39.

[78] Sompallae R, Hofmann O, Maher CA, Gedye C, Behren A, Vitezic M, et al. A comprehensive promoter landscape identifies a novel promoter for CD133 in restricted tissues, cancers, and stem cells. Front Genet 2013;4:209.

[79] Tabu K, Bizen N, Taga T, Tanaka S. Gene regulation of prominin-1 (CD133) in normal and cancerous tissues. Adv Exp Med Biol 2013;777:73–85.

[80] Sallustio F, Serino G, Costantino V, Curci C, Cox SN, De Palma G, et al. miR-1915 and miR-1225-5p regulate the expression of CD133, PAX2 and TLR2 in adult renal progenitor cells. PLoS One 2013;8:e68296.

[81] Cheng M, Yang L, Yang R, Yang X, Deng J, Yu B, et al. A microRNA-135a/b binding polymorphism in CD133 confers decreased risk and favorable prognosis of lung cancer in Chinese by reducing CD133 expression. Carcinogenesis 2013;34:2292–9.

[82] Shen WW, Zeng Z, Zhu WX, Fu GH. MiR-142-3p functions as a tumor suppressor by targeting CD133, ABCG2, and Lgr5 in colon cancer cells. J Mol Med (Berl) 2013;91:989–1000.

[83] Corbeil D, Marzesco AM, Wilsch-Bräuninger M, Huttner WB. The intriguing links between prominin-1 (CD133), cholesterol-based membrane microdomains, remodeling of apical plasma membrane protrusions, extracellular membrane particles, and (neuro)epithelial cell differentiation. FEBS Lett 2010;584:1659–64.

[84] Guemez-Gamboa A, Coufal NG, Gleeson JG. Primary cilia in the developing and mature brain. Neuron 2014;82:511–21.

[85] Chen CT, Ettinger AW, Huttner WB, Doxsey SJ. Resurrecting remnants: the lives of post-mitotic midbodies. Trends Cell Biol 2013;23:118–28.

[86] Kosodo Y, Röper K, Haubensak W, Marzesco AM, Corbeil D, Huttner WB. Asymmetric distribution of the apical plasma membrane during neurogenic divisions of mammalian neuroepithelial cells. EMBO J 2004;23:2314–24.

[87] Marzesco AM, Janich P, Wilsch-Bräuninger M, Dubreuil V, Langenfeld K, Corbeil D, et al. Release of extracellular membrane particles carrying the stem cell marker prominin-1 (CD133) from neural progenitors and other epithelial cells. J Cell Sci 2005;118:2849–58.

[88] Sykes AM, Huttner WB. Prominin-1 (CD133) and the cell biology of neural progenitors and their progeny. Adv Exp Med Biol 2013;777:89–98.

[89] Ettinger AW, Wilsch-Bräuninger M, Marzesco AM, Bickle M, Lohmann A, Maliga Z, et al. Proliferating versus differentiating stem and cancer cells exhibit distinct midbody-release behaviour. Nat Commun 2011;2:503.

[90] Marimpietri D, Petretto A, Raffaghello L, Pezzolo A, Gagliani C, Tacchetti C, et al. Proteome profiling of neuroblastoma-derived exosomes reveal the expression of proteins potentially involved in tumor progression. PLoS One 2013;8:e75054.

[91] Rappa G, Mercapide J, Anzanello F, Le TT, Johlfs MG, Fiscus RR, et al. Wnt interaction and extracellular release of prominin-1/CD133 in human malignant melanoma cells. Exp Cell Res 2013;319:810–9.

[92] Chen H, Luo Z, Dong L, Tan Y, Yang J, Feng G, et al. CD133/prominin-1-mediated autophagy and glucose uptake beneficial for hepatoma cell survival. PLoS One 2013;8:e56878.

[93] Karbanová J, Laco J, Marzseco AM, Janich P, Voborníková M, Mokrý J, Fargeas CA, Huttner WB, Corbeil D. Human PROMININ-1 (CD133) is detected in both neoplastic and non-neoplastic salivary gland diseases and released into saliva in a ubiquitinated form. PLoS One 2014;9:e98927.

[94] Rappa G, Mercapide J, Anzanello F, Pope RM, Lorico A. Biochemical and biological characterization of exosomes containing prominin-1/CD133. Mol Cancer 2013;12:62.

[95] Feliciano DM, Zhang S, Nasrallah CM, Lisgo SN, Bordey A. Embryonic cerebrospinal fluid nanovesicles carry evolutionarily conserved molecules and promote neural stem cell amplification. PLoS One 2014;9:e88810.

[96] Huttner HB, Corbeil D, Thirmeyer C, Coras R, Kohrmann M, Mauer C, et al. Increased membrane shedding – indicated by an elevation of CD133-enriched membrane particles – into the CSF in partial epilepsy. Epilepsy Res 2012;99:101–6.

[97] Marzesco AM. Prominin-1-containing membrane vesicles: origins, formation, and utility. Adv Exp Med Biol 2013;777:41–54.

[98] Wodarz A, Huttner WB. Asymmetric cell division during neurogenesis in *Drosophila* and vertebrates. Mech Dev 2003;120:1297–309.

[99] Götz M, Huttner WB. The cell biology of neurogenesis. Nat Rev Mol Cell Biol 2005;6:777–88.

[100] Paridaen JT, Huttner WB. Neurogenesis during development of the vertebrate central nervous system. EMBO Rep 2014;15:351–64.

[101] Fonseca AV, Bauer N, Corbeil D. The stem cell marker CD133 meets the endosomal compartment – new insights into the cell division of hematopoietic stem cells. Blood Cells Mol Dis 2008;41:194–5.

[102] Lathia JD, Hitomi M, Gallagher J, Gadani SP, Adkins J, Vasanji A, et al. Distribution of CD133 reveals glioma stem cells self-renew through symmetric and asymmetric cell divisions. Cell Death Dis 2011;2:e200.

[103] Pine SR, Ryan BM, Varticovski L, Robles AI, Harris CC. Microenvironmental modulation of asymmetric cell division in human lung cancer cells. Proc Natl Acad Sci USA 2010;107:2195–200.

[104] Redmer T, Welte Y, Behrens D, Fichtner I, Przybilla D, Wruck W, et al. The nerve growth factor receptor CD271 is crucial to maintain tumorigenicity and stem-like properties of melanoma cells. PLoS One 2014;9:e92596.

[105] Lubanska D, Market-Velker BA, deCarvalho AC, Mikkelsen T, Fidalgo da Silva E, Porter LA. The cyclin-like protein Spy1 regulates growth and division characteristics of the CD133+ population in human glioma. Cancer Cell 2014;25:64–76.

[106] Barraud P, Stott S, Mollgard K, Parmar M, Bjorklund A. In vitro characterization of a human neural progenitor cell coexpressing SSEA4 and CD133. J Neurosci Res 2007;85:250–9.

[107] Corti S, Nizzardo M, Nardini M, Donadoni C, Locatelli F, Papadimitriou D, et al. Isolation and characterization of murine neural stem/progenitor cells based on prominin-1 expression. Exp Neurol 2007;205:547–62.

[108] Pfenninger CV, Roschupkina T, Hertwig F, Kottwitz D, Englund E, Bengzon J, et al. CD133 is not present on neurogenic astrocytes in the adult subventricular zone, but on embryonic neural stem cells, ependymal cells, and glioblastoma cells. Cancer Res 2007;67:5727–36.

[109] Mirzadeh Z, Merkle FT, Soriano-Navarro M, Garcia-Verdugo JM, Alvarez-Buylla A. Neural stem cells confer unique pinwheel architecture to the ventricular surface in neurogenic regions of the adult brain. Cell Stem Cell 2008;3:265–78.

[110] Pinto L, Mader MT, Irmler M, Gentilini M, Santoni F, Drechsel D, et al. Prospective isolation of functionally distinct radial glial subtypes—lineage and transcriptome analysis. Mol Cell Neurosci 2008;38:15–42.

[111] Peh GS, Lang RJ, Pera MF, Hawes SM. CD133 expression by neural progenitors derived from human embryonic stem cells and its use for their prospective isolation. Stem Cells Dev 2009;18:269–82.

[112] Beckervordersandforth R, Tripathi P, Ninkovic J, Bayam E, Lepier A, Stempfhuber B, et al. In vivo fate mapping and expression analysis reveals molecular hallmarks of prospectively isolated adult neural stem cells. Cell Stem Cell 2010;7:744–58.

[113] Beckervordersandforth R, Deshpande A, Schaffner I, Huttner HB, Lepier A, Lie DC, et al. In vivo targeting of adult neural stem cells in the dentate gyrus by a split-cre approach. Stem Cell Reports 2014;2:153–62.

[114] Codega P, Silva-Vargas V, Paul A, Maldonado-Soto AR, Deleo AM, Pastrana E, et al. Prospective identification and purification of quiescent adult neural stem cells from their in vivo niche. Neuron 2014;82:545–59.

[115] Artavanis-Tsakonas S, Matsuno K, Fortini ME. Notch signaling. Science 1995;268:225–32.

[116] Chitnis AB. The role of notch in lateral inhibition and cell fate specification. Mol Cell Neurosci 1995;6:311–21.

[117] Haubensak W, Attardo A, Denk W, Huttner WB. Neurons arise in the basal neuroepithelium of the early mammalian telencephalon: a major site of neurogenesis. Proc Natl Acad Sci USA 2004;101:3196–201.

[118] Fietz SA, Kelava I, Vogt J, Wilsch-Bräuninger M, Stenzel D, Fish JL, et al. OSVZ progenitors of human and ferret neocortex are epithelial-like and expand by integrin signaling. Nat Neurosci 2010;13:690–9.

[119] Florio M, Huttner WB. Neural progenitors, neurogenesis and the evolution of the neocortex. Development 2014;141:2182–94.

[120] Menn B, Garcia-Verdugo JM, Yaschine C, Gonzalez-Perez O, Rowitch D, Alvarez-Buylla A. Origin of oligodendrocytes in the subventricular zone of the adult brain. J Neurosci 2006;26:7907–18.

[121] Susaki K, Kaneko J, Yamano Y, Nakamura K, Inami W, Yoshikawa T, et al. Musashi-1, an RNA-binding protein, is indispensable for survival of photoreceptors. Exp Eye Res 2009;88:347–55.

[122] Sun Y, Kong W, Falk A, Hu J, Zhou L, Pollard S, et al. CD133 (prominin) negative human neural stem cells are clonogenic and tripotent. PLoS One 2009;4:e5498.

[123] Silva-Vargas V, Crouch EE, Doetsch F. Adult neural stem cells and their niche: a dynamic duo during homeostasis, regeneration, and aging. Curr Opin Neurobiol 2013;23:935–42.

[124] Doetsch F, Caille I, Lim DA, Garcia-Verdugo JM, Alvarez-Buylla A. Subventricular zone astrocytes are neural stem cells in the adult mammalian brain. Cell 1999;97:703–16.

[125] Tramontin AD, Garcia-Verdugo JM, Lim DA, Alvarez-Buylla A. Postnatal development of radial glia and the ventricular zone (VZ): a continuum of the neural stem cell compartment. Cereb Cortex 2003;13:580–7.

[126] Merkle FT, Tramontin AD, Garcia-Verdugo JM, Alvarez-Buylla A. Radial glia give rise to adult neural stem cells in the subventricular zone. Proc Natl Acad Sci USA 2004;101:17528–32.

[127] Chojnacki AK, Mak GK, Weiss S. Identity crisis for adult periventricular neural stem cells: subventricular zone astrocytes, ependymal cells or both? Nat Rev Neurosci 2009;10:153–63.

[128] Carlen M, Meletis K, Goritz C, Darsalia V, Evergren E, Tanigaki K, et al. Forebrain ependymal cells are notch-dependent and generate neuroblasts and astrocytes after stroke. Nat Neurosci 2009;12:259–67.

[129] Meletis K, Barnabe-Heider F, Carlen M, Evergren E, Tomilin N, Shupliakov O, et al. Spinal cord injury reveals multilineage differentiation of ependymal cells. PLoS Biol 2008;6:e182.

[130] Oyarce K, Nualart F. Unconventional neurogenic niches and neurogenesis modulation by vitamins. J Stem Cell Res Ther 2014;4:1–11.

[131] Tanaka EM, Ferretti P. Considering the evolution of regeneration in the central nervous system. Nat Rev Neurosci 2009;10:713–23.

[132] Chapouton P, Jagasia R, Bally-Cuif L. Adult neurogenesis in non-mammalian vertebrates. Bioessays 2007;29:745–57.

[133] Adolf B, Chapouton P, Lam CS, Topp S, Tannhauser B, Strahle U, et al. Conserved and acquired features of adult neurogenesis in the zebrafish telencephalon. Dev Biol 2006;295:278–93.

[134] Pellegrini E, Mouriec K, Anglade I, Menuet A, Le Page Y, Gueguen MM, et al. Identification of aromatase-positive radial glial cells as progenitor cells in the ventricular layer of the forebrain in zebrafish. J Comp Neurol 2007;501:150–67.

[135] März M, Chapouton P, Diotel N, Vaillant C, Hesl B, Takamiya M, et al. Heterogeneity in progenitor cell subtypes in the ventricular zone of the zebrafish adult telencephalon. Glia 2010;58:870–88.

[136] Reimer MM, Sorensen I, Kuscha V, Frank RE, Liu C, Becker CG, et al. Motor neuron regeneration in adult zebrafish. J Neurosci 2008;28:8510–6.

[137] März M, Schmidt R, Rastegar S, Strähle U. Expression of the transcription factor Olig2 in proliferating cells in the adult zebrafish telencephalon. Dev Dyn 2010;239:3336–49.

[138] Roth G, Nishikawa KC, Naujoks-Manteuffel C, Schmidt A, Wake DB. Paedomorphosis and simplification in the nervous system of salamanders. Brain Behav Evol 1993;42:137–70.

[139] Richter W, Kranz D. Autoradiographic investigations on postnatal proliferative activity of the matrix-zones of the brain in the trout (*Salmo irideus*) (author's transl). Z Mikrosk Anat Forsch 1981;95:491–520.

[140] McHedlishvili L, Epperlein HH, Telzerow A, Tanaka EM. A clonal analysis of neural progenitors during axolotl spinal cord regeneration reveals evidence for both spatially restricted and multipotent progenitors. Development 2007;134:2083–93.

[141] McHedlishvili L, Mazurov V, Grassme KS, Goehler K, Robl B, Tazaki A, et al. Reconstitution of the central and peripheral nervous system during salamander tail regeneration. Proc Natl Acad Sci USA 2012;109:E2258–66.

[142] Schnapp E, Kragl M, Rubin L, Tanaka EM. Hedgehog signaling controls dorsoventral patterning, blastema cell proliferation and cartilage induction during axolotl tail regeneration. Development 2005;132:3243–53.

[143] O'Hara CM, Chernoff EA. Growth factor modulation of injury-reactive ependymal cell proliferation and migration. Tissue Cell 1994;26:599–611.

[144] Echeverri K, Tanaka EM. Ectoderm to mesoderm lineage switching during axolotl tail regeneration. Science 2002;298:1993–6.

[145] Walder S, Zhang F, Ferretti P. Up-regulation of neural stem cell markers suggests the occurrence of dedifferentiation in regenerating spinal cord. Dev Genes Evol 2003;213:625–30.

[146] Reimer MM, Kuscha V, Wyatt C, Sorensen I, Frank RE, Knuwer M, et al. Sonic hedgehog is a polarized signal for motor neuron regeneration in adult zebrafish. J Neurosci 2009;29:15073–82.

[147] Dessaud E, McMahon AP, Briscoe J. Pattern formation in the vertebrate neural tube: a sonic hedgehog morphogen-regulated transcriptional network. Development 2008;135:2489–503.

[148] Takanaga H, Tsuchida-Straeten N, Nishide K, Watanabe A, Aburatani H, Kondo T. Gli2 is a novel regulator of sox2 expression in telencephalic neuroepithelial cells. Stem Cells 2009;27:165–74.

[149] Schiapparelli P, Shahi MH, Enguita-German M, Johnsen JI, Kogner P, Lazcoz P, et al. Inhibition of the sonic hedgehog pathway by cyplopamine reduces the CD133+/CD15+ cell compartment and the in vitro tumorigenic capability of neuroblastoma cells. Cancer Lett 2011;310:222–31.

[150] Rohatgi R, Milenkovic L, Scott MP. Patched1 regulates hedgehog signaling at the primary cilium. Science 2007;317:372–6.

[151] Murdoch JN, Copp AJ. The relationship between sonic hedgehog signaling, cilia, and neural tube defects. Birth Defects Res A Clin Mol Teratol 2010;88:633–52.

[152] Bonnet D, Dick JE. Human acute myeloid leukemia is organized as a hierarchy that originates from a primitive hematopoietic cell. Nat Med 1997;3:730–7.

[153] Bao S, Wu Q, McLendon RE, Hao Y, Shi Q, Hjelmeland AB, et al. Glioma stem cells promote radioresistance by preferential activation of the DNA damage response. Nature 2006;444:756–60.

[154] Liu G, Yuan X, Zeng Z, Tunici P, Ng H, Abdulkadir IR, et al. Analysis of gene expression and chemoresistance of CD133+ cancer stem cells in glioblastoma. Mol Cancer 2006;5:67.

[155] Blazek ER, Foutch JL, Maki G. Daoy medulloblastoma cells that express CD133 are radioresistant relative to CD133− cells, and the CD133+ sector is enlarged by hypoxia. Int J Radiat Oncol Biol Phys 2007;67:1–5.

[156] Yi L, Zhou ZH, Ping YF, Chen JH, Yao XH, Feng H, et al. Isolation and characterization of stem cell-like precursor cells from primary human anaplastic oligoastrocytoma. Mod Pathol 2007;20:1061–8.

[157] Chiou SH, Kao CL, Chen YW, Chien CS, Hung SC, Lo JF, et al. Identification of CD133-positive radioresistant cells in atypical teratoid/rhabdoid tumor. PLoS One 2008;3:e2090.

[158] Donovan LK, Potter NE, Warr T, Pilkington GJ. A prominin-1-rich pediatric glioblastoma: biologic behavior is determined by oxygen tension-modulated CD133 expression but not accompanied by underlying molecular profiles. Transl Oncol 2012;5:141–54.

[159] Sartelet H, Imbriglio T, Nyalendo C, Haddad E, Annabi B, Duval M, et al. CD133 expression is associated with poor outcome in neuroblastoma via chemoresistance mediated by the Akt pathway. Histopathology 2012;60:1144–55.

[160] Chen R, Nishimura MC, Bumbaca SM, Kharbanda S, Forrest WF, Kasman IM, et al. A hierarchy of self-renewing tumor-initiating cell types in glioblastoma. Cancer Cell 2010;17:362–75.

[161] Yao J, Zhang T, Ren J, Yu M, Wu G. Effect of CD133/prominin-1 antisense oligodeoxynucleotide on in vitro growth characteristics of Huh-7 human hepatocarcinoma cells and U251 human glioma cells. Oncol Rep 2009;22:781–7.

[162] Takenobu H, Shimozato O, Nakamura T, Ochiai H, Yamaguchi Y, Ohira M, et al. CD133 suppresses neuroblastoma cell differentiation via signal pathway modification. Oncogene 2011;30:97–105.

[163] Knoepfler PS, Cheng PF, Eisenman RN. N-myc is essential during neurogenesis for the rapid expansion of progenitor cell populations and the inhibition of neuronal differentiation. Genes Dev 2002;16:2699–712.

[164] Cotterman R, Knoepfler PS. N-Myc regulates expression of pluripotency genes in neuroblastoma including *lif, klf2, klf4,* and *lin28b.* PLoS One 2009;4:e5799.

[165] Izumi H, Kaneko Y. Evidence of asymmetric cell division and centrosome inheritance in human neuroblastoma cells. Proc Natl Acad Sci USA 2012;109:18048–53.

[166] Nguyen LV, Vanner R, Dirks P, Eaves CJ. Cancer stem cells: an evolving concept. Nat Rev Cancer 2012;12:133–43.

[167] Ding BS, James D, Iyer R, Falciatori I, Hambardzumyan D, Wang S, et al. Prominin 1/CD133 endothelium sustains growth of proneural glioma. PLoS One 2013;8:e62150.

[168] Jászai J, Fargeas CA, Florek M, Huttner WB, Corbeil D. Focus on molecules: prominin-1 (CD133). Exp Eye Res 2007;85:585–6.

[169] Gurudev N, Florek M, Corbeil D, Knust E. Prominent role of prominin in the retina. Adv Exp Med Biol 2013;777:55–71.

[170] Yang Z, Chen Y, Lillo C, Chien J, Yu Z, Michaelides M, et al. Mutant prominin 1 found in patients with macular degeneration disrupts photoreceptor disk morphogenesis in mice. J Clin Invest 2008;118:2908–16.

[171] Han Z, Anderson DW, Papermaster DS. Prominin-1 localizes to the open rims of outer segment lamellae in *Xenopus laevis* rod and cone photoreceptors. Invest Ophthalmol Vis Sci 2012;53:361–73.

[172] Zelhof AC, Hardy RW, Becker A, Zuker CS. Transforming the architecture of compound eyes. Nature 2006;443:696–9.

[173] Gurudev N, Yuan M, Knust E. Chaoptin, prominin, eyes shut and crumbs form a genetic network controlling the apical compartment of *Drosophila* photoreceptor cells. Biol Open 2014;3:332–41.

[174] Nie J, Mahato S, Mustill W, Tipping C, Bhattacharya SS, Zelhof AC. Cross species analysis of prominin reveals a conserved cellular role in invertebrate and vertebrate photoreceptor cells. Dev Biol 2012;371:312–20.

Chapter 11

Fundamentals of NCAM Expression, Function, and Regulation of Alternative Splicing in Neuronal Differentiation

Ana Fiszbein[1], Ignacio E. Schor[1,2] and Alberto R. Kornblihtt[1]

[1]*Laboratorio de Fisiología y Biología Molecular, Departamento de Fisiología, Biología Molecular y Celular, IFIBYNE-CONICET, Facultad de Ciencias Exactas y Naturales, Universidad de Buenos Aires, Buenos Aires, Argentina;* [2]*European Molecular Biology Laboratory, Heidelberg, Germany*

11.1 INTRODUCTION

The neural cell adhesion molecule (NCAM) is a member of the immunoglobulin protein superfamily implicated in cell recognition and cell–cell adhesion through a homophilic, Ca^{2+}-independent, binding mechanism. NCAM is thought to play an important role in cell migration, synaptogenesis, memory formation, and other fundamental functions that are discussed in this chapter. NCAM was first identified as a cell surface glycoprotein [1]. Originally, it was described as a cell adhesion molecule from the neural retina of chick embryos, being intimately involved in the initial formation of cell–cell bonds during aggregation of embryonic retinal cells [2]. However, over time, it became clear that NCAM is essential for many other cellular processes, including the promotion of neuronal plasticity, and its expression has been reported on the surface of most cells in the central and peripheral nervous systems, as well as in heart, gonads, and muscle [3]. NCAM can also be used to enrich for neuronal fraction in human embryonic stem cell cultures because, for the later stages of stem cell differentiation, an antibody targeting human-specific NCAM was shown to monitor neural specification in vitro and in vivo. Fluorescence-activated cell sorting (FACS)-purified neuronal cells could be transplanted and survived in the brain of a rodent model [4]. Furthermore, NCAM/CD56 can be considered a natural killer (NK) cell marker. Human NK cells can be subdivided into two major subsets, the NCAM/CD56 (bright) NK cell subset being the one that constitutes the majority of NK cells in secondary lymphoid tissues [5]. It has been shown that NCAM/CD56, in addition to being a marker for neuronal and NK cells, is also a marker for muscle cells [6] and it is expressed in a small subset of bone marrow CD271[bright]

mesenchymal cells [7]. In the literature, many different NCAM antibodies have been reported, including antibodies that recognize specific isoforms [8].

In view of the fact that NCAM has many protein isoforms resulting from alternative splicing and that these variants have been reported to play different roles [4], the study of NCAM alternative splicing regulation becomes fundamental to understanding its specific functions. In this chapter, we focus on the role of chromatin structure in the regulation of NCAM alternative splicing during neuronal differentiation. First, we briefly review the features of NCAM alternative isoforms and specifically the changes in pattern during neuronal differentiation. Second, we describe the molecular mechanisms of alternative splicing regulation, discuss how chromatin structure affects splicing decisions in different models, and elaborate on the alternative splicing regulation of NCAM during neuronal differentiation.

11.2 NCAM GENE AND ALTERNATIVE SPLICING ISOFORMS

The NCAM gene is conserved in many species, including chimpanzee, dog, cow, mouse, rat, zebrafish, and fruit fly, and orthologs have been reported in 78 organisms. One of the most prominent is the *Drosophila* Down syndrome cell adhesion molecule (Dscam) gene, which can generate 38,016 isoforms by the alternative splicing of 95 variable exons [9].

Human NCAM is encoded by a single-copy gene in chromosome 11 that spans more than 314 kb and contains 19 major exons as well as six additional smaller exons [10]. In mice NCAM is encoded by a single gene located

Neural Surface Antigens. http://dx.doi.org/10.1016/B978-0-12-800781-5.00011-6

in chromosome 9 that contains 20 exons. A set of six additional small exons is usually described separately [11]. Alternative splicing gives rise to at least 20–30 mRNA isoforms whose encoded polypeptides can also be modified by posttranslational mechanisms [12]. The three more abundant protein isoforms are named according to their apparent molecular weight (NCAM180, NCAM140, and NCAM120). NCAM180 is a single-pass transmembrane protein generated from exons 1 to 19 with the exclusion of exon 15. NCAM140 differs from NCAM180 only by the absence of exon 18, which causes the shortening of the cytoplasmic domain (Figure 11.1). On the other hand, NCAM120 results from transcription of the first 15 exons. Inclusion of exon 15 in NCAM120 generates an alternative polyadenylation site causing the total absence of a cytoplasmic domain in the mature protein (Figure 11.1). Despite the absence of a hydrophobic anchoring peptide, NCAM120 is linked to the cell surface, as are NCAM180 and 140, by a glycosylphosphatidylinositol intermediate [13]. A further degree of variability is added when a 30-nucleotide exon is included between exons 7 and 8. This exon is called VASE

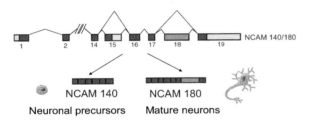

FIGURE 11.1 Molecular features of NCAM. Schema illustrating the identifiable domains and some posttranslational modifications of different NCAM isoforms. NCAM180 and NCAM140 are transmembrane proteins that differ only in a small portion of the intracellular domain. NCAM120 contains the extracellular domain linked to the membrane through an anchor.

FIGURE 11.2 Alternative splicing gives rise to NCAM180 and NCAM140. Two main isoforms of NCAM are characterized by different cytoplasmic domains generated by alternative splicing of exon 18. NCAM140 expression is more abundant in neuronal precursors, whereas NCAM180 is typical of mature neurons. Exons are represented by rectangles and introns by lines.

and when present in the mature protein it could modulate NCAM binding properties. Last, there are 4 short exons located between exons 12 and 13 that can give rise to different combinations that encode a region in the extracellular domain of the molecule known as the muscle-specific domain [14]. When one of these small exons containing a stop codon is included into the mature isoform, it gives rise to a truncated secreted form of the NCAM protein [15]. Soluble forms of NCAM can also be generated by the enzymatic excision of NCAM120 from the glycosylphosphatidylinositol anchor [16] or by the proteolytic cleavage of the extracellular part of NCAM molecules [17].

During neuronal differentiation from neuronal precursors to mature neurons both in vivo and in vitro, the alternative splicing pattern of NCAM changes, favoring the expression of NCAM180 to the detriment of NCAM140 (Figure 11.2). As a consequence of this splicing regulation, NCAM140 expression is more abundant in glia and in neuronal precursors, where it is homogeneously distributed in the cell membrane and favors neurite growth. Conversely, NCAM180 is more specific of mature neurons and is primarily restricted to the postsynaptic densities of neurons [18]. Moreover, NCAM180 is enriched in cell-to-cell contacts and contributes to organized stable and mature synapses [19–21]. Thus, the modulation of NCAM alternative splicing during neuronal differentiation has important functional consequences and allows the utilization of NCAM180 as a marker for mature neural cells.

Not only during neuronal differentiation does NCAM alternative splicing change, favoring the expression of specific isoforms. Data suggest that the expression of different isoforms of NCAM influences the progression and survival of patients who suffer from numerous malignant neoplasms. Moreover, NCAM140 was shown to be exclusively expressed in a number of highly malignant neoplasms,

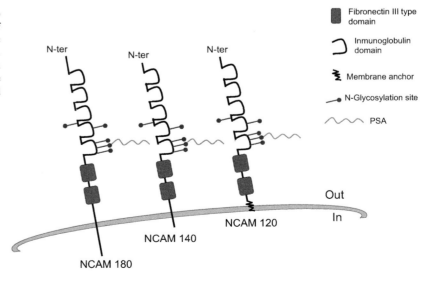

in which it induces antiapoptotic/proliferative pathways and affects calcium-dependent kinases that are relevant for tumorigenesis [8].

11.3 MOLECULAR STRUCTURE AND FUNCTION OF NCAM

NCAM proteins are synthesized in the endoplasmic reticulum as glycoproteins that can undergo various posttranslational modifications before reaching the cell surface. Once placed on the cell membrane, the extracellular region of NCAM comprises five immunoglobulin and two fibronectin type III domains (Figure 11.1). Structural studies of the N-terminal domain of NCAM revealed that the β structure of the first two immunoglobulin domains is responsible for the homophilic recognition that participates in cell–cell adhesion, characterized by the formation of antiparallel dimers between two cells [22]. NCAM proteins are able to interact among themselves by homophilic contacts, as well as with other molecules, forming heterophilic bonds. The most important heterophilic interactions take place through NCAM immunoglobulin domains binding to other adhesion molecules such as L1 and the neuron–glia cell adhesion molecule [23,24]. Other types of heterophilic interactions are illustrated by the connections between NCAM fibronectin type III domains and growth factor receptors such as the fibroblast growth factor receptor FGFR1 [25–27].

One of the most important posttranslational modifications of NCAM is the addition of long homopolymers of sialic residues in the nervous system. This polysialic acid (PSA) carbohydrate is added to the NCAM extracellular portion during its transit through the Golgi apparatus. In mammals, PSA is a large and simple molecule based on linear homopolymers of α2,8-linked N-acetylneuraminic acid, with the monomer number varying from 8 up to 100 [28,29]. Levels of polysialylation are high in the embryo and decrease during development to a minimum until most of the NCAM in the adult brain lacks PSA [30]. Once attached to NCAM, PSA interferes with cell–cell adhesion by reducing the contact force between molecules. Treatments with the enzyme neuroaminidase, which specifically eliminates PSA from NCAM–PSA, revealed that the PSA moiety plays key roles in nerve fasciculation, axon branching, and formation of synapses [31–33]. The synthesis of PSA is catalyzed by two different polysialyltransferases, ST8Sia II and IV. The first has a key role during development and the second plays an important role in neural plasticity in adults [34–36]. Biosynthesis of NCAM–PSA is regulated by cell activation [37] such as electrical activity in axons and can also be affected by alterations in the intracellular levels of Ca^{2+} that can induce exocytosis/endocytosis of NCAM–PSA [38,39]. Moreover, it is known that NCAM contains heparin-binding domains that are also essential for cell–cell adhesion [40]. Furthermore, NCAM also undergoes other regulatory posttranslational modifications such as phosphorylation, O- and N-glycosylation, sulfation, and palmitoylation [12].

The distribution of NCAM and NCAM–PSA is not permanent and can vary between developmental stages and tissues. Of particular interest is the expression pattern of NCAM–PSA in the nerve cell. It was found to be highly expressed in axons and dendrites before cell–cell adhesions are formed, but appears to be reduced once mature synapses are established [41]. Although the levels of PSA are dramatically reduced in the adult brain, NCAM–PSA is specifically retained in those adult brain regions that display a high degree of structural plasticity. Based on these findings, NCAM has been proposed as a plasticity-promoting molecule in the adult nervous system [42].

NCAM plays a crucial role in determining synapse formation and stability when expressed post- but not presynaptically [42]. Studies with knockout mice for NCAM [43] or for the enzymes responsible for transferring PSA [44] revealed its essential role in the hippocampus-controlled learning progression in adult mice and its implication in long-term, but not short-term, plasticity. Specifically, NCAM-deficient mice show impaired survival of newly generated hippocampal neurons [45]. Furthermore, NCAM–PSA is known to have an essential role in axonal regeneration in the peripheral nervous system [31] and there is evidence that NCAM–PSA could in some cases hamper brain repair [46]. In the absence of functional NCAM it has been demonstrated that a reduced number of new cells become engaged in a neuronal destiny [47]. Moreover, the addition of soluble NCAM to neuronal precursors in culture was shown to induce their differentiation into mature neurons, suggesting that NCAM promotes neuronal differentiation and that the mechanism implicated could be related to the binding to cell surface receptors [48]. An involvement of the mitogen-activated protein kinase kinase and its downstream kinase, the extracellular signal-regulated kinase, in the neural differentiation process has been suggested [49]. Consistently, it has been demonstrated that the addition of a synthetic ligand that blocks NCAM interactions prevents cell differentiation in culture [50].

The ectopic expression of NCAM in skeletal muscle revealed an increase in size and complexity of neuromuscular synapses without affecting any functional mechanism [51]. Altogether, this evidence shows that expression of NCAM–PSA during development somehow contributes to the establishment of specific patterns of synaptic connections. The current hypothesis on the role of synaptic NCAM–PSA states that nonpolysialylated NCAM isoforms promote stability, whereas NCAM–PSA favors plasticity. It appears that polysialylation directs the differentiation of adult neuroblasts by controlling NCAM interactions and it is therefore now clear that, by modifying the molecular properties of NCAM, PSA plays a crucial role in postnatal neurogenesis [52].

It was shown that neurons expressing NCAM–PSA constitute a subpopulation of interneurons. Moreover, in the adult central nervous system, NCAM–PSA also appears to be expressed in inhibitory neurons, because they express markers associated with interneuronal neurites or inhibitory synapses [53]. The presence of NCAM–PSA in a considerable proportion of interneurons and the alteration of its expression by various intrinsic and extrinsic factors indicate that this molecule must have an important function in the physiology of these inhibitory neurons. The interneurons expressing NCAM–PSA show reduced structural features and synaptic connectivity compared to other interneurons lacking this molecule, suggesting a role in the control of the connectivity of these inhibitory neurons [54]. As NCAM–PSA is a good candidate for mediating putative structural changes in interneurons, it has been proposed to be involved in various psychiatric disorders, including schizophrenia and major depression [55,56].

As mentioned above, NCAM180 differs from NCAM140 only by the presence of exon 18. Inclusion of exon 18 and the consequent addition of an intracellular domain functionally allow the protein to interact with cytoskeletal components through spectrin adapter proteins, reducing their mobility and fixing their location in the mature synapses [57,58]. This evidence suggests a crucial role for NCAM180 in mature synapse establishment. More specifically, inactivation of NCAM180 causes a retraction of the growth cones of some neurons without affecting neurite extension [59]. All these data suggest that NCAM and NCAM–PSA could regulate specific mechanisms that control the growth of axons and dendrites as well as the establishment of synaptic contacts [60].

In a 2014 study, NCAM functions were related to the cellular prion protein during neuronal differentiation. The authors found that cellular prion protein (CD230) is required for the neural stem/precursor cells' response to NCAM-induced-neuronal differentiation [61].

11.4 ALTERNATIVE SPLICING REGULATION

To understand the regulation of NCAM alternative splicing patterns during neuronal differentiation we briefly introduce here the molecular basis of alternative splicing and its various levels of regulation. There is no doubt that alternative splicing is one of the major contributors to protein diversity in metazoans. Among pre-mRNA-processing mechanisms, alternative splicing is fundamental to explain a high biological complexity from a limited number of genes, largely expanding the coding capacities of genetic information. Alternative splicing provides an extensive level of gene expression control that acts in parallel with the well-established transcriptional regulation. Furthermore, the number of protein isoforms generated by alternative splicing is particularly important in the nervous system and often these isoforms have different functions and regulatory features.

Before presenting the ways in which alternative splicing operates, we will describe essential features of the splicing process. Splicing is the mechanism of processing pre-mRNAs by the elimination of introns and the joining of exons. The term introns refers to noncoding interspersed sequences that in most cases are not included in the mature mRNA. In contrast, exons are the sequences that remain once the introns have been removed. Most times, exons are included in the mature mRNA and introns excluded, but alternative splicing sets the exception. The boundaries of exons and introns are marked by the 5′ splice site at the 5′ end of the intron (splice donor site) and the 3′ splice site at the 3′ end of the intron (splice acceptor site). Both splice sites and the polypyrimidine track near the downstream end of the intron involve defined consensus sequences [62]. Components of a large ribonucleoprotein complex known as the spliceosome recognize these particular sites, assemble specifically, and catalyze the excision of introns and ligation of the flanking exons. The spliceosome comprises five small nuclear ribonucleoproteins (U1, U2, U4, U5, and U6 snRNPs) and a set of auxiliary proteins. Assembly and catalytic activation of the spliceosome is not a random process, but is a complex one requiring the sequential pairing and unpairing of specific snRNAs aided by spliceosomal and nonspliceosomal proteins [63]. It is known that 90% of the pre-mRNA is removed as introns, whereas just about 10% represents exonic sequences.

If we take into account all the exons present in a genome, we will verify that most exons are constitutive and they are always spliced in or included in the final mRNA in all studied conditions. However, the splicing reaction that assembles eukaryotic mRNAs from their much longer precursors provides a uniquely versatile means of genetic regulation because of variations in the splicing choices. In other words, exon usage is often alternative and even introns can became part of mature mRNA when these parts are not removed from the precursor. Alternative splicing has been estimated to affect 95% of human protein-coding genes [64]. Most of the alternative exons encode protein segments and their alternate use allows the generation of multiple proteins from a single gene, which increases the coding potential of the genome. Many gene transcripts have multiple splicing patterns and some have thousands [65]. Furthermore, many genes show multiple alternative splicing events, creating a much more complex scenario and a large number of isoforms. Those isoforms may have different chemical and physiological activities because they differ in their peptide sequence.

There are many different modes of alternative splicing that can be combined with the use of alternative promoters to generate alternative 5′-terminal exons and, with the use of alternative polyadenylation sites, to give rise to alternative 3′-terminal exons. One mode of alternative splicing involves the use of alternative 5′ or 3′ splice sites. In other cases, as mentioned before, an intronic sequence can be retained and take part in the mature mRNA. The most common mode of alternative splicing is a regulated exon, also known as a cassette exon, which is sometimes included in and sometimes excluded from the mature mRNA, NCAM exon 18 serving as a salient example. A particular and noticeable mode of alternative splicing involves two or more cassette exons, usually arranged in tandem, that display mutually exclusive inclusion into the mature mRNA, meaning that when one of the exons is included, the other or other ones are excluded. The most common examples of the last mode are the mutually exclusive splicing of α-tropomyosin exons 2 and 3 [66] and several clusters of Dscam [9].

Factors that alter spliceosome assembly will also affect alternative splicing choices. For instance, the secondary structure of a pre-mRNA can alter the ability of the spliceosome to gain access to the splice sites and thus affect its usage. There are also multiple protein factors that bind to the pre-mRNA and enhance or repress spliceosome assembly at various steps. It is well known that even small changes in the rates of spliceosome assembly can give rise to important changes in the alternative splicing patterns.

In addition, it is now widely accepted that most splicing or splicing commitment takes place cotranscriptionally. The first evidence of cotranscriptional splicing came from the electron microscopy visualization of *Drosophila* embryo nascent transcripts, in which intron pre-mRNA loops are detected before transcription is completed [67]. However, the actual splicing catalysis cannot be strictly cotranscriptional for all introns. For example, if every single intron were processed before the following one was transcribed, alternative splicing would not exist. It is believed that some introns could be processed cotranscriptionally and others could be spliced posttranscriptionally and it is not known whether the same intron is always processed in the same way.

The cotranscriptionality opens an additional layer of splicing regulation based on the regulation of transcription. Alternative splicing could then be affected by all factors that alter transcription. One of the factors involved in transcription regulation is the RNA polymerase II elongation rate. Studies have revealed many factors implicated in regulation of the elongation rate, having paid special attention to the general initiation factors TFIIE, TFIIF, and TFIIH, as well as the C-terminal domain of the largest subunit of polymerase II [68]. The kinetic model proposes an explanation of how the transcription rate could be coupled to alternative splicing choices. This model presents the case in which the 3′ splice site of the intron upstream of a cassette exon is weaker compared to the 3′ splice site of the intron downstream. A fast polymerase II would expose both 3′ sites to the spliceosome components almost simultaneously, leading to a competition between them. This competition will favor the use of the stronger 3′ splice site of the intron downstream of the cassette exon. In contrast, a slower polymerase II would allow more time for spliceosome assembly to the first transcribed

weak 3′ splice site of the intron upstream of the cassette exon, committing it to splicing despite its weakness. Moreover, because many splicing regulatory proteins are recruited by the polymerase II machinery itself, the transcription elongation rate would affect the action of these *trans*-acting regulators as well. In conclusion, the kinetic model explains how low elongation rates or transcriptional pausing regulate the inclusion of alternative exons, whereas rapid elongation rates or the absence of transcriptional pausing favor its exclusion. A direct demonstration that transcription elongation affects alternative splicing in human cells was provided by the use of a mutant form of RNA polymerase II with a point mutation that confers on the enzyme a lower elongation rate. The slow polymerase stimulated the inclusion of the fibronectin EDI exon, confirming the inverse correlation between elongation rate and inclusion of this alternative exon [69].

It has been demonstrated that promoters can control alternative splicing patterns. The original observation of the promoter effect was seen using reporter minigenes for the alternatively spliced cassette exon of fibronectin, E33 or EDI, under the control of different promoters [70]. The effects show that the change in the alternative splicing choices between the constructs used must depend on some properties conferred by promoters to the transcription machinery. It has been proposed as a possible explanation that the promoter might recruit splicing factors to the nascent mRNA. However, other explanations take into account the possible role of promoters affecting the polymerase II elongation rate.

11.5 CHROMATIN STRUCTURE REGULATES ALTERNATIVE SPLICING

More recently, chromatin structure has emerged as a novel potential alternative splicing regulator. As transcription occurs through chromatin it is evident that any particular factor that affects its structure will alter the transcription process and thus eventually could affect alternative splicing. The first evidence comes from Adami and Babiss [71], who proposed that splicing could be modulated by chromatin after the observation that two copies of a viral gene in the same nucleus were spliced differentially [71]. They showed how two identical adenoviral genomes inserted into the same nucleus have different alternative splicing patterns at different stages of infection. Other evidence comes from treatments with drugs affecting the chromatin structure. For instance, treatment of cells with trichostatin A (TSA), a potent inhibitor of histone deacetylation, favors a cassette exon skipping [72]. This finding is consistent with the kinetic model, as TSA generates histone hyperacetylation, facilitating the passage of the transcribing polymerase. Similarly, it has been shown that DNA methylation at internal regions of a gene provokes a closed chromatin structure and reduces the efficiency of transcription elongation [73].

Moreover, Luco et al. [74] discovered an adaptor system of splicing factor recruitment to specific histone marks involved in alternative splicing regulation, Ameyar et al. [75] uncovered a new model for the regulation of alternative splicing in which Argonaute proteins couple RNA polymerase II elongation to chromatin modification, and, in a more physiological context, results from Saint-Andre et al. [76] support the role of intragenic repressive histone marks in alternative splicing regulation in the case of the *CD*44 gene. This work provides evidence that chromatin could function as a matrix displaying the immature transcripts to the spliceosomes [77].

Some newer evidence points out that chromatin structure influences splicing choices, but splicing can also actively modulate the pattern of histone modification in chromatin. For instance, the distribution of histone H3 lysine 36 trimethylation (H3K36me3), a well-established repressive chromatin mark, was found to be enriched on exons of active genes. Moreover, the same repressive mark positively correlates with the inclusion levels of alternative exons and with constitutive exons instead of alternative exons [78,79]. In conclusion, these reports suggest that in actively transcribed genes, transcription and alternative splicing feed back to chromatin structure.

11.6 BACKGROUND IN NCAM ALTERNATIVE SPLICING REGULATION

The length of NCAM cassette exon 18 (E18), with 801 nucleotides in the mouse, is greatly above the average length of vertebrate exons (from 100 to 150 nucleotides). It is well known that the increase in exon length hinders exon definition [80] and that the presence of a weak 5′ splice site contributes to the formation of an alternative exon and, therefore, its ability to be regulated. One of the first studies on the subject shows that when E18 is deleted splicing regulation is affected [81]. Moreover, some evidence suggests the implication of more than a single regulatory element in E18 [81]. These observations suggest that E18 splicing regulation is complex and that regulatory elements in other exons or even intronic elements could be involved. In fact, exon 17, which is constitutive in all studied conditions, regulates the recognition of the weak 5′ splice site upstream of E18 and binds to a conserved family of proteins involved in RNA splicing (SR proteins) [82]. Other studies emphasize the formation of internal RNA secondary structures that affect the use of this weak 5′ splice site, regulating its inclusion on the mature mRNA [83]. These works highlight that changes at the transcriptional level could affect not only the recruitment of accessory proteins, but also the time at which the regulatory sequences are transcribed, and thus affect significantly the inclusion or exclusion of this alternative exon.

More recently, our laboratory investigated the basis of NCAM alternative splicing regulation by an extracellular signal: membrane depolarization of neuronal cells [84]. The authors treated the murine neuroblastoma cell line N2a with increasing concentrations of extracellular KCl and assessed the splicing patterns of NCAM using radioactive semiquantitative RT-PCR. They showed that depolarization of N2a cells with 40 or 60 mM KCl induced E18 skipping. The effect is specific, as incubation with similar concentrations of NaCl does not affect NCAM alternative splicing patterns and, in the same way, alternative splicing of other cassette exon events is not affected by KCl treatment. To test if NCAM exon 18 splicing is affected by polymerase II elongation rate, the authors cotransfected N2a neuroblastoma cells with an NCAM splicing reporter minigene, harboring exons 17 to 19 and the introns between, and plasmids expressing a mutant "slow" polymerase II or a wild-type polymerase II. They found that when transcription is driven by the slow polymerase II, inclusion of NCAM exon 18 is increased, indicating that exon 18 splicing is affected by elongation and that its regulation can be interpreted in light of the kinetic model. To investigate if NCAM exon 18 splicing could be modulated by a change in the histone acetylation patterns, they performed native chromatin immunoprecipitation with antibodies recognizing acetylated histones, using primers distributed along the gene to evaluate enrichment by quantitative PCR. The patterns of H3 lysine 9 (H3K9ac) and H4 acetylation in untreated cells were comparable and low around E18. However, H3K9ac increased specifically in that region in response to depolarization, indicating that a localized change in intragenic chromatin structure was triggered by the treatment. These changes affecting the chromatin structure did not alter the gene promoter region, suggesting that transcription initiation was not affected. The more relaxed chromatin configuration around E18 caused by membrane depolarization correlated with skipping of the E18 during pre-mRNA processing by a mechanism that presumably involves enhanced transcriptional elongation. If the decrease in NCAM exon 18 inclusion in response to depolarization is caused by a local opening of chromatin, a derived hypothesis would be that inducing histone acetylation and chromatin opening directly should mimic the depolarization effect. To test this, the authors treated cells with the histone deacetylase inhibitor TSA to verify that E18 skipping was enhanced, as happens with depolarization, confirming the hypothesis.

11.7 REGULATION OF NCAM ALTERNATIVE SPLICING BY CHROMATIN CHANGES IN NEURONAL DIFFERENTIATION

The results from our lab showing that NCAM E18 is sensitive to chromatin structure and polymerase II elongation rate made the study of this splicing event a good system to assess the influence of chromatin structure in alternative splicing regulation.

Some years later, we published another analysis of the regulation of NCAM alternative splicing by chromatin structure, but this time during neuronal differentiation [85]. Using a previously established neuronal differentiation model, we incubated N2a cells in a medium with low serum and high dimethyl sulfoxide concentration. As expected from previous work, we showed that during the differentiation of N2a cells into mature neurons inclusion of NCAM E18 in the mature mRNA increases. This upregulation correlated with the acquisition of the histone heterochromatin marks H3K9me2 and H3K27me3 along the NCAM gene body (Figure 11.3), with a concomitant reduction in RNA polymerase II elongation rate at the region of E18. To address these results we used a validated elongation assay based on analyzing the kinetics of pre-mRNA levels after releasing undifferentiated or differentiated N2a cells from a transcriptional blockage by 5,6-dichloro-1-β-D-ribofuranosylbenzimidazole [86]. We, then, demonstrated that the change in E18 splicing is reverted by treatments with drugs that promote chromatin relaxation, showing that the general repressive chromatin conformation being deployed during differentiation inside the NCAM gene (including the alternatively spliced regions) plays an important role in the NCAM alternative splicing regulation. To prove that a chromatin change occurring in the NCAM gene is sufficient to promote exon 18 inclusion, we made use of a previously described methodology called TGS-AS, for transcriptional gene silencing-regulated alternative splicing [87]. This methodology uses small interfering RNAs targeting intronic sequences downstream of an alternative exon to trigger the deposition of heterochromatin marks in the target region and thus regulate alternative splicing patterns. Using this approach, we showed that small interfering RNAs targeting the intron downstream of E18 of NCAM regulate the splicing of that exon by creating silencing marks at the target site, which act as roadblocks to polymerase II elongation. These results

indicate that local TGS elicited by specific intronic siRNAs at a particular region of the NCAM gene duplicates the effects of differentiation on E18 splicing and confirm that the chromatin change is able to cause the change in the splicing pattern. Our results complete a picture in which differentiation promotes intragenic chromatin silencing, localized decrease in elongation, and higher E18 inclusion.

Taken together, the results of the two described reports of our lab [84,85] support a general model of "alternative chromatin–alternative splicing" (Figure 11.3) in which cells subjected to two different stimuli (membrane potential depolarization or differentiation signals) can modulate the chromatin surrounding a single alternative splicing event in opposite ways to have opposite effects on localized polymerase II elongation. In this way, this mechanism gives rise to two different alternative splicing patterns, each of them characteristic of a functional and differentiation status of the neuronal cells.

11.8 CONCLUSIONS

In this chapter we presented NCAM as a member of the immunoglobulin protein superfamily. We reviewed the features of NCAM alternative mRNA and protein variants and focused on the two isoforms generated by alternative splicing of exon 18, whose inclusion in the mature mRNA is upregulated during neuronal differentiation. We highlighted the key roles of NCAM in cell migration, synaptogenesis, neuronal differentiation, and plasticity, giving special importance to those roles controlled by the addition of polysialic acid to NCAM molecules. Current evidence indicates that chromatin structure is critical to modulating NCAM alternative splicing choices through changes in transcription elongation rate. This control mechanism, described in our laboratory, has been fundamental to supporting a general model of regulation that we named "alternative chromatin–alternative splicing."

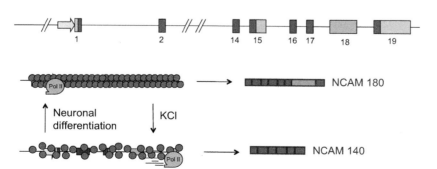

FIGURE 11.3 Alternative chromatin–alternative splicing model. In mature neurons chromatin basally adopts a closed structure (middle) in the region of NCAM exon 18. Upon depolarization with high extracellular KCl concentration, chromatin relaxation is detected in this region (bottom). Concomitantly, RNA polymerase II (Pol II) processivity increases and less time is available for recognition of the weak exon, giving a greater proportion of NCAM140. In contrast, during neuronal differentiation, chromatin adopts a more closed structure that reduces RNA polymerase II processivity and increases NCAM180 expression.

ACKNOWLEDGMENTS

This work was supported by grants to A.R.K. from the Agencia Nacional de Promoción de Ciencia y Tecnología of Argentina (ANPCYT) and the University of Buenos Aires. A.F. is recipient of a fellowship, and I.E.S and A.R.K. are career investigators from the Consejo Nacional de Investigaciones Científicas y Técnicas of Argentina (CONICET). A.R.K. is a senior international research scholar of the Howard Hughes Medical Institute.

REFERENCES

[1] Rutishauser U, Thiery JP, Brackenbury R, Sela BA, Edelman GM. Mechanisms of adhesion among cells from neural tissues of the chick embryo. Proc Natl Acad Sci USA 1976;73:577–81.

[2] Thiery JP, Brackenbury R, Rutishauser U, Edelman GM. Characterization of a cell adhesion molecule from neural retina. II. Purification and adhesion among neural cells of the chick embryo. J Biol Chem 1977;252:6841–5.

[3] Andersson AM, Olsen M, Zhernosekov D, Gaardsvoll H, Krog L, Linnemann D, et al. Age-related changes in expression of the neural cell adhesion molecule in skeletal muscle: a comparative study of newborn, adult and aged rats. Biochem J 1993;290:641–8.

[4] Pruszak J, Sonntag K, Aung MH, Sanchez-Pernaute R, Isacson O. Markers and methods for cell sorting of human embryonic stem cell-derived neural cell populations. Stem Cells 2007;25(9):2257–68.

[5] Poli A, Michel T, Thérésine M, Andrès E, Hentges F, Zimmer J. CD56bright natural killer (NK) cells: an important NK cell subset. Immunology 2009;126(4):458–65.

[6] Zola H, Swart B, Cholson I, Voss E. CD56. Leukocyte and stromal cell molecules—the CD markers. New Jersey: Wiley-Liss; 2007. p. 138–139.

[7] Buhring HJ, Battula VL, Treml S, Schewe B, Kanz L, Vogel W. Novel markers for the prospective isolation of human MSC. Ann NY Acad Sci 2007;1106:262–71.

[8] Gattenlöhner S, Stühmer T, Leich E, Reinhard M, Etschmann B, Völker HU, et al. Specific detection of CD56 (NCAM) isoforms for the identification of aggressive malignant neoplasms with progressive development. Am J Pathol 2009;174(4):1160–71.

[9] Graveley BR, Kaur A, Gunning D, Zipursky SL, Rowen L, Clemens JC. The organization and evolution of the dipteran and hymenopteran down syndrome cell adhesion molecule (Dscam) genes. RNA 2004; 10(10):1499–506.

[10] Walsh FS, Putt W, Dickson JG, Quinn CA, Cox RD, Webb M, et al. Human N-CAM gene: mapping to chromosome 11 by analysis of somatic cell hybrids with mouse and human cDNA probes. Brain Res 1986;387:197–200.

[11] Goridis C, Brunet JF. NCAM: structural diversity, function and regulation of expression. Semin Cell Biol 1992;3:189–97.

[12] Kolkova K. Biosynthesis of NCAM. Adv Exp Med Biol 2010; 663:213–25.

[13] Edelman GM, Crossin KL. Cell adhesion molecules: implications for a molecular histology. Annu Rev Biochem 1991;60:155–90.

[14] Byeon MK, Sugi Y, Markwald RR, Hoffman S. NCAM polypeptides in heart development: association with Z discs of forms that contain the muscle-specific domain. J Cell Biol 1995;128(1–2):209–21.

[15] Walmod PS, Kolkova K, Berezin V, Bock E. Zippers make signals: NCAM-mediated molecular interactions and signal transduction. Neurochem Res 2004;29:2015–35.

[16] He HT, Barbet J, Chaix JC, Goridis C. Phosphatidylinositol is involved in the membrane attachment of NCAM-120, the smallest component of the neural cell adhesion molecule. EMBO J 1986;5:2489–94.

[17] Hinkle CL, Diestel S, Lieberman J, Maness PF. Metalloprotease-induced ectodomain shedding of neural cell adhesion molecule (NCAM). J Neurobiol 2006;66:1378–95.

[18] Gibbons AS, Thomas EA, Dean B. Regional and duration of illness differences in the alteration of NCAM-180 mRNA expression within the cortex of subjects with schizophrenia. Schizophr Res 2009;112(1–3):65–71.

[19] Doherty P, Rimon G, Mann DA, Walsh FS. Alternative splicing of the cytoplasmic domain of neural cell adhesion molecule alters its ability to act as a substrate for neurite outgrowth. J Neurochem 1992;58:2338–41.

[20] Persohn E, Schachner M. Immunohistological localization of the neural adhesion molecules L1 and N-CAM in the developing hippocampus of the mouse. J Neurocytol 1990;19:807–19.

[21] Sytnyk V, Leshchyns'ka I, Nikonenko AG, Schachner M. NCAM promotes assembly and activity-dependent remodeling of the postsynaptic signaling complex. J Cell Biol 2006;174:1071–85.

[22] Kasper C, Rasmussen H, Kastrup JS, Ikemizu S, Jones EY, Berezin V, et al. Structural basis of cell-cell adhesion by NCAM. Nat Struct Biol 2000;5:389–93.

[23] Horstkorte R, Schachner M, Magyar JP, Vorherr T, Schmitz B. The fourth immunoglobulin-like domain of NCAM contains a carbohydrate recognition domain for oligomannosidic glycans implicated in association with L1 and neurite outgrowth. J Cell Biol 1993;121:1409–21.

[24] Brümmendorf T, Rathjen FG. Structure/function relationships of axon-associated adhesion receptors of the immunoglobulin superfamily. Curr Opin Neurobiol 1996;6:584–93.

[25] Kiselyov VV, Skladchikova G, Hinsby AM, Jensen PH, Kulahin N, Soroka V, et al. Structural basis for a direct interaction between FGFR1 and NCAM and evidence for a regulatory role of ATP. Structure 2003;11:691–701.

[26] Doherty P, Walsh F. CAM-FGF receptor interactions: a model for axonal growth. Mol Cell Neurosci 1996;8:99–111.

[27] Kiselyov VV. NCAM and the FGF-Receptor. Adv Exp Med Biol 2010;663:67–79.

[28] Rougon G. Structure, metabolism and cell biology of polysialic acids. Eur J Cell Biol 1993;61(2):197–207.

[29] Mühlenhoff M, Eckhardt M, Gerardy-Schahn R. Polysialic acid: three-dimensional structure, biosynthesis and function. Curr Opin Struct Biol 1998;8(5):558–64.

[30] Finne J, Finne U, Deagostini-Bazin H, Goridis C. Ocurrende of alpha 2-8 linked polysialosyl units in a neural cell adheseion molecule. Biochem Biophys Res Commun 1983;112(2):482–7.

[31] Rutishauser U, Landmesser L. Polysialic acid in the vertebrate nervous system: a promotor of plasticity in cell–cell interactions. Trends Neurosci 1996;19(10):422–7.

[32] Tang J, Landmesser L. Reduction of intramuscular nerve branching and synaptogenesis is correlated with decreased motoneuron survival. J Neurosci 1993;13(7):3095–103.

[33] Tang J, Landmesser L, Rutishauser U. Polysialic acid influences specific pathfinding by avian motoneurons. Neuron 1992;8(6):1031–44.

[34] Angata K, Nakayama J, Fredette B, Chong K, Ranscht B, Fukuda M. Human STX polysialyltransferase forms the embryonic form of the neural cell adhesion molecule. Tissue–specific expression, neurite outgrowth, and chromosomal localization in comparison with another polysialyltransferase, PST. J Biol Chem 1997;272(11):7182–90.

[35] Hildebrandt H, Becker C, Murau M, Gerardy–Schahn R, Rahmann H. Heterogeneous expression of the polysialyltransferases ST8Sia II and ST8Sia IV during postnatal rat brain development. J Neurochem 1998;71(6):2339–48.

[36] Eckhardt M, Bukalo O, Chazal G, Wang L, Goridis C, Schachner M, et al. Mice deficient in the polysialyltransferase ST8SiaIV/PST–1 allow discrimination of the roles of neural cell adhesion molecule protein and polysialic acid in neural development and synaptic plasticity. J Neurosci 2000;20(14):5234–44.

[37] Kiss JZ, Wang C, Olive S, Rougon G, Lang J, Baetens D, et al. Activity-dependent mobilization of the adhesion molecule polysialic NCAM to the cell surface of neurons and endocrine cells. EMBO J 1994;13(22):5284–92.

[38] Wang C, Pralong WF, Schulz MF, Rougon G, Aubry JM, Pagliusi S, et al. Functional N-methyl-D-aspartate receptors in O-2A glial precursor cells: a critical role in regulating polysialic acid-neural cell adhesion molecule expression and cell migration. J Cell Biol 1996;135(6 Pt 1):1565–81.

[39] Bouzioukh F, Tell F, Rougon G, Jean A. Dual effects of NMDA receptor activation on polysialylated neural cell adhesion molecule expression during brainstem postnatal development. Eur J Neurosci 2001;14(8):1194–202.

[40] Cole GJ, Loewy A, Glaser L. Neuronal cell-cell adhesion depends on interactions of N-CAM with heparin-like molecules. Nature 1986; 320:445–7.

[41] Bruses JL, Chauvet N, Rubio ME, Rutishauser U. Polysialic acid and the formation of oculomotor synapses on chick ciliary neurons. J Comp Neurol 2002;446:244–56.

[42] Gascon E, Vutskits L, Kiss JZ. Polysialic acid-neural cell adhesion molecule in brain plasticity: from synapses to integration of new neurons. Brain Res Rev 2007;56(1):101–18.

[43] Cremer H, Chazal G, Carleton A, Goridis C, Vincent JD, Lledo PM. Long-term but not short-term plasticity at mossy fiber synapses is impaired in neural cell adhesion molecule-deficient mice. Proc Natl Acad Sci USA 1998;95:13242–7.

[44] Angata K, Long JM, Bukalo O, Lee W, Dityatev A, Wynshaw–Boris A, et al. Sialyltransferase ST8Sia–II assembles a subset of polysialic acid that directs hippocampal axonal targeting and promotes fear behavior. J Biol Chem 2004;279(31):32603–13.

[45] Aonurm-Helm A, Jurgenson M, Zharkovsky T, Sonn K, Berezin V, Bock E, et al. Depression-like behaviour in neural cell adhesion molecule (NCAM)-deficient mice and its reversal by an NCAM-derived peptide, FGL. Eur J Neurosci 2008;28:1618–28.

[46] Bonfati L. PSA-NCAM in mammalian structural plasticity and neurogenesis. Prog Neurobiol 2006;80(3):129–64.

[47] Vutskits L, Gascon E, Zgraggen E, Kiss JZ. The polysialylated neural cell adhesion molecule promotes neurogenesis in vitro. Neurochem Res 2006;31:215–25.

[48] Amoureux MC, Cunningham BA, Edelman GM, Crossin KL. N-CAM binding inhibits the proliferation of hippocampal progenitor cells and promotes their differentiation to a neuronal phenotype. J Neurosci 2000;20:3631–40.

[49] Kim BW, Son H. Neural cell adhesion molecule (NCAM) induces neuronal phenotype acquisition in dominant negative MEK1-expressing hippocampal neural progenitor cells. Exp Mol Med 2006;38(6):732–8.

[50] Röckle I, Seidenfaden R, Weinhold B, Mühlenhoff M, Gerardy-Schahn R, Hildebrandt H. Polysialic acid controls NCAM-induced differentiation of neuronal precursors into calretinin-positive olfactory bulb interneurons. Dev Neurobiol 2008;68:1170–84.

[51] Walsh FS, Hobbs C, Wells DJ, Slater CR, Fazeli S. Ectopic expression of NCAM in skeletal muscle of transgenic mice results in terminal sprouting at the neuromuscular junction and altered structure but not function. Mol Cell Neurosci 2000;15(3):244–61.

[52] Dallérac G, Rampon C, Doyère V. NCAM function in the adult brain: lessons from mimetic peptides and therapeutic potential. Neurochem Res 2013;38(6):1163–73.

[53] Gomez-Climent MA, Guirado R, Castillo-Gomez E, Varea E, Gutierrez-Mecinas M, Gilabert-Juan J, et al. The polysialylated form of the neural cell adhesion molecule (PSA-NCAM) is expressed in a subpopulation of mature cortical interneurons characterized by reduced structural features and connectivity. Cereb Cortex 2011;21:1028–41.

[54] Nacher J, Guirado R, Castillo-Gómez E. Structural plasticity of interneurons in the adult brain: role of PSA-NCAM and implications for psychiatric disorders. Neurochem Res 2005;38(6):1122–33.

[55] Castren E. Is mood chemistry? Nat Rev Neurosci 2005;6:241–6.

[56] Lewis DA, Gonzalez-Burgos G. Neuroplasticity of neocortical circuits in schizophrenia. Neuropsychopharmacology 2008;33:141–65.

[57] Pollerberg GE, Schachner M, Davoust J. Differentiation state-dependent surface mobilities of two forms of the neural cell adhesion molecule. Nature 1986;324(6096):462–5.

[58] Pollerberg GE, Burridge K, Krebs KE, Goodman SR, Schachner M. The 180-kD component of the neural cell adhesion molecule N-CAM is involved in cell-cell contact and cytoskeleton-membrane interactions. Cell Tissue Res 1987;250(1):227–36.

[59] Takei K, Chan TA, Wang FS, Deng H, Rutishauser U, Jay DJ. The neural cell adhesion molecules L1 and NCAM-180 act in different steps of neurite outgrowth. J Neurosci 1999;19(21):9469–79.

[60] Kiss JZ, Troncoso E, Djebbara Z, Vutskits L, Muller D. The role of neural cell adhesion molecules in plasticity and repair. Brain Res Rev 2001;36(2001):175–84.

[61] Prodromidou K, Papastefanaki F, Sklaviadis T, Matsas R. Functional cross-talk between the cellular prion protein and the neural cell adhesion molecule NCAM is critical for neuronal differentiation of neural stem/precursor cells. Stem Cells 2014;32(6):1674–87.

[62] Burge CB, Tuschl T, Sharp PA. Splicing of precursors to mRNAs by the spliceosomes. RNA World 1999:525–60.

[63] Black DL, Grabowski PJ. Alternative pre-mRNA splicing and neuronal function. Prog Mol Subcell Biol 2003;31:187–216.

[64] Kornblihtt AR, Schor IE, Alló M, Dujardin G, Petrillo E, Muñoz MJ. Alternative splicing: a pivotal step between eukaryotic transcription and translation. Nat Rev Mol Cell Biol 2013;14(3):153–65.

[65] Black DL. Protein diversity from alternative splicing: a challenge for bioinformatics and post-genome biology. Cell 2000;103:367–70.

[66] Smith CW, Nadal-Ginard B. Mutually exclusive splicing of alpha-tropomyosin exons enforced by an unusual lariat branch point location: implications for constitutive splicing. Cell 1989; 56(5):749–58.

[67] Beyer AL, Osheim YN. Splice site selection, rate of splicing and alternative splicing on nascent transcripts. Genes Dev 1988;2(6):754–65.

[68] Conaway JW, Shilatifard A, Dvir A, Conaway RC. Control of elongation by RNA polymerase II. TIBS 2000;25(8):375–80.

[69] de la Mata M, Alonso CR, Kadener S, Fededa JP, Blaustein M, Pelisch F, et al. A slow RNA polymerase II affects alternative splicing in vivo. Mol Cell 2003;12(2):525–32.

[70] Cramer P, Pesce CG, Baralle FE, Kornblihtt AR. Functional association between promoter structure and transcript alternative splicing. Proc Natl Acad Sci USA 1997;94(21):11456–60.

[71] Adami G, Babiss LE. DNA template effect on RNA splicing: two copies of the same gene in the same nucleus are processed differently. EMBO J 1991;10:3457–65.

[72] Nogues G, Kadener S, Cramer P, Bentley D, Kornblihtt AR. Transcriptional activators differ in their abilities to control alternative splicing. J Biol Chem 2002;277(45):43110–4.

[73] Lorincz MC, Dickerson DR, Schmitt M, Groudine M. Intragenic DNA methylation alters chromatin structure and elongation efficiency in mammalian cells. Nat Struct Mol Biol 2004;11(11):1068–75.

[74] Luco RF, Pan Q, Tominaga K, Blencowe BJ, Pereira-Smith OM, Misteli T. Regulation of alternative splicing by histone modifications. Science 2010;327(5968):996–1000.

[75] Ameyar-Zazoua M, Rachez C, Souidi M, Robin P, Fritsch L, Young R, et al. Argonaute proteins couple chromatin silencing to alternative splicing. Nat Struct Mol Biol 2012;19(10):998–1004.

[76] Saint-Andre V, Batsche E, Rachez C, Muchardt C. Histone H3 lysine 9 trimethylation and HP1gamma favor inclusion of alternative exons. Nat Struct Mol Biol 2011;18(3):337–44.

[77] Allemand E, Batsché E, Muchardt C. Splicing, transcription, and chromatin: a ménage à trois. Curr Opin Genet Dev 2008;18(2):145–51.

[78] Kolasinska-Zwierz P, Down T, Latorre I, Liu T, Liu XS, Ahringer J. Differential chromatin marking of introns and expressed exons by H3K36me3. Nat Genet 2009;41(3):376–81.

[79] Andersson R, Enroth S, Rada-Iglesias A, Wadelius C, Komorowski J. Nucleosomes are well positioned in exons and carry characteristic histone modifications. Genome Res 2009;19(10):1732–41.

[80] Berget SM. Exon recognition in vertebrate splicing. J Biol Chem 1995;270(6):2411–4.

[81] Tacke R, Goridis C. Alternative splicing in the neural cell adhesion molecule pre-mRNA: regulation of exon 18 skipping depends on the 5′-splice site. Genes Dev 1991;5(8):1416–29.

[82] Côté J, Simard MJ, Chabot B. An element in the 5′ common exon of the NCAM alternative splicing unit interacts with SR proteins and modulates 5′ splice site selection. Nucleic Acids Res 1999;27(12):2529–37.

[83] Côté J, Chabot B. Natural base-pairing interactions between 5′ splice site and branch site sequences affect mammalian 5′ splice site selection. RNA 1997;3(11):1248–61.

[84] Schor IE, Rascovan N, Pelisch F, Alló M, Kornblihtt AR. Neuronal cell depolarization induces intragenic chromatin modifications affecting NCAM alternative splicing. Proc Natl Acad Sci USA 2009;106(11):4325–30.

[85] Schor IE, Fiszbein A, Petrillo E, Kornblihtt AR. Intragenic epigenetic changes modulate NCAM alternative splicing in neuronal differentiation. EMBO J 2013;32(16):2264–74.

[86] Singh J, Padgett RA. Rates of in situ transcription and splicing in large human genes. Nat Struct Mol Biol 2009;16(11):1128–33.

[87] Alló M, Buggiano V, Fededa JP, Petrillo E, Schor I, de la Mata M, et al. Control of alternative splicing through siRNA-mediated transcriptional gene silencing. Nat Struct Mol Biol 2009;16(7):717–24.

Chapter 12

Role of the Clustered Protocadherins in Promoting Neuronal Diversity and Function

Takeshi Yagi

KOKORO-Biology Group, Laboratories for Integrated Biology, Graduate School of Frontier Biosciences, Osaka University, Suita, Osaka, Japan; Core Research for Evolutional Science and Technology (CREST), Japan Science and Technology Agency, Japan

12.1 INTRODUCTION

In complex biological systems, diverse cell-surface antigens play significant roles in maintaining cell-specific functions. In the immune system, immunoglobulins (Igs) and T cell receptors (TCRs) are differentially expressed in B and T cells, and their activation leads to primary and secondary immune responses. Their diversity is developmentally generated by random somatic DNA rearrangements, specific expression, heteromeric combination, and antigen-induced proliferation [1,2]. In the olfactory nervous system, diverse odorant receptors (ORs) are uniquely expressed in individual olfactory sensory neurons and induce selective odorant recognition. Whereas several hundred to over 1000 ORs are expressed in the olfactory sensory neurons [3], only one functional OR is expressed in each olfactory sensory neuron and induces specific axonal targeting to the appropriate glomerulus in the olfactory bulb. Thus, ORs play significant roles in building the complex neural networks that enable animals to recognize an enormous number of odors.

In vertebrate brains, the clustered protocadherins (Pcdhs) were first identified as cadherin-related neuronal receptors (CNRs) [4–8]. These proteins are predominantly expressed in the nervous system, and experiments using cerebellar Purkinje cells showed that almost all of them are stochastically and combinatorially expressed in individual neurons. The clustered *Pcdh* members are organized in gene clusters and undergo stochastic allelic expression in individual neurons [9–11]. Each clustered Pcdh protein interacts with different family members to form cell-surface tetramers possessing homophilic binding activity [12]. Following homophilic binding, the clustered Pcdh proteins form intracellular protein complexes to induce signal transduction and receptor proteolysis [13,14]. In mice, the clustered Pcdhs play various roles in neuronal survival, axonal targeting, dendrite arborization, synaptogenesis, cortical activity, and learning and memory and are predicted to play significant roles in the assembly of neuronal networks [6,15–20]. In human populations, the clustered *Pcdh* genes have many polymorphisms [21,22] and are known susceptibility genes for certain neurological disorders [23–25]. Their expression is regulated by epigenetic modifications, including DNA methylation [26]. These epigenetic modifications may occur in response to early environmental stresses, such as reduced maternal care [27], leading to changes in brain activity. The molecular characterization of the clustered Pcdhs is reviewed in the next few sections and is summarized in Figure 12.1.

12.2 THE CADHERIN SUPERFAMILY AND THE CLUSTERED Pcdhs

The cadherin superfamily was first identified in studies aimed at isolating calcium-responsive cell-adhesion molecules (reviewed in Ref. [28]) and is defined by the presence of cadherin sequence repeats in the extracellular (EC) domain. The cadherin superfamily is divided into several subgroups, including the classical type I and closely related type II cadherins, the desmosomal cadherins, and the clustered and nonclustered Pcdhs. In vertebrate genomes, there are more than 100 different cadherin-related genes [16].

The *Pcdh* genes were first discovered in an effort to isolate additional cadherin repeat-containing genes using degenerate PCR [29]. In vertebrates, the clustered Pcdhs comprise a major family within the cadherin superfamily. Interestingly in teleost fish genomes, there are over 100 distinct clustered *Pcdh* genes [30,31]. The clustered Pcdhs are divided into three gene clusters: the *Pcdh-α* (*CNR*), *Pcdh-β*, and *Pcdh-γ* families, consisting of approximately 60 genes in mammals [4,5,32–34], and the nonclustered *Pcdhs* consist of 13 members in mammals: four in the *Pcdh-δ1*, five

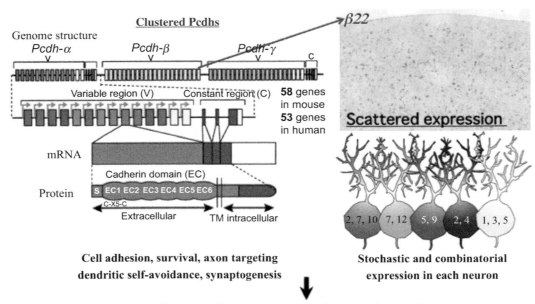

FIGURE 12.1 The clustered Pcdh gene family. The genomic organization of the clustered *Pcdh* genes is shown. Fifty-eight mouse and 53 human genes are organized into *Pcdh-α*, *Pcdh-β*, and *Pcdh-γ* clusters arranged in tandem. The *Pcdh-α* and -*γ* clusters contain many variable (V) and common constant (C) region exons, whereas the *Pcdh-β* cluster contains only variable region exons. Each mature mRNA encoding a *Pcdh-α* or -*γ* isoform is produced from a variable region exon spliced to a set of constant region exons, which encode the intracellular tail. Each variable region exon of all the *Pcdh* members encodes an extracellular (EC), transmembrane (TM), and intracellular domain. All clustered Pcdh proteins contain a signal peptide (S) with six EC domains. The EC1 domain contains a Cys-$(X)_5$-Cys (C-X5-C) motif, which is involved in tetramer formation among heteromeric clustered Pcdh proteins. Most clustered *Pcdh* members (except for the five C-type clustered *Pcdh* members) are expressed in a scattered pattern (shown by *Pcdh-β22* expression in the cerebral cortex). Stochastic and combinatorial expression of the genes is observed in individual neurons.

in the *Pcdh-δ2*, and four in other *Pcdh* subgroups [35]. In mammals, the nonclustered *Pcdh* genes are scattered throughout the genomes, whereas, in contrast, the clustered *Pcdh* genes lie in tandem arrays, encompassing ~1 Mb on mouse chromosome 18 and human chromosome 5q31 [5,33,34].

12.3 GENOMIC STRUCTURES OF THE CLUSTERED *Pcdh* GENES

The clustered *Pcdh* genes are exclusively found in vertebrates (Figure 12.1) [32]. Within the clusters, over 50 long "variable" exons, which encode the N-terminal extracellular cadherin domain, a transmembrane domain, and a short cytoplasmic domain, are arranged in tandem. Both the *Pcdh-α* and the *Pcdh-γ* gene clusters are present in all sequenced vertebrate genomes, and each has a genomic organization similar to that of the *Ig* and *TCR* gene clusters, in that each cluster contains variable region exons, described above, and "constant" region exons, which encode a shared cytoplasmic tail (Figure 12.1). In each case, the variable region exon is spliced to the constant region exons, resulting in the generation of Pcdh-α or Pcdh-γ isoforms with many different N-terminal extracellular domains and identical cytoplasmic tails [4,5]. The cytoplasmic tail binds directly

to the nonreceptor protein tyrosine kinases focal adhesion kinase (FAK) and proline-rich tyrosine kinase (PYK2) [13]. Similar to the signaling induced by membrane-bound Ig and TCRs, different extracellular interactions between clustered Pcdh proteins are translated to a common cytoplasmic signaling pathway. In contrast, the *Pcdh-β* gene cluster does not contain constant region exons; thus each *Pcdh-β* gene is represented by a single exon. The *Pcdh-β* cluster is not found in teleost fishes, including zebrafish, medaka, and fugu, but seems to have arisen separately in coelacanths [32,36]. In mice, a total of 58 genes are arranged in the *Pcdh-α*, *Pcdh-β*, and *Pcdh-γ* gene clusters, which contain 14, 22, and 22 members, respectively. In humans, a total of 53 *Pcdh-α*, *Pcdh-β*, and *Pcdh-γ* members have been identified. Phylogenetic analyses show that the Pcdh isoforms encoded by each gene cluster are highly homologous to one another; however, a notable exception is a set of five isoforms encoded by the *Pcdh-α* (*αC1* and *αC2*) and *Pcdh-γ* (*γC3*, *γC4*, and *γC5*) clusters, which exhibit more similarity with one another than with the other isoforms of their respective clusters. These divergent C-type *Pcdh* members are located near the constant exons in the *Pcdh-α* or *Pcdh-γ* gene clusters [32].

A phylogenetic analysis of the clustered *Pcdh* variable and constant exons revealed orthologous relationships among the human, rat, and mouse sequences [31,34]. However, these relationships do not extend to the variable exons of the chicken, zebrafish, and coelacanth [30,36,37]. In contrast, the constant region exons and the CGCT conserved sequence element (CSE) in their promoters are highly conserved across species. The identification of species- and/or gene-cluster-specific homologous sequences among the variable exons within the human, mouse, chicken, zebrafish, and coelacanth *Pcdh* genes [30,31,34,36,37] suggested that these regions underwent sequence homogenization. It is also reported that the clustered *Pcdh* family has been particularly prone to gene conversion events through the course of evolution [31]. These results indicate that the evolution of clustered *Pcdh* genes is driven by a combination of lineage-specific duplication of variable exons, region-restricted gene conversion within each species, and adaptive variation [31]. In mice, many single-nucleotide polymorphisms (SNPs) have been identified in the *Pcdh-α* gene clusters from both wild-derived and laboratory strains [38]. Synonymous SNPs (silent nucleotide substitutions) are concentrated in the EC1 and EC5 regions, which have high nucleotide homology among *Pcdh-α* members. Gene conversion events associated with mouse specification have been shown to occur in these regions [38]. In humans, many SNPs, including those that cause amino acid changes leading to extensive linkage disequilibrium, are observed in the *Pcdh-α* gene cluster [21,22]. Several intermediate-frequency variants in the *Pcdh-α* gene cluster of certain human populations have been identified and suggested to represent a signature of balancing selection. A 16.7-kb deletion that truncates *Pcdh-α8* and removes both *Pcdh-α9* and *Pcdh-α10* is found in Europeans at a frequency of 11% [21]. These features of molecular diversity in the human population suggest that the clustered *Pcdh* genes may be involved in generating brain function variation. Furthermore, SNPs in the *Pcdh-α* genes are also associated with autism and bipolar diseases [21,23–25].

12.4 GENE EXPRESSION OF THE CLUSTERED *Pcdhs*

The unique genomic organization of clustered *Pcdhs* contributes to their differential and combinatorial expression in individual neurons. The *Pcdh-α* and *Pcdh-γ* transcripts are expressed from a promoter located upstream of each variable exon [39,40], and selective transcription of each clustered *Pcdh* member is achieved by "promoter choice" followed by alternative *cis*-splicing of the transcript. The CSE is conserved in most promoters within the three gene clusters and plays a role in the clustered *Pcdh* gene expression. Selective expression of the clustered *Pcdh* transcripts was first demonstrated in specific cultured cell lines [39,41] and was shown to be correlated with the state of DNA methylation. The promoters of actively transcribed

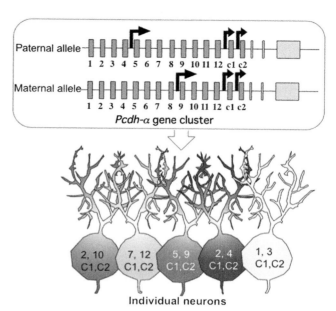

FIGURE 12.2 Monoallelic, biallelic, and combinatorial expression of the clustered *Pcdh-α* variable region exons in individual neurons. The expression levels of *Pcdh-α* transcripts from each allele are regulated independently. Broad arrows indicate strongly expressed exons. Exons *α1* to *α12* undergo monoallelic, stochastic expression, whereas *αC1* and *αC2* are constitutively expressed from both alleles in individual neurons.

genes are hypomethylated, whereas those of silent genes are hypermethylated. Thus, the DNA methylation pattern of the promoter region is involved in regulating *Pcdh* gene expression.

Clustered *Pcdh* transcripts are expressed in the nervous system and are highly expressed in the developing brain [42]. During neuronal maturation, *Pcdh-α* expression is highly localized to axons and is dramatically downregulated by myelination, similar to the expression of GAP43 and L1 [42]. *Pcdh-γ* is widely expressed in neurons, glia, and the choroid plexus [43], whereas *Pcdh-β* is expressed specifically in neurons and glias [10].

Most of the clustered *Pcdh* members (except for the five C-type *Pcdh* isoforms) are expressed in a scattered pattern (Figure 12.1) throughout the brain [9–11]. In addition, translation analysis using single Purkinje cells from interspecies F1 hybrid mice revealed that most *Pcdh* family members (except for the five C-type members) are expressed stochastically and monoallelically. Although the gene clusters of both alleles are transcriptionally active (Figure 12.2) [9–11,42,44], promoter selection occurs independently in both alleles. The C-type *Pcdh* transcripts are constitutively expressed from both alleles in individual neurons [9,11].

The stochastic expression of the *Pcdh-α* transcripts depends on the number of variable exons in the *Pcdh-α* cluster. The deletion of variable exons *α2* to *α11* in one allele (from a total of 12) induces an increased expression of the remaining variable exons *α1* and *α12*, compared

FIGURE 12.3 Monoallelic, biallelic, and combinatorial expression in the *Pcdh-α* (*α2* to *α11*) deletion mutant. In wild-type mice, *Pcdh-α1* shows a scattered mRNA expression pattern in Purkinje cells (left) and cortical neurons (right). Each individual neuron expresses a distinct, but random, set of *Pcdh-α* (*α1* to *α12*) members and expresses *αC1* and *αC2* constitutively. In the deletion mutant, the *Pcdh-α1* expression frequency is upregulated in both Purkinje cells and cortical neurons. Each individual neuron expresses either *α1* or *α12*, or both *α1* and *α12*, and expresses *αC1* and *αC2* constitutively [44].

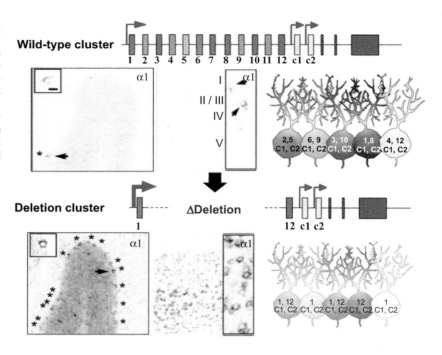

to their expression in the wild-type allele. That is, in the wild-type allele *α1* and *α12* are expressed rarely and stochastically, whereas in the deletion allele they are highly expressed in individual neurons (Figure 12.3) [44]. Interestingly, the total expression level of *Pcdh-α* transcripts (based on the expression of the common constant region) in the gene cluster is maintained even in the deletion allele [44]. The same level of expression is also maintained in cells containing a duplication of *Pcdh-α* genes. Studies of several mouse lines containing deletions or duplications within the *Pcdh-α* gene cluster show that the frequency of a variable isoform's stochastic expression depends on the number of variable exons in the cluster and that the total expression level is maintained by *cis*-regulatory elements outside of the variable exons. Thus, the stochastic expression of *Pcdh-α* isoforms is analogous to the results of throwing dice.

In contrast to the variable region exons, *αC1* and *αC2* in the *Pcdh-α* cluster are constitutively expressed in both wild-type alleles of individual neurons. Studies in which variable exons were deleted in the *Pcdh-α* gene cluster showed that the constitutive expression of *αC1* and *αC2* requires their location at the most 3′ position of the cluster. Normally, the 3′-most variable exon (*α10*) exhibits stochastic expression. However, when the position of *α10* is shifted to a more 3′ location in a mutant allele containing a deletion encompassing *α11* to *αC2*, the expression pattern of this exon switches from stochastic to constitutive and biallelic, similar to that of *αC2* in the wild-type allele [44]. Thus, the position of the constant region exon in the *Pcdh-α* cluster plays a role in its constitutive expression.

As a result of the stochastic expression of the clustered *Pcdh* genes, it is estimated that individual neurons express approximately two of the 12 *Pcdh-α*, approximately four

of the 22 *Pcdh-β*, and approximately four of the 19 *Pcdh-γ* genes. Therefore, the total number of possible combinations is 78 (2 from the 12 *Pcdh-α* cluster) × 26,796 (4 from the 22 *Pcdh-β* cluster) × 14,706 (4 from the 19 *Pcdh-γ* cluster) = 30,736,834,128. Thus, approximately 3×10^{10} variations are possible in each individual neuron [7].

12.5 GENE REGULATION OF THE CLUSTERED *Pcdhs*

Except for the C-type members, over 50 of the clustered *Pcdh* genes are stochastically and differentially expressed from each allele in individual neurons. Based on their sequence conservation and DNase I hypersensitivity, *cis*-regulatory elements have been identified in the genomic region of the *Pcdh* gene clusters and shown to have independent functions for each gene cluster [39,40]. For example, the DNase I hypersensitivity site HS5-1 located between the *Pcdh-α* and the *Pcdh-β* clusters is involved in regulating the expression of *Pcdh-α6* to -*α12* and -*αC1* (Figures 12.4 and 12.5) [46,47]. The HS16-20 site (known as the cluster control region or CCR), located downstream of the *Pcdh-γ* cluster, is involved in the stochastic and allelic expression of the *Pcdh-β* gene cluster (Figure 12.5) [47]. Deletion of the CCR dramatically reduces the expression of the *Pcdh-β* gene, but has a much smaller effect on *Pcdh-γ* gene expression and has no effect on the *Pcdh-α* gene expression in mice [47], thus demonstrating that the CCR is an independent long-range *cis*-element that regulates only *Pcdh-β* gene expression (Figure 12.5).

The CCCTC-binding factor (CTCF), a zinc finger transcription factor, regulates the stochastic and allelic

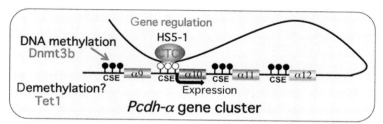

FIGURE 12.4 Regulatory mechanism leading to stochastic expression of the clustered *Pcdh* genes. Each variable region exon is independently transcribed from its own promoter, each of which possesses a conserved sequence element (CSE). The CTCF protein binds to *cis*-enhancer elements (HS5-1 of *Pcdh-α*) and active promoter regions by DNA looping. The promoter regions are methylated by Dnmt3b during early embryonic stages [26]. Strongly methylated promoters show reduced activity. Tet1, a DNA hydroxylase, binds to *Pcdh* gene clusters and may play a role in DNA demethylation.

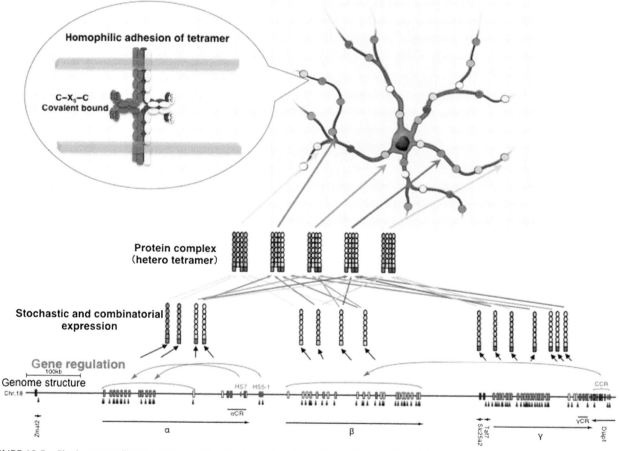

FIGURE 12.5 Single neuron diversity determined by the stochastic and combinatorial expression of clustered *Pcdh*s and the combinatorial production of heteromeric *cis*-tetramers. Approximately 10 *Pcdh-α*, *Pcdh-β*, and *Pcdh-γ* proteins (excluding the C-type isoforms) are stochastically expressed, and five of the C-type isoforms are constitutively expressed from the clustered *Pcdh* alleles. The stochastic expression is regulated by the *cis*-elements, HS5-1 in the *Pcdh-α* cluster and CCR in the *Pcdh-β* cluster. HS7 also regulates *Pcdh-α* cluster transcription, but functions independent of CTCF. The HS5-1 and CCR motifs are located downstream of the *Pcdh-α* and *Pcdh-γ* cluster, respectively. The stochastic expression is also regulated by the transcription factor CTCF and the Rad21 cohesion complex. Black triangles show CTCF-binding sites [45]. The expressed Pcdh isoforms form random heteromeric *cis*-tetramers. The Cys-(X)$_5$-Cys motif, which is conserved among all clustered Pcdh and nonclustered Pcdh-δ2 proteins, plays a role in forming the *cis*-tetramers. The same heteromeric *cis*-tetramers can bind homophilically between cell surfaces. Genetic analyses in mice showed that these interactions influence dendritic arborization. (*This figure is modified from previously published figures [7,8].*)

expression of the clustered *Pcdh*s in individual neurons, but does not regulate the constitutive and biallelic expression of *αC2*, *γC3*, *γC4*, and *γC5* [48]. Although *αC1* is a constitutively expressed C-type member, its expression can be regulated by CTCF [48]. Interestingly, the *αC1* promoter

contains a CTCF-binding CSE sequence. CTCF-binding sites are also found in all of the stochastically expressed clustered *Pcdh* promoters and in HS5-1 and HS16-20 (Figures 12.4 and 12.5) [49–51]. The HS5-1 site also functions as transcriptional silencer in nonneuronal cells by binding to the RE1-silencing transcription factor (REST, also known as NRSF) receptor complex [49].

In neuroblastoma cell lines, CTCF and the nuclear phosphoprotein Rad21 (a subunit of the cohesion complex involved in sister chromatid cohesion during mitosis) bind directly to the CSE of transcriptionally active *Pcdh-α* promoters and the conserved downstream sequence [51]. The HS5-1 enhancer also binds CTCF and Rad21 [51]. However, the HS7 enhancer, which is located in the intron of the *Pcdh-α* constant region, lacks a CTCF-binding site and binds only to Rad21. It is known that CTCF mediates enhancer and promoter interactions by DNA looping [52]. In long-range regulation, the CTCF-mediated DNA looping between a promoter and an enhancer accounts for the stochastic "promoter choice" of clustered *Pcdh* genes in individual neurons (Figure 12.4). Interestingly, differentiated excitatory neurons lacking CTCF exhibit abnormal dendritic arborization and decreased numbers of synapses, but show no effects on synapse maturation or the amplitude of excitatory synaptic currents [48]. In addition, differentiated excitatory neurons in conditional CTCF-knockout mice exhibit a somatosensory map defect involving a loss of barrel structure in the cortex, suggesting that CTCF plays a role in the final refinement of functional neural circuits during postnatal brain development [48]. Many genes exhibit altered expression in the CTCF-deficient cortex. Among them, the stochastically expressed clustered *Pcdh* genes are particularly sensitive to the CTCF defect, exhibiting a marked reduction in expression. A similar phenotype affecting barrel formation in the cortex is also observed in clustered *Pcdh* loss-of-function mutants and in *Pcdh-β* CCR deletion mutant mice [47]. In the latter case, all of the *Pcdh-β* isoforms except *β1* are strongly downregulated, and *Pcdh-γ* expression is also downregulated owing to a *lacZ* gene insertion in the 3′ terminus of the *Pcdh-γ* constant region. Taken together these results show that the stochastic expression of the clustered *Pcdh* genes in individual neurons influences functional map formation in the cortex [48]; however, the underlying mechanisms remain to be determined.

Pcdh expression in individual neurons is epigenetically controlled by DNA methylation [26,39,41]. In two distinct mouse cell lines that expressed different combinations of *Pcdh-α* isoforms, it is observed that the DNA methylation level of each *Pcdh-α* promoter was negatively correlated with its level of activity [41]. In addition, treatment of cells with 5-azacytidine, which induces DNA demethylation, led to increased *Pcdh-α* gene transcription. In the mouse brain, the promoters of the constitutively expressed *Pcdh-αC1* and -*αC2* are hypomethylated, whereas the promoters of stochastically expressed clustered *Pcdh* members exhibit various mosaic methylation patterns [41]. Hypermethylation of CpG islands in the variable exons of *Pcdh-α* and *Pcdh-β* is a strong determinant of poor prognosis in a large number of

neuroblastomas [53]. However, examination of the tumor samples indicates that the hypermethylation of *Pcdh-β* genes does not lead to the downregulation of *Pcdh-β* transcripts, suggesting that *Pcdh-β* gene expression may not be strongly dependent on DNA methylation or linked to neuroblastoma prognosis [53].

The promoters of the stochastically expressed clustered *Pcdh* genes are differentially methylated by the de novo DNA methyltransferase Dnmt3b in early embryogenesis, whereas those of the constitutively expressed *αC1* and *αC2* are not [26]. Consistent with this finding, the 3′-most variable exon in the *Pcdh-α* deletion allele, which exhibits a constitutive and biallelic expression pattern similar to that of *Pcdh−αC2* in the wild-type allele, is also not methylated by Dnmt3b. These results suggest that DNA methylation of the clustered *Pcdh* promoter regions is necessary to inhibit their transcription. Single Dnmt3b-deficient neurons in chimeric mice, generated by injecting Dnmt3b-deficient induced pluripotent stem cells into wild-type embryos, express increased numbers of clustered *Pcdh* isoforms [26]. Thus, DNA methylation by Dnmt3b at early embryonic stages regulates the probability of stochastic expression of clustered *Pcdh* isoforms [26]. Furthermore, Dnmt3b-deficient Purkinje cells exhibit abnormal dendritic arborization, suggesting that Dnmt3b may contribute to the mechanisms for establishing neuronal identity and fine neural network formation, and that Dnmt3b deficiency may lead to the development of human recessive diseases, such as the immunodeficiency, centromere instability, and facial anomalies syndrome.

Smchd1 (structural maintenance of chromosomes hinge domain-containing 1) is a protein with homology to the SMC family of proteins involved in chromosomal condensation and cohesion [54]. Smchd1 plays an important role in CpG island methylation on the inactive X chromosome and in the stable silencing of some X-linked genes in females. Smchd1 is also required for the CpG island methylation and silencing of the *Pcdh-α* and *Pcdh-β* clusters [55]. Interestingly, normally inactive X-chromosome genes that are active in Smchd1-deficient embryos are not strongly affected in Dnmt3b-deficient female embryos, in which CpG island methylation is lost [56]. Taken together, these findings suggest that region-specific DNA methylation in the *Pcdh-α* cluster may involve the cooperative activities of both Smchd1 and Dnmt3b.

The DNA-methylation patterns of multiple clustered *Pcdh* genes are modified in response to variations in maternal care (licking and grooming behaviors) in mouse offspring [27]. Poor maternal care is correlated with changes in methylation and histone H3 lysine-9 acetylation and gene expression patterns across a 7-Mb region surrounding the glucocorticoid receptor (NR3C1) gene. The three *Pcdh* clusters are located in this region on rat chromosome 18, and hypermethylation of multiple *Pcdh* genes is found in the offspring of low-maternal-care mothers [27]. Similar results are reported in humans. In fact, hypermethylation across the clustered *Pcdh* genes is observed in hippocampal samples from suicide victims with a history of severe child abuse [57], further evidence that the clustered *Pcdh* genes may be involved in environmental stress responses. The Tet1 protein, a DNA hydroxylase that converts 5-methylcytosine to 5-hydroxymethylcytosine during DNA demethylation, also binds to the clustered *Pcdh* genes [58]. These DNA-methylation studies suggest that the clustered *Pcdh* genes may be important mediators of neural circuit formation that respond to environmental stress during brain development.

12.6 CELL ADHESION ACTIVITY OF THE CLUSTERED Pcdhs

Classical cadherins mediate calcium-dependent cell–cell adhesion via *trans*-dimerization between their extracellular domains. Because cadherins can undergo homophilic, but not heterophilic, binding, specific cadherin expression in a cell induces selective cell–cell adhesion and interactions with other cells expressing the same cadherin. Classical cadherins also assemble intercellular adherence junctions through *cis*-clustering [29]. Similar to classical cadherins, most Pcdh proteins mediate intercellular adhesion. Unlike classical cadherins, however, the homophilic binding activity of clustered Pcdh proteins is not dependent on extracellular calcium. In addition, whereas the binding activity of classical cadherins is mediated by their EC1 domain (Figure 12.1), the binding specificity of clustered Pcdh proteins is determined by their EC2 and EC3 domains (Figure 12.5) [12], which are highly divergent among clustered Pcdh members. Whereas classical cadherins bind to one another as monomers, it has been suggested that the unit of Pcdh homophilic *trans* interactions is a *cis*-tetramer. Pcdh-γ isoforms have been shown to form heteromultimeric *cis*-tetramers that undergo homophilic binding [12]. Similarly, Pcdh-α and Pcdh-γ isoforms form multimeric complexes *in cis* [59], and Pcdh-β proteins may also form multimeric complexes with Pcdh-α and Pcdh-γ proteins (Figure 12.5) [14,60,61]. The disulfide-bonded Cys-$(X)_5$-Cys sequence in the EC1 domain, conserved among all clustered Pcdhs, plays an important role in their tetramer formation. The structural characterization of the Cys-$(X)_5$-Cys motif shows that it is exposed to solvent when in a loop conformation [62]. Interestingly, the Cys-$(X)_5$-Cys motif is also conserved in the EC1 domains of the nonclustered Pcdh-δ2 isoforms [62]. Thus, the multimerization of stochastically expressed Pcdhs in a single neuron could provide a large number of distinct homophilic interaction units, potentially increasing the cell-surface diversity for cellular interactions (Figure 12.5) [7,8,12]. As an example, if two distinct clustered Pcdh isoforms are expressed in a single cell, six different combinatorial *cis*-tetramers

can be expressed on the cell surface. If one isoform is expressed on both of two cells, only one of the six tetramers will be present on both cell surfaces, resulting in weak cell–cell aggregation [7]. However, the formation of multiple heteromeric *cis*-tetramers of the clustered Pcdhs (and most likely the Pcdh-δ2 isoforms) has the potential to provide an exponential increase in the number of protein complexes expressed at the cell surface, thus increasing the number of commonly expressed proteins between any two cells, leading to stronger cell–cell interactions.

In estimation, 15 clustered Pcdh members (i.e., 10 randomly expressed, 2 Pcdh-α, 4 Pcdh-β, and 4 Pcdh-γ, and 5 constitutively expressed, αC1, αC2, γC3, γC4, and γC5) are possibly expressed on an individual neuron. Random combinations provide 12,720 15-member sets of Pcdhs and 15^4 (50,625) possible tetramers [7]. The probability of matching tetramers between a pair of distinct neurons exponentially decreases by increasing the number of isoforms [7]. Therefore, the tetramer formation of the clustered Pcdh members could result in an exponential diversity of functional units on the cell surface between any given pair of neurons [7].

12.7 CELL SIGNALING MEDIATED BY THE CLUSTERED Pcdhs

In addition to the cell–cell adhesion caused by homophilic tetramer interactions, clustered Pcdh proteins can also interact with the classical cadherins (N- and R-cadherin), Pcdh17 [60], and the receptor tyrosine kinase Ret [61]. Ret binds Pcdh-α and -γ proteins in differentiated neuroblastoma cells and is required both for their stabilization and for the differentiation-induced phosphorylation of their intracellular domains [61]. In addition, glial cell line-derived neurotrophic factor, a Ret ligand, induces the phosphorylation of the Pcdhs, which are in turn required for stabilization of the activated Ret. Thus, the Pcdh proteins and Ret form a functional signaling complex [61]. Similarly, Pcdh-γC5 specifically binds to the γ-aminobutyric acid A (GABA-A) receptor, and this interaction is required for the stabilization and maintenance of GABAergic synapses in cultured hippocampal neurons [63]. In addition, most Pcdh-α proteins can bind to β-integrin via the RGD motif in their EC1 domain [62,64]. The RGD motif is highly conserved among Pcdh-α proteins, but is absent from the constitutively expressed Pcdh-αC1 and -αC2 proteins.

Clustered Pcdhs also mediate intracellular signaling by interacting with other proteins. Although the intracellular domains of the Pcdh-α and Pcdh-γ proteins differ from each other, they are highly conserved among vertebrates, suggesting they have a conserved biological function. The cytoplasmic tails of both Pcdhs can bind to the nonreceptor tyrosine kinases FAK and PYK2 [13]. These protein interactions are implicated in the cell survival and dendritic patterning defects observed in both Pcdh-α- and

Pcdh-γ-deficient neurons [13,65,66]. The intracellular domains of the Pcdh-γ proteins also bind to PDCD10 (programmed cell death 10), also known as CCM3, the protein involved in cerebral cavernous malformations in humans. In the developing chicken spinal cord, PCDC10 is required for Pcdh-γ depletion-induced apoptosis. Moreover, the overexpression and membrane localization of PCDC10 are sufficient to induce neuronal apoptosis [67].

Pcdh-α and Pcdh-γ proteins are proteolytically processed by the disintegrin and metalloprotease ADAM10 and γ-secretase complex, resulting in the generation of intracellular fragments that are soluble in the cytoplasm. Interestingly, the combinatorial expression of Pcdh-α and Pcdh-γ proteins can inhibit their proteolytic processing [14], suggesting that heteromeric protein complexes containing both Pcdhs are stable on the cell surface. In addition, many cell-surface antigens are regulated by their trafficking in cells. Pcdh-α and Pcdh-γ proteins are abundantly localized to intracellular organelles, with relatively low expression on the cell surface [59,68]. The control of Pcdh-γ trafficking is dependent on elements in its intracellular domains; deletion of the cytoplasmic tail significantly increases the surface delivery of the Pcdh-γ proteins [12,69].

12.8 ROLES OF CLUSTERED Pcdhs IN THE NERVOUS SYSTEM

Genetic manipulations of the clustered Pcdhs in mice have revealed multiple roles for these proteins in the developing nervous system. Mice lacking Pcdh-α are viable and fertile with no visible abnormalities, but exhibit abnormal learning and memory functions [70] and show abnormal cortical activities in response to visual and sensory stimulation [71,72]. In addition, the precise targeting of olfactory and serotonergic axonal terminals is impaired in Pcdh-α-deficient mice [73,74]. These neurological defects are also found in mice deficient in the Pcdh-α constant region [70,73,74], indicating that their shared cytoplasmic tail may be required for proper Pcdh-α function. In addition, a single constitutively expressed Pcdh-α1 isoform rescued mice containing a deletion encompassing Pcdh-α2 to -αC2, resulting in the normal axonal coalescence of olfactory sensory neurons [75], suggesting that intracellular Pcdh-α function, rather than isoform diversity, may be more important for this role of the clustered Pcdhs. Consistent with this suggestion, Pcdh-α proteins have been shown to regulate dendritic and spine development in hippocampal neurons by modulating the activity of FAK and PYK2 [66]. The FAK/PYK2 kinases activate small GTPases such as Ras homolog gene family member A and Ras-related C3 botulinum toxin substrate 1, leading to the induction of cytoskeletal reorganization.

The loss of Pcdh-γ leads to neonatal death with neurological defects induced by massive neuronal apoptosis

[76]. In both the spinal cord and the retina, the neuronal apoptosis is accompanied by reduced numbers of synapses, suggesting that the neuronal cell death is induced by synaptic loss. This possibility is supported by the observation that synaptic loss in the spinal cord is also observed in Pcdh-γ-deficient mice crossed onto a Bax-deficient background, in which neuronal apoptosis is genetically blocked [77]. However, both the neuronal and the synaptic losses in the retina are rescued in the Bax-deficient background [78].

The phenotype of neuronal apoptosis in the Pcdh-γ-deficient mice is also observed in mutant mice harboring a deletion encompassing γC3 to γC5 [78]. The triple-γC-deficient mice also die shortly after birth and exhibit neuronal apoptosis and synaptic loss in the spinal cord and retina. However, genetically blocking apoptosis by crossing these mice onto a Bax-deficient background can rescue the neonatal lethality [78], suggesting that the triple-γC isoforms play a significant role in neuronal survival and that the other Pcdh-γ proteins may be important in postnatal neuronal development.

Pcdh-γ conditional knockout experiments revealed a role for this Pcdh family in mediating neurite self-avoidance during dendrite formation in retinal starburst amacrine cells and cerebellar Purkinje neurons [79]. The Pcdh-γ family also modulates this process during dendritic formation in cortical neurons and granule cells [80,81]. In addition, Pcdh-γ proteins play a role in mediating astrocyte–neuron contact during synaptogenesis in the spinal cord and are also involved in the maturation of postnatally generated olfactory granule cells [80,81]. In cortical neurons, Pcdh-γ proteins inhibit FAK, which downregulates the downstream signaling pathway involving protein kinase C, phospholipase C, and myristoylated alanine-rich C-kinase substrate. This pathway activation promotes dendritic arborization [80]. Taken together, these results support the possibility that the clustered Pcdhs play significant roles in the assembly of neural networks in the nervous system. However, the molecular mechanisms by which the clustered Pcdhs expressed in individual neurons contribute to the assembly of neural networks are not completely understood.

12.9 CONCLUSIONS

Genetic manipulations of the clustered *Pcdhs* in mice have gradually revealed their functions in axonal targeting, dendritic arborization, dendritic self-avoidance, and synapse formation. The clustered *Pcdhs* provide neuronal diversity through their stochastic and combinatorial expression in individual neurons. This diversity among cells is generated by DNA looping in the chromatin structure and by epigenetic regulation involving DNA methylation in the genome during early embryogenesis. In addition, tetramer formation involving the various clustered Pcdh proteins provides specific cell-adhesion activities and exponential neuronal diversity at the cell surface.

Physiological studies in the mammalian brain have revealed that local networks form complex topological neuronal ensembles in which there are differences at the individual neuron level [82–88]. For example, specific local connectivities and similar physiological responses to visual stimuli are observed among clonally related excitatory cortical neurons [89–92]. The molecular features of the clustered Pcdhs, which are diverse neural surface antigens, provide a fascinating mechanism underlying the assembly of complex neural networks in the brain.

REFERENCES

[1] Tonegawa S. Somatic generation of antibody diversity. Nature 1983;302:575–81.

[2] Lieber MR. The mechanism of V(D)J recombination: a balance of diversity, specificity, and stability. Cell 1992;70:873–6.

[3] Buck L, Axel R. A novel multigene family may encode odorant receptors: a molecular basis for odor recognition. Cell 1991;65:175–87.

[4] Kohmura N, Senzaki K, Hamada S, Kai N, Yasuda R, Watanabe M, et al. Diversity revealed by a novel family of cadherins expressed in neurons at a synaptic complex. Neuron 1998;20:1137–51.

[5] Wu Q, Maniatis T. A striking organization of a large family of human neural cadherin-like cell adhesion genes. Cell 1999;97:779–90.

[6] Yagi T. Clustered protocadherin family. Dev Growth Differ 2008;50:S131–40.

[7] Yagi T. Molecular codes for neuronal individuality and cell assembly in the brain. Front Mol Neurosci 2012;5:45.

[8] Yagi T. Genetic basis of neuronal individuality in the mammalian brain. J Neurogenet 2013;27:97–105.

[9] Esumi S, Kakazu N, Taguchi Y, Hirayama T, Sasaki A, Hirabayashi T, et al. Monoallelic yet combinatorial expression of variable exons of the protocadherin-alpha gene cluster in single neurons. Nat Genet 2005;37:171–6.

[10] Hirano K, Kaneko R, Izawa T, Kawaguchi M, Kitsukawa T, Yagi T. Single-neuron diversity generated by Protocadherin-beta cluster in mouse central and peripheral nervous systems. Front Mol Neurosci 2012;5:90.

[11] Kaneko R, Kato H, Kawamura Y, Esumi S, Hirayama T, Hirabayashi T, et al. Allelic gene regulation of Pcdh-alpha and Pcdh-gamma clusters involving both monoallelic and biallelic expression in single Purkinje cells. J Biol Chem 2006;281:30551–60.

[12] Schreiner D, Weiner JA. Combinatorial homophilic interaction between gamma-protocadherin multimers greatly expands the molecular diversity of cell adhesion. Proc Natl Acad Sci USA 2010;107:14893–8.

[13] Chen J, Lu Y, Meng S, Han MH, Lin C, Wang X. alpha- and gamma-Protocadherins negatively regulate PYK2. J Biol Chem 2009;284:2880–90.

[14] Bonn S, Seeburg PH, Schwarz MK. Combinatorial expression of alpha- and gamma-protocadherins alters their presenilin-dependent processing. Mol Cell Biol 2007;27:4121–32.

[15] Yagi T, Takeichi M. Cadherin superfamily genes: functions, genomic organization, and neurologic diversity. Genes Dev 2000;14:1169–80.

[16] Morishita H, Yagi T. Protocadherin family: diversity, structure, and function. Curr Opin Cell Biol 2007;19:584–92.

[17] Zipursky SL, Sanes JR. Chemoaffinity revisited: dscams, protocadherins, and neural circuit assembly. Cell 2010;143:343–53.

[18] Hirayama T, Yagi T. Clustered protocadherins and neuronal diversity. Prog Mol Biol Transl Sci 2013;116:145–67.

[19] Weiner JA, Jontes JD. Protocadherins, not prototypical: a complex tale of their interactions, expression, and functions. Front Mol Neurosci 2013;6:4.

[20] Chen WV, Maniatis T. Clustered protocadherins. Development 2013;140:3297–302.

[21] Noonan JP, Li J, Nguyen L, Caoile C, Dickson M, Grimwood J, et al. Extensive linkage disequilibrium, a common 16.7-kilobase deletion, and evidence of balancing selection in the human protocadherin alpha cluster. Am J Hum Genet 2003;72:621–35.

[22] Miki R, Hattori K, Taguchi Y, Tada MN, Isosaka T, Hidaka Y, et al. Identification and characterization of coding single-nucleotide polymorphisms within human protocadherin-alpha and -beta gene clusters. Gene 2005;349:1–14.

[23] Anitha A, Thanseem I, Nakamura K, Yamada K, Iwayama Y, Toyota T, et al. Protocadherin alpha (PCDHA) as a novel susceptibility gene for autism. J Psychiatr Neurosci 2012;37:120058.

[24] Redies C, Hertel N, Hubner CA. Cadherins and neuropsychiatric disorders. Brain Res 2012;1470:130–44.

[25] Iossifov I, Ronemus M, Levy D, Wang Z, Hakker I, Rosenbaum J, et al. De novo gene disruptions in children on the autistic spectrum. Neuron 2012;74:285–99.

[26] Toyoda S, Kawaguchi M, Kobayashi T, Tarusawa E, Toyama T, Okano M, et al. Developmental epigenetic modification regulates stochastic expression of clustered protocadherin genes, generating single neuron diversity. Neuron 2014;82:94–108.

[27] McGowan PO, Suderman M, Sasaki A, Huang TC, Hallett M, Meaney MJ, et al. Broad epigenetic signature of maternal care in the brain of adult rats. PLoS One 2011;6:e14739.

[28] Takeichi M. The cadherin superfamily in neuronal connections and interactions. Nat Rev Neurosci 2007;8:11–20.

[29] Sano K, Tanihara H, Heimark RL, Obata S, Davidson M, St John T, et al. Protocadherins: a large family of cadherin-related molecules in central nervous system. EMBO J 1993;12:2249–56.

[30] Tada MN, Senzaki K, Tai Y, Morishita H, Tanaka YZ, Murata Y, et al. Genomic organization and transcripts of the zebrafish protocadherin genes. Gene 2004;340:197–211.

[31] Noonan JP, Grimwood J, Schmutz J, Dickson M, Myers RM. Gene conversion and the evolution of protocadherin gene cluster diversity. Genome Res 2004;14:354–66.

[32] Hirayama T, Yagi T. The role and expression of the protocadherin-alpha clusters in the CNS. Curr Opin Neurobiol 2006;16:336–42.

[33] Sugino H, Hamada S, Yasuda R, Tuji A, Matsuda Y, Fujita M, et al. Genomic organization of the family of CNR cadherin genes in mice and humans. Genomics 2000;63:75–87.

[34] Wu Q, Zhang T, Cheng JF, Kim Y, Grimwood J, Schmutz J, et al. Comparative DNA sequence analysis of mouse and human protocadherin gene clusters. Genome Res 2001;11:389–404.

[35] Kim SY, Yasuda S, Tanaka H, Yamagata K, Kim H. Non-clustered protocadherin. Cell Adh Migr 2011;5:97–105.

[36] Noonan JP, Grimwood J, Danke J, Schmutz J, Dickson M, Amemiya CT, et al. Coelacanth genome sequence reveals the evolutionary history of vertebrate genes. Genome Res 2004;14:2397–405.

[37] Sugino H, Yanase H, Hamada S, Kurokawa K, Asakawa S, Shimizu N, et al. Distinct genomic sequence of the CNR/Pcdhalpha genes in chicken. Biochem Biophys Res Commun 2004;316:437–45.

[38] Taguchi Y, Koide T, Shiroishi T, Yagi T. Molecular evolution of cadherin-related neuronal receptor/protocadherin(alpha) (CNR/Pcdh(alpha)) gene cluster in mus musculus subspecies. Mol Biol Evol 2005;22:1433–43.

[39] Tasic B, Nabholz CE, Baldwin KK, Kim Y, Rueckert EH, Ribich SA, et al. Promoter choice determines splice site selection in protocadherin alpha and gamma pre-mRNA splicing. Mol Cell 2002;10:21–33.

[40] Wang X, Su H, Bradley A. Molecular mechanisms governing Pcdh-gamma gene expression: evidence for a multiple promoter and cis-alternative splicing model. Genes Dev 2002;16:1890–905.

[41] Kawaguchi M, Toyama T, Kaneko R, Hirayama T, Kawamura Y, Yagi T. Relationship between DNA methylation states and transcription of individual isoforms encoded by the protocadherin-alpha gene cluster. J Biol Chem 2008;283:12064–75.

[42] Morishita H, Murata Y, Esumi S, Hamada S, Yagi T. CNR/Pcdhalpha family in subplate neurons, and developing cortical connectivity. Neuroreport 2004;15:2595–9.

[43] Lobas MA, Helsper L, Vernon CG, Schreiner D, Zhang Y, Holtzman MJ, et al. Molecular heterogeneity in the choroid plexus epithelium: the 22-member gamma-protocadherin family is differentially expressed, apically localized, and implicated in CSF regulation. J Neurochem 2012;120:913–27.

[44] Noguchi Y, Hirabayashi T, Katori S, Kawamura Y, Sanbo M, Hirabayashi M, et al. Total expression and dual gene-regulatory mechanisms maintained in deletions and duplications of the Pcdha cluster. J Biol Chem 2009;284:32002–14.

[45] Handoko L, Xu H, Li G, Ngan CY, Chew E, Schnapp M, et al. CTCF-mediated functional chromatin interactome in pluripotent cells. Nat Genet 2011;43:630–8.

[46] Ribich S, Tasic B, Maniatis T. Identification of long-range regulatory elements in the protocadherin-alpha gene cluster. Proc Natl Acad Sci USA 2006;103:19719–24.

[47] Yokota S, Hirayama T, Hirano K, Kaneko R, Toyoda S, Kawamura Y, et al. Identification of the cluster control region for the protocadherin-beta genes located beyond the protocadherin-gamma cluster. J Biol Chem 2011;286:31885–95.

[48] Hirayama T, Tarusawa E, Yoshimura Y, Galjart N, Yagi T. CTCF is required for neural development and stochastic expression of clustered Pcdh genes in neurons. Cell Rep 2012;2:345–57.

[49] Kehayova P, Monahan K, Chen W, Maniatis T. Regulatory elements required for the activation and repression of the protocadherin-alpha gene cluster. Proc Natl Acad Sci USA 2011;108:17195–200.

[50] Golan-Mashiach M, Grunspan M, Emmanuel R, Gibbs-Bar L, Dikstein R, Shapiro E. Identification of CTCF as a master regulator of the clustered protocadherin genes. Nucleic Acids Res 2012;40:3378–91.

[51] Monahan K, Rudnick ND, Kehayova PD, Pauli F, Newberry KM, Myers RM, et al. Role of CCCTC binding factor (CTCF) and cohesin in the generation of single-cell diversity of protocadherin-alpha gene expression. Proc Natl Acad Sci USA 2012;109:9125–30.

[52] Merkenschlager M, Odom DT. CTCF and cohesin: linking gene regulatory elements with their targets. Cell 2013;152:1285–97.

[53] Abe M, Ohira M, Kaneda A, Yagi Y, Yamamoto S, Kitano Y, et al. CpG island methylator phenotype is a strong determinant of poor prognosis in neuroblastomas. Cancer Res 2005;65:828–34.

[54] Hirano T. At the heart of the chromosome: SMC proteins in action. Nat Rev Mol Cell Biol 2006;7:311–22.

[55] Gendrel AV, Tang YA, Suzuki M, Godwin J, Nesterova TB, Greally JM, et al. Epigenetic functions of smchd1 repress gene clusters on the inactive X chromosome and on autosomes. Mol Cell Biol 2013;33:3150–65.

[56] Horsthemke B, Wagstaff J. Mechanisms of imprinting of the Prader-Willi/Angelman region. Am J Med Genet A 2008;146A:2041–52.

[57] Suderman M, McGowan PO, Sasaki A, Huang TC, Hallett MT, Meaney MJ, et al. Conserved epigenetic sensitivity to early life experience in the rat and human hippocampus. Proc Natl Acad Sci USA 2012;109(Suppl. 2):17266–72.

[58] Xu Y, Wu F, Tan L, Kong L, Xiong L, Deng J, et al. Genome-wide regulation of 5hmC, 5mC, and gene expression by Tet1 hydroxylase in mouse embryonic stem cells. Mol Cell 2011;42:451–64.

[59] Murata Y, Hamada S, Morishita H, Mutoh T, Yagi T. Interaction with protocadherin-gamma regulates the cell surface expression of protocadherin-alpha. J Biol Chem 2004;279:49508–16.

[60] Han MH, Lin C, Meng S, Wang X. Proteomics analysis reveals overlapping functions of clustered protocadherins. Mol Cell Proteomics 2010;9:71–83.

[61] Schalm SS, Ballif BA, Buchanan SM, Phillips GR, Maniatis T. Phosphorylation of protocadherin proteins by the receptor tyrosine kinase Ret. Proc Natl Acad Sci USA 2010;107:13894–9.

[62] Morishita H, Umitsu M, Murata Y, Shibata N, Udaka K, Higuchi Y, et al. Structure of the cadherin-related neuronal receptor/protocadherin-alpha first extracellular cadherin domain reveals diversity across cadherin families. J Biol Chem 2006;281:33650–63.

[63] Li Y, Xiao H, Chiou TT, Jin H, Bonhomme B, Miralles CP, et al. Molecular and functional interaction between protocadherin-gammaC5 and GABAA receptors. J Neurosci 2012;32:11780–97.

[64] Mutoh T, Hamada S, Senzaki K, Murata Y, Yagi T. Cadherin-related neuronal receptor 1 (CNR1) has cell adhesion activity with beta1 integrin mediated through the RGD site of CNR1. Exp Cell Res 2004;294:494–508.

[65] Garrett AM, Schreiner D, Lobas MA, Weiner JA. gamma-protocadherins control cortical dendrite arborization by regulating the activity of a FAK/PKC/MARCKS signaling pathway. Neuron 2012;74:269–76.

[66] Suo L, Lu H, Ying G, Capecchi MR, Wu Q. Protocadherin clusters and cell adhesion kinase regulate dendrite complexity through Rho GTPase. J Mol Cell Biol 2012;4:362–76.

[67] Lin C, Meng S, Zhu T, Wang X. PDCD10/CCM3 acts downstream of {gamma}-protocadherins to regulate neuronal survival. J Biol Chem 2010;285:41675–85.

[68] Phillips GR, Tanaka H, Frank M, Elste A, Fidler L, Benson DL, et al. Gamma-protocadherins are targeted to subsets of synapses and intracellular organelles in neurons. J Neurosci 2003;23:5096–104.

[69] Fernandez-Monreal M, Kang S, Phillips GR. Gamma-protocadherin homophilic interaction and intracellular trafficking is controlled by the cytoplasmic domain in neurons. Mol Cell Neurosci 2009;40:344–53.

[70] Fukuda E, Hamada S, Hasegawa S, Katori S, Sanbo M, Miyakawa T, et al. Down-regulation of protocadherin-alpha A isoforms in mice changes contextual fear conditioning and spatial working memory. Eur J Neurosci 2008;28:1362–76.

[71] Yamashita H, Chen S, Komagata S, Hishida R, Iwasato T, Itohara S, et al. Restoration of contralateral representation in the mouse somatosensory cortex after crossing nerve transfer. PLoS One 2012;7:e35676.

[72] Yoshitake K, Tsukano H, Tohmi M, Komagata S, Hishida R, Yagi T, et al. Visual map shifts based on whisker-guided cues in the young mouse visual cortex. Cell Rep 2013;5:1365–74.

[73] Hasegawa S, Hamada S, Kumode Y, Esumi S, Katori S, Fukuda E, et al. The protocadherin-alpha family is involved in axonal coalescence of olfactory sensory neurons into glomeruli of the olfactory bulb in mouse. Mol Cell Neurosci 2008;38:66–79.

[74] Katori S, Hamada S, Noguchi Y, Fukuda E, Yamamoto T, Yamamoto H, et al. Protocadherin-alpha family is required for serotonergic projections to appropriately innervate target brain areas. J Neurosci 2009;29:9137–47.

[75] Hasegawa S, Hirabayashi T, Kondo T, Inoue K, Esumi S, Okayama A, et al. Constitutively expressed protocadherin-alpha regulates the coalescence and elimination of homotypic olfactory axons through its cytoplasmic region. Front Mol Neurosci 2012;5:97.

[76] Wang X, Weiner JA, Levi S, Craig AM, Bradley A, Sanes JR. Gamma protocadherins are required for survival of spinal interneurons. Neuron 2002b;36:843–54.

[77] Weiner JA, Wang X, Tapia JC, Sanes JR. Gamma protocadherins are required for synaptic development in the spinal cord. Proc Natl Acad Sci USA 2005;102:8–14.

[78] Chen WV, Alvarez FJ, Lefebvre JL, Friedman B, Nwakeze C, Geiman E, et al. Functional significance of isoform diversification in the protocadherin gamma gene cluster. Neuron 2012;75:402–9.

[79] Lefebvre JL, Kostadinov D, Chen WV, Maniatis T, Sanes JR. Protocadherins mediate dendritic self-avoidance in the mammalian nervous system. Nature 2012;488:517–21.

[80] Garrett AM, Weiner JA. Control of CNS synapse development by {gamma}-protocadherin-mediated astrocyte-neuron contact. J Neurosci 2009;29:11723–31.

[81] Ledderose J, Dieter S, Schwarz MK. Maturation of postnatally generated olfactory bulb granule cells depends on functional gamma-protocadherin expression. Sci Rep 2013;3:1514.

[82] Markram H. A network of tufted layer 5 pyramidal neurons. Cereb Cortex 1997;7:523–33.

[83] Kalisman N, Silberberg G, Markram H. The neocortical microcircuit as a tabula rasa. Proc Natl Acad Sci USA 2005;102:880–5.

[84] Song S, Sjostrom PJ, Reigl M, Nelson S, Chklovskii DB. Highly nonrandom features of synaptic connectivity in local cortical circuits. PLoS Biol 2005;3:e68.

[85] Yoshimura Y, Callaway EM. Fine-scale specificity of cortical networks depends on inhibitory cell type and connectivity. Nat Neurosci 2005;8:1552–9.

[86] Koulakov AA, Hromadka T, Zador AM. Correlated connectivity and the distribution of firing rates in the neocortex. J Neurosci 2009;29:3685–94.

[87] Yassin L, Benedetti BL, Jouhanneau JS, Wen JA, Poulet JF, Barth AL. An embedded subnetwork of highly active neurons in the neocortex. Neuron 2010;68:1043–50.

[88] Perin R, Berger TK, Markram H. A synaptic organizing principle for cortical neuronal groups. Proc Natl Acad Sci USA 2011;108:5419–24.

[89] Yu YC, Bultje RS, Wang X, Shi SH. Specific synapses develop preferentially among sister excitatory neurons in the neocortex. Nature 2009;458:501–4.

[90] Yu YC, He S, Chen S, Fu Y, Brown KN, Yao XH, et al. Preferential electrical coupling regulates neocortical lineage-dependent microcircuit assembly. Nature 2012;486:113–7.

[91] Li Y, Lu H, Cheng PL, Ge S, Xu H, Shi SH, et al. Clonally related visual cortical neurons show similar stimulus feature selectivity. Nature 2012;486:118–21.

[92] Ohtsuki G, Nishiyama M, Yoshida T, Murakami T, Histed M, Lois C, et al. Similarity of visual selectivity among clonally related neurons in visual cortex. Neuron 2012;75:65–72.

Chapter 13

β1-Integrin Function and Interplay during Enteric Nervous System Development

Sylvie Dufour[1,2,3], Florence Broders-Bondon[1] and Nadège Bondurand[2,3]

[1]Institut Curie/CNRS UMR144, Paris, France; [2]INSERM U955, IMRB, Equipe 6, Créteil, France; [3]Faculté de Médecine, Université Paris Est, Créteil, France

13.1 INTRODUCTION

The enteric nervous system (ENS) is the part of the peripheral nervous system (PNS) that controls the peristaltic and secretory activity of the gut wall (for review: Refs. [1–3]). It consists of a large number of neurons and glial cells organized into a network of interconnected ganglia distributed along the length of the gut. The ENS is organized into two concentric plexuses: the myenteric plexus, located between the longitudinal and the circular smooth muscles, and the submucosal plexus, located between the circular smooth muscle and the mucosa (Figure 13.1(A) and (B)).

The ENS is derived from two subpopulations of neural crest cells (NCCs) [4]. The ontogenesis of NCCs is a crucial morphogenetic process during early embryogenesis. These cells are initially located in the dorsal part of the neural tube and are categorized into cranial, vagal, trunk, and sacral populations depending on their position along the anteroposterior axis [5]. The four populations of NCCs give rise to all the melanocytes and, depending on their axial level, NCCs also produce several other derivatives. The cranial NCCs give rise to the craniofacial structures, teeth, cornea, and cranial ganglia. The trunk NCCs produce adrenal medulla as well as the dorsal root and sympathetic ganglia. The vagal NCCs are at the origin of the cardiac septum and smooth muscle cells of great vessels. They also produce the enteric neural crest cells (ENCCs) that contribute to most of the ENS, whereas the sacral NCCs produce the ENCCs contributing to the postumbilical ENS.

After delamination from the dorsal part of the neural tube, ENCCs move ventromedially toward the developing gut, entering its rostral extremity (foregut) at embryonic day (E) 9.5 in mice, and migrate in a rostrocaudal direction to colonize the entire length of the gut. A population of ENCCs has been shown to follow an alternative migration "route," crossing from the midgut to the hindgut via the mesentery during a period of development in which these gut regions are transiently juxtaposed, with these cells subsequently making a major contribution to the hindgut ENS [6].

ENCCs adopt various modes of migration, depending on the region of the gut invaded. An analysis of the dynamics of ENCC migration has, indeed, revealed complex patterns of cell movement at the migratory front that appear to differ between different portions: chains of ENCCs form in the linear parts of the gut (the ileum and hindgut), whereas the proportion of isolated cells is higher in the cecum and proximal hindgut [7–10]. These differences in the mode of ENCC migration are poorly understood, but suggestive of the existence of various molecular mechanisms of ENCC adhesion and progression along the gut rostrocaudal axis, with cells encountering successive regions with different extracellular matrix (ECM) compositions and growth, signaling, and differentiation states. During colonization, ENCCs also encounter various signals modulating their proliferation, acting as guidance or chemoattractive cues, inhibiting or stimulating commitment to the glial or neuronal lineage.

Analyses of knockout and conditional mutant mice have implicated cell adhesion molecules of the cadherin family and of the superfamily of immunoglobulins (CAM–Ig family), together with β1-integrins, in ENS ontogenesis. Integrins are the major family of ECM adhesion receptors and the interaction with their ligands elicits various signaling cascades that regulate adhesion, migration, survival, or differentiation and gene expression [11]. Integrins are heterodimeric receptors composed of one α and one β chain. In mammals, 24 integrin heterodimers comprise several subfamilies with various ligand specificities. The β1-integrins (CD29 antigen) constitute the largest subfamily, playing a key role, together with the αV-integrin (CD51) subfamily, in cell–ECM interactions. Most integrins can bind to several ligands. The association between one α and one β chain forms an ectodomain that conveys ligand recognition. The cytoplasmic domains are essential for integrin function in

Neural Surface Antigens. http://dx.doi.org/10.1016/B978-0-12-800781-5.00013-X

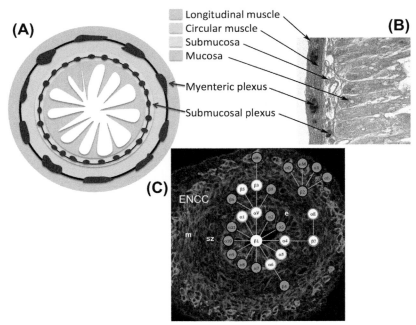

FIGURE 13.1 (A) Schematic representation of the transverse section of the adult small intestine. It shows the organization of the enteric nervous system (ENS) in two plexuses, the myenteric plexus and the submucosal plexus. The myenteric plexus is located between the longitudinal (green) and the circular (blue) smooth muscle layer. The submucosal plexus is located in the submucosa (orange). The mucosa and its lumen are represented in gray and white, respectively. *(Adapted from Ref. [3].)* (B) Transverse section of adult mouse gut at postnatal day 28 stained with hematoxylin and eosin to reveal tissue layers. The myenteric plexus and submucosal plexus are colored in blue (X-gal staining for β-galactosidase activity thanks to the use of the Ht-PA-Cre transgenic mouse model allowing one to target reporter gene expression in the ENS). (C) Schematic representation of integrin repertoire expressed by enteric neural crest cells (ENCCs). The integrin family is composed of 18 α and 8 β subunits (represented by circles) and the association of one α and one β subunit (represented by lines) forms 24 heterodimers. The integrin subunits expressed by ENCCs are highlighted in white, with the α subunits that bind laminins and collagens in yellow circles, and heterodimers that recognize fibronectin are highlighted with blue lines and the RGD-specific α subunits with blue circles. The background picture shows a transverse section through embryonic mouse midgut at embryonic day 12.5 immunostained for β1-integrin subunit (green) and p75ntr (blue; CD271, a marker of ENCCs, with glial and neuronal ENCC derivatives expressing high and low levels, respectively). Please refer to Chapter 7 by T. Glaser for the CD271 function in neural development. The β1 subunit is ubiquitously expressed in the gut wall with higher levels in the muscle layer, underneath the ENCC layer (blue). m, muscle layer; sz, submucosal zone; e, epithelium. (For interpretation of the references to color in this figure legend, the reader is referred to the online version of this book.)

recruiting regulators, adaptor proteins, and proteins with enzymatic activity and controlling integrin activation state and signaling [12,13].

13.2 FUNCTIONS OF THE ECM AND INTEGRINS DURING ENS DEVELOPMENT

13.2.1 ECM and Integrin Expression during ENS Development

The ECM plays a crucial role during embryonic development [14,15]. During their colonization of the developing gut, ENCCs encounter various ECM components, including fibronectin, collagens, laminins, vitronectin, tenascin-C and tenascin-W, hyaluronic acid, heparan sulfate, and chondroitin sulfate proteoglycans [16–19]. The surrounding cells secrete most of these ECM components, but some of them, such as tenascin-C, are produced by ENCCs (in avians). Fibronectin, vitronectin, collagen-IV, and laminin isoform 1 (laminin-1, containing the laminin α1 chain) have a broad spatial distribution during the period of ENCC migration [20]. Fibronectin is expressed at higher levels in the cecum and hindgut in mice. Tenascin-C is a multidomain protein with a dual role in promoting or inhibiting cell adhesion. Tenascin-C is weakly expressed in the mouse gut mesenchyme as a whole, but with much stronger expression in these two regions of the gut. The pattern of tenascin-C expression varies with species. In avians, unlike mice, tenascin-C is absent from the cecum before the arrival of the ENCCs, its expression coinciding with the colonization of the cecum by the vagal NCCs secreting this protein [21]. The laminin α5 chain is expressed during ENCC migration [22]. Laminin-1 and collagen-IV are also abundant in the basal laminae and around the vessels. By contrast, the laminin α2 and α4 chains and tenascin-W have also been detected in gut tissue, but only after colonization by the ENCCs [23,24].

In adults, heparan sulfate proteoglycan, collagen-IV, and laminins are located around the myenteric ganglia and are in close contact with enteric glial cells, whereas neurons

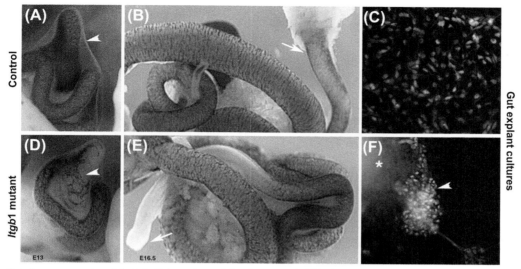

FIGURE 13.2 Enteric nervous system defects in β1-integrin conditional mutants. At E13, the colonization of the gut by enteric neural crest cells (blue staining) in controls has reached the hindgut (A), whereas in the *Itgb1* conditional mutant it is delayed and has reached the proximal cecum (arrowhead, (D)). At E16.5, control enteric neural crest cells have reached the rectum (arrow, (B)), whereas the mutant cells have stopped in the middle of the hindgut (arrow, (E)). In in vitro gut explant cultures, enteric progenitor (green) and neuronal derivatives (red) are well spread and dispersed (C), whereas the mutant cells are aggregated (arrowhead, (F)) close to the gut explant (*, blue staining). (For interpretation of the references to color in this figure legend, the reader is referred to the online version of this book.)

have a more central location. The abnormal expression of laminin isoforms, collagen-IV, and tenascin-C in the gut tissue has been demonstrated in human diseases involving ENS defects, such as Hirschsprung disease (HSCR), and in mouse models of ENS diseases [25–27]. HSCR is a neurocristopathy affecting the ENS characterized by the absence of enteric ganglia along a variable length of the intestine, causing intestinal obstruction in neonates and severe constipation in adults [3].

The composition of the ECM also regulates the fate of enteric progenitors. In vitro, laminin-1 promotes the development of mouse enteric neurons [28]. The culture of rabbit enteric progenitors on surfaces coated with polylysine or a mixture of collagen-I and -IV promotes the differentiation of ENCCs into glia, whereas the presence of laminin or heparan sulfate promotes their differentiation into neurons [29].

During development, the ENCCs produce several integrin chains (Figure 13.1(C)), including α4 (CD49d), α5 (CD49e), α6 (CD49f), αV (CD51), β1 (CD29), β3 (CD61), and β5 (CD29) [14,22,30] (Figure 13.1). αE (CD103) and β7 and low levels of α1 (CD49a) chain mRNAs have been detected in enteric progenitors, but the corresponding proteins were not detected [30]. In adults, enteric ganglia express α5 [31]. High levels of α4 or β1 at the cell surface are used as a criterion for the purification of enteric progenitors from embryonic or adult dissociated gut tissues [30,32,33]. β1-Integrins play crucial roles in the regulation of neural stem cell survival, maintenance, and fate [34].

13.2.2 Integrin Function during ENCC Migration and ENS Development

Constitutive knockout of the *Itgb1* gene, encoding the β1-integrin chain, is embryo-lethal at peri-implantation stages in mice [35]. Thus, the analysis of β1-integrin function during ENS development requires using genetic tools leading to invalidation of *Itgb1* specifically in NCCs. Among these tools, the Ht-PA-Cre transgenic mouse model [36] allows the targeting of Cre recombinase expression into migratory NCCs. By crossing Ht-PA-Cre mice carrying one null *Itgb1* allele with mice having two *Itgb1* floxed alleles it was possible to produce conditional mutant embryos with *Itgb1*-null NCCs at the stage of their escape from the neural tube. The conditional invalidation of *Itgb1* in NCCs leads to several embryonic defects affecting the PNS [37], including a delayed migration of Schwann cells along axons, abnormal spinal nerve arborization and morphology, defective Schwann cell and sensory axon segregation, and retarded maturation in neuromuscular synaptogenesis. Conditional mutants also exhibit severe alterations during the ENS development (Figure 13.2), resulting in an abnormal organization of the enteric ganglia network and aganglionosis of the distal colon [38], yielding a phenotype resembling that of HSCR. Thus, β1-integrins play a crucial role in PNS and ENS ontogenesis.

In the ENS, *Itgb1*-mutant NCCs lack the members of the β1-integrin subfamily but they are not affected in the expression of the β3- and β5-integrins, which can still promote ENCC interaction with the ECM. However, a dynamic analysis of ENCC behavior in the cecum revealed a severe adhesion and migratory defect, characterized by slow loco-motion in abnormal directions and a lack of persistence of cell movement [20]. β1-Integrins also regulate cell survival and fate. However, conditional *Itgb1* mutants display no obvious defect of neuronal or glial cell content and do not have abnormally small numbers of ENCCs in the colonized zone of the embryonic gut. Thus, β1-integrins mostly contribute to ENS ontogenesis by regulating ENCC adhesion and migration in the gut environment.

The defects observed in the cecum may reflect an inappropriate response of the ENCCs to the specific ECM composition of this gut region. In vitro, ENCCs adhere to and migrate well on ECM components [21,38,39]. ECM components expressed in the cecum region, including tenascin-C, fibronectin, and vitronectin, can individually support ENCC adhesion and migration [21,40–43]. Tenascin-C upregulates the migration of ENCCs on fibronectin [21], but has the opposite effect in the presence of vitronectin [20]. The major integrin controlling adhesion to fibronectin is α5β1, whereas the αV-integrins regulate adhesion to vitronectin and fibronectin but also to many other ECM components, including tenascin-C, collagens, fibrinogen, and thrombospondin. *Itgb1* mutant ENCC do not respond positively to fibronectin and have a higher colonization potential in gut environments with lower levels of tenascin-C [20]. Thus, β1-integrins are required to overcome the negative response to tenascin-C and to promote the fibronectin-dependent stimulation of ENCC adhesion and migration in the developing gut. β1-Integrins are also required to mediate the interaction of ENCCs with other ECM components because *Itgb1* mutant ENCCs display defective spreading and migration on ECM gel consisting mostly of laminin-1, collagen-IV, and heparan sulfate proteoglycan [38].

The inhibition of glioblastoma cell adhesion by tenascin-C can be counteracted, in an α5β1-integrin- and syndecan-4-dependent manner, by the stimulation of the endothelin receptor B (EDNRB) [44]. EDNRB is a G-protein-coupled receptor that binds to endothelin (EDN)-1 and EDN-3. ENCCs express EDNRB, and EDN-3 is expressed in the mesenchyme during ENS ontogenesis. EDN-3/EDNRB signaling plays a crucial role during ENS development, promoting enteric progenitor maintenance and migration while inhibiting neuronal differentiation [45–47]. The possible role of EDNRB signaling in rescuing the inhibition of ENCC adhesion due to tenascin-C remains to be established.

It has been suggested that β1-integrins drive ENS development by mediating the interaction of ENCCs with endothelial cells and vessel-associated ECM [39]. Indeed, the migration of NCCs and endothelial cells is controlled by these ECM components, and the embryonic vasculature transiently supports NCC migration in the trunk [48]. In vitro studies showed that ENCCs actively migrate on human umbilical vein endothelial cell monolayers.

Antibodies directed against β1-integrins block this process. In addition, ENCC colonization of the avian hindgut is inhibited in ex vivo perturbation experiments, affecting endothelial development [39]. However, these conclusions have been called into question by another study making use of adequate mouse models, targeted fluorescent markers in ENCCs and endothelial cells, and three-dimensional reconstructions of their respective locations and the cellular network organization along the rostrocaudal axis of the gut [49].

13.3 EFFECTORS OF THE INTEGRIN SIGNALING PATHWAY DURING ENS DEVELOPMENT

Integrins are activated in response to ECM binding or inside-out signaling. They cluster, recruit scaffold and signaling proteins to their cytoplasmic tail in adhesion complexes known as focal adhesions (FAs), and activate signaling cascades regulating many aspects of cell behavior. The integrin-based adhesion process starts with the formation of nascent adhesions, the smallest integrin-containing adhesion structures to form FAs connected to the actomyosin cytoskeleton. Proteomic and dynamic analyses of the protein content of integrin adhesions have revealed the complexity of these structures [50]. Up to 150 proteins have been identified in the integrin adhesome [51], and these proteins are thought to initiate and contribute to integrin-dependent signaling. Although the role of β1-integrins during ENCC migration and ENS ontogenesis was demonstrated some years ago, the underlying signaling pathways remain poorly understood.

13.3.1 Phactr-4 and Integrin Signaling

PHACTR-4 (phosphatase and actin regulator 4) has been identified as a new player in ENCC migration, blocking β1-integrin signaling. The PHACTR family has four members (PHACTR-1 to -4), which regulate protein phosphatase 1 (PP1) activity and can bind G-actin via their three RPEL motifs (core sequence RPxxxEL) [52]. Disruption of the *Phactr4* gene in mice leads to an HSCR-like phenotype, with a hypoganglionic phenotype and defects of network organization. Using gut explants and time-lapse video-microscopy, Zhang and coworkers [53] showed that *Phactr4*[humdy], a missense mutation in the PP1 motif of PHACTR-4, leads to ENCC collective migration defects, with an inability to form chains or to migrate in the caudal direction, leading to hypogangliosis in the hindgut. ENCCs are more randomly distributed, often well separated from other cells, and highly motile. Thus, PHACTR-4 is required to retain ENCCs in chains and to guide the direction of migration to facilitate complete innervation of the gut. Defects of ENCC directional migration are rescued by the inhibition of RHO/RHO kinase (ROCK) or integrin function.

One of the roles of PHACTR-4 is the modulation of integrin trafficking, redistributing integrin at the membrane and contributing to the formation of new adhesions at the leading edge. PHACTR-4 is located in the lamellipodium, FAs, cytoplasm, and nucleolus, suggesting multiple roles in the regulation of cellular processes [53]. It has been suggested that PHACTR-4 acts as a scaffold at FAs. Phactr-4 connects actin and PP1 to integrin signaling and to the activity of the actin-severing protein cofilin, thereby regulating cytoskeletal dynamics, directed protrusion growth, and their stabilization. PHACTR-4 antagonizes β1-integrin signaling through the ROCK pathway. PHACTR-4 mutants display a misregulation of PP1 activity with a lack of cofilin activation. Another parameter controlling migration is cell polarity. The polarity machinery includes the Par complex (Par3, Par6, and atypical protein kinase C), which can activate RHO guanosine 5'-triphosphatase (GTPase) signaling and coordinate the reorientation of the centrosome and vesicle trafficking. Several lines of evidence suggest that PHACTR-4 regulates polarity through the Par complex or through phosphatidylinositol diphosphate and phosphatidylinositol triphosphate signaling [54].

A detailed commentary and perspective on the roles of PHACTR-4 and integrins in the ENS has been published [55]. The results of Zhang and coworkers [53] also raise many new questions relating to, for example, the regulation of β1-integrin by PHACTR-4 during ENCC migration at various stages of development.

13.3.2 Role of Phosphatase and Tensin Homolog

Integrin signaling can be downregulated by phosphatase and tensin homolog (PTEN) [56], a protein with sequence similarity to protein tyrosine phosphatases and tensin (a component of fibrillar adhesions). PTEN regulates the phosphatidylinositol 3-kinase (PI3K)/AKT (also known as protein kinase B) pathway [57] and dephosphorylates the focal adhesion kinase (FAK), both important elements of the integrin-signaling cascades, with consequences for cell adhesion, spread, and migration and for the organization of the actin cytoskeleton [58]. PTEN also regulates cell apoptosis, growth, and invasion. It is almost ubiquitously produced early in embryogenesis, its expression being gradually restricted during development [59]. However, migratory ENCCs do not express PTEN. This protein is first produced by a subset of neuronal derivatives at E15, and by E17.5 at the latest, all enteric neurons are positive for this protein. This suggests that PTEN probably exerts its effects at a stage of ENS development occurring after ENCC migration [60]. Indeed, the Tyr::Cre-mediated deletion of PTEN in a subset of vagal NCCs [61] leads to hyperplasia and hypertrophy of the enteric ganglia. In the conditional mutant, the ganglia exhibit higher proportions of enteric progenitors and glial

derivatives than control ganglia. This indicates a role for PTEN at later stages of ENS development, in the regulation of self-renewal and of the fate of enteric progenitors through the modulation of PI3K/Akt and mitogen-activated protein kinase/extracellular signal-regulated kinase signaling. The putative effects of PTEN on ENCC adhesion and integrin-dependent signaling remain to be investigated.

13.3.3 Roles of Integrin-Linked Kinase and Pinch

The cytoplasmic proteins integrin-linked kinase (ILK), Pinch, and parvin form the complex (IPP) at integrin adhesion sites. They are important regulators of integrin-mediated signaling [62,63]. The formation of the IPP complex precedes integrin-mediated localization to cell adhesion sites, cell spreading, and FA assembly [64,65]. The IPP complex serves as a link between integrins and receptor tyrosine kinases and GTPase signaling pathways. A loss of ILK or Pinch leads to embryonic lethality at the peri-implantation stage in mice (E5.5–6.5) [66,67].

ILK is a serine/threonine protein kinase that binds the cytoplasmic tail of β1-, β3-, or β5-integrins directly or indirectly, by binding to kindlin. Parvin can bind F-actin and contributes to the function of ILK in integrin–actin interactions [63]. ILK-deficient mouse cells have impairments in actin reorganization, FA maturation, fibrillar adhesion, and fibronectin (FN) matrix deposition. They cannot transmit forces to the ECM and migrate. Conditional *Ilk*-knockout studies have pointed to its role in regulating adhesion, spreading, and migration of embryonic cells; neurite extension; cell proliferation; and F-actin remodeling during histogenesis [68–70]. NCCs express ILK, and the Wnt1::Cre-dependent depletion of ILK in NCCs prior to their escape from the neural tube leads to embryonic death, with a defective NCC migration to the pharyngeal arches and cardiovascular defects [71,72]. However, possible effects on ENS ontogenesis have not been investigated. The precise role of ILK in the integrin-mediated adhesion, migration, and cytoskeletal dynamics of ENCCs should be determined, to provide greater insight into the way in which these cells interpret ECM signals for efficient gut colonization.

The adaptor proteins Pinch-1 and Pinch-2 have overlapping expression profiles during development but have different functions. Pinch-1 induces cell migration and spreading and mediates survival, whereas Pinch-2 is involved in mediating the Pinch-1/ILK interaction but cannot transduce integrin-mediated signals that control cell spreading and migration. The Wnt1::Cre-dependent invalidation of Pinch-1 leads to perinatal lethality with impaired development of cranial and cardiac NCC derivatives [73]. Mutant NCCs display altered smooth muscle differentiation and increased apoptosis but lack any FA defects. Pinch-1 is restricted to the nuclei of NCCs, suggesting that it may act as a shuttle/signaling protein or as a transcription factor (TF). In mutants, Pinch-1 leads to downregulation of transforming growth factor β (TGF-β) signaling and decreases Smad phosphorylation [73], suggesting an essential role for Pinch-1 in NCC development, mediated via TGF-β.

The Wnt1::Cre-dependent invalidation of Pinch-1 produced a transient delay in the gut colonization by ENCCs at the entry of the cecum in 16% of E12.5 embryos, subsequently recovering at later stages (our unpublished results). If Pinch-1 is depleted using the Ht-PA::Cre mouse model [36] the colonization of the gut is not impaired (our unpublished results), suggesting that Pinch-1 is dispensable for ENCC migration. Alternatively, Pinch-1 loss might be rescued by Pinch-2, but it is not possible to test this hypothesis, because at this writing there are no antibodies capable of discriminating between the two isoforms. Further studies are required to elucidate the Pinch effect in ENCC differentiation and survival.

13.3.4 Role of FAK

FAK is a key component of the signaling cascades elicited by extracellular signals. It is activated by integrin-mediated adhesion and lies at the crossroads between the integrin and the growth factor receptor signaling cascades. FAK plays a crucial role in cell proliferation, differentiation, and survival and regulates migration [74]. The constitutive knockout of FAK results in the death of mouse embryos at E8.5, owing to the impairment of mesoderm development as a result of severe actin-cytoskeletal disorganization and cell migration defects [75]. Wnt1/Cre-dependent FAK deletion results in perinatal death [76]. The mutants present cardiovascular defects, mostly due to impairments in NCC differentiation in smooth muscle. Oligopeptide kinase assay revealed that FAK is a direct substrate of the receptor tyrosine kinase RET (rearranged during transfection). FAK is activated upon stimulation of a human breast cancer cell line with the RET ligand glial cell line-derived neurotrophic factor (GDNF) [77]. FAK become rapidly activated upon EDN-3 treatment of astrocytes, which express EDNRB [78]. RET/GDNF and EDNRB/EDN-3 are key players in ENS ontogenesis (see below). However, the possible role of FAK in ENS development has not been investigated.

13.4 β1-INTEGRIN CROSSTALK WITH ENS GENES AND N-CADHERIN DURING ENS DEVELOPMENT

Several molecules have been implicated in ENS development in both animal models and humans, including the receptor RET and its ligand GDNF, the G-coupled receptor EDNRB and its ligand EDN-3, and the bone morphogenetic protein (BMP) signaling pathway. In addition, various transcriptional regulators play important roles during ENS

ontogeny, including SOX10 (Sry-like HMG box 10), ZEB2 (zinc-finger E-box-binding homeobox 2), paired-like homeobox 2B, and the HAND2 member of the heart and neural crest derivatives subclass of basic helix–loop–helix factors [2,3]. The ENS defects observed in *Itgb1* mutants suggest that there may be some interplay between β1-integrins and the products of some ENS genes.

13.4.1 Interplay between β1-Integrins and SOX10

The possible synergy between *Sox10* and *Itgb1* has been analyzed in double-mutant studies. *Sox10;Itgb1* double mutants have been reported to display more severe ENS defects than single mutants [79], with larger meshwork size and a lower density of ENCCs, indicating that the organization of the ENS is modulated by the interplay between β1-integrins and SOX10. At E14.5, when the gut is fully colonized by ENCCs in control embryos, the migratory front of these cells in *Sox10^{lacZ/+}* and *Itgb1* conditional mutants is located in the cecum and the middle of the colon, respectively, whereas this front is located more proximally in double mutants. Time-lapse imaging in ex vivo double-mutant gut cultures revealed slow ENCC locomotion, with lower levels of persistence and alterations in the direction of migration. Interestingly, *Sox10* haploinsufficiency leads to an enhancement of ENCC–fibronectin adhesion properties, with larger FAs and higher levels of activated β1-integrin than in control cells, which may reflect a difference in turnover rate potentially accounting for the altered migratory properties of ENCCs in the mutant and highlighting an essential function of SOX10 in the control of β1-integrin-dependent ENCC behavior [79]. Additional studies have been performed to decipher the underlying molecular mechanism. As SOX10 regulates the expression of several genes, including *Ednrb*, *L1cam*, *Ret*, and *Mpz*, during the development of the ENS and other NCC derivatives [80–83], we first investigated the impact of this TF on *Itgb1* expression, but found no effects [79]. By contrast, the expression and/or activation of β1-integrins was decreased by SOX10 overexpression in cell lines and increased in *Sox10* heterozygous mutants (Figure 13.3); an observation that could partially explain the origin of the severe ENS defects observed in *Sox10;Itgb1* double mutants [79] (Figure 13.3).

13.4.2 β1-Integrins and GDNF/RET Crosstalk

GDNF belongs to the TGF-β family of GDNF-family ligands (GFLs). GDNF, its receptor RET, and the coreceptor GFRα1 (GDNF-family receptor α1) are essential for ENS development. Their signaling pathway promotes the survival, proliferation, and migration of enteric progenitors and the differentiation of neurons [3]. The other GFLs, neurturin, artemin, and persephin, are also expressed during

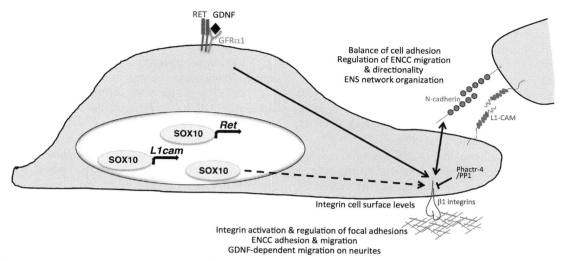

FIGURE 13.3 Interplay between adhesion molecules, SOX10, and RET/glial cell line-derived neurotrophic factor (GDNF) during enteric nervous system development. SOX10 is known to control *L1cam* and *Ret* expression. It does not regulate *Itgb1* expression but modulates integrin chain expression at the enteric neural crest cell (ENCC) surface. GDNF and SOX10 stimulate activation of β1-integrins and focal adhesion development in ENCCs, whereas PHACTR-4 inhibits β1-integrin signaling. ENS, enteric nervous system.

or after ENS development [84–86]. GFRα2, which binds neurturin, is expressed during ENS development but at later stages than GFRα1 [87]. GFRα4, the main coreceptor for persephin, is expressed by enteric progenitors [88].

The signaling cascades elicited by GDNF/RET involve kinases and scaffold proteins, such as FAK, p130cas, and paxillin, and they modulate the actin cytoskeleton via RHO GTPases [89]. Indeed, GDNF signaling and integrins act together to promote pancreatic cancer cell invasion and metastasis [90]. GDNF modulates integrin expression and RET activates β1-integrins [91]. On collagens and fibronectin, GDNF stimulates the adhesion and migration of cancer cells (neural-derived cancer cells and papillary thyroid carcinoma cells) via β1- and β3-integrin-dependent mechanisms. In addition, coimmunoprecipitation experiments performed on dopaminergic neurons in the substantia nigra of mammalian brain have revealed a possible interaction between GFRα1 and β1-integrins [92], suggesting the possible existence of a RET-independent pathway in GDNF-mediated responses. Sensory neurons isolated from dorsal root ganglia exhibit both a RET-dependent and a RET-independent pathway in response to GDNF-family ligands, but β1-integrins are not required for the RET-independent response to GDNF [93].

The *Itgb1* conditional mutant ENCCs lacking β1-integrins exhibit defective migration in the cecum compared to control cells [20]. At this stage, GDNF is present in this region of the gut. Mutant ENCCs express RET and GFRα1 and can respond to GDNF stimulation to grow neurites in 3D-collagen gut cultures. However, the migration of neuron cell bodies and glial progenitors along these neurites is abolished in the mutant [38], suggesting that β1-integrins contribute to the GDNF-dependent control of ENCC behavior. GDNF

is required for ENCCs to follow the alternative migration "route" via the mesentery to colonize the hindgut [6]. This process could be altered in the *Itgb1* mutant cells. As reported for cancer cells, GDNF stimulation increases β1-integrin activation in enteric progenitors (N. Bondurand and S. Dufour, unpublished data), suggesting that crosstalk between RET and β1-integrins occurs during ENS development.

Goto and coworkers used the expression of fluorescence resonance energy transfer (FRET) biosensors and the targeted expression of a fluorescent reporter to ENCCs to investigate the distribution of the activity of signaling molecules, including cAMP-dependent protein kinase A (PKA) and Rac1, during early ENS development [94]. GDNF and EDN-3 have opposite effects on PKA. GDNF inhibits PKA activity, whereas EDN-3 stimulates this activity. In ENCCs, the activities of PKA and RAC1 (ras-related C3 botulinum toxin substrate 1) are inversely and positively correlated, respectively, with cell velocity. This suggests that PKA acts as a molecular switch, regulating RAC1-dependent ENCC migration, but the link between PKA and RAC1 was not determined.

13.4.3 β1-Integrins and N-Cadherin Crosstalk

β1-Integrins are probably involved in remodeling the initial ENS network after gut colonization by ENCCs. Indeed, these cells lacking β1-integrins display greater aggregation, leading to a disorganization of the ENS in newborn mice, characterized by larger ganglia and a looser meshwork, suggesting a modulation of the intercellular adhesion properties of the ENCC. Cell surface receptors of the cadherin and CAM–Ig superfamilies are the major receptors regulating intercellular adhesion. Please refer to Chapter 12 by

T. Yagi and Chapter 11 by A. Fiszbein and colleagues for the roles of cadherins and N-CAM, respectively, in neural development. Mouse ENCCs express N-cadherin (CD325) and cadherins 11, 9, and 6, as well as protocadherins α1, α4, α10, and α15 [38,95,96]. ENCCs express N-cadherin at low levels during migration and at higher levels, together with N-CAM (CD56) and PSA-NCAM, during ENCC aggregation into ganglia, whereas Ng-CAM/L1-CAM (CD171) levels decrease during this process [97].

The loss of N-cadherin in ENCCs transiently delays colonization in the developing gut at E12.5 [98]. Mutant ENCCs display a disturbance of directionality and of the chain-like pattern of migration. The double depletion of N-cadherin and β1-integrin in ENCCs leads to embryonic lethality, with severe defects of ENS development, suggesting that there is cooperation between these two molecules during ENCC migration. However, in the colonized region of the gut, the abnormal network observed in β1-integrin mutants is partially rescued in the double mutant, indicating that collective ENCC migration and the organization of the ENS network require a correct balance between cell–ECM adhesion and cell–cell adhesion, controlled by β1-integrins and N-cadherin, respectively [98] (Figure 13.4). The adhesion receptor repertoire and crosstalk between integrins and cadherins regulate the strength of intercellular adhesion and traction force at cell–ECM adhesion sites [99,100], undoubtedly contributing to the control of ENS development [101].

Adhesion receptors, including cadherins and integrins, are also components of various signaling pathways governing cell behavior, including proliferation and cytoskeletal dynamics [102,103]. However, the molecular mechanisms involved in their crosstalk during ENCC migration have yet to be elucidated. The small GTPase Rap1 may be involved, as this protein has been reported to regulate endosome signaling and membrane trafficking pathways, orchestrating the delivery of cadherins and integrins to adhesion sites at the plasma membrane [104]. The GTPase-activating protein (GAP) of Rap1 has been shown to interact with RET and to downregulate GDNF-mediated neurite outgrowth [105].

Reactive oxygen species (ROS) generated by integrin activation may influence the adhesive functions of cadherins [106]. It would be interesting to evaluate the roles of Rap1, its GAP, and ROS in ENS development. The function of adhesion receptors can be regulated by interaction *in cis* with growth factor receptors. The disruption of these interactions when adhesion receptors are depleted could, therefore, modify the signaling cascades and the response to growth factors. Furthermore, cadherin and CAM–Ig levels can be altered by shedding and cleavage of functional

FIGURE 13.4 Function of β1-integrins and interplay with N-cadherin and enteric nervous system effectors in enteric neural crest cells (ENCCs). (A) Schematic representation of the impact of β1-integrin and N-cadherin depletion in ENCCs, which modifies the equilibrium between intercellular and cell–extracellular matrix (ECM) adhesion. When both are depleted, ENCC adhesion properties are reduced but the adhesion balance is rescued. (B) Schematic representation of the impact of β1-integrins and N-cadherin depletion in ENCC (represented as gray ovals) migratory behavior in the gut mesenchyme (represented by orange background). Top left panel: In the control, ENCCs migrate as thin chains (*) with some advanced isolated cells at the front (arrow) or cells exchanging between chains or migrating circumferentially to populate the gut tissue. Left to right upper panels: The β1-integrin loss strongly perturbs the chain-like mode of migration of ENCCs in the gut wall and favors the formation of clusters (arrowhead) and thick static chains (*) in the proximal hindgut. Top to bottom panels: The N-cadherin loss slightly disorganizes the chains (*) and directionality of ENCCs and transiently influences the caudal progression of the migratory front. Bottom right panel: The double mutation disrupts both the formation of chains and the ENCC adhesion and migration toward the caudal part of the gut. Double-mutant ENCCs are rounder and are mostly isolated (arrow) compared to control conditions.

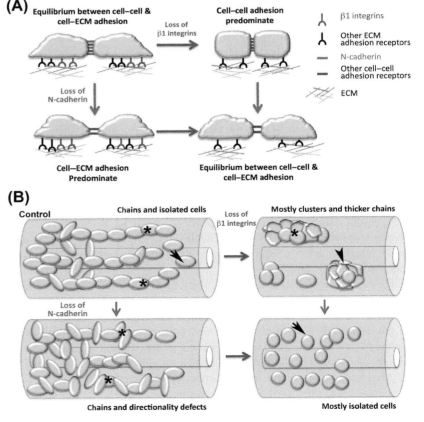

domains by metalloproteases and secretases, which modulate cell behavior [107]. Such cleaved fragments may affect the behavior of ENCCs, but this remains to be demonstrated.

13.5 CONTROL OF INTEGRIN GENE EXPRESSION

The expression of integrins must be tightly controlled, to ensure their correct action. Very little is currently known about the regulation of integrin gene expression during ENS development. However, the organization of the promoters/regulatory elements and the TFs binding to them have been described in other cell types, during development or in cancers. We describe some of these data here, focusing on the transcriptional and posttranscriptional control of integrins expressed during ENS development, in particular the *ITGA4*, *ITGA5*, *ITGA6*, *ITGAV*, *ITGB1*, and *ITGB3* genes, encoding CD49d, CD49e, CD49f, CD51, CD29, and CD61, respectively.

13.5.1 Organization of 5′ Regions/Promoters

Most of the promoter regions of these genes lack TATA and CCAAT boxes, but are GC-rich. This is the case for *ITGA4*, which maps to chromosome 2q31–q32; *ITGA5*, on human chromosome 12q13; *ITGAV*, which maps to chromosome 2q31; *ITGB3*, on chromosome 17; and the two promoter regions of *ITGB1* identified on human chromosome 10. The *ITGA6* promoter is the only promoter of an integrin gene expressed in the ENS containing a TATA box, located 22 nucleotides upstream of the primary transcription initiation site. Like the other promoters, it is also GC-rich (for review see Ref. [108]). TFs binding to the core promoters of these genes are redundant, suggesting that similar regulatory mechanisms could be used to couple integrin gene transcription to cellular differentiation/activation state in different cellular backgrounds.

13.5.2 Control of Integrin Gene Expression by SP1 and Interacting TFs

Proximal specificity protein 1 (SP1) binding sites are required for basal and regulated activities of promoters (see for example Refs. [109–112]). SP1 binding sites have been identified within the promoters or 5′ regions of *ITGB1*, *ITGB3*, *ITGA4*, *ITGA5*, and *ITGA6* [108,113–115].

SP1 sites are often located close to binding sites for activator proteins (AP-1/AP-2), Ets (E-twenty-six) family TFs, or nuclear factor 1 (NF1), and the coordinated activation or repression of integrin genes by these factors has been described. Both AP-1 and SP1 activate the *ITGA5* promoter, but NF1 has been shown to be a potent repressor [116,117]. Interestingly, fibronectin can either promote or repress *ITGA5* promoter activity in a cell-dependent

manner, through different effects on the ratios of these TFs [108,116]. *ITGA6* regulation seems to involve similar mechanisms. The negative influence of NF1 appears to be potentiated by the degradation of SP1 when cells are grown on laminin [118]. The proteolytic cleavage of SP1 or SP3 by laminin has been reported [119].

FOSL1 (Fos-like 1, also named Fra1) represses the transcription of *ITGAV* and *ITGB3* by binding, together with JunD, via SP1, this regulation being required for endothelial assembly into vessel structures [120]. ZEB2, a member of the deltaEF1 family of two-handed zinc-finger factors, has been shown to induce the expression of *ITGA5* and vimentin by interacting with and activating SP1, suggesting that cooperation between these two TFs constitutes a new mechanism of mesenchymal gene activation during epithelial–mesenchymal transition [121]. Conversely, ZEB TFs seem to repress the expression of *ITGA4* [122]. The c-Myb sites within this promoter mediate a balance between repression by ZEB and transcriptional activation by c-Myb/Ets. Another regulatory loop, between ZEB2 and β1-integrins, has been described [123]. In some breast cancer cells, the blocking or silencing of β1-integrins induces a switch from collective to single-cell migration. In these cells, β1-integrins enhance TGF-β signaling, leading to a shift in the balance between miR200 and ZEB2 [123]. It remains unclear whether one or several of these regulatory mechanisms involving ZEB2 and integrins occur during ENS development.

13.5.3 Control of Integrin Gene Expression by other TFs

In vivo chromatin immunoprecipitation experiments have demonstrated the binding of E2F1 to the *ITGA6* promoter and of CCAAT enhancer-binding proteins (C/EBPs) to the *ITGA5* promoter [119,124]. The C/EBPs upregulate *ITGA5* transcription in HepG2 cells, but they repress the activity of the *ITGA5* promoter in keratinocytes. Given the broad range of TFs known to interact physically with C/EBPs (SP1, acute myeloid leukemia 1, c-Myb, nuclear factor κB (NF-κB), c-Jun, or E47), the C/EBP-mediated repression of integrin promoters in keratinocytes may involve interference with the functional activity of TFs with binding sites adjacent to those for C/EBP. FoxC2 (forkhead box C2), a member of the forkhead/winged-helix TF family, the expression of which can be upregulated by BMP2, binds directly to the *ITGB1* promoter to activate its expression, thereby promoting osteoblastogenesis [125]. Several FoxC2 binding sites have also been described in *ITGB3* regulatory elements, with a potential role in the regulation of angiogenesis [126]. The HOX (homeobox-containing) TFs HOXD3, HOXD1, and HOXA10 have been shown to regulate *ITGB3*, *ITGA5*, and *ITGB1*, inducing their expression during angiogenesis or in the endometrium, in response to estrogen and progesterone [127–129]. Retinoblastoma binding protein 2 has also been shown to regulate *ITGB*1 and to rescue proliferation or migration/invasion in lung cancer cells [130].

Binding sites for the paired-box factor 6 (PAX6) are present in the *ITGA*4, *ITGA*5, and *ITGB*1 promoters. PAX6 has been shown to alter the expression of *ITGA*4 in corneal epithelial cells, by interacting with at least three different regions of the promoter, to exert positive or negative effects [131,132]. A near-perfect target site for the microphthalmia TF (MITF) overlaps with one of the PAX6 binding sites. MITF (which has been reported to regulate α4 chain expression in mast cells [133]) interacts with its binding site and may prevent either the binding or the transactivating effect of PAX6 [132].

NF-κB binding sites have been identified in the regulatory regions of *ITGA*6, *ITGAV*, and *ITGB*1 (for review see Refs. [108,134]). Data have highlighted the existence of an EDN–integrin axis regulating chemotactic interactions between macrophages and endothelial cells [135]. The authors showed that EDN-1 induced Ets-like kinase-1, signal transducer and activator 3, and the phosphorylation of NF-κB, leading to the activation of *ITGB*1 and *ITGAV* expression through direct binding to their promoters. It remains to be determined whether a similar regulatory mechanism involving EDN-3 and *ITGB*1 occurs during ENS development.

Finally, chromatin immunoprecipitation experiments coupled to high-throughput sequencing technology and microarray analyses have also revealed that the regulatory regions of integrin genes or their mRNAs are directly or indirectly controlled by SOX factors. Consistent with this observation, we have recently highlighted a strong genetic interaction between *Sox*10 and *Itgb*1 during ENS development. However, as stated above, our data suggest that *Itgb*1 is not under the regulation of SOX10 [79]. As several of these TFs (or members of their families) are expressed in ENCCs, their involvement in *Itgb*1 regulation should be considered.

13.5.4 Posttranscriptional Control of Integrin Expression by MicroRNAs

Integrins also seem to be subject to posttranscriptional regulation by microRNAs (miRNAs), small, single-stranded, noncoding RNAs in cancer and neural development. Some of these molecules have been shown to downregulate β1-integrin by interacting with its 3′ untranslated region [136,137], and the number of reports on the regulation of integrin mRNAs in this manner is increasing exponentially (for review, Ref. [138]).

Interestingly, NCC development seems to be under the regulation of miRNAs [139]. The loss of the endonuclease Dicer in NCCs has been shown to lead to ENS defects [140]. Indeed, conditional ablation of *Dicer* in the NCC lineage does not prevent the initial induction, migration, and differentiation of NCCs. However, as development progresses,

NCC derivatives, including ENS cells and melanocytes, are lost through apoptosis. Once the miRNAs expressed in the ENS have been identified, it will be of great interest to determine their role in regulating integrin gene expression.

13.6 CONCLUSION

Integrins play a crucial role in cell adhesion, migration, and mechanosensing [141], and effectors of integrin signaling are found at the crossroads of many cytoplasmic cascades, including those downstream of growth factor receptors. More detailed analyses of integrin signaling during ENS ontogenesis and elucidation of the way in which enteric progenitors integrate signals elicited by the chemical and physical properties of their environment will improve our understanding of ENS ontogenesis. In addition to cytoplasmic signaling, integrins are regulated by various TFs. Data are emerging for the ENS, but the transcriptional and posttranscriptional regulation of integrins remains largely unknown and should be the focus of future studies.

ENS defects are linked to the abnormal proliferation, survival, and differentiation of enteric progenitors or to alterations in the migration of ENCCs. Resections of the aganglionic bowel and supportive care have decreased the mortality and morbidity of HSCR patients, but progress in the molecular and cell biology of enteric progenitors and advances in transplantation techniques are favoring the development of therapies based on cell replacement strategies [142–144]. As integrins play a key role in regulating stem cell fate, a better knowledge of the interplay between ENS genes and integrin signaling and mechanosensing should provide us with a more complete understanding of the mechanisms controlling the fate and behavior of enteric progenitors.

REFERENCES

[1] Sasselli V, Pachnis V, Burns AJ. The enteric nervous system. Dev Biol 2012;366(1):64–73.

[2] Lake JI, Heuckeroth RO. Enteric nervous system development: migration, differentiation, and disease. Am J Physiol Gastrointest Liver Physiol 2013;305(1):G1–24.

[3] Heanue TA, Pachnis V. Enteric nervous system development and Hirschsprung's disease: advances in genetic and stem cell studies. Nat Rev Neurosci 2007;8(6):466–79.

[4] Le Douarin NM, Teillet MA. The migration of neural crest cells to the wall of the digestive tract in avian embryo. J Embryol Exp Morphol 1973;30(1):31–48.

[5] Le Douarin NM, Kalcheim C. The neural crest. In: Bard BL, Barlow PW, Kirk DL, editors. Developmental and cell biology series. 2nd ed. Cambridge: Cambridge University Press; 1999.

[6] Nishiyama C, et al. Trans-mesenteric neural crest cells are the principal source of the colonic enteric nervous system. Nat Neurosci 2012;15(9):1211–8.

[7] Young HM, et al. Dynamics of neural crest-derived cell migration in the embryonic mouse gut. Dev Biol 2004;270(2):455–73.

[8] Druckenbrod NR, Epstein ML. The pattern of neural crest advance in the cecum and colon. Dev Biol 2005;287(1):125–33.

[9] Druckenbrod NR, Epstein ML. Behavior of enteric neural crest-derived cells varies with respect to the migratory wavefront. Dev Dyn 2007;236(1):84–92.

[10] Young HM, et al. Colonizing while migrating: how do individual enteric neural crest cells behave? BMC Biol 2014;12(1):23.

[11] Hynes RO. Integrins: bidirectional, allosteric signaling machines. Cell 2002;110(6):673–87.

[12] Campbell ID, Humphries MJ. Integrin structure, activation, and interactions. Cold Spring Harb Perspect Biol 2011;3(3).

[13] Morse EM, Brahme NN, Calderwood DA. Integrin cytoplasmic tail interactions. Biochemistry 2014;53(5):810–20.

[14] Daley WP, Yamada KM. ECM-modulated cellular dynamics as a driving force for tissue morphogenesis. Curr Opin Genet Dev 2013;23(4):408–14.

[15] Rozario T, De Simone DW. The extracellular matrix in development and morphogenesis: a dynamic view. Dev Biol 2010;341(1):126–40.

[16] Newgreen DF, Hartley L. Extracellular matrix and adhesive molecules in the early development of the gut and its innervation in normal and spotting lethal rat embryos. Acta Anat (Basel) 1995;154(4):243–60.

[17] Simon-Assmann P, et al. Extracellular matrix components in intestinal development. Experientia 1995;51(9–10):883–900.

[18] Rauch U, Schafer KH. The extracellular matrix and its role in cell migration and development of the enteric nervous system. Eur J Pediatr Surg 2003;13(3):158–62.

[19] Breau MA, Dufour S. Biological development of the enteric nervous system. In: Núñez RN, López-Alonso M, editors. "Hirschsprung's disease: diagnosis and treatment." Congenital disorders – laboratory and clinical research series. New York: Nova; 2009. pp. 15–44.

[20] Breau MA, et al. Beta1 integrins are required for the invasion of the caecum and proximal hindgut by enteric neural crest cells. Development 2009;136(16):2791–801.

[21] Akbareian SE, et al. Enteric neural crest-derived cells promote their migration by modifying their microenvironment through tenascin-C production. Dev Biol 2013;382(2):446–56.

[22] Bolcato-Bellemin AL, et al. Laminin alpha5 chain is required for intestinal smooth muscle development. Dev Biol 2003;260(2):376–90.

[23] Lefebvre O, et al. Developmental expression and cellular origin of the laminin alpha2, alpha4, and alpha5 chains in the intestine. Dev Biol 1999;210(1):135–50.

[24] Scherberich A, Murine tenascin W. A novel mammalian tenascin expressed in kidney and at sites of bone and smooth muscle development. J Cell Sci 2004;117(4):571–81.

[25] Payette RF, et al. Accumulation of components of basal laminae: association with the failure of neural crest cells to colonize the presumptive aganglionic bowel of ls/ls mutant mice. Dev Biol 1988;125(2):341–60.

[26] Parikh DH, et al. Abnormalities in the distribution of laminin and collagen type IV in Hirschsprung's disease. Gastroenterology 1992;102(4 Pt 1):1236–41.

[27] Alpy F, et al. The expression pattern of laminin isoforms in Hirschsprung disease reveals a distal peripheral nerve differentiation. Hum Pathol 2005;36(10):1055–65.

[28] Chalazonitis A, et al. The alpha1 subunit of laminin-1 promotes the development of neurons by interacting with LBP110 expressed by neural crest-derived cells immunoselected from the fetal mouse gut. J Neurobiol 1997;33(2):118–38.

[29] Raghavan S, Gilmont RR, Bitar KN. Neuroglial differentiation of adult enteric neuronal progenitor cells as a function of extracellular matrix composition. Biomaterials 2013;34(28):6649–58.

[30] Iwashita T, et al. Hirschsprung disease is linked to defects in neural crest stem cell function. Science 2003;301(5635):972–6.

[31] Ikawa H, et al. Impaired expression of neural cell adhesion molecule L1 in the extrinsic nerve fibers in Hirschsprung's disease. J Pediatr Surg 1997;32(4):542–5.

[32] Bixby S, et al. Cell-intrinsic differences between stem cells from different regions of the peripheral nervous system regulate the generation of neural diversity. Neuron 2002;35(4):643–56.

[33] Kruger GM, et al. Neural crest stem cells persist in the adult gut but undergo changes in self-renewal, neuronal subtype potential, and factor responsiveness. Neuron 2002;35(4):657–69.

[34] Campos LS. β1 integrins and neural stem cells: making sense of the extracellular environment. BioEssays 2005;27(7):698–707.

[35] Fässler R, Meyer M. Consequence of lack of β1 integrin gene expression in mice. Genes Dev 1995;9:1896–908.

[36] Pietri T, et al. The human tissue plasminogen activator-Cre mouse: a new tool for targeting specifically neural crest cells and their derivatives in vivo. Dev Biol 2003;259(1):176–87.

[37] Pietri T, et al. Conditional beta1-integrin gene deletion in neural crest cells causes severe developmental alterations of the peripheral nervous system. Development 2004;131(16):3871–83.

[38] Breau MA, et al. Lack of beta1 integrins in enteric neural crest cells leads to a Hirschsprung-like phenotype. Development 2006;133(9):1725–34.

[39] Nagy N, et al. Endothelial cells promote migration and proliferation of enteric neural crest cells via β1 integrin signaling. Dev Biol 2009;330(2):263–72.

[40] Dufour S, et al. Attachment, spreading and locomotion of avian neural crest cells are mediated by multiple adhesion sites on fibronectin molecules. EMBO J 1988;7:2661–71.

[41] Fujimoto T, et al. A study of the extracellular matrix protein as the migration pathway of neural crest cells in the gut: analysis in human embryos with special reference to the pathogenesis of Hirschsprung's disease. J Pediatr Surg 1989;24(6):550–6.

[42] Halfter W, Chiquet-Ehrismann R, Tucker RP. The effect of tenascin and embryonic basal lamina on the behavior and morphology of neural crest cells in vitro. Dev Biol 1989;132(1):14–25.

[43] Tucker RP. Abnormal neural crest cell migration after the in vivo knockdown of tenascin-C expression with morpholino antisense oligonucleotides. Dev Dyn 2001;222(1):115–9.

[44] Lange K, et al. Endothelin receptor type B counteracts tenascin-C-induced endothelin receptor type A-dependent focal adhesion and actin stress fiber disorganization. Cancer Res 2007;67(13):6163–73.

[45] Barlow A, de Graaff E, Pachnis V. Enteric nervous system progenitors are coordinately controlled by the G protein-coupled receptor EDNRB and the receptor tyrosine kinase RET. Neuron 2003;40(5):905–16.

[46] Lee H. The endothelin receptor-B is required for the migration of neural crest-derived melanocyte and enteric neuron precursors. Dev Biol 2003;259(1):162–75.

[47] Bondurand N, et al. Maintenance of mammalian enteric nervous system progenitors by SOX10 and endothelin 3 signalling. Development 2006;133(10):2075–86.

[48] Spence SG, Poole TJ. Developing blood vessels and associated extracellular matrix as substrates for neural crest migration in Japanese quail, Coturnix coturnix japonica. Int J Dev Biol 1994;38(1):85–98.

[49] Delalande JM, et al. Vascularisation is not necessary for gut colonisation by enteric neural crest cells. Dev Biol 2014;385(2):220–9.

[50] Wolfenson H, Lavelin I, Geiger B. Dynamic regulation of the structure and functions of integrin adhesions. Dev Cell 2013;24(5):447–58.

[51] Zaidel-Bar R, Geiger B. The switchable integrin adhesome. J Cell Sci 2010;123(9):1385–8.

[52] Allen PB, et al. Phactrs 1–4: a family of protein phosphatase 1 and actin regulatory proteins. Proc Natl Acad Sci USA 2004;101(18):7187–92.

[53] Zhang Y, Kim TH, Niswander L. Phactr4 regulates directional migration of enteric neural crest through PP1, integrin signaling, and cofilin activity. Genes Dev 2012;26(1):69–81.

[54] Zhang Y, Niswander L. Phactr4: a new integrin modulator required for directional migration of enteric neural crest cells. Cell Adh Migr 2012;6(5):419–23.

[55] Sun Z, Fassler R. A firm grip does not always pay off: a new Phact(r) 4 integrin signaling. Genes Dev 2012;26(1):1–5.

[56] Tamura M, et al. PTEN gene and integrin signaling in cancer. J Natl Cancer Inst 1999;91(21):1820–8.

[57] Katso R, et al. Cellular function of phosphoinositide 3-kinases: implications for development, homeostasis, and cancer. Annu Rev Cell Dev Biol 2001;17:615–75.

[58] Tamura M, et al. Inhibition of cell migration, spreading, and focal adhesions by tumor suppressor PTEN. Science 1998;280(5369):1614–7.

[59] Podsypanina K, et al. Mutation of Pten/Mmac1 in mice causes neoplasia in multiple organ systems. Proc Natl Acad Sci USA 1999;96(4):1563–8.

[60] Puig I, et al. Deletion of Pten in the mouse enteric nervous system induces ganglioneuromatosis and mimics intestinal pseudoobstruction. J Clin Invest 2009;119(12):3586–96.

[61] Puig I, et al. The tyrosinase promoter is active in a subset of vagal neural crest cells during early development in mice. Pigment Cell Melanoma Res 2009;22(3):331–4.

[62] Legate KR, et al. ILK, PINCH and parvin: the tIPP of integrin signalling. Nat Rev Mol Cell Biol 2005;7(1):20–31.

[63] Ghatak S, Morgner J, Wickstrom SA. ILK: a pseudokinase with a unique function in the integrin-actin linkage. Biochem Soc Trans 2013;41(4):995–1001.

[64] Zhang Y, et al. Assembly of the PINCH-ILK-CH-ILKBP complex precedes and is essential for localization of each component to cell-matrix adhesion sites. J Cell Sci 2002;115(Pt 24):4777–86.

[65] Wickstrom SA, et al. The ILK/PINCH/parvin complex: the kinase is dead, long live the pseudokinase! EMBO J 2010;29(2):281–91.

[66] Sakai T. Integrin-linked kinase (ILK) is required for polarizing the epiblast, cell adhesion, and controlling actin accumulation. Genes Dev 2003;17(7):926–40.

[67] Liang X, et al. PINCH1 plays an essential role in early murine embryonic development but is dispensable in ventricular cardiomyocytes. Mol Cell Biol 2005;25(8):3056–62.

[68] Grashoff C, et al. Integrin-linked kinase regulates chondrocyte shape and proliferation. EMBO Rep 2003;4(4):432–8.

[69] Friedrich EB, et al. Integrin-linked kinase regulates endothelial cell survival and vascular development. Mol Cell Biol 2004;24(18):8134–44.

[70] Ishii T, Satoh E, Nishimura M. Integrin-linked kinase controls neurite outgrowth in N1E-115 neuroblastoma cells. J Biol Chem 2001;276(46):42994–3003.

[71] Arnold TD, Zang K, Vallejo-Illarramendi A. Deletion of integrin-linked kinase from neural crest cells in mice results in aortic aneurysms and embryonic lethality. Dis Model Mech 2013;6(5):1205–12.

[72] Dai X, et al. Requirement for integrin-linked kinase in neural crest migration and differentiation and outflow tract morphogenesis. BMC Biol 2013;11:107.

[73] Liang X, et al. Pinch1 is required for Normal development of cranial and cardiac neural crest-derived structures. Circ Res 2007;100(4):527–35.

[74] Mitra SK, Schlaepfer DD. Integrin-regulated FAK–Src signaling in normal and cancer cells. Curr Opin Cell Biol 2006;18(5):516–23.

[75] Ilic D, et al. Reduced cell motility and enhanced focal adhesion contact formation in cells from FAK-deficient mice. Nature 1995;377(6549):539–44.

[76] Vallejo-Illarramendi A, Zang K, Reichardt LF. Focal adhesion kinase is required for neural crest cell morphogenesis during mouse cardiovascular development. J Clin Invest 2009;119(8):2218–30.

[77] Plaza-Menacho I, et al. Focal adhesion kinase (FAK) binds RET kinase via its FERM domain, priming a direct and reciprocal RET-FAK transactivation mechanism. J Biol Chem 2011;286(19):17292–302.

[78] Koyama Y, et al. Endothelins increase tyrosine phosphorylation of astrocytic focal adhesion kinase and paxillin accompanied by their association with cytoskeletal components. Neuroscience 2000;101(1):219–27.

[79] Watanabe Y, et al. Sox10 and Itgb1 interaction in enteric neural crest cell migration. Dev Biol 2013;379(1):92–106.

[80] Lang D, et al. Pax3 is required for enteric ganglia formation and functions with Sox10 to modulate expression of c-ret. J Clin Invest 2000;106(8):963–71.

[81] Peirano RI, et al. Protein zero gene expression is regulated by the glial transcription factor Sox10. Mol Cell Biol 2000;20(9):3198–209.

[82] Zhu L, et al. Spatiotemporal regulation of endothelin receptor-B by SOX10 in neural crest–derived enteric neuron precursors. Nat Genet 2004;36(7):732–7.

[83] Wallace AS, et al. L1cam acts as a modifier gene during enteric nervous system development. Neurobiol Dis 2010;40(3):622–33.

[84] Heuckeroth RO, et al. Neurturin and GDNF promote proliferation and survival of enteric neuron and glial progenitors in vitro. Dev Biol 1998;200(1):116–29.

[85] Maruccio L, et al. The development of avian enteric nervous system: distribution of artemin immunoreactivity. Acta Histochem 2008;110(2):163–71. Epub November 26, 2007.

[86] Yan H, et al. Neural cells in the esophagus respond to glial cell line-derived neurotrophic factor and neurturin, and are RET-dependent. Dev Biol 2004;272(1):118–33.

[87] Shepherd IT, Pietsch J, Elworthy S, Kelsh RN, and Raible DW. Roles for GFRalpha1 receptors in zebrafish enteric nervous system development. Development 2004;131:241–249.

[88] Ruiz-Ferrer M, et al. Novel mutations at RET ligand genes preventing receptor activation are associated to Hirschsprung's disease. J Mol Med Berl 2011;89(5):471–80.

[89] Murakami H, et al. Rho-dependent and -independent tyrosine phosphorylation of focal adhesion kinase, paxillin and p130Cas mediated by Ret kinase. Oncogene 1999;18(11):1975–82.

[90] Funahashi H, et al. The role of glial cell line-derived neurotrophic factor (GDNF) and integrins for invasion and metastasis in human pancreatic cancer cells. J Surg Oncol 2005;91(1):77–83.

[91] Cockburn JG, et al. RET-mediated cell adhesion and migration require multiple integrin subunits. J Clin Endocrinol Metab 2010;95(11):E342–6.

[92] Cao JP, et al. Integrin beta1 is involved in the signaling of glial cell line-derived neurotrophic factor. J Comp Neurol 2008;509(2):203–10.

[93] Schmutzler BS, et al. Ret-dependent and Ret-independent mechanisms of Gfl-induced sensitization. Mol Pain 2011;7:22.

[94] Goto A, et al. GDNF and endothelin 3 regulate migration of enteric neural crest-derived cells via protein kinase A and Rac1. J Neurosci 2013;33(11):4901–12.

[95] Heanue TA, Pachnis V. Expression profiling the developing mammalian enteric nervous system identifies marker and candidate Hirschsprung disease genes. Proc Natl Acad Sci USA 2006;103(18):6919–24.

[96] Vohra BPS, et al. Differential gene expression and functional analysis implicate novel mechanisms in enteric nervous system precursor migration and neuritogenesis. Dev Biol 2006;298(1):259–71.

[97] Hackett-Jones EJ, et al. On the role of differential adhesion in gangliogenesis in the enteric nervous system. J Theor Biol 2011;287:148–59.

[98] Broders-Bondon F, et al. N-cadherin and beta1-integrins cooperate during the development of the enteric nervous system. Dev Biol 2012;364(2):178–91.

[99] Martinez-Rico C, et al. Integrins stimulate E-cadherin-mediated intercellular adhesion by regulating Src-kinase activation and actomyosin contractility. J Cell Sci 2010;123(Pt 5):712–22.

[100] Jasaitis A, et al. E-Cadherin-Dependent stimulation of traction force at focal adhesions via the Src and PI3K signaling pathways. Biophys J 2012;103(2):175–84.

[101] Newgreen DF, et al. Simple rules for a "simple" nervous system? Molecular and biomathematical approaches to enteric nervous system formation and malformation. Dev Biol 2013;382(1):305–19.

[102] Epifano C, Perez-Moreno M. Crossroads of integrins and cadherins in epithelia and stroma remodeling. Cell Adh Migr 2012;6(3):261–73.

[103] McKeown SJ, Wallace AS, Anderson RB. Expression and function of cell adhesion molecules during neural crest migration. Dev Biol 2013;373(2):244–57.

[104] Retta SF, Balzac F, Avolio M. Rap1: a turnabout for the crosstalk between cadherins and integrins. Eur J Cell Biol 2006;85(3–4):283–93.

[105] Jiao L, et al. Rap1GAP interacts with RET and suppresses GDNF-induced neurite outgrowth. Cell Res 2011;21(2):327–37.

[106] Goitre L, et al. Molecular crosstalk between integrins and cadherins: do Reactive oxygen species Set the Talk? J Signal Transduction 2012;2012:807682.

[107] Cavallaro U, Dejana E. Adhesion molecule signalling: not always a sticky business. Nat Rev Mol Cell Biol 2011;12(3):189–97.

[108] Vigneault F, et al. Control of integrin genes expression in the eye. Prog Retin Eye Res 2007;26(2):99–161.

[109] Chen HM, et al. The Sp1 transcription factor binds the CD11b promoter specifically in myeloid cells in vivo and is essential for myeloid-specific promoter activity. J Biol Chem 1993;268(11):8230–9.

[110] Lopez-Rodriguez C, et al. Identification of Sp1-binding sites in the CD11c (p150,95 alpha) and CD11a (LFA-1 alpha) integrin subunit promoters and their involvement in the tissue-specific expression of CD11c. Eur J Immunol 1995;25(12):3496–503.

[111] Ye J, et al. Sp1 binding plays a critical role in Erb-B2- and v-ras-mediated downregulation of alpha2-integrin expression in human mammary epithelial cells. Mol Cell Biol 1996;16(11):6178–89.

[112] Zutter MM, Ryan EE, Painter AD. Binding of phosphorylated Sp1 protein to tandem Sp1 binding sites regulates alpha2 integrin gene core promoter activity. Blood 1997;90(2):678–89.

[113] Birkenmeier TM, et al. The alpha 5 beta 1 fibronectin receptor. Characterization of the alpha 5 gene promoter. J Biol Chem 1991;266(30):20544–9.

[114] Cervella P, et al. Human beta 1-integrin gene expression is regulated by two promoter regions. J Biol Chem 1993;268(7):5148–55.

[115] Nishida K, et al. Identification of regulatory elements of human alpha 6 integrin subunit gene. Biochem Biophys Res Commun 1997;241(2):258–63.

[116] Gingras ME, et al. Differential binding of the transcription factors Sp1, AP-1, and NFI to the promoter of the human alpha5 integrin gene dictates its transcriptional activity. Invest Ophthalmol Vis Sci 2009;50(1):57–67.

[117] Landreville S, et al. Suppression of alpha5 gene expression is closely related to the tumorigenic properties of uveal melanoma cell lines. Pigment Cell Melanoma Res 2011;24(4):643–55.

[118] Gaudreault M, et al. Transcriptional regulation of the human alpha6 integrin gene by the transcription factor NFI during corneal wound healing. Invest Ophthalmol Vis Sci 2008;49(9):3758–67.

[119] Gaudreault M, et al. Laminin reduces expression of the human alpha6 integrin subunit gene by altering the level of the transcription factors Sp1 and Sp3. Invest Ophthalmol Vis Sci 2007;48(8):3490–505.

[120] Evellin S, et al. FOSL1 controls the assembly of endothelial cells into capillary tubes by direct repression of alphav and beta3 integrin transcription. Mol Cell Biol 2013;33(6):1198–209.

[121] Nam EH, et al. ZEB2 upregulates integrin alpha5 expression through cooperation with Sp1 to induce invasion during epithelial-mesenchymal transition of human cancer cells. Carcinogenesis 2012;33(3):563–71.

[122] Postigo AA, et al. c-Myb and Ets proteins synergize to overcome transcriptional repression by ZEB. EMBO J 1997;16(13):3924–34.

[123] Truong HH, et al. Beta1 integrin inhibition elicits a prometastatic switch through the TGFbeta-miR-200-ZEB network in e-cadherin-positive triple-negative breast Cancer. Sci Signal 2014;7(312):ra15.

[124] Corbi AL, Jensen UB, Watt FM. The alpha2 and alpha5 integrin genes: identification of transcription factors that regulate promoter activity in epidermal keratinocytes. FEBS Lett 2000;474(2–3):201–7.

[125] Park SJ, et al. The forkhead transcription factor Foxc2 promotes osteoblastogenesis via up-regulation of integrin beta1 expression. Bone 2011;49(3):428–38.

[126] Hayashi H, et al. The Foxc2 transcription factor regulates angiogenesis via induction of integrin beta3 expression. J Biol Chem 2008;283(35):23791–800.

[127] Daftary GS, et al. Direct regulation of beta3-integrin subunit gene expression by HOXA10 in endometrial cells. Mol Endocrinol 2002;16(3):571–9.

[128] Boudreau NJ, Varner JA. The homeobox transcription factor Hox D3 promotes integrin alpha5beta1 expression and function during angiogenesis. J Biol Chem 2004;279(6):4862–8.

[129] Park H, et al. Homeobox D1 regulates angiogenic functions of endothelial cells via integrin beta1 expression. Biochem Biophys Res Commun 2011;408(1):186–92.

[130] Teng YC, et al. Histone demethylase RBP2 promotes lung tumorigenesis and cancer metastasis. Cancer Res 2013;73(15):4711–21.

[131] Duncan MK, et al. Overexpression of PAX6(5a) in lens fiber cells results in cataract and upregulation of (alpha)5(beta)1 integrin expression. J Cell Sci 2000;113(Pt 18):3173–85.

[132] Zaniolo K, et al. Expression of the alpha4 integrin subunit gene promoter is modulated by the transcription factor Pax-6 in corneal epithelial cells. Invest Ophthalmol Vis Sci 2004;45(6):1692–704.

[133] Kim DK, et al. Impaired expression of integrin alpha-4 subunit in cultured mast cells derived from mutant mice of mi/mi genotype. Blood 1998;92(6):1973–80.

[134] Sharma HW, et al. A DNA motif present in alpha V integrin promoter exhibits dual binding preference to distinct transcription factors. Anticancer Res 1995;15(5B):1857–67.

[135] Chen CC, et al. The endothelin-integrin axis is involved in macrophage-induced breast Cancer cell chemotactic interactions with endothelial cells. J Biol Chem 2014;289(14):10029–44.

[136] Hunt S, et al. MicroRNA-124 suppresses oral squamous cell carcinoma motility by targeting ITGB1. FEBS Lett 2011;585(1):187–92.

[137] Zha R, et al. Genome-wide screening identified that miR-134 acts as a metastasis suppressor by targeting integrin beta1 in hepatocellular carcinoma. PloS One 2014;9(2):e87665.

[138] Rutnam ZJ, Wight TN, Yang BB. miRNAs regulate expression and function of extracellular matrix molecules. Matrix Biol 2013;32(2):74–85.

[139] Strobl-Mazzulla PH, Marini M, Buzzi A. Epigenetic landscape and miRNA involvement during neural crest development. Dev Dyn 2012;241(12):1849–56.

[140] Huang T, et al. Wnt1-cre-mediated conditional loss of Dicer results in malformation of the midbrain and cerebellum and failure of neural crest and dopaminergic differentiation in mice. J Mol Cell Biol 2010;2(3):152–63.

[141] Schiller HB, Fassler R. Mechanosensitivity and compositional dynamics of cell-matrix adhesions. EMBO Rep 2013;14(6):509–19.

[142] Hotta R, et al. Stem cells for GI motility disorders. Curr Opin Pharmacol 2011;11(6):617–23.

[143] Hotta R, et al. Transplanted progenitors generate functional enteric neurons in the postnatal colon. J Clin Invest 2013;123(3):1182–91.

[144] Burns AJ, Thapar N. Neural stem cell therapies for enteric nervous system disorders. Nat Rev Gastroenterol Hepatol 2014;11(5):317–28.

Chapter 14

Neural Flow Cytometry – A Historical Account from a Personal Perspective

Henry J. Klassen

University of California, Irvine, CA, USA

Neural progenitor cells are critical to central nervous system (CNS) development in all vertebrates and, indeed, constitute the very contingent of cells that become both the neuronal and macroglial elements of this elaborate structure. The transition from immature progenitor cells of neural lineage to the resident cells of the mature CNS is achieved through many rounds of mitosis, as well as a complex series of interrelated cellular migrations and interactions involving the extension of processes and formation of physiological connections as the cells differentiate. While it was long thought that the exceedingly complex process of neural development in mammals represented a progressive march toward the formation of a rigidly structured organ with little capacity for self-repair, as vividly evidenced by the generally poor outcome from neurological trauma in humans, work over the past several decades has revealed dramatic surprises that may have considerable clinical relevance. In particular, it may be possible to restore meaningful levels of neurological function through the reintroduction of neural progenitor cells to the diseased human CNS via cell transplantation [1].

My own interest in neural progenitor cells (NPCs) dates back to the early era of work with these cells and followed from my preceding studies involving the transplantation of fetal neural tissue in rats. Those earlier studies, performed as a graduate student in Ray Lund's laboratory, then at the University of Pittsburgh, examined the functional capabilities of retinal tissue grafts transplanted to neonatal albino (Sprague Dawley) rats. It had already been shown that a fetal retina placed in the rat brain could extend neural processes that selectively innervated the retino-recipient nuclei of the host brainstem following optic deafferentation [2]. In turn, I asked the question as to whether these projections were functional. Using the pupillary light reflex as an assay, I showed that illumination of the grafts was sufficient to drive a brisk pupilloconstriction response in the host eye by way of graft–host connectivity in the olivary pretectal nucleus [3–5].

This result had favorable implications for the reestablishment of neural connectivity in the mammalian CNS;

however, a number of caveats subsequently emerged. First of all, Lucia Galli-Resta, also working in Ray Lund's lab, showed that the projection to the host lacked the topographic organization needed for spatial vision [6]. In addition, I went on to show that the formation of new graft-host connections in adult rats—although possible—was notably more difficult to achieve [4]. Furthermore, it was clear from the outset that the intracranial retinal graft model was good for proof-of-principle but did not represent a potential treatment strategy. A breakthrough of sorts was needed to propel neural repair from a provocative lab result to a strategy with potential clinical relevance. That breakthrough turned out to be the ability to isolate and culture neural progenitor cells.

Upon completing clinical training in ophthalmology, I became aware of the work of Masayo Takahashi, then a research fellow in the lab of Fred H. Gage at the Salk Institute. She had transplanted neural progenitor cells to the eyes of neonatal rats and found surprising levels of integration of these cells into the immature retina [7]. This seemed to provide a new cell-based tool for pursuing the overarching goal of retinal reconstruction; however, there were at least two major challenges to clinical translation identified by this work. First, neonatal rats are at a developmental stage roughly equivalent to the fetal human and therefore not representative of a currently treatable patient demographic. Unfortunately, the profound donor cell integration seen in newborn rats was not replicated in adult recipients. A separate highly problematic challenge was that, although the differentiating hippocampal progenitors appeared to replicate a wide range of retinal neurons (included expression of some cell-appropriate markers), the glaring exception was the critical photoreceptor cell type (in this case, rods) that was the source of the visual deficits seen in many retinopathic conditions. Nevertheless, the clinical potential of the neural progenitor cell type had seized my attention and I was determined to confront the remaining challenges.

The challenge of achieving integration of neural progenitors into the mature mammalian retina turned out

Neural Surface Antigens. http://dx.doi.org/10.1016/B978-0-12-800781-5.00014-1

not to be a problem after all. Although such progenitors do not migrate into the mature retina of healthy recipients, in experiments with Gage and Mike Young, we showed that these cells migrated extensively into the degenerating retina of the Royal College of Surgeons (RCS) rat [8]. This result was gratifying from a translational perspective, since donor cell selectivity for targeting a diseased tissue was in reality a remarkably helpful property. However, the second challenge, namely generating photoreceptors from brain-derived progenitors, remained unsolved. It appeared to be the case that hippocampal, adult brain-derived progenitors were developmentally restricted such that they had lost the ability to reliably generate these retina-specific cells.

To address this, it was necessary to derive progenitor cells from the immature neural retina. The resulting retinal progenitor cells (RPCs) from mice were capable of differentiation into photoreceptor cells and in vivo cell replacement [9]. Even with this achieved, there remained the translational challenge of deriving human RPCs and conducting the full gamut of preclinical studies, an effort that continues up to the present. But now that I have provided the rationale and background story, I would like to consider in greater detail our original work exploring surface markers on neural progenitor populations from the brain and, later, the retina.

14.1 SURFACE MARKERS ON NEURAL PROGENITOR CELLS

Surface markers hold a prominent role in stem cell studies for a number of reasons, which include not only the role played by these molecules in basic cellular physiology but also various practical aspects of working with these cells. These include methods for isolation, proliferation, and identification of cells before and after differentiation. For instance, the very ability to study neural stem and progenitor cells in culture without inadvertently inducing differentiation (as when using serum), or without immortalization (which alters the natural properties of the cells), hinges on the use of recombinant growth factors to maintain cellular mitosis and viability [10]. In this way, the use of the factors EGF and/or bFGF allows for passive selection of NPCs under serum-free conditions, and therefore the presence of the associated receptors (e.g., EGFR) on NPCs is not particularly surprising [11]. The now classic marker for neural progenitor cells is the cytoskeletal protein nestin, its very name derived from "neural stem cell protein" [12]. Although arguably not perfect in its specificity, from a practical standpoint nestin expression remains extremely helpful, if not essential, to the identification of these cell types, and is particularly useful when the cells are known to be of neural tissue origin. Various exceptions not withstanding, the major utilitarian challenge related to nestin as a marker, from a cell product standpoint, is that it is intracellular and

therefore problematic to assess in living cells. This also means that it is not very useful for prospective cell selection. In my work, a major motivation for looking at surface markers on NPCs was the possibility of identifying epitopes that could be used for specific identification, and hopefully active selection, of this cell type.

This strategy had been well established by the Weissman lab when they identified and used the surface marker CD34 to select hematopoietic progenitor cells from mixed samples. The approach has been so successful that it remains in current clinical use, and these cells are routinely referred to as "CD34 cells." I was interested in pursing this avenue with neural progenitors and teamed up with an experienced cord blood group at Children's Hospital of Orange County (CHOC). At this point in time, the application of flow cytometry to cells of neural lineage was relatively novel, with little in the way of publications, particularly with respect to human cells, with a notable exception [13].

Fluorescence activated cell sorting (FACS, colloquially "flow") requires that the cell product be passed in through a tiny nozzle at high speed, one cell at a time, while a laser beam interacts with each passing cell (for details please see Chapter 2 of this volume). Without going into detail here, the samples being routinely processed in the flow cytometry core at CHOC were of hematological origin and, as such, were naturally in the form of a single-cell suspension (once the potential for clotting was eliminated). Neural progenitors, on the other hand, are comparably sticky, adherent cells that either form a monolayer in culture or cluster into tight aggregates referred to as *neurospheres*. Therefore, methodologically, an immediate challenge for us was the preparation of neural progenitors into a viable single-cell suspension that could be run as samples for FACS, without losing target surface antigens in the process. Short-term enzymatic treatment plus quench turned out to be rapid and effective. We prefer trypsin; however, protocols vary considerably among labs. With practice, the problem of inadvertently lysing cells and creating a DNA "gumball" can be avoided. An alternate approach is to use a nonenzymatic dissociation buffer.

While comprehensive screening panels are available today, the experimental design we used at the time was the shotgun approach in which we cultured large quantities of vendor-sourced human NPCs and then tested every single antihuman surface marker in the repertoire of the flow (FACS) service at CHOC. This antibody selection was very much directed toward known markers of hematological and oncological significance, therefore the hits were relatively few and far between; however, there were a number of positive results that ultimately proved of considerable interest.

Arguably the most exciting of the findings was in the presence of sugar moieties as epitopes. These included CD15 and GD2 ganglioside [14]. There were several considerations that made these targets exciting. First, they would not have been identified using the typical transcriptome screening, thus justifying the labor-intensive FACS-based approach. Second, these kinds of nonprotein moieties were known to play a role in neural development, as evidenced by the known distinction between neural cell adhesion molecule (NCAM) and polysialated (PSA) NCAM, the later frequently associated with immaturity in a neural lineage population (e.g., [15]). Also, being poorly unrecognized up to that point, they had potential as cell selection tools not yet in the public domain.

14.2 CD15

The CD15 marker, also known as the Lewis X (Le^x) antigen or SSEA-1, had previously been identified as important to neural development in seminal studies by the late Paul Patterson of the California Institute of Technology (Caltech), using the novel FORSE-1 antibody, which was eventually determined to recognize the same antigen [16]. CD15 is a fucosyl moiety (specifically, 3-fucosly-N-acetyllactosamine) that is a posttranslational modification of some surface proteins. The enzyme responsible for this process is fucosyltransferase 4 (FUT4). The same fucosyl moiety can also be found on glycolipids and proteoglycans, and is thought to play a role in adhesion, migration, chemotaxis, immune recognition, and phagocytosis.

In the clinical hematology-oncology setting, CD15 is an important marker for neoplastic Reed Sternberg cells, hence useful for the diagnosis of Hodgkin's lymphoma. Using this same antibody to probe neural progenitor cells using FACS, we found substantial expression of this marker on their surface [14]. Because CD15 is a carbohydrate, instead of a strictly protein epitope, the same antibodies might also be used to verify expression of the marker on murine neural progenitor cells, hence coming full circle back to Professor Patterson's earlier neurodevelopmental studies in rodents.

During the course of our work, we learned that Alexandra Capela, then working with Sally Temple, was also investigating this epitope on neural stem cells [17]. CD15 is now widely recognized as being expressed on immature cells, including a range of CNS progenitors [18], including those from the retina [19].

14.3 CD133

Of note, during a similar timeframe, the more widely recognized marker CD133 was identified elsewhere by another group using FACS, as reported by Nobuko Uchida and colleagues, working for Stem Cells Inc. [20]. This work led directly to a patented method of neural stem cell isolation currently used to generate clinical product, as used by that company in current trials in diseases of the brain, retina, and spinal cord (http://www.stemcellsinc.com/Therapeutic-Programs/Overview.htm). This marker also has proven useful in human neural stem cell (NSC) identification and isolation, and may enter clinical routines in the context of CNS tumor subsets such as medulloblastoma or glioblastoma (see Chapters 17 and 18).

14.4 GD2 GANGLIOSIDE

During the process of testing numerous antibodies of clinical relevance to heme-onc, we quickly came across positivity for a nonprotein epitope, namely GD2 ganglioside [14]. Gangliosides are sialyated glycosphingolipids composed of ceramide and an oligosaccharide (see Chapter 13). These molecules are found in the plasma membrane, concentrated in lipid rafts (e.g., [21]). Gangliosides were first discovered in the brain, which is a particularly rich source of these molecules. Functionally, they appear to contribute to signal transduction. GD2, in particular, has two sialic acid (NANA) groups and is used clinically as a marker for circulating neuroblastoma cells. It is also found on melanoma cells [22].

Using immunocytochemistry (ICC), we saw very distinctive labeling of human retinal progenitor cells with this marker, exhibiting a spotlike surface pattern suggesting the previously reported concentration in lipid rafts [19]. These cells were not obtained from aborted tissue, but from developmentally more mature tissue obtained from premature infants. Unfortunately, the particular antibody that we used in these studies has since become unavailable, and alternate products and cells have yielded more variable results. At this point, the role and specificity of GD2 in neural development remains relatively unexplored and deserves further investigation, although there is a report in the literature in which GD2 expression was associated with upregulation of the proneuronal transcription factors MASH1 and neurogenin1 [23].

14.5 TETRASPANINS: CD9 AND CD81

In terms of surface proteins, it was not particularly surprising that NCAM was identified on neural progenitors, but it was more interesting that relatively novel (at the time) tetraspanins were represented, both on the murine and human neural progenitors. The specific tetraspanins identified were CD9 and CD81 [14], and although both were present on progenitors from both species, the mouse cells showed higher CD9 expression, while the human cells showed higher expression of CD81. As will become evident, this was not to be the only time that we found species-specific differences and variations in marker expression on the equivalent CNS progenitor cell types.

As the name implies, tetraspanins (or tetraspans) have four transmembrane domains, thus share a distinct property of crossing the plasma membrane four times. The sequences tend to be highly conserved, and both the N- and C-termini are intracellular, while two unequal loops extend extracellularly. The tetraspanins are thought to bind to adjacent membrane proteins, including each other, and to play a role in adhesion, migration, and proliferation [24], i.e., typical neural progenitor behaviors, when properly regulated, as seen with the previous markers discussed.

14.6 MHC

Of particular interest was the expression on neural progenitors of surface antigens comprising the major histocompatibility complex (MHC), because these molecules play a critical role in graft rejection, particularly the direct involvement of class II antigens in initiating type IV (cell-mediated) hypersensitivity reactions. Therefore, MHC expression might be predictive of immune tolerance and survival of neural progenitors as allografts. This story proved to be complex and in some ways varied, although that might have been expected, based on the range of seemingly conflicting data present in the prior literature.

When we started this work, there was a degree of controversy regarding the expression of MHC by cells of the CNS. Without getting into the specifics, my impression of the general consensus was that neurons, and most glia, do not generally express MHC molecules, at least under normal conditions in vivo. This basic notion is at least consistent with (but not necessarily casual to) the so-called immune privilege exhibited by the brain [25] and routinely seen in the sustained survival of embryonic neural tissue transplants to that CNS compartment ([26]; JAMA). On the other hand, there were not infrequent reports suggesting MHC expression in the CNS, including during development (e.g., [27,28]). What we found by examining cultured CNS progenitors was a somewhat complex story, consistent and reproducible for a given cell type under strictly defined conditions, yet highly variable among conditions and species.

The first cells we examined with FACS were the adult hippocampal progenitor cells (AHPCs) from Rusty Gage's lab. These rat brain progenitors consistently expressed low but well-defined levels of MHC class I without detectable class II [29]. Because of the key role played by the recognition of class II by host CD4+ cells in graft rejection, the lack of class II expression was at least consistent with the immune tolerance shown for these cells as allografts to the vitreous cavity of allogeneic recipients [7,8]. The expression of MHC class I was curious in light of the prior reports suggesting lack of MHC in the developing brain, but the low level of the expression seemed at least vaguely consistent with the general lack of immunogenicity exhibited by the cells following allo-transplantation.

We next examined neural progenitors from fetal human brain. Unlike the rat progenitors, these cells were strongly positive for MHC class I, with no detectable MHC class II [14].

As part of that study, we compared the human cells to mouse brain progenitors and found that the murine cells also did not express class II but, as if to make matters more interesting, they did not express detectable class I either. This striking difference versus the human, rat, and mouse cells in terms of class I (high versus low versus nil) suggested to us that perhaps some of the confusion on the literature surrounding MHC expression in the nervous system might relate to underappreciated differences between species. Clearly, the topic deserves further attention.

But here it is important to remember that all of our data were generated from cell cultures, whereas much of the preceding work had been performed on histological preparations. Of course it is always prudent to view in vitro data as distinct from in vivo, until proven otherwise, and this turned out to be important in this instance as well. That said, in vitro work in this instance is quite relevant in its own right, defining as it does the MHC expression characteristics of a potential therapeutic cellular agent. In fact, the results on cultured cells were not random and quite consistent for a given cell type of a particular species, under standard proliferation conditions in culture.

To experimentally model a pro-inflammatory microenvironment, we exposed the cultured cells to species-specific interferon gamma and then reexamined them for changes in MHC expression. This treatment was sufficient to induce expression of MHC antigens in all cells tested, regardless of primary CNS source tissue or species. Both class I and class II expression could be induced in this manner, albeit not always equally, and this induction was reversible upon withdrawal of interferon gamma. For cells with elevated class I at baseline, like human and rat, the expression rose to higher levels in response to the interferon stimulation, but then would progressively revert to baseline once the interferon was no longer present in the medium. For the mouse progenitor cells, MHC expression also reverted to baseline, in that case nil [30].

These studies showed that baseline MHC class I expression reflected a stable set point for progenitors of a given species, yet was dynamically modulated in response to certain environmental cues like interferon gamma. Although these data were obtained in vitro, they suggested that inflammatory cues present in the cellular microenvironment could influence MHC expression, either in culture or post-transplantation, thus raising potential concerns for methods of neural progenitor manufacturing, as well as adding to the numerous sources of variability reported for MHC expression in the CNS.

At this point, the above research had begun to provide some practical insights into potential clinical relevance. An example of a manufacturing concern would be the use of LIF in cell culture protocols, since LIF can also stimulate MHC expression by neural progenitor cells (P Schwartz, S Huhn; personal communication). CNS progenitor cells exposed to LIF can have elevated class II expression and therefore present an elevated rejection hazard following transplantation. This could necessitate the use of immune suppression in patients who might not otherwise require it. An example of a treatment-related concern would be the transplantation of neural progenitor product to an actively inflamed site, such as multiple sclerosis. The survival of transplanted neural progenitors might be compromised by the induction of MHC class II in response to pro-inflammatory cytokines present in inflamed host tissue.

14.7 FAS/CD95

Another marker that we discovered on the surface of neural progenitors was Fas, also known as CD95, the apoptosis antigen 1, or the "death receptor," is a surface molecule of the tumor necrosis factor receptor (TNFR) superfamily. When stimulated by Fas ligand (FasL; CD95L), Fas triggers a major apoptotic pathway leading to cell death. Fas is known to play an important role in the immune system and has been implicated in tumor growth (see Chapter 4). Interestingly, the reported tumor-associated activity of Fas may be complex and varied, such that expression of Fas on mouse cells could tend to be tumor promoting [31], whereas on human cells it is thought to be tumor suppressing, although not invariably [32].

Interestingly, the expression pattern of Fas on neural progenitor cells very closely mirrored the levels of MHC class I across the three species studied, i.e., high for human, low for rat, and nil for mouse [33]. In that way, Fas (as well as MHC class I) expression may correlate with positive tumor suppression activity, in a species-dependent manner, in these actively proliferating progenitor cell populations. Again, correlation does not necessarily imply causation, and more work will be needed to explore these possibilities. Overall, the role of Fas on neural progenitors has received limited attention since our initial observations; however, an interesting paper reported that stimulation of this "death receptor" actually promoted cell survival and neuronal specification [34] (see Chapter 4). It seems that ongoing exploration of the role of Fas on neural progenitor cells could reveal additional surprises.

14.8 RETINAL PROGENITOR CELLS

In addition to the above work examining surface marker expression on brain-derived neural progenitors, we have taken a particularly strong interest in the derivation of similar progenitors from the immature neural retina. In particular, we have examined the same markers on cultured retinal progenitor cells from mouse [9] and human [19] tissue. The findings from these studies show that retinal progenitors from a given species display essentially identical expression patterns as their brain-derived counterparts for the markers

described in detail above, specifically including MHC class I and Fas. In addition, retinal progenitors do not express MHC class II unless stimulated to do so.

This is not to imply that retinal and other CNS progenitor cells are identical. The populations can be distinguished based on differentiation patterns. Retinal specification leads to an increased propensity toward photoreceptor differentiation and expression of proteins such as recoverin and rhodopsin [9], whereas progenitors from extraretinal loci of the neuraxis are capable of giving rise to oligodendrocytes, a glial cell of nonretinal origin. However, the classic surface marker profile, particularly including MHC antigens, generally looks quite similar, on a species-specific basis.

We have generated initial lots of hRPC product for use in proposed clinical trials for the blinding orphan disease retinitis pigmentosa (RP). Based on the foundation laid down by the work described above, the process currently employed involves the use of flow cytometery to characterize marker expression by the cells. Multiple markers are examined and results to date show a high level of reproducibility between lots. The manufacturing of hRPC product for human use poses additional challenges that need to be met, both initially and over the longer term, as the likelihood of increased demand leads to the necessity of scale-up and standardization of production methods [35].

14.9 FURTHER STUDIES OF SURFACE MARKERS

Flow cytometry is a powerful, but labor- and reagent-intensive, method for examining surface markers in vitro. The use of other techniques can greatly expedite the exploration of surface marker profiles, specifically including ICC but also PCR and microarray analysis. Whole-genome microarray can provide a rapid assessment of surface markers expressed by a population of cells, with the caveats that various epigenetic and post-translational modifications, some of which are quite relevant, can be missed entirely. Also, the above RNA-based methods are not particularly helpful in identifying subpopulations of cells, since there is only a single readout for the cellular population as a whole. Therefore, transcriptome-based techniques are very powerful for initial screening; however, FACS remains the gold standard for intensive characterization of surface markers expressed within a cell population. However, the hopeful notion that there might exist a single specific surface marker for each stage or cell type during development is not supported by available evidence.

14.10 CONCLUSIONS

Our early studies of neural progenitors using FACS helped to define the expression landscape of a number of important surface markers, including MHC antigens, and identify certain commonalities and differences. In general, it emerged that marker expression for neural progenitors from brain and retina was similar for a given species. It was also seen that neural progenitors from either tissue, and from multiple species, did not express MHC class II in culture, under standard proliferation conditions. This commonality is of interest in the context of the immune tolerance exhibited by these cells as allografts.

In addition, we found that relative levels of some surface markers, such as tetraspanins CD9 and CD81, varied among species. Most dramatically, we found that MHC class I and Fas varied considerably among mammalian species, and expression levels of these markers could be dynamically modified by treatment with certain cytokines. Together, these findings show that cultured neural progenitor populations exhibit stable, well-defined, and reproducible expression patterns for these particular markers, and that these profiles may therefore be useful for characterization of cultured cell products. It will be interesting to see if interspecies differences in MHC class I and Fas correspond to phylogenetic considerations, or if in vitro cell data correspond to developmental expression patterns of these markers in vivo.

In the broader view, we now have a firm initial grip on the expression of certain key markers by neural progenitors; however, much remains to be learned. Nevertheless, various types of neural progenitor cell products are currently being manufactured for testing in humans, and a number of early trials are already underway, with more about to be initiated. The results of the initial bench work with stem and progenitor cells performed over that last several decades are, hopefully, now in the process of achieving clinical relevance.

REFERENCES

[1] Brüstle O, McKay RD. Neuronal progenitors as tools for cell replacement in the nervous system. Curr Opin Neurobiol 1996;6(5):688–95.

[2] McLoon SC, Lund RD. Development of fetal retina, tectum, and cortex transplanted to the superior colliculus of adult rats. Comp Neurol 1983;217(4):376–89.

[3] Klassen H, Lund RD. Retinal transplants can drive a pupillary reflex in host rat brains. Proc Natl Acad Sci USA 1987;84:6958–60.

[4] Klassen H, Lund RD. Retinal graft-mediated pupillary responses in rats: restoration of a reflex function in the mature mammalian brain. J Neurosci 1990;10(2):578–87.

[5] Klassen H, Lund RD. Parameters of retinal graft-mediated responses are related to underlying target innervation. Brain Res 1990;533(2):181–91.

[6] Galli L, Rao K, Lund RD. Transplanted rat retinae do not project in a topographic fashion on the host tectum. Exp Brain Res 1989;74(2):427–30.

[7] Takahashi M, Palmer TD, Takahashi J, Gage FH. Widespread integration and survival of adult-derived neural progenitor cells in the developing optic retina. Mol Cell Neurosci 1998;12(6):340–8.

[8] Young MJ, Ray J, Whiteley SJ, Klassen H, Gage FH. Neuronal differentiation and morphological integration of hippocampal progenitor cells transplanted to the retina of immature and mature dystrophic rats. Mol Cell Neurosci 2000;16:197–205.

[9] Klassen HJ, Ng TF, Kurimoto Y, Kirov I, Shatos M, Coffey P, et al. Multipotent retinal progenitors express developmental markers, differentiate into retinal neurons, and preserve light-mediated behavior. Invest Ophthalmol Vis Sci 2004;45:4167–73.

[10] Reynolds BA, Weiss S. Generation of neurons and astrocytes from isolated cells of the adult mammalian central nervous system. Science 1992;255(5052):1707–10.

[11] Schwartz PH, Bryant PJ, Fuja TJ, Su H, O'Dowd DK, Klassen H. Isolation and characterization of neural progenitor cells from postmortem human cortex. J Neurosci Res 2003;74(6):838–51.

[12] Lendahl U, Zimmerman LB, McKay RD. CNS stem cells express a new class of intermediate filament protein. Cell 1990;60(4):585–95.

[13] Hulspas R, Quesenberry PJ. Characterization of neurosphere cell phenotypes by flow cytometry. Cytometry 2000;40(3):245–50.

[14] Klassen H, Schwartz MR, Bailey AH, Young MJ. Surface markers expressed by multipotent human and mouse neural progenitor cells include tetraspanins and non-protein epitopes. Neurosci Lett 2001;312(3):180–2.

[15] Szele FG, Chesselet MF. Cortical lesions induce an increase in cell number and PSA-NCAM expression in the subventricular zone of adult rats. J Comp Neurol 1996;368(3):439–54.

[16] Allendoerfer KL, Magnani JL, Patterson PH. FORSE-1, an antibody that labels regionally restricted subpopulations of progenitor cells in the embryonic central nervous system, recognizes the Le(x) carbohydrate on a proteoglycan and two glycolipid antigens. Mol Cell Neurosci 1995;6(4):381–95.

[17] Capela A, Temple S. LeX/ssea-1 is expressed by adult mouse CNS stem cells, identifying them as nonependymal. Neuron 2002;35(5):865–75.

[18] Yanagisawa M, Yu RK. The expression and functions of glycoconjugates in neural stem cells. Glycobiology 2007;17(7):57R–74R.

[19] Klassen H, Ziaeian B, Kirov II, Young MJ, Schwartz PH. Isolation of retinal progenitor cells from post-mortem human tissue and comparison with autologous brain progenitors. J Neurosci Res 2004;77(3):334–43.

[20] Uchida N, Buck DW, He D, Reitsma MJ, Masek M, Phan TV, et al. Direct isolation of human central nervous system stem cells. Proc Natl Acad Sci USA 2000;97(26):14720–5.

[21] Ohmi Y, Tajima O, Ohkawa Y, Yamauchi Y, Sugiura Y, Furukawa K, et al. Gangliosides are essential in the protection of inflammation and neurodegeneration via maintenance of lipid rafts: elucidation by a series of ganglioside-deficient mutant mice. J Neurochem 2011;116(5):926–35.

[22] Dobrenkov K, Cheung NK. GD2-Targeted immunotherapy and radioimmunotherapy. Semin Oncol 2014;41(5):589–612.

[23] Jin HJ, Nam HY, Bae YK, Kim SY, Im IR, Oh W, et al. GD2 expression is closely associated with neuronal differentiation of human umbilical cord blood-derived mesenchymal stem cells. Cell Mol Life Sci 2010;67(11):1845–58.

[24] Hemler ME. Tetraspanin proteins promote multiple cancer stages. Nat Rev Cancer 2014;14(1):49–60.

[25] Wenkel H, Streilein JW, Young MJ. Systemic immune deviation in the brain that does not depend on the integrity of the blood–brain barrier. J Immunol 2000;164(10):5125–31.

[26] Gill 3rd TJ, Lund RD. Implantation of tissue into the brain. An immunologic perspective. JAMA 1989;261(18):2674–6.

[27] Raedler E, Raedler A. Developmental modulation of neuronal cell surface determinants. Bibl Anat 1986;27:61–130.

[28] Huh GS, Boulanger LM, Du H, Riquelme PA, Brotz TM, Shatz CJ. Functional requirement for class I MHC in CNS development and plasticity. Science 2000;290(5499):2155–9.

[29] Klassen H, Imfeld KL, Ray J, Young MJ, Gage FH, Berman MA. The immunological properties of adult hippocampal progenitor cells. Vis Res 2003;43(8):947–56.

[30] Hori J, Ng TF, Shatos M, Klassen H, Streilein JW, Young MJ. Neural progenitor cells lack immunogenicity and resist destruction as allografts. Stem Cells 2003;21(4):405–16.

[31] Chen L, Park SM, Tumanov AV, Hau A, Sawada K, Feig C, et al. CD95 promotes tumour growth. Nature 2010;465(7297):492–6.

[32] Fouqué A, Debure L, Legembre P. The CD95/CD95L signaling pathway: a role in carcinogenesis. Biochim Biophys Acta 2014;1846(1):130–41.

[33] Klassen H. Transplantation of cultured progenitor cells to the mammalian retina. Expert Opin Biol Ther 2006;6(5):443–51.

[34] Corsini NS, Sancho-Martinez I, Laudenklos S, Glagow D, Kumar S, Letellier E, et al. The death receptor CD95 activates adult neural stem cells for working memory formation and brain repair. Cell Stem Cell 2009;5(2):178–90.

[35] Sheu J, Klassen H, Bauer G. Cellular manufacturing for clinical applications. Dev Ophthalmol 2014;53:178–88.

Chapter 15

Multimarker Flow Cytometric Characterization, Isolation and Differentiation of Neural Stem Cells and Progenitors of the Normal and Injured Mouse Subventricular Zone

Krista D. Buono[1,2], Matthew T. Goodus[1], Lisamarie Moore[1], Amber N. Ziegler[1] and Steven W. Levison[1]

[1]Department of Neurology and Neuroscience, New Jersey Medical School, Rutgers University-New Jersey Medical School, Newark, NJ, USA;

[2]ICON Central Laboratories,123 Smith Street, Farmingdale, NY

15.1 INTRODUCTION

Studies performed over the past 20 years on neural stem cells and progenitors have opened up new possibilities for treating neurological diseases and for regenerating the central nervous system (CNS) after injury. But our understanding of the types of stem cells and progenitors that produce the CNS and then persist into adulthood has lagged behind studies on other organs in part due to the complexity of the neural precursor (NP) pool but also due to the lack of widely accepted methods to isolate and analyze NPs. Until recently, there have been no cell surface markers known to reside on specific types of neural precursors that could be used for flow cytometric or fluorescence activated cell sorting (FACS) studies. Indeed, there is still no unique surface epitope identified to be exclusively expressed by neural stem cells (NSCs). Therefore, investigators have used cell surface markers in combination to begin to establish the variety of neural precursors that reside in the brain's germinal zones and to understand how they produce new neurons and glia. In this chapter, we will review progress made in discerning the different types of precursors that reside within the brain as assessed by flow cytometry and FACS. Based upon our trials and errors in establishing a multimarker flow cytometry panel, we will discuss the technical considerations that complicate the application of flow cytometry and FACS to studies of the developing and adult nervous tissues. We predict that with additional refinements to the current methods being used, neuroscientists will make great strides in understanding the complexities of the brain's germinal matrices and their precursors.

15.2 NEURAL STEM CELLS AND PROGENITORS

There are many biologically distinct types of stem cells. Stem cells are characterized by their "potentiality," which refers to the number of distinct cell types that a stem cell can produce. For example, the inner cell mass of the blastocyst comprises embryonic stem cells (ESCs), which are pluripotent cells because they have the capacity to produce all of the cells derived from any of the three embryonic germ layers. Most relevant to the work presented in this chapter are *multipotent stem cells*. Multipotent stem cells, also known as *somatic stem cells* or *adult stem cells*, are found throughout the body after embryonic development and are limited in the differentiated cell types that they can produce. They persist to provide a continuous supply of new cells in tissues where there is extensive cellular turnover such as the skin, gut, and hematopoietic system. Adult stem cells also provide a reserve for the replacement of cells in the event of disease and injury.

There are specific characteristics that a cell should fulfill to be classified as a stem cell. It must: (1) be quiescent or slowly dividing; (2) be capable of indefinite self-renewal; and (3) have the ability to generate all of the cell types of the organ or tissue in which it resides. The nature of this definition presents a challenge in identifying stem cells in vivo—their identity must be verified through confirmation of their functional capabilities. In fact, there is controversy on whether the brain after birth has NSCs or whether the putative stem cells are long-lived primitive progenitors. This controversy is based on the inability of the NSCs to generate all of the cells within the brain, particularly each different type of neuron [1], and the fact that the numbers

Neural Surface Antigens. http://dx.doi.org/10.1016/B978-0-12-800781-5.00015-3

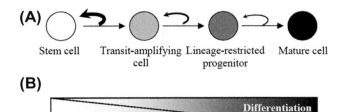

FIGURE 15.1 Two models of a stem cell developmental restriction.
(A) Simplified depiction of discreet phase model of cell development. Arrow width represents self-renewal capacity of the cell. (B) Continuum model of cell development depicting gradual loss of proliferation potential in favor of differentiated functionality.

of putative NSCs decline with age [2,3]. It has been counter-argued that NSCs surely do exist but that they have not been properly identified.

Through unique patterns of cell divisions, stem cells generate pools of rapidly dividing cells with finite lifespans called *transit-amplifying precursors* or *intermediate progenitors*. Progenitors are defined as cells possessing: (1) short cell cycles, (2) limited self renewal, and (3) limited potentiality. As these cells divide and mature, transit-amplifying cells gradually give rise to committed progenitors (lineage-restricted precursors) that mature into functional noncycling cells (Figure 15.1).

As reviewed in Chapter 1, the CNS is established from a spatially defined region of embryonic ectoderm known as *neuroectoderm*, which undergoes significant growth and becomes specified dorsally and ventrally to produce distinct sets of precursors. The most primitive cells persist in a region defined as the ventricular zone (VZ). These NPs will produce almost all of the cells of the adult brain and spinal cord [4–7]. Not long after the first neurons emerge from the VZ, a second proliferative population becomes discernable at the border between the VZ and the cell sparse intermediate zone. Whereas the proliferative fraction and prominence of the VZ declines rapidly, the subventricular zone (SVZ) cell population expands exponentially during the latter third of prenatal development. In the E16 mouse, over 90% of the SVZ cell population is dividing, whereas the majority of the cells in the VZ are leaving the cell cycle [8]. Thus, as the VZ regresses, the SVZ achieves prominence, peaking in size during early postnatal development [9–11].

In the prenatal forebrain, the germinal cells in the SVZ are positioned adjacent to the VZ and are easily recognizable. The SVZ is densely populated and its cells can be identified as far caudally as the third and fourth ventricles. The SVZ is not prominent in more caudal regions of the brain. Unlike VZ cells, SVZ cells show no evidence of interkinetic nuclear movements during the cell cycle, and their processes lack any topographic orientation [8]. In the mouse, the SVZ peaks during the first week after birth, after

which the SVZ begins to decrease in size [9]. However, in all mammals, a residual, mitotically active SVZ persists into adulthood.

15.3 FLOW CYTOMETRIC STUDIES ON NEURAL PRECURSORS – TECHNICAL CONSIDERATIONS

Certainly, deciding which phenotypic markers will be analyzed is critical for successfully implementing flow cytometry or FACS, but there are other technical aspects that are just as important for the ultimate success or failure of a flow cytometry study on neural precursors derived from primary tissues. Other variables that need to be optimized include methods for tissue dissociation, excluding dead cells and debris, and designing and implementing a reproducible gating strategy.

The CNS is not easily digested into a single cell suspension, which is critical for flow cytometric and FACS analyses. Neuroscientists use an assortment of nonenzymatic and enzymatic methods for tissue dissociation. Enzymes that are typically used include papain, trypsin, Accutase, and collagenase, with or without dispase or other neutral proteases. As these are proteases and as many of the antigens used to differentiate among neural precursors are proteins, the composition of the enzyme solution chosen for tissue dissociation can greatly affect the outcome of a flow cytometry study. For example, it was not initially appreciated that CD24 is sensitive to enzymatic degradation [12,13]. Several studies reported that NSCs and primitive progenitors express low levels of CD24 [12–14], whereas others have reported that NSCs are CD24 negative [15,16]. But a review of the methods used for tissue dissociation is enlightening. In those studies that reported that CD24 was absent, papain was used to dissociate the cells into a single-cell suspension. By contrast, those studies that reported low levels of CD24 used more gentle enzymes, such as collagenase or collagenase plus dispase (in Liberase; [12,13]) or collagenase plus hyaluronidase [14]. Therefore, one must be careful to use the least destructive enzymes to preserve these cell surface antigens and to avoid false negative results when performing flow cytometric studies. On the other hand, when a gentler enzyme is selected, more rigorous mechanical dissociation methods are necessary, which may destroy sensitive cells, leave clumps of intact tissue, and produce a significant amount of debris. Thus, to successfully integrate flow cytometry or FACS into a neuroscience research program, one must wisely choose the most appropriate enzyme mixture, and optimize the concentration used and digestion time based on the species, region of CNS, stage of maturation, and surface antigens being analyzed (see Chapter 2).

Unfortunately, cells are destroyed and debris is created during the tissue dissociation process even when gentle dissociation methods are employed. This can confound

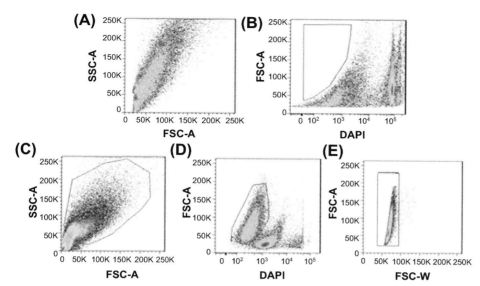

FIGURE 15.2   Flow cytometric gating profile of debris, dead cells, and doublets. (A,B) Cells were heated to 95 °C for 3 to 5 min to produce a 100% dead cell population. The cells were then filtered and then placed on ice. (C–E) Plots of dissociated spheres labeled with isotype controls and 4′,6-diamidino-2-phenylindole (DAPI). (A) Dead cell scatter. (B) Dead cell FSC-A versus DAPI; gate represents the absence of live cells. (C) Gate 1: Live cell scatter: SSC-A versus FSC-A. (D) Gate 2: Live cell gating with DAPI exclusion: FSC-A versus DAPI. Gate is of live cells while debris and dead cells are to the right of gate, respectively. (E) Gate 3: Single cell gating to exclude doublets: FSC-A versus FSC-W. These gates were then applied to SVZ samples.

analyses of flow cytometric data because dead cells tend to bind antibodies nonspecifically. A number of density gradient methods have been applied prior to flow cytometry, including Ficoll, Percoll, or sucrose density gradients [15]. We have tried all of these methods, but noted that density centrifugation can increase the extent of cell death and, in our experience, does not sufficiently decrease the amount of debris. Therefore, we currently use large volume washes to help remove debris.

In the absence of physical methods to reduce debris, other techniques must be employed to exclude dead cells and debris from live cells using the flow cytometer. One means by which living cells can be identified is by including viability dyes in the biomarker panels. To date, several DNA intercalating dyes have been used, including 4′,6-diamidino-2-phenylindole [DAPI], propidium iodide [PI] and 7-amino-actinomycin D [7AAD]. Investigators also have used Annexin V to exclude dead cells. Each of the dyes has technical shortcomings, the major one being the instability of the stained cells, especially after fixation and permeabilization. For example, 7AAD has a low binding affinity for DNA. Therefore, cells should be analyzed within an hour after staining. To circumvent the limitations of the DNA binding dyes, amine-reactive viability dyes have been developed for dead cell exclusion. Including a viability dye in a biomarker panel is relatively simple, and improves sensitivity, reproducibility, and accuracy in separating the dead cell population and debris from the live cells [17]. Unlike the DNA-binding dyes, amine-reactive dyes are stable after fixation, and the cells can be rendered permeable to assess intracellular proteins. Their stability provides investigators with flexibility in performing flow cytometric studies, which is especially desirable when using a shared flow cytometer, where instrument time

may be limited. This allows the flow cytometric analysis to be postponed for one or two days after completing the staining. However, in practice we have found that large amounts of debris will interfere with the use of amine-reactive dyes; therefore, DNA-binding dyes are still valuable, especially when analyzing cells directly from the brain [18]. Other considerations that need to be taken into consideration are discussed in Chapter 2. In Figure 15.2, we have provided an example of the gating strategy that we employ to exclude debris, dead cells, and doublets from our flow cytometry analyses.

With the optimization of the flow cytometry method and data analysis, there are two other essential factors that need to be considered when applying the flow cytometry phenotyping results to FACS on nervous tissues. These essential factors are the type of sheath fluid used in sorting neural cells and the media into which these cells are collected. A common sheath fluid is 1X PBS without Mg^{2+} and Ca^{2+}; however, from our experience, this sheath fluid is not optimal for sorting neural precursors, especially NSCs. Our laboratory and other laboratories have found that the survival rate for rare and fragile cells is poor, and cells do not often survive beyond 24 h using a simple ionic sheath fluid. However, using the solution used to isolate the cells for the sheath fluid, i.e., a solution containing 0.6% dextrose and 0.5–2% bovine serum albumin (BSA) will greatly increase the viability and survival of these cells, allowing functional assays to be carried out [13]. The other critical factor in successfully propagating the sorted cells is to optimize the medium for each NP. We found while characterizing the sorted NPs that their growth factor requirements differed [13], which is logical, since they express different growth factor receptors and rely upon the milieu and neighboring cells both in vivo and in vitro to aid in their survival.

15.4 FLOW CYTOMETRY STUDIES OF THE FETAL MOUSE FOREBRAIN

When developmental neurobiologists first became interested in identifying and characterizing the neural stem cell of the CNS, a number of cell surface markers had been identified to be expressed by neural precursors (Table 15.1); however, none were readily available to clearly define those precursors that possessed the features of NSCs. Consequently, several groups used negative selection to enrich for the most primitive neural precursors. Maric et al., in 2003, assembled a cocktail of neuronal markers that included cholera toxin B and tetanus toxin, and neuroglial markers that included A2B5, JONES, O4, and O1, and markers for other cell types that included CD44, Ran-2, CD11b, and CD45 [19]. At that time, there were no available markers for immature astrocytes, which contributed to their decision to sort neural precursors from the early embryonic CNS (e13).

Negative selection on embryonic neural precursors (as well as adult neural precursors) was also employed by Rietze et al., in 2003, to purify mouse NSCs. They used the neurosphere assay as a read-out and found that sorted cells that were greater than 12 µm in diameter by forward scatter and that were negative for both peanut agglutinin and CD24, comprised 63% of the total sphere forming cells. All of the cells in this population were also nestin+ and were self-renewing and multipotent. These cells also failed to express CD30, CD34, CD90.2, CD24, CD117, CD135, and CD31, thus were antigenically unique from other stem cells, most particularly, hematopoietic stem cells. These neural precursors

also differed from the HSCs in that they were bright for Hoechst 33,342, whereas HSCs possess ABC transporters that pump this dye out of their cytoplasm [20].

15.5 STUDIES USING FLOW CYTOMETRY ON THE NEONATAL SVZ

Taking advantage of studies performed on HSCs, Capela et al. (2002) assessed the expression of SSEA-1 (CD-15 or LeX) on the neural precursors of the neonatal SVZ and reported that LeX+ cells shared several features of stem cell. LeX expression in the brain is tightly regulated by the enzyme alpha1, 3-fucosyltransferase 9 (Fut9) [34]. LeX labeled approximately 4% of total SVZ cells, and LeX expression was not coincident with cells expressing ependymal or neuroblast markers, although a fraction of LeX+ cells colabeled with GFAP+ (glial fibrillary acidic protein+). Using BrdU and LeX double labeling, LeX+/GFAP− cells retained BrdU label for over two weeks post-BrdU treatment. Furthermore, FACS purified LeX+ cells made abundant neurospheres (one out of four LeX+ cells), while LeX− cells rarely made neurospheres [16].

Panchision and coworkers combined observations made by earlier investigators to combine CD133, CD-15, CD24, A2B5, and PSA-NCAM to produce a panel that enabled them to identify and enrich for four separate sets of neural precursors from the E13.5 and P2 mouse SVZ. They established that there were precursors that were multipotent and could produce neurons, astrocytes, and oligodendrocytes, that there were precursors that were bipotential and either

TABLE 15.1 Markers Used to Classify Neural Precursors by Flow Cytometry

Marker	Other Names/Specificity	Expression	Citation
CD24	Heat stable antigen (HSA)	Multiple neural precursors	[19]
PNA	Peanut agglutinin	Oligodendrocytes and myelin	[20–22]
CD133	Prominin-1	Stem cells and ependymal cells	[14,23,24]
CD15	Lewis-X, SSEA-1	Neural precursors, some immature astrocytes	[16]
NG2	NG2 proteoglycan	Multipotential progenitors, bipotential and unipotential progenitors	[25,26]
CD140a	PDGFRa	Multipotential, bipotential, and unipotential progenitors	[27–29]
O4	Sulfatide and POA	Late OPCs	[30]
A2B5	Complex gangliosides	Early OPCs and immature NPs	[31]
Cholera toxin B subunit	GM1 ganglioside	Neurons	[19]
Tetanus toxin fragment C	Complex gangliosides	Neurons, oligodendrocyte progenitors	[19]
GLAST	Glutamate transporter	NSCs, astrocytes, subsets of oligodendrocytes	[32,33]

produced neurons and oligodendrocytes, or astrocytes and oligodendrocytes, and that there were precursors that only produced neurons [12].

Extending Panchision's studies Buono et al. (2012) combined CD133 and LeX with CD24, and then added two intermediate progenitor antigens CD140a and NG2, with the prediction that the NSCs would be CD133+/Lex+/NG2−/CD140a− and CD24−. Based on studies by Panchision et al. (2007) we chose to dissociate the tissue using Liberase-DH, which is a gentle enzyme mixture that increased dissociation efficiency, viability, and cell surface antigen retention (particularly CD133 and CD24) [12]. As all of the NPs expressed CD24, CD24 was eliminated from the panel. Debris, dead cells, and doublets were excluded from analyses using DAPI and forward scatter gates. Nonspecific staining was established using isotype controls. A two-gate strategy was then employed to parse the cells into subsets, where the cells were first gated on CD140a versus NG2 and then gated for CD133 versus LeX (Figure 15.3). With this strategy, eight phenotypically defined subsets could be identified. To determine the developmental potential of these eight subpopulations, NPs propagated as neurospheres were separated by FACS, plated onto poly-D-lysine and laminin coated chamber slides at low density, and expanded with growth factors. Initially, all of the subpopulations

were plated with neurosphere-conditioned medium supplemented with EGF and FGF-2. However, some precursors could not be expanded using these growth factors. Since some of these precursors expressed CD140a, which is the receptor for platelet-derived growth factor (PDGF), Buono et al. in 2012 established that two of these precursors could be expanded using the PDGF and FGF-2 (Table 15.2).

However, even using four cell surface markers, our FACS analyses have suggested that there is an even greater diversity of precursors within the subsets that we have defined. Therefore, to further parse the NPs that reside within the SVZ, we have expanded the four-marker panel by adding an additional three cell surface markers recognized by the antibodies O4, A2B5, and antiglutamate aspartate transporter (GLAST) (Table 15.3).

A2B5 was added, as it is found on O-2A progenitors and is retained by those precursors that can differentiate into type 2 astrocytes. The O4 antibody was added, as it is an extraordinarily useful marker for cells in the oligodendrocyte lineage [35]. GLAST or excitatory amino acid transporter 1 (EAAT1) is a glutamate transporter that is expressed by immature astrocytes, NSCs, and some oligodendrocyte progenitors [21,36]. Using this new seven-color panel, NSCs should be CD133+/Lex+/NG2−/CD140a−/A2B5−/O4−/GLAST+. The addition of these three makers

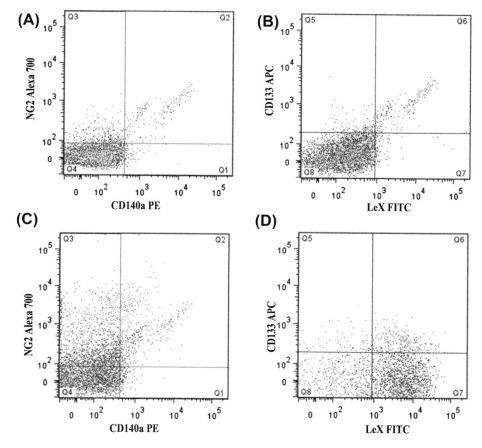

FIGURE 15.3 **Flow cytometric isotype controls and profiles of SVZ neural precursors from 8-month old animals.** (A) Isotype controls of CD140a-PE versus NG2-Alexa 700 and (B) CD133-APC versus LeX-FITC. (C) CD140a-PE versus NG2-Alexa 700 and (D) Q4 from CD140a-PE versus NG2-Alexa 700 plotted for CD133-APC versus LeX-FITC in order to show NSCs (Q6) and MP1 (Q7). Both C and D show example shifts in cell populations for each marker relative to isotype control gates. Cells are from one adult mouse (bilateral SVZs).

TABLE 15.2 Neural Precursor Subsets of the Neonatal Mouse as Defined Using Multipanel Flow Cytometry and FACS Analyses

Antigenic Profile	Developmental Potential	Growth Factor Responsiveness	Designation
CD133+LeX+NG2–CD140a–	Multipotential	EGF and FGF-2	NSC
CD133–LeX+NG2–CD140a–	Multipotential	EGF and FGF-2	MP1
CD133+LeX+NG2+CD140a–	Multipotential	EGF and FGF-2	MP2
CD133–LeX+NG2+CD140a–	Bipotential	EGF and FGF-2	BNAP/GRP1
CD133+LeX+NG2+CD140a+	Multipotential	PDGF and FGF-2 and 0.5% FBS	MP4
CD133–LeX–NG2+CD140a–	Multipotential/Bipotential	PDGF and FGF-2	MP3/GRP2
CD133–LeX+NG2+CD140a+	Multipotential	PDGF and FGF-2	PFMP
CD133–LeX-NG2+CD140a+	Bipotential	PDGF and FGF-2	GRP3

Table Summarizes the Cell Surface Phenotype, Growth Factor Responsiveness and Designation Based Upon Studies by Buono et al., 2012.

TABLE 15.3 Antibody Panels Used for 7-Color Flow Cytometry

Antibody	Antibody	Source	Secondary Antibody	LSRII Configuration BP Filter: Laser
CD133-APC	13A4	eBioscience		660/20: 633 nm
LeX-V450	MMA	BD Bioscience		440/40: 407 nm
CD140a-PE	APA5	BioLegend		575/26: 488 nm
A2B5-Alexa 488	A2B5-105	Millipore		530/30: 488 nm
NG2	Rabbit Polyclonal	Millipore	Alexa 700	710/20: 633 nm
GLAST-biotin	ACSA-1	Milltenyi	Strep-APC-eFluor 780	780/60: 633 nm
O4	Mouse IgM	Supernatant	PerCp-eFluor 710	695/40: 488 nm
DAPI		Sigma		440/40: 350 nm

has allowed us to further divide the SVZ into more than 75 sets of precursors. As this is unmanageable, we have reduced the numbers of groups by only presenting the data from the eight distinctly defined populations that we have previously characterized (Table 15.2), adding their expression of GLAST and A2B5 (Table 15.4).

To determine potentiality, Buono et al. (2012) differentiated the colonies formed from these NPs. Triple immunostaining for βIII-tubulin (Tuj1, for immature neurons), glial fibrillary acidic protein (GFAP) (for astrocytes), and O4 (for oligodendrocytes) revealed that five of the eight subpopulations were multipotential precursors (MPs), yielding neurons, oligodendrocytes, and astrocytes, whereas two subsets were bipotential. In separate studies, we sorted NPs directly from the P11 SVZ and found that seven out of eight of these precursors can be obtained (the MP4 cells were an insignificant population at P11). We found that the progeny that each

SVZ precursor produced was the same, with the exception that NPs with the cell surface phenotype of GRP2s when sorted directly from the SVZ differentiated into astrocytes, oligodendrocytes and neurons, suggesting that this class of precursors comprised two sets of precursors—one that was bipotential and another that was multipotential.

Figure 15.4 depicts a putative lineage scheme constructed from our flow cytometry analyses and cell sorting results. In this lineage diagram, we have relied both upon surface antigen expression and the clonal analyses performed to propose lineage relationships. In addition, we depict four precursors that could not be separated using the four-color panel, but whose existence was revealed by the clonal analyses. For example, the colonies formed by the Lex+/NG2+NPs contained either neurons and astrocytes or astrocytes and oligodendrocytes, which suggested that there are two NPs that are phenotypically similar within this defined population. Similarly, cells that

TABLE 15.4 Antigenic Features of Nine Neural Precursors Defined Using a Seven Antigen Flow Panel

Antigenic Profile	Designation	% of Total	% Glast+ of Total	% A2B5+ of Total
CD133+LeX+NG2–CD140a–	NSC	0.31±0.12	0.21±0.09	0.08±0.02
CD133–LeX+NG2–CD140a–	MP1	1.01±0.41	0.71±0.30	0.13±0.03
CD133+LeX+NG2+CD140a–	MP2	6.12±2.01	5.36±1.81	3.10±1.34
CD133–LeX–NG2+CD140a–	MP3/GRP2	25±4.54	5.87±1.36	5.68±1.59
CD133+LeX+NG2+CD140a+	MP4	3.42±1.68	2.72±1.37	2.98±1.46
CD133–LeX+NG2+CD140a+	PFMP	3.97±0.73	2.86±0.52	3.13±0.57
CD133–LeX+NG2+CD140a–	BNAP/GRP1	9.25±2.34	7.11±1.79	2.99±1.04
CD133–LeX–NG2+CD140a+	GRP3	4.18±0.62	2.06±0.57	2.45±0.57
CD133–LeX–NG2–CD140a–O4+	Late OPCs	0.10±0.04	0.01±0.001	0.02±0.01
	Others	44.48		

were NG2 only are designated as either MP3 NPs or GRP2s. Further studies will be required to validate the existence of these less well-characterized precursors; however, the new data depicted in Table 15.4 provide additional support that these subpopulations exist and that they may be isolated and characterized using the seven-color flow panel.

Neural precursors from the postnatal day one mouse SVZ were analyzed by flow cytometry using a panel of seven antigens. Values represent the percentage of each designated cell type out of the total analyzed population. Since both O4 and A2B5 are mouse IgM isotypes, cells were stained with O4 and then anti-IgM secondary antibody. The cells were then blocked with unconjugated IgMs to occupy any potential binding sites on the secondary anti-IgM before adding other conjugated primary IgMs (LeX and A2B5).

15.6 STUDIES USING FLOW CYTOMETRY ON THE ADULT SVZ

Rietze and Bartlett (2001) were the first investigators to use FACS to isolate NSCs from the adult SVZ. They gated on a population of cells that were larger than 12 μm in diameter and used a combination of CD24 and peanut agglutinin (PNA) and sorted for the negatively stained cells, under the assumption that the NSCs would be low for both of these markers. They found that 80% of the cells in these fractions could form neurospheres. However, this method only isolated a subpopulation of the total NSCs. With the lack of suitable cell surface markers to identify NSCs, several groups attempted to use other physical properties of the cells such as forward scatter, side scatter, autofluorescence, and Hoechst dye exclusion [37–39]. However, these approaches lacked the selectivity needed to enrich for NSCs.

To increase the specificity of their analyses, investigators began to combine FACS with the use of genetically

engineered mice that ex pressed a fluorescent reporter driven by a promoter that was enriched in NSCs. Murayama et al.(2002) used a nestin enhancer-EGFP expressing genetically engineered mouse (GEM) to enable the researchers to follow the NSCs. They found that the putative NSCs varied in their properties from embryonic into the adult brain, and they failed to find a correspondence between the cells that corresponded to nestin enhancer GFP+ cells and the side population based on Hoechst exclusion. Mignone et al. (2004) produced a separate line of nestin enhancer-driven EGFP GEMs to analyze NPs and isolated EGFP+ cells using FACS; however, beyond establishing that they could produce neurospheres, no further characterization was performed [40]. Barraud produced a GFP-expressing mouse where the GFP was driven off of the Sox-1 promoter. The researchers established that these cells could be purified using FACS from E13 brains and transplanted into newborn rat brains, where they generated neurons and glia [21].

Pastrana et al. (2009) used a similar approach, but used GEMs where GFP was driven off of the human GFAP promoter, as studies by Mignone et al. (2004) and Doetsch et al. (2003) had shown that stem cells of the SVZ expressed GFAP [41]. Using the hGFAP-GFP mice, together with fluorescently tagged EGF-ligand and antibodies to CD24, Pastrana et al. (2009) separated adult NSCs from transit-amplifying cells and resident astrocytes by flow cytometry [15]. While this approach is useful for adult mice, neonatal NSCs do not express GFAP since the transition from GFAP- to GFAP+ NSCs occurs during the second week of life; thus, these mice have limited utility for studies of embryonic or neonatal NSCs [42]. Furthermore, those neonatal NPs that express hGFAP produce progeny with diverse cell fates across different regions of the developing forebrain [43]. These data support the view that adult NSCs and neonatal NSCs cannot be identified and isolated using the same

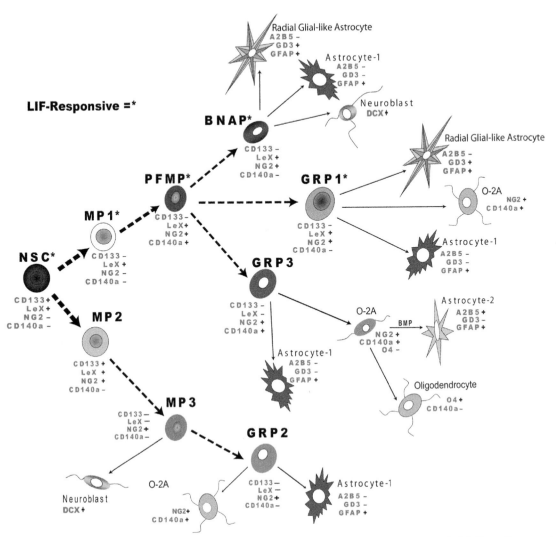

FIGURE 15.4 Neonatal Mouse SVZ Neural Precursors as defined using the four-color neural precursor panel. This figure illustrates the complexity of the neonatal SVZ and proposes a new lineage scheme constructed from our flow cytometry analysis and cell-sorting results. This schematic suggests that there are two coincident, but separate, multipotential progenitor cell lineages: CD133-/LeX/NG2 versus CD133+Lex+/NG2+. The width of the arrows depicts the continual loss of potential to self-renew as the precursors mature. It remains possible that cells can reverse their differentiation states and that there may be more than one NSC within CD133+LeX+ population.

approach, and they substantiate the view that the neonatal germinal zones are significantly more heterogeneous than the adult SVZ.

15.7 USING FLOW CYTOMETRY TO EVALUATE EFFECTS OF CYTOKINES AND GROWTH FACTORS ON NEURAL PRECURSORS

The powerful capability of flow cytometry will enable neuroscientists to gain insights into how cytokines and growth factors affect the proliferation and differentiation of these cells. Studies using propidium iodide or carboxyfluorescein diacetate succinimidyl ester (CFSE) have been used for

many years to study cell cycle kinetics of neural precursors [44], but more sophisticated analyses are beginning to be performed. Using flow cytometry with multimarker panels for neural precursors and oligodendrocyte progenitors, Buono et al. (2012) demonstrated that LIF and CNTF exerted different effects. LIF specifically promoted the growth and maintenance of three phenotypically distinct multipotent precursors; NSCs, MP1s, and PFMPs, and one mixed population of bipotential progenitors (BNAP/GRP1s), while simultaneously repressing the expansion of a separate set of LIF-nonresponsive progenitors. LIF significantly decreased MP2s, MP4s, GRP3s, and GRP2s. By contrast, the related cytokine, CNTF, only modestly increased the proportion of NSCs, but more strongly increased the percentage of

PFMPs. Using a multimarker panel comprising CD133/LeX/CD140a/NG2/O4 LIF was shown to increase the production of immature oligodendrocytes (CD133–CD140a–O4+), whereas CNTF had no effect on the production of premature or immature oligodendrocytes.

We also have used flow cytometry to assess the independent effects of IGF-II and IGF-II analogs on SVZ neural precursors. Using the multipanel markers CD133, LeX, CD140a, and NG2 for neural precursors, Ziegler et al. (2014) examined the effects of IGF-II and an IGF-II analog on cultured neurospheres [45]. Like LIF, IGF-II and the IGF-II analog increased the percentage of NSCs. However, no effect was observed on MP1, MP2, MP3, or MP4 populations. In contrast to LIF, IGF-II and the IGF-II analog reduced the PFMP population. While results from these studies were reflected in the standard neurosphere assay and limiting dilution analysis and transplantation studies (Ziegler stem cells), the flow cytometric analysis revealed more specific information that could not be gained from the other culture assays, and is a higher throughput and more timely analysis compared with the transplantation studies. This flow cytometry panel complements standard techniques in the field, which can aid investigators in the choice of other assays to choose to analyze the effects of various cytokines on stem and progenitor cells. With in vitro culture studies, careful attention must be paid to the culture technique to ensure reproducibility. Experiments must follow identical timelines for media changes, time from last medium change until cell collection and cell density, otherwise these differences will be reflected in the data.

15.8 USING FLOW CYTOMETRY TO EVALUATE EFFECTS OF NEUROLOGICAL INJURIES AND DISEASES ON NEURAL PRECURSORS

Flow cytometry may also be used to evaluate how neural precursors respond to injury and disease. A robust flow cytometric assay will enable investigators to establish with great precision which NP populations are expanding and/or contracting, and which signals are coordinating these shifts to occur. We recently used the multimarker panel described above to evaluate the effects of neonatal hypoxia-ischemia (H-I) and a focal brain injury on the precursors of the SVZ. In our previous studies on the effects of H-I we had shown that this injury induces SVZ cell proliferation. Using the neurosphere assay, we obtained data suggesting that H-I was inducing an expansion in the numbers of NSCs. However, much to our surprise, when we performed H-I multimarker panel to assess which NPs were increased by H-I, we discovered that NSCs (CD133+LeX+NG2–CD140a–) were significantly diminished as a consequence of H-I, whereas there was an expansion of intermediate progenitors; the MP2s and three types of glial restricted precursors [46].

Whereas the NSCs did not increase after neonatal H-I, they do increase as a consequence of a focal neocortical injury. For these studies the SVZs from 14-day old mice were analyzed by flow cytometry at 24 h after producing a controlled cortical impact (CCI) injury to the somatosensory cortex. In response to the cortical injury, there were robust changes within the SVZ that included a proliferation of putative stem cells and progenitors [47]. To gain more detailed insights into which neural precursors were responding to this injury we performed flow cytometry using the four-color flow panel described earlier. This analysis revealed a three-fold increase in the NSCs in the SVZ ipsilateral to the injury when compared to the SVZ in the contralateral hemisphere or to sham-operated and naive controls. CCI also produced a bilateral increase of the MP2s, which were increased four times in the ipsilateral SVZ and 2.5 times in the contralateral SVZ compared to the sham-operated and naive controls. There was also a 2.5-fold increase in the ipsilateral and a 1.5-fold increase in the contralateral SVZ of the GRP2s/MP3s versus the contralateral hemisphere and controls. The percentages of MP1 cells and PFMPs were both decreased 2.5-fold ($p < 0.01$ and $P < 0.05$, respectively) in the ipsilateral hemisphere compared to the contralateral SVZ and controls [47].

Thus, a different set of SVZ precursors responded to CCI versus H-I. Notably, we found an increase in the proportion of NSCs after CCI injury, whereas these cells became less abundant after neonatal H-I [13]. Furthermore, there was a significant expansion of glial precursors after H-I, whereas these precursors are not as significantly amplified after CCI. However, one similarity is the increase in the MP3/GRP2 cells. Interestingly, these cells are EGF responsive precursors, suggesting that increases in either EGF ligands or the EGFR could be responsible for the amplification of these precursors. These examples illustrate the type of information that can be gleaned using this flow panel, demonstrating that this method can be used reliably and with great sensitivity and resolution to compare and contrast SVZ responses to multiple types of injuries and/or treatments.

The four-marker flow panel that we have characterized provides a large amount of data on the relative proportions of NSCs as well as seven multipotential and bipotential progenitors. However, it only provides information about the relative percentages of each of the eight populations examined. It does not establish whether these changes are the result of proliferation after injury or due to changes in cell surface profiles. To address this issue, we have enhanced the flow cytometry protocol to include the incorporation of 5-ethylnyl-2′-deoxyuridine (EdU) as an index of cell proliferation. EdU is a synthetic nucleoside that is taken up by actively dividing cells that are in S-phase of the cell cycle and incorporated into their DNA. The incorporated EdU is visualized using Click-It chemistry. The Click-It reaction is a copper-catalyzed reaction where the alkyne within the EdU nucleoside reacts with a fluorescent dye-labeled picolyl azide, forming a stable covalent bond. The small size of the azide reagent allows efficient access to the DNA and florescent labeling of the EdU without the need for harsh cell treatments. For our studies, we have found that an intraperitoneal injection of 100 mg/kg of EdU in saline, injected twice, beginning 4 h before sacrifice or addition of 1 μg/ml EdU to cultured neurospheres strongly labels proliferating neural precursors. To determine by flow cytometry which neural precursors are proliferating, we dissociated the neural precursors as described above, stained the cells for surface antigens, fixed the cells and then after a brief wash, incubated the cells with a saponin-based permeabilization reagent for 15 min. The cells were then incubated in the Click-It reaction cocktail for 30 min. The cells were then washed twice in phosphate buffer and were then analyzed by flow cytometry (Chen et al., unpublished data).

15.9 USING FLOW CYTOMETRY TO BETTER UNDERSTAND GLIOBLASTOMA

Chapters 17 and 18 describe the utility of flow cytometry to characterize the cell types that comprise neural tumors. As they discuss, glioblastoma (GBM) is a heterogeneous glial tumor found most frequently in the adult cerebral cortex and represents the majority of diagnosed brain tumors with an extremely high mortality rate. Some studies suggest that this cancer arises from resident NSCs, whereas others indicate that the tumor-initiating cells are progenitors that reside within the brain parenchyma. To date, definitive evidence is lacking to establish either source as the exclusive "cell of origin" in GBM. Since our studies revealed the existence of multiple antigenically defined progenitors within the SVZ that are self-renewing and potentially tumor-initiating cells, we have used the eight-color panel described above to classify the types of cells that reside within an experimentally induced end-stage GBM tumor. With this approach, we have established that there is considerable heterogeneity within the GBM. In particular, using a mouse model where PDGF is overexpressed and tumor suppressors PTEN and P53 are excised, the majority of the tumor initiating cells are CD133-, Lex-(low%), NG2+, CD140a+, GLAST-(low%), A2B5+/-, and O4+/-. Whereas the tumors formed in this model reveal the sensitivities of OPCs to certain genetic and epigenetic aberrations, changes in other growth factors, such as EGF or FGF, may provoke the transformation of NSCs, which we would predict would possess a completely different antigenic phenotype. As studies using flow cytometry evolve, investigators will be able to choose from an ever-expanding list of markers and gather useful and valuable data from both clinical and animal model tumor samples. A reasonable prediction is that by phenotypically defining the composition of a tumor, important insights can be obtained that will lead to more effective and specific clinical therapies.

15.10 CONCLUSION

Our understanding of the CNS, like many other fields, has expanded exponentially over the last several decades. One of the greatest advancements made is the discovery that the brain is not fixed, but is indeed plastic. The concept of neurogenesis has provided hope that, while brain injury has long been considered irreparable, this might not actually be the case. Utilizing a powerful tool like flow cytometry, we can expand our understanding of the mechanisms that regulate neurogenesis and provide a novel selection method to isolate NSCs as well as specific subsets of their progeny for experiments that previously were not possible. This discovery will provide an excellent basis for the elucidation of never-before-seen factors that regulate the possible regenerative behavior of NSCs and their progeny in the mammalian CNS. These studies will not only be informative for understanding neural development and basic biology, but they also will tremendously impact the translation of basic science discoveries into novel therapeutics for nervous system regeneration and repair. Thus, with continued diligence, scientists will be able to show that the vision of brain regeneration offered by Ramon y Cajal long ago is no longer a dream, but a reality [48].

Clearly, many more studies will be required to fully understand the intrinsic properties of each precursor type and how they respond to environmental cues. Using different growth factors to maintain and expand the isolated NPs, future studies will be able to gain insights into which extrinsic factors (small molecules, growth factors, ECM) are required for NSC maintenance and expansion and, similarly, which environmental signals are required by the intermediate progenitors. The power of isolating phenotypically defined NP populations by FACS will also lead to the understanding of their intrinsic mechanisms of self-renewal or fate specialization. Most of all, the data will extend our knowledge of neural development. These types of studies also will provide neuroscientists with a new platform to study regenerative mechanisms in the CNS. For example, one can determine which factors a particular population of precursors most responds to, and then use this knowledge to promote the self-renewal, proliferation, or differentiation of a single cell type. Future studies will also illuminate which signaling cascades and transcription factors drive these behaviors. Having this knowledge, cell transplantation or in vivo regenerative paradigms may be tailored to a specific brain region, with the proper precursor being selected and manipulated. In the future, we will know the mitogens and survival signals that need to be provided, thus we can enhance CNS regenerative responses to restore neurological function to individuals who have a sustained a brain or spinal cord injury or lost function due to a degenerative disease.

REFERENCES

[1] Merkle FT, Mirzadeh Z, Alvarez-Buylla A. Mosaic organization of neural stem cells in the adult brain. Science 2007;317(5836): 381–4.

[2] Maslov AY, et al. Neural stem cell detection, characterization, and age-related changes in the subventricular zone of mice. J Neurosci 2004;24(7):1726–33.

[3] Klein C, Fishell G. Neural stem cells: progenitors or panacea? Dev Neurosci 2004;26(2–4):82–92.

[4] Sauer FC. Mitosis in the neural tube. J Comp Neurol 1935;62:377–405.

[5] Misson JP, et al. The alignment of migrating neural cells in relation to the murine neopallial radial glial fiber system. Cereb Cortex 1991;1(3):221–9.

[6] Rakic P. Guidance of neurons migrating to the fetal monkey neocortex. Brain Res 1971;33(2):471–6.

[7] Angevine JB, Sidman RL. Autoadiographic study of cell migration during histogenesis of cerebral cortex in the mouse. Nature 1961;192:766–8.

[8] Takahashi T, Nowakowski RS, Caviness Jr VS. Early ontogeny of the secondary proliferative population of the embryonic murine cerebral wall. J Neurosci 1995;15(9):6058–68.

[9] Thomaidou D, et al. Apoptosis and its relation to the cell cycle in the developing cerebral cortex. J Neurosci 1997;17:1075–85.

[10] Bayer SA, Altman J. Neocortical development. New York: Raven Press; 1991.

[11] Lewis PD, Lai M. Cell generation in the subependymal layer of the rat brain during the early postnatal period. Brain Res 1974;77:520–5.

[12] Panchision DM, et al. Optimized flow cytometric analysis of central nervous system tissue reveals novel functional relationships among cells expressing CD133, CD15, and CD24. Stem Cells 2007;25(6):1560–70.

[13] Buono KD, et al. Leukemia inhibitory factor is essential for subventricular zone neural stem cell and progenitor homeostasis as revealed by a novel flow cytometric analysis. Dev Neurosci 2012;34(5):449–62.

[14] Uchida N, et al. Direct isolation of human central nervous system stem cells. Proc Natl Acad Sci USA 2000;97(26):14720–5.

[15] Pastrana E, Cheng LC, Doetsch F. Simultaneous prospective purification of adult subventricular zone neural stem cells and their progeny. Proc Natl Acad Sci USA 2009;106(15):6387–92.

[16] Capela A, Temple S. LeX/ssea-1 is expressed by adult mouse CNS stem cells, identifying them as nonependymal. Neuron 2002; 35(5):865–75.

[17] Perfetto SP, et al. Amine-reactive dyes for dead cell discrimination in fixed samples. Curr Protoc Cytom 2010. [Chapter 9]: p. Unit 9 34.

[18] Maecker HT, Trotter J. Flow cytometry controls, instrument setup, and the determination of positivity. Cytom A 2006;69(9):1037–42.

[19] Maric D, et al. Prospective cell sorting of embryonic rat neural stem cells and neuronal and glial progenitors reveals selective effects of basic fibroblast growth factor and epidermal growth factor on self-renewal and differentiation. J Neurosci 2003;23(1):240–51.

[20] Rietze RL, et al. Purification of a pluripotent neural stem cell from the adult mouse brain. Nature 2001;412(6848):736–9.

[21] Barraud P, et al. Isolation and characterization of neural precursor cells from the Sox1-GFP reporter mouse. Eur J Neurosci 2005; 22(7):1555–69.

[22] Kitada M, Kuroda Y, Dezawa M. Lectins as a tool for detecting neural stem/progenitor cells in the adult mouse brain. Anat Rec Hob 2011;294(2):305–21.

[23] Coskun V, et al. CD133+ neural stem cells in the ependyma of mammalian postnatal forebrain. Proc Natl Acad Sci USA 2008; 105(3):1026–31.

[24] Sawamoto K, et al. Visualization, direct isolation, and transplantation of midbrain dopaminergic neurons. Proc Natl Acad Sci USA 2001;98(11):6423–8.

[25] Belachew S, et al. Postnatal NG2 proteoglycan-expressing progenitor cells are intrinsically multipotent and generate functional neurons. J Cell Biol 2003;161(1):169–86.

[26] Jablonska B, et al. Cdk2 is critical for proliferation and self-renewal of neural progenitor cells in the adult subventricular zone. J Cell Biol 2007;179(6):1231–45.

[27] Zhang SC. Defining glial cells during CNS development. Nat Rev Neurosci 2001;2(11):840–3.

[28] Chojnacki A, Weiss S. Isolation of a novel platelet-derived growth factor-responsive precursor from the embryonic ventral forebrain. J Neurosci 2004;24(48):10888–99.

[29] Rivers LE, et al. PDGFRA/NG2 glia generate myelinating oligodendrocytes and piriform projection neurons in adult mice. Nat Neurosci 2008;11(12):1392–401.

[30] Bansal R, Stefansson K, Pfeiffer SE. Proligodendroblast antigen (POA), a developmental antigen expressed by A007/O4-positive oligodendrocyte progenitors prior to the appearance of sulfatide and galactocerebroside. J Neurochem 1992;58:2221–9.

[31] Schnitzer J, Schachner M. Cell type specificity of a neural cell surface antigen recognized by the monoclonal antibody A2B5. Cell Tissue Res 1982;224:625–36.

[32] Kaneko Y, et al. Musashi1: an evolutionarily conserved marker for CNS progenitor cells including neural stem cells. Dev Neurosci 2000;22(1–2):139–53.

[33] Regan MR, et al. Variations in promoter activity reveal a differential expression and physiology of glutamate transporters by glia in the developing and mature CNS. J Neurosci 2007;27(25): 6607–19.

[34] Nishihara S, et al. Alpha1,3-fucosyltransferase IX (Fut9) determines Lewis X expression in brain. Glycobiology 2003;13(6):445–55.

[35] Sommer I, Schachner M. Monoclonal antibodies (O1 to O4) to oligodendrocyte cell surfaces: an immunocytological study in the central nervous system. Dev Biol 1981;83(2):311–27.

[36] Domercq M, et al. Excitotoxic oligodendrocyte death and axonal damage induced by glutamate transporter inhibition. Glia 2005; 52(1):36–46.

[37] McLaren FH, et al. Analysis of neural stem cells by flow cytometry: cellular differentiation modifies patterns of MHC expression. J Neuroimmunol 2001;112(1–2):35–46.

[38] Murayama A, et al. Flow cytometric analysis of neural stem cells in the developing and adult mouse brain. J Neurosci Res 2002;69(6):837–47.

[39] Mouthon MA, et al. Neural stem cells from mouse forebrain are contained in a population distinct from the 'side population'. J Neurochem 2006;99(3):807–17.

[40] Mignone JL, et al. Neural stem and progenitor cells in nestin-GFP transgenic mice. J Comp Neurol 2004;469(3):311–24.

[41] Doetsch F. The glial identity of neural stem cells. Nat Neurosci 2003;6(11):1127–34.

[42] Merkle FT, et al. Radial glia give rise to adult neural stem cells in the subventricular zone. Proc Natl Acad Sci USA 2004.

[43] Ganat YM, et al. Early postnatal astroglial cells produce multilineage precursors and neural stem cells in vivo. J Neurosci 2006;26(33):8609–21.

[44] Groszer M, et al. Negative regulation of neural stem/progenitor cell proliferation by the Pten tumor suppressor gene in vivo. Science 2001;294(5549):2186–9.

[45] Ziegler AN, et al. Insulin-like growth factor-II (IGF-II) and IGF-II analogs with enhanced insulin receptor-a binding affinity promote neural stem cell expansion. J Biol Chem 2014;289(8):4626–33.

[46] Buono KD. Analyses of mouse neural precursor responses to leukemia inhibitor factor and hypoxia/ischemia, in graduate school of biomedical sciences. Newark: University of Medicine and Dentistry of New Jersey; 2011. pp. 1–200.

[47] Goodus, M, et al. Neural stem cells of the immature, but not the mature, subventricular zone mount a robust regenerative response to traumatic brain injury. Dev Neurosci. http://dx.doi.org/10.1159/000367784.

[48] Cajal SR, DeFelipe J, Jones EG. Cajal's degeneration and regeneration of the nervous system. Oxford University Press; 1991.

Chapter 16

Multiparameter Flow Cytometry Applications for Analyzing and Isolating Neural Cell Populations Derived from Human Pluripotent Stem Cells

Nil Emre, Jason G. Vidal, Christopher Boyce, Lissette Wilensky, Mirko Corselli and Christian T. Carson
BD Biosciences, La Jolla, CA, USA

16.1 INTRODUCTION

Multiparameter flow cytometry has been instrumental in the field of immunology and has evolved to meet the demands of this biological field for over 30 years [1]. Flow cytometry can be used to identify numerous cell populations based on their unique cell surface phenotype, size, and light scattering properties. The current state of the art permits researchers to interrogate heterogeneous cell cultures at single-cell resolution utilizing hundreds of verified fluorophore-conjugated antibodies to cell surface markers, transcription factors, cytokines, and phosphorylated proteins. These reagents can be used for or combined with other assays to measure cell cycle, cell proliferation, apoptosis, signaling, and senescence to provide an exquisite analysis of all cells in a heterogeneous cell population. This knowledge base, available technology, and reagent pool are indispensable in the field of hematopoiesis but have not been fully utilized in other biological fields.

Stem cell biologists working with pluripotent stem cells (PSCs) struggle with obstacles similar to those of hematopoietic biologists with respect to the heterogeneous nature of their cell cultures. There are numerous methods to differentiate human embryonic stem cells (hESCs) and human induced pluripotent stem cells (hiPSCs) to neural cells using spontaneous differentiation, chemical induction, or mouse stromal feeder cells [2–7]. These approaches often lead to heterogeneous cell cultures composed of cells from multiple stages of differentiation and from different germ layers. Moreover, there is variability in the differentiation potential of PSC lines, which encumbers downstream assays that require near-pure cell populations for comparative analysis such as in vitro assays, microarrays, and transplantations.

In this chapter we review the various cell sorting signatures and strategies currently employed for isolating neural cells derived from PSCs by fluorescence-activated cell sorting (FACS). In addition, we review how these cell surface signatures were discovered and developed to provide insights into how to identify new signatures for important neural cell types. We also describe the various neural differentiation methods relevant to each cell sorting strategy and illustrate how these combined approaches have been instrumental in both in vitro and in vivo disease models. Finally, we show how flow cytometry has been employed to quantify heterogeneous neural differentiation cultures.

16.2 METHODS FOR NEURAL DIFFERENTIATION OF HUMAN PLURIPOTENT STEM CELLS

There are many neural induction methods that lead to cell cultures enriched for neural ectoderm and the description of these methods is important as many of the cell signatures described are in part tailored to the differentiation method used. All of these methods have their advantages depending on the end goal.

Serum-free embryoid body (SFEB) culture utilizes embryoid body (EB) formation in suspension and subsequent culture in a serum-free culture medium containing N2 and B27 supplements. EBs are ultimately adhered onto matrix-coated plates and neural ectoderm becomes evident as neural rosettes. At this stage neural rosettes can be harvested manually and enzymatically separated to yield a monolayer cell culture of neural stem cells (NSCs) that can self-renew for many passages and are capable of differentiating into neurons and glia upon withdrawal of growth factors [8]. Alternatively, the NSC isolation step can be omitted by allowing the adherent SFEB cultures to spontaneously differentiate in neural induction culture medium, which results in a more heterogeneous cell culture.

Neural Surface Antigens. http://dx.doi.org/10.1016/B978-0-12-800781-5.00016-5

Stromal-derived induction activity (SDIA) utilizes mouse stromal feeders (MS5 or PA6) for neural induction [2,3,5–7]. Human PSCs are plated in clumps or as single cells onto stromal feeder layers and allowed to propagate and differentiate to heterogeneous cultures containing cells from all germ layers but enriched for neural ectoderm. Two of the limitations of mouse stromal feeder layers is the contamination of the cell cultures with mouse cells and the batch-to-batch variability of these cells for neural induction.

Dual Similar to Mothers Against Decapentaplegic (SMAD) inhibition utilizes a combination of recombinant Noggin and/or small-molecule inhibitors of SMAD signaling to induce neurogenesis of human PSCs [4]. This highly effective method can be performed on PSCs grown in monolayers or in combination with SFEB and SDIA.

16.3 CELL SURFACE SIGNATURES FOR NEURAL CELL ISOLATION FROM PLURIPOTENT STEM CELLS

Cell surface cluster of differentiation (CD) markers have enabled the field of hematopoiesis and our understanding of blood-borne diseases [1]. We are just beginning to understand the cell surface signatures that define cell types of other tissues, including the nervous system [9]. Many of the strategies for isolating neural cells derived from human PSCs by FACS are derived from our knowledge of stem cell surface markers expressed in vivo on NSCs in humans and rodents [10–14]. Many groups have utilized CD133 and CD15 to isolate prospective NSCs from differentiating human PSC cultures [15,16]. Peh et al. reported the isolation of neurosphere-forming NSCs from neural differentiation cultures of hESCs based on the expression of CD133, CD15, and GCTM-2 [15]. Golebiewska et al. used $CD133^+CD45^-CD34^-$ to isolate NSCs from differentiating hESCs [16]. Many groups have utilized engineered promoter reporter lines for isolating NSCs and neurons from differentiating cultures of hESCs [17,18]. However, many of these promoters are leaky, which precludes achieving optimal purity of cells, and engineering numerous PSC lines is time-consuming and not practical for therapeutic applications.

Pruszak et al. [19] surveyed hESC SDIA neural differentiation cultures with a number of antibodies to cell surface markers to characterize the surface antigens present during neural differentiation of hESCs [19]. This was the first study that characterized changes in cell surface marker expression over the course of neural differentiation of hESCs. In this study, the authors reported the isolation by FACS of enriched populations of neurons using NCAM (CD56) and that these enriched neurons could survive in vivo after transplantation in rodent brains without tumor formation. Importantly, by utilizing a 100-μm nozzle and low sheath pressure (20–25 psi), they were able to achieve viability comparable to that of unsorted controls and therefore illustrated for the first time that viable neurons derived from hESCs could be isolated by FACS in a robust assay.

This team further continued their studies and performed a biased combinatorial antibody screen of hESC SDIA neural differentiation cultures. Pruszak et al. [20] identified three distinct cell populations based on expression of CD15, CD24, and CD29 [20]. $CD15^+CD29^{high}CD24^{low}$ defined cells that expressed NSC and proliferation markers and formed neuroepithelial tumors after transplantation in rodent brains. $CD15^-CD29^{high}CD24^{low}$ yielded neural crest-like and mesenchymal cells that expressed neural crest and mesenchymal markers. This cell population formed neural crest tumors upon transplantation in rodent brains. $CD15^-CD29^{low}CD24^{high}$ identified a population of neurons and neuroblasts that had reduced expression of proliferation markers and immature neural markers Pax6 and FORSE-1 and elevated expression of neuronal markers Tuj1 and MAP2. This cell population yielded highly pure neuronal cultures and led to tumor-free neuronal grafts in rodent brains. Ultimately, this work led to a novel surface antigen code for neural differentiation cultures of hESCs.

Yuan et al. [21] performed a flow cytometry and image-based unbiased screen of 190 antibodies to cell surface markers to elucidate cell surface signatures of NSCs, neurons, and glia derived from SFEB neural induction cultures [21]. From a flow-based screen of SFEB hESC neural induction cultures they identified $CD184^+CD271^-CD44^-CD24^+$ as a subpopulation of NSCs that could be expanded in culture for many passages and be subsequently differentiated to a mixed cell population of neurons and glia in vitro. $CD184^+CD271^-CD44^-CD24^+$ NSCs also resulted in tumor-free grafts in rat spinal cords and differentiated to both neurons and astrocytes. This sorting method was validated in numerous hESC and hiPSC lines, illustrating its versatility as an alternative to manually isolating neural rosettes from neural induction cultures and serially passaging to create NSC lines.

From an image-based screen of purified and differentiated NSCs Yuan et al. [21] identified $CD184^-CD44^-CD24^+CD15^{low}$ as a panneuronal signature [21]. Neurons sorted using this method were highly pure, were viable in culture for many weeks, and could fire action potentials when depolarized. Figure 16.1 illustrates the use of this method. From the same image-based screen, this group identified $CD184^+CD44^+$ as a population of immature replicating glia cells from differentiated NSC cultures. Upon isolation, this cell population expressed nestin, and after many weeks in culture in astrocyte differentiation medium, a portion of these cells expressed the astrocyte marker glial fibrillary acidic protein (GFAP). This work revealed novel signatures for the isolation of neural cells from SFEB cultures by FACS. This work also demonstrated the utility of unbiased antibody-based surface marker

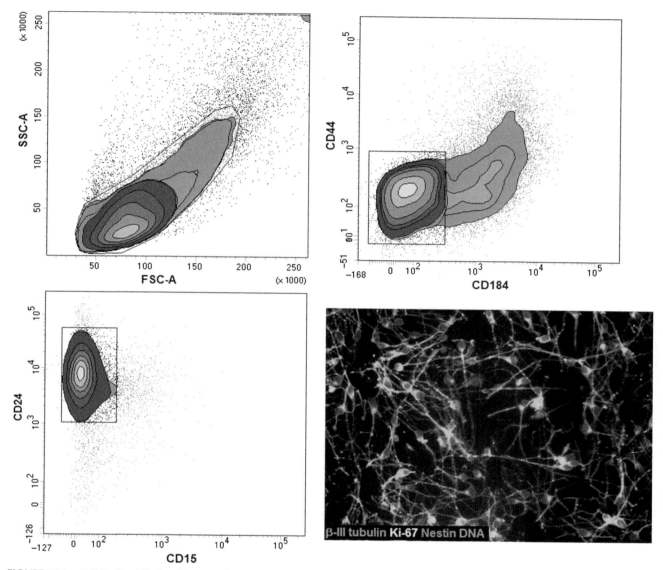

FIGURE 16.1 Cell Surface Marker Signature for Isolating Neurons. NSC cultures were differentiated as described in Yuan et al. [21] and stained with fluorophore-conjugated antibodies against CD44, CD184, CD24, and CD15. Cells were identified by their light-scattering properties and then analyzed for CD44 and CD184 expression. The CD44−CD184− cells were then analyzed for CD24 and CD15 expression. *Contour plots*: Sorting strategy for the isolation of CD44−CD184−CD24highCD15dim/− neurons. *Microscope image*: Immunofluorescence analysis of sorted neurons 1 week after the sort. The cells were stained with β-tubulin III, Ki-67, and nestin.

screening for identifying cell surface signatures for isolating differentiated cell types from human PSCs.

In addition to focusing on central nervous system lineages, many groups have developed signatures for isolating neural crest cells. Lee et al. [22] described a method for isolating neural crest cells from differentiating hESC cultures using dual SMAD inhibition or SDIA [22]. The HNK-1+p75+ (CD57+CD271+) signature identified a cell subpopulation of neural crest capable of differentiating toward autonomic and sensory neurons, Schwann cells, myofibroblasts, and mesenchymal cell types. Similarly, Zhou et al. [23] reported the isolation by FACS of

frizzled-3+cadherin-11+ cranial neural crest-like cells derived from differentiating hESCs, capable of maintaining multipotent differentiation potential [23].

In summary, there has been much effort to develop cell surface signatures and sorting methods to isolate neural cell types from both the central nervous system and neural crest lineages. Both candidate marker approaches and unbiased screens have proven to be effective methods for identifying cell surface signatures of cells of interest. Whereas we are starting to understand the surface phenotypes for many neural cell types, such as NSCs, neural crest, and neurons, the surface signatures for specific neuronal cell types remain elusive. In 2014, the surface marker CORIN was shown to be effective for FACS-based enrichment of midbrain dopaminergic neuronal progenitor cells [24]. There is a lot of promise for using both RNA expression analysis and unbiased cell surface marker screening to further elucidate cell surface signatures of specific neural subtypes.

16.4 NEURAL APPLICATIONS FOR MULTIPARAMETER FLOW CYTOMETRY

16.4.1 Neural Fate Determination

One of the key limitations to obtaining relevant neural cell types for drug discovery and therapeutic platforms is the ability to efficiently and reproducibly obtain large numbers of cells. Studies to understand culture conditions and regulatory pathways that are involved in neural cell fate determination can help to address this problem of scalability. The ability to obtain pure neural cell populations through sorting to characterize and optimize culture conditions and the use of defined cell surface markers to delineate cell differentiation status aid in the optimization of neural differentiation schemes.

Optimal matrix compositions are crucial for recapitulating in vitro the conditions of neural development in vivo. To mimic in vivo conditions, decellularized porcine brain matrix was examined for its ability to support neuronal cell growth [25]. Purified populations of neurons (CD184$^-$CD44$^-$CD15^{low}CD24$^+$) [21] were obtained from hiPSCs and plated on decellularized porcine brain matrix. Consistently pure populations of neurons allowed for the comparison of neuronal morphology on various matrices. Interestingly, neurons were able to form a higher degree of complex dendritic processes on the decellularized porcine brain matrix than on traditional substrates.

The identification of novel neural fate determinants not only increases the understanding of neural development but also can help in the optimization of differentiation schemes. To examine the involvement of microRNAs (miRNAs) in neural differentiation regulation, hESC-derived NSCs and neurons were screened for the expression of miRNAs [26]. The differentiated neurons were a heterogeneous cell population, and the presence of other cell types in the culture had the potential to perturb the miRNA profiling. To eliminate contaminating cell artifacts, a combinatorial cell surface signature, CD15$^-$CD29^{low}CD24^{high} [20], was used to obtain a pure population of neurons for miRNA profiling. From the expression analysis, multiple miRNAs were examined for their potential to modulate neural lineage commitment, and miR7 was found to be a potential regulator of neurogenesis.

Engineered reporter cell lines are often used to track lineage fates as well as to optimize differentiation protocols. An mCherry nestin hESC line was derived to monitor neural

cell fate [27] and the expression of nestin was compared to known cell surface signatures of NSCs [21]. Cells that expressed a high level of mCherry nestin coincided with a CD184$^+$CD44$^-$CD24$^+$ as well as a CD271$^-$CD44$^-$CD24$^+$ cell population. Sorting for CD184$^+$CD44$^-$CD24$^+$ enriched to a greater degree for NSC markers such as Sox1, Sox2, or Pax6 in comparison to sorting with nestin^{high} cells alone, indicating that nestin expression alone is not sufficient to purify NSCs.

Cortical interneurons are an important subclass of neurons because their dysfunction has consequences in multiple neuropsychiatric diseases such as schizophrenia and autism. An Nkx2.1 green fluorescent protein (GFP) hESC line was used to optimize culture conditions for obtaining a high proportion of Nkx2.1-positive, ventral prosencephalic progenitor cells [28]. Further studies with different culture conditions resulted in three distinct ventral forebrain precursor cell populations. Nkx2.1-positive cells were tested for their maturation in vivo on human cortical neuron feeders, which were derived from hESCs and sorted to purity using the CD44$^-$CD184$^-$CD24$^+$ signature [21]. Coculture systems can allow for the maturation of cell types in vitro and it is important for the feeder systems to be defined both to better understand neural fate and eventually to be able to model disease states.

16.4.2 In vitro Disease Modeling

Pluripotent stem cells can be utilized as model systems in recapitulating disease states in vitro. To better compare and characterize the phenotypes of cell populations, it is becoming increasingly important to purify cell types of interest for comparison and to also interrogate expression levels of particular markers in phenotyping disease states. Multiparameter flow cytometry-based sorting and analysis applications have enabled many assays for in vitro modeling of neurological disorders.

Alzheimer disease is challenging to model because of the inability to obtain live neurons from patients. Furthermore, the sporadic form of the disease, which is the predominant form, is particularly challenging because of a lack of appropriate models to study the disease. To investigate if hiPSC technology could be utilized to model both familial and sporadic forms of Alzheimer disease, hiPSCs were generated and compared to nondemented age-matched controls (NDCs) [29]. To examine the efficiency of neural differentiation between NDC and Alzheimer disease patient-derived hiPSC lines, multiparameter flow cytometry was utilized to quantitate the efficiency of NSC generation using a CD184$^+$CD15$^+$CD44$^-$CD271$^-$ cell surface signature, whereas neuronal cultures that were derived from purified and expanded NSCs were quantitated using a CD24$^+$CD44$^-$CD184$^-$ cell surface signature [21]. There were no significant differences in the differentiation ability of the hiPSC lines. Using the same respective cell surface signatures, NSCs and neurons were sorted to purity and, in particular, the neurons were subjected to further analyses [29]. The ability to obtain consistently pure neuronal cell populations for downstream analyses was an important milestone that enabled the comparison of phenotypes between the various patient hiPSC-derived neurons. Interestingly, purified neurons derived from hiPSCs from two familial cases and one case of sporadic Alzheimer disease exhibited higher levels of three key markers of Alzheimer disease. The increased levels of amyloid-β(1–40), a GSK-3β, and p-Tau to total Tau ratio seen in the neurons are significant because Alzheimer disease can take decades to develop in an individual and now can be modeled in vitro in a relatively short period of time.

Dominant missense mutations of presenilin 1 (PS1) are common in familial Alzheimer disease. PS1 is part of the γ-secretase protease complex that is involved in the cleavage of APP to amyloid-β, which is the major component of amyloid plaques. To address whether PS1 mutations result in a gain or loss of function, a hiPSC line whose genome had been sequenced was targeted with a series of mutations using TALEN genome editing [30]. Phenotypically normal as well as altered hiPSC were differentiated and CD184$^+$CD44$^-$CD271$^-$CD24$^+$ NSCs [21] were purified. Expanded NSCs were differentiated to neurons and purified using a CD24$^+$CD44$^-$CD184$^-$ cell surface signature. A comparison of surface marker expression between the different genotypes demonstrated that there were no differences in the percentages of derived NSC or neurons between the groups. Further characterization of purified neurons revealed a complex etiology in which PS1 mutations do not lead to a simple loss of function in familial Alzheimer disease.

In neurodegenerative disorders, Parkinson disease is second only to Alzheimer disease in terms of occurrence. Heterozygous mutations in the *GBA1* gene, which encodes the lysosomal enzyme β-glucocerebrosidase, are associated with a higher risk of Parkinson disease, and homozygous mutations in *GBA1* cause Gaucher disease. HiPSC were generated from Parkinson disease and Gaucher disease patients with a *GBA1* mutation [31]. From the differentiated hiPSCs, neuronal cell populations were enriched using a CD24^{high}CD29$^-$CD184$^-$CD15$^-$ cell surface signature (based on [20] and [21]). In comparing the Parkinson disease and Gaucher disease purified neurons to healthy controls, multiple defects were observed. These defects included decreased β-glucocerebrosidase levels, an increase in glucosylceramide and α-synuclein, lysosomal and autophagic deficiencies, as well as calcium homeostasis defects. In addition, a comparison of the ganglioside profile of the various cell types, including unsorted neurons, revealed that purified neurons displayed a ganglioside profile that more closely mimicked the adult human brain. This study again highlights the utility of obtaining highly pure cell populations for downstream phenotypic analysis.

Through whole exome sequencing of two families presenting with autism, a null mutation in the branched-chain ketoacid dehydrogenase kinase (BCKDK) gene was identified [32]. To model the disease in vitro, hiPSCs were generated and cells were differentiated into NSCs. CD184$^+$CD24$^+$CD44$^-$CD271$^-$ NSCs [21] were purified by cell sorting to reduce variability and enable comparison between BCKDK mutant and wild-type cells. There was no difference in either the morphology or the proliferative and neural differentiation capacities of the cells. BCKDK is involved in the catabolism of branched-chain amino acids (BCAAs) and NSCs exposed to lower levels of BCAAs showed no change in cell survival or proliferation. This suggested against a cell autonomous role of BCKDK in the disease. Amino acid transporters in the blood–brain barrier can change in response to amino acid availability, and examination of knockout BCKDK mice revealed an imbalance in brain amino acid profiles, possibly caused by the low concentration of plasma BCAAs. In addition, dietary supplementation of BCAA to the BCKDK-deficient mice ameliorated neurological symptoms.

Niemann Pick type C1 (NPC1) is a fatal progressive lysosomal neurodegenerative disease in which cholesterol and glycolipids accumulate in the lysosomal compartment owing to a loss-of-function mutation of NPC1. This lysosomal accumulation has a particularly damaging effect on neurons. An hESC knockdown (KD) model of NPCs was generated to characterize the neuronal defect seen in the disease [33]. Using a CD184$^+$CD15$^+$CD44$^-$CD271$^-$ signature [21], NSCs were successfully isolated from wild-type but not from KD NPC1 hESCs. To effectively generate an NPC1 model, wild-type purified NSCs were transduced with a short hairpin RNA to decrease levels of NPC1. The NPC1-deficient NSCs were further differentiated into neurons; neurons were characterized by autophagy defects, leading to an accumulation of mitochondrial fragments in neurons. The novel autophagy defect seen in neurons for NPC1 now opens a new path toward the treatment of NPC1 disease symptoms.

Pure hiPSC-derived neural cell types not only enable disease modeling but also are a powerful tool to screen for small-molecule regulators that could potentially ameliorate disease phenotypes. Familial dysautonomia (FD) is a fatal neurodegenerative disease that is caused by reduced levels of I$\kappa\beta$ kinase complex-associated protein (IKBKAP). To model FD, patient-derived hiPSCs were differentiated into five lineages representative of the three germ lineages and purified by flow cytometry using lineage-specific markers, and levels of IKBKAP were characterized [34]. Interestingly, HNK-1$^+$ neural crest precursors had significantly low IKBKAP transcript levels and comparative transcriptome analysis of HNK-1$^+$ purified cells revealed deficiencies in genes involved in neurogenesis and neuronal differentiation. Small molecules that lead to an increase in IKBKAP levels could potentially lessen the impact of FD. To screen for potential regulators of IKBKAP levels, FD-derived hiPSC HNK-1$^+$ (CD57$^+$) neural crest progenitors were screened for small-molecule regulators [35]. Eight potential small molecules were found to increase IKBKAP levels and one potential small molecule, SKF-86466, lessened the disease-specific phenotype of autonomic neuronal marker loss. These studies not only show the utility of hiPSCs for disease modeling but also highlight the utility of flow cytometry to obtain pure cell populations for downstream analysis such as drug screening for therapeutic intervention.

16.5 CELL TRANSPLANTATION

Pluripotent stem cells are a potentially useful source of multiple therapeutically relevant cell types for cell therapy applications. Prior to the realization of cell replacement therapy, effective methods to demonstrate not only efficacy but also safety need to be addressed. One step in this process is the use of in vivo models to demonstrate engraftment safety and efficacy. Multiparameter flow cytometry has been applied to both neural cell analysis and neural cell sorting to better understand and define therapeutically relevant neural cell populations.

The utility of hESC-derived neural cell populations as a source for cell replacement therapies was examined in a rodent and a small-mammal model [36]. CD184$^+$CD44$^-$CD271$^-$CD24$^+$CD15$^+$ NSCs [21] derived from hESCs were sorted and further expanded prior to transplantation. The sorted and cultured NSCs were analyzed using multiparameter intracellular flow cytometry to confirm purity after sorting and expansion as well as assess for the presence of any undifferentiated hESCs. This homogeneous cell population was then transplanted into spinal cords of spinal ischemia-injured rats or immunosuppressed minipigs and the cells were evaluated for maturation and survival. The maturation and proliferation of grafted cells were similar to those of human fetal spinal cord-derived neural precursors and suggest that hESC-derived NSCs that have been sorted to purity represent a viable source for cell therapy for neurodegenerative disorders. Another study demonstrated the successful transplantation of differentiated Parkinson disease patient-derived hiPSCs into rodent models [37]. In addition to the transplantation of the total population of differentiated cells, cells were purified prior to transplantation. Specifically, cells were sorted using NCAM (CD56) as a positive marker and survival was monitored. Both studies demonstrated the ability to use purified neural cell populations for engraftment studies.

Attaining safe immunosuppression upon engraftment is also a key step in cell therapy enablement. As part of a study focusing on the application of varying tacrolimus (TAC) formulations as an immunosuppressant in spinal trauma-injured and SOD1$^+$ amyotrophic lateral sclerosis rats, NSCs were evaluated in addition to human fetal spinal cord-derived stem cells [38]. NSCs were sorted using a CD184$^+$CD44$^-$CD271$^-$CD24$^+$ cell surface signature [21], expanded, and frozen prior to implantation. SOD1$^+$ rats as well as rats with spinal trauma were immunosuppressed through the use of a TAC pellet and NSCs were implanted and successful long-term engraftment and maturation were achieved. Increased levels of TAC demonstrated less T cell infiltration into the area of engraftment.

Although the maturation and survival of neural cell populations at the site of implantation as well as effective immunosuppressive methodologies are important, another key milestone in cell therapy is the ability of cells to extend beyond the site of implantation and to aid in functional recovery. Rats that had undergone spinal cord transection were coengrafted with rat embryo spinal cord-derived or human fetal spinal cord-derived NSCs along with a growth factor cocktail containing fibrin matrix [39]. Engrafted cells differentiated into neurons, extended axons over long distances, and contributed to a functional improvement in locomotion. hESC-derived NSCs were sorted using a defined cell surface signature [21] and subsequently engrafted into sites of spinal injury with a growth factor cocktail-containing matrix. Similar to other cell types examined, axons were able to extend out from the lesion over a long distance. It is significant that in this spinal cord injury model, early neural cell types from multiple sources and species were able to integrate and, in addition, axonal growth and functional connectivity were achieved.

16.6 THE DETECTION OF INTRACELLULAR ANTIGENS BY FLOW CYTOMETRY

The detection of intracellular antigens by flow cytometry is the ability to detect cytokines, transcription factors, phosphorylated proteins, structural proteins, or any other proteins inside a cell by utilizing fluorescence-labeled antibodies or dyes and a flow cytometer. Intracellular flow cytometry (IFC) presents many challenges not present when assaying cell surface expression by flow cytometry. Because IFC requires protein cross-linking (fixation) within the cell and subsequent permeabilization of the cell membrane, there are different ways to accomplish these requirements: Traditionally cells are cross-linked using an aldehyde (additive, noncoagulant fixative), including formaldehyde, paraformaldehyde, and glutaraldehyde, or by an alcohol (denaturing, coagulating fixative), specifically ethanol or methanol. After the fixation of a cell, a permeabilization step must follow, or be in combination with fixation, to allow for the entrance of an antibody into the cell, especially in the case of the aldehyde fixatives. Permeabilization methods can vary widely as well. These include the use of nonionic detergents such as saponin, Triton X-100, and Tween 20; ionic detergents such as sodium dodecyl sulfate in a buffered solution; and also alcohols, specifically various concentrations of methanol [40–42]. The cellular compartment containing the epitope of interest, i.e., the cytoplasm, the nucleus, or sub-organelle structures such as the nucleolus, can help guide the user in choosing methodology. Additionally, the type of fluorophore and the sequence in which antibody staining is performed is also important. Specifically, the hydrodynamic volume (size) and structural flexibility of the fluorophore and the resistance to damage by solvents can contribute to protocol design input and differences in the resolution detected of the antibody–antigen–fluorophore. Detergents or solvents, or high-temperature incubations in either, can contribute to poor cell recovery, limiting epitope detection and adding to the difficulty in resolving specific cell phenotypes in heterogeneous samples. Yields from sample processing techniques can vary greatly with some methods and users, but with careful attention to specific handling methods, cell loss can be limited. Vendors sell buffer systems for IFC for specific applications, which are usually designed around a given assay. It may be desirable to develop a custom system or select a commercially available option that is verified in a specific application.

Multiplexing cell surface staining combined with IFC is also a powerful tool and can be accomplished, but has its own set of challenges [43]. Depending on the epitope and the fixation and/or permeabilization method, many surface epitopes can be changed or destroyed, eliminating the antibody binding capacity of a specific antibody. There are two main variations of the protocol to stain both surface and intracellular epitopes: (1) stain with the surface antibodies and then proceed with the fixation and permeabilization as per the IFC protocol or (2) perform the fixation and permeabilization portion of the IFC protocol first and then stain with antibodies against both surface and intracellular epitopes. There are caveats to both protocols. With the first option, it is imperative to ensure that the fluorophore being used is compatible with the fixation/permeabilization protocol. Some fluorophores, mainly the large proteins (e.g., PE, APC), will be denatured after long exposure times in alcohols. In the second option it is imperative to ensure that the epitope will not be damaged or altered upon fixation and permeabilization. Regardless of the protocol used, fixation and permeabilization result in cell death, hence the need for surface marker signatures to identify and purify viable and functional cells of interest.

16.7 THE UTILITY OF FLOW CYTOMETRY FOR ANALYZING HUMAN PSC-DERIVED NEURAL CELLS

The ability of flow cytometry to analyze cells at the single-cell level is a powerful tool compared to assays such as ELISA or Western blot. Combining this with the ability to analyze epitopes intracellularly, especially in the situation of analyzing master regulator genes such as Nanog for

hESCs or Sox1 for NSCs, makes IFC a very powerful tool for quantitating heterogeneous differentiation cultures. Yuan et al. [21] utilized IFC to quantitate the expression of nestin, Sox1, Sox2, Oct3/4, and the proliferation marker Ki-67 in both heterogeneous neural cell cultures and sorted populations from these cultures [21]. Analyzing the cells for the expression of these proteins allowed for the quantitation of the efficiency of their differentiation and the purity of a FACS-derived population of neurons from their differentiation cultures. Kakinohana et al. [36] used IFC to verify their NSC cultures for the expression of Sox1 and nestin and also the lack of expression of Nanog before implantation studies [36]. Chambers et al. [44] utilized IFC to compare alternative differentiation protocols for converting hPSCs into nociceptors [44]. Analysis by IFC is a powerful tool to study neural biology as there are few surface markers known in the field to delineate the hierarchy of neural cell types.

IFC has also been used as a tool to help with the discovery of surface phenotypes. Turac et al. [45] stained various cell lines that are capable of neuronal differentiation with 10 surface markers and subsequently costained, by IFC, with the neural markers doublecortin (DCX), tyrosine hydroxylase (TH), or GFAP to discover prospective cell surface signatures for neural subtypes that are defined by the expression of DCX, TH, or GFAP. They then used the prospective neuron cell surface signature CD49f$^-$CD200^{high} to sort a population of postmitotic neurons from hiPSC neural differentiation cultures [45].

There are also other multiparameter assays that utilize IFC for the detection of cytokines or phosphorylated proteins and to interrogate aspects of the cell cycle. Such an assay is shown in Figure 16.2. Here a neuron differentiation culture from NSCs was analyzed for cycling cells by the incorporation of BrdU and the expression of Sox1 at days 0, 8, and 21 of differentiation. Cells were pulsed with BrdU (10 μM) for 1 h immediately before being harvested for analysis and then analyzed for BrdU incorporation and Sox1 expression. Cycling cells were limited to the Sox1$^+$ cell population on days 0 and 8, whereas there were no more cycling cells at day 21.

16.8 CELL SURFACE MARKER SCREENING APPLICATIONS FOR NONNEURAL STEM CELL POPULATIONS

The ability to use multiparameter flow cytometry to both analyze and purify heterogeneous cell populations has wide applications to other fields of stem cell biology. An inherent challenge that comes with heterogeneity is also a paucity of known cell surface markers for particular stem cell types of interest. Oftentimes the first step to purifying a cell population is performing cell surface marker screens to discover the cell surface signature of the cell population of interest.

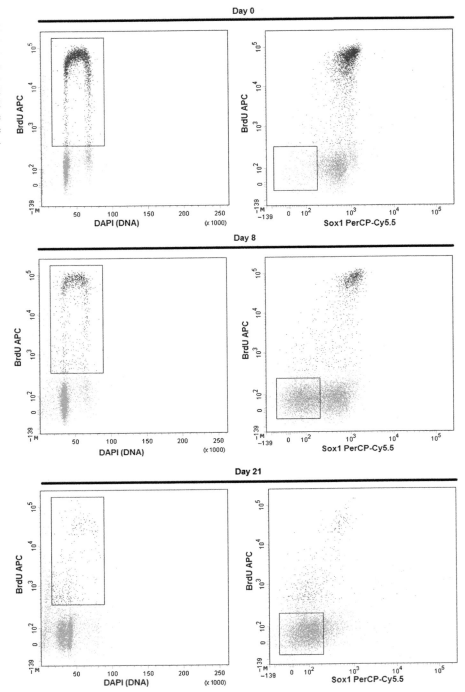

FIGURE 16.2 Intracellular Flow Cytometric Analysis of Differentiating NSCs. NSC cultures were differentiated as described in Yuan et al. [21] and then analyzed at day 0, day 8, and day 21. Cells were pulsed with BrdU (10 mM) for 1 h immediately before being harvested for analysis and then analyzed for BrdU incorporation and Sox1 expression. BrdU$^+$ cycling cells were limited to the Sox1$^+$ cell population. The gate on the Sox1$^-$ population identifies prospective neurons.

16.9 PLURIPOTENT STEM CELLS AND THEIR DERIVATIVES

To fully realize the therapeutic potential of pluripotent stem cell biology, cell types of interest need to be well characterized and purified. For cell therapy applications, it is important to be able to identify and remove undifferentiated cell types that might be able to form teratomas. Screening both pluripotent cells and cells in an early stage of differentiation with a large cell surface library led to the identification of a novel cell surface marker, SSEA-5, the expression of which correlated with teratoma formation potential [46]. When SSEA-5$^+$ cells were removed from a culture, the teratoma potential of differentiated cultures decreased. To completely remove teratoma potential, multiparameter flow cytometry was utilized in which SSEA-5 was combined with two additional positive cell surface markers, either CD9 and CD90 or CD50 and CD200.

In addition to the removal of cell types that could cause teratomas, it is also important to identify and purify differentiated cell types from pluripotent cells. A cell surface marker screening strategy was used to identify cell surface markers present on pancreatic cells derived from pluripotent stem cells [47]. Cell surface markers that showed potentially interesting staining patterns were further explored. A combination of cell surface and IFC analysis was used to compare cell populations that were delineated by transcription factors (chromogranin A, NKX6.1, and PDX1) to define the particular cell subtypes of interest. Upon further characterization, endocrine cells displayed a CD200$^+$CD318$^+$ cell surface signature, whereas pancreatic endoderm cells were enriched in a CD142$^+$ cell population.

Cell surface marker screening has also been used to delineate cell surface markers that are present on cardiomyocytes derived from PSCs. By combining cell surface marker library screening with an intracellular marker of cardiomyocytes, cTNT, a multiparameter flow cytometric screening strategy was utilized to discover novel markers of cardiomyocytes [48]. Vascular cell adhesion molecule (VCAM1, CD106) was identified as a novel cell surface marker for cardiomyocytes. VCAM1 expression was mutually exclusive with the expression of markers of pluripotency (TRA-1-60), endothelium (CD144), and vascular pericytes (CD140b). From a screen of hESC-derived cardiomyocytes using an hESC Nkx2.5 GFP reporter cell line, SIRPA (CD172a) was identified as a cell surface marker expressed on cardiomyocytes [49]. Using a combination of surface and intracellular flow cytometry, SIRPA$^+$ cells were shown to coexpress a key cardiac intracellular marker, cTNT. Additional nonmyocyte markers, CD31, CD90, CD140B, and CD49a, were also identified from the screen and were utilized to enrich for cardiomyocytes by depletion.

16.10 ADULT STEM CELLS

Mesenchymal stromal cells (MSCs) are heterogeneous cells comprising stem cells with regenerative medicine potential. Because of their highly heterogeneous nature, comprehensive characterization of their phenotype is key to their application for cell therapy. Because the sources of MSC can vary, it is important to understand their particular identity. Cultured adipose stromal cells (ASCs) from multiple donors were interrogated for their cell surface signature, and specific subsets were further examined using multiparameter flow cytometry [50]. The cell surface signature as defined by the International Society for Cellular Therapy as the minimal criterion for cultured MSCs [51] was confirmed for the cultured ASCs from the five donors. Donor-specific variances were seen for certain markers such as CD34 and CD36 [50]. To further characterize subpopulations, multicolor

flow cytometry analysis for CD34 and CD36 expression, in combination with CD90, CD73, and CD105, known MSC-positive markers, was performed. Along with donor-specific variations, source-specific variation can also occur. Subcutaneous and visceral fat-derived ASCs obtained from the same donor were interrogated by cell surface marker profiling [52]. The initial profiling was an imaging screen and subsequent markers were confirmed and validated by flow cytometry. CD10 was predominantly expressed in subcutaneous fat ASCs, and sorting of CD10^{high} cells yielded cells with a higher capacity to differentiate. Visceral fat ASCs, which have a lower adipogenic differentiative capacity than subcutaneous fat ASCs, expressed higher levels of CD200, and significantly, CD200^{high} cells, in comparison to CD200^{low} cells, had a lower differentiative capacity.

In addition to understanding cell surface phenotypes, it is also significant to examine if different subsets of MSCs have differing differentiation capacities. To examine if a particular cell surface marker signature of MSCs could be indicative of differentiation capacity, mouse bone marrow-derived MSCs were screened for their cell surface marker signature [53]. An initial screen was performed on MSCs and the markers that were positive or heterogeneously expressed were also screened on clonal populations of MSCs that had variable osteogenic and adipogenic potential. CD200 was identified as a marker of clones with higher osteogenic potential, whereas SSEA-4 marked cells with higher adipogenic potential. CD140a was expressed in adipogenic cells but was not predictive of a lack of osteogenic potential.

Cultured human corneal endothelial cells (HCECs) are a potential therapeutic cell source for visual blindness from endothelial dysfunction. One of the main challenges is that isolated HCECs are often contaminated with stromal fibroblasts and the cell surface markers that would distinguish HCECs from stromal fibroblasts are not well characterized. Gene expression profiling in combination with an imaging-based cell surface marker screening strategy was utilized to identify GPC4 and CD200 as cell surface markers that mark HCECs but not fibroblasts [54]. The use of either CD200 or GPC4 to stain and effectively sort HCECs from stromal fibroblasts was also demonstrated.

16.11 CANCER STEM CELLS

Similarly, cancer cell populations can be interrogated through flow cytometric-based approaches to enable the discovery of new markers. In two flow cytometry-based cell surface marker screens [55,56], a fluorescent cell barcoding technique was applied to increase throughput and decrease interexperiment variability. Cell barcoding is a method by which different cell populations are stained with a distinguishing dye or a distinguishing concentration of dye and then pooled together prior to antibody labeling; cells can subsequently be deconvoluted based on fluorescence and/or the levels of their particular distinguishing dye [57]. Two primary adenocarcinoma (SW480 and HCT116) cell lines and one metastatic (SW620) colon cancer cell line were bar coded and analyzed for cell surface marker expression using a flow cytometric-based cell surface marker screening strategy [56]. In comparing cell surface marker expression from primary versus metastatic colon cancer cell lines, multiple cell surface markers were found to be higher in expression level in the metastatic cancer cell line, including CD10. These new markers can be further studied for their disease-state-specific expression.

The cell surface phenotypes of a stem cell such as the osteosarcoma (OS) cell line 3-aminobenzamide (3AB)-OS and the parental, differentiated cell line MG63 were profiled by flow cytometry [58]. Cell surface markers that were differentially expressed, homogeneously expressed, and homogeneously not expressed were identified, and differentially expressed markers were further characterized by ingenuity pathway analysis. The MAPK signaling pathway was activated in the 3AB-OS cell line in comparison to the MG63 cells, highlighting a potential target for further investigation.

A flow cytometric-based cell surface marker screen of six patient-derived glioblastomas in which cell barcoding was also used led to the identification of a novel cell surface marker, JAM-A, which plays a key role in cancer stem cell adhesion [55]. Cell adhesion is an important mechanism by which a cancer stem cell interacts with its niche. A better understanding of the cell surface signature of cancer stem cell populations with respect to adhesion molecule expression could enable therapeutic intervention. The glioblastoma screen was novel in that it was targeted toward finding biologically relevant cell adhesion surface markers expressed on glioblastoma cancer stem cells. Normal and neoplastic stem cells share adhesion mechanisms but JAM-A is a unique marker in that targeting of JAM-A did not negatively affect normal stem cells, making targeting of JAM-A a promising therapeutic application.

16.12 CONCLUSIONS AND FUTURE CONSIDERATIONS

The purification and analysis of stem cell cultures and their derivatives using multiparameter flow cytometry are a key element in realizing the potential of stem cell biology. The isolation of pure populations of cell types often requires knowledge of the cell surface signature of the particular cell type of interest. Both unbiased and biased cell surface marker screening strategies using either imaging or flow cytometry platforms can aid in understanding the cell surface signature of a cell type of interest. For stem cell lineages, transcription factors or other intracellular markers that demarcate a particular lineage are often known.

When either a cell reporter line is used or intracellular lineage markers amenable to flow cytometry exist, cell surface marker screens can be multiplexed with the intracellular markers to obtain rich datasets. To map the cell surface phenotype of particular neuronal subtypes, one can imagine performing a multicolor cell surface marker screen in which multiple neuronal subtype-specific intracellular markers are used. Cellular function-related markers, such as Ki-67, could also easily be added to screens to understand the proliferative capacity of the various cell types being screened. Additionally, fluorescent cell barcoding techniques can be employed in multiparameter cell surface marker screening to increase throughput and decrease variability. All in all, from multiparameter flow cytometry screens, incredibly informational datasets can be obtained and further investigated.

Both intracellular and cell surface marker signatures are invaluable to better characterize and serve as quantitative quality control tools in regenerative medicine. Flow cytometry is a particularly advantageous quality control method because of the quantitative nature of the results, especially compared to imaging and Western blot applications. In addition, multiparameter flow allows for multiple marker interrogation of one sample. As stem cell-based therapies move to becoming a reality, the ability to quantitate and monitor specific cell populations is going to become increasingly important.

Defined cell surface signatures also enable the purification of cell types. Homogeneous cell types of interest are important for drug screening models in that contaminating cell types can often perturb drug responses. In cell therapy applications, contaminating cell types should not be introduced into a patient, so understanding both the cell surface signature of the cell type of interest and the contaminating cells is key to designing proper cell sorting surface marker panels. Finally, as the fields of genomics and personalized medicine become more prevalent, the ability to isolate pure cell populations for subsequent characterization is increasingly important.

REFERENCES

[1] Herzenberg LA, et al. The history and future of the fluorescence activated cell sorter and flow cytometry: a view from Stanford. Clin Chem 2002;48(10):1819–27.

[2] Itsykson P, et al. Derivation of neural precursors from human embryonic stem cells in the presence of noggin. Mol Cell Neurosci 2005;30(1):24–36.

[3] Zeng X, et al. Dopaminergic differentiation of human embryonic stem cells. Stem Cells 2004;22(6):925–40.

[4] Chambers SM, et al. Highly efficient neural conversion of human ES and iPS cells by dual inhibition of SMAD signaling. Nat Biotechnol 2009;27(3):275–80.

[5] Reubinoff BE, et al. Neural progenitors from human embryonic stem cells. Nat Biotechnol 2001;19(12):1134–40.

[6] Perrier AL, et al. Derivation of midbrain dopamine neurons from human embryonic stem cells. Proc Natl Acad Sci USA 2004;101(34): 12543–8.

[7] Wu H, et al. Integrative genomic and functional analyses reveal neuronal subtype differentiation bias in human embryonic stem cell lines. Proc Natl Acad Sci USA 2007;104(34):13821–6.

[8] Yeo GW, et al. Alternative splicing events identified in human embryonic stem cells and neural progenitors. PLoS Comput Biol 2007;3(10):1951–67.

[9] Carson CT, Aigner S, Gage FH. Stem cells: the good, bad and barely in control. Nat Med 2006;12(11):1237–8.

[10] Roy NS, et al. Promoter-targeted selection and isolation of neural progenitor cells from the adult human ventricular zone. J Neurosci Res 2000;59(3):321–31.

[11] Maric D, et al. Prospective cell sorting of embryonic rat neural stem cells and neuronal and glial progenitors reveals selective effects of basic fibroblast growth factor and epidermal growth factor on self-renewal and differentiation. J Neurosci 2003;23(1): 240–51.

[12] Rietze RL, et al. Purification of a pluripotent neural stem cell from the adult mouse brain. Nature 2001;412(6848):736–9.

[13] Panchision DM, et al. Optimized flow cytometric analysis of central nervous system tissue reveals novel functional relationships among cells expressing CD133, CD15, and CD24. Stem Cells 2007;25(6): 1560–70.

[14] Keyoung HM, et al. High-yield selection and extraction of two promoter-defined phenotypes of neural stem cells from the fetal human brain. Nat Biotechnol 2001;19(9):843–50.

[15] Peh GS, et al. CD133 expression by neural progenitors derived from human embryonic stem cells and its use for their prospective isolation. Stem Cells Dev 2009;18(2):269–82.

[16] Golebiewska A, et al. Epigenetic landscaping during hESC differentiation to neural cells. Stem Cells 2009;27(6):1298–308.

[17] Chung S, et al. Genetic selection of sox1GFP-expressing neural precursors removes residual tumorigenic pluripotent stem cells and attenuates tumor formation after transplantation. J Neurochem 2006;97(5):1467–80.

[18] Hedlund E, et al. Selection of embryonic stem cell-derived enhanced green fluorescent protein-positive dopamine neurons using the tyrosine hydroxylase promoter is confounded by reporter gene expression in immature cell populations. Stem Cells 2007;25(5): 1126–35.

[19] Pruszak J, et al. Markers and methods for cell sorting of human embryonic stem cell-derived neural cell populations. Stem Cells 2007;25(9): 2257–68.

[20] Pruszak J, et al. CD15, CD24 and CD29 define a surface biomarker code for neural lineage differentiation of stem cells. Stem Cells 2009;27(12):2928–40.

[21] Yuan SH, et al. Cell-surface marker signatures for the isolation of neural stem cells, glia and neurons derived from human pluripotent stem cells. PLoS One 2011;6(3):e17540.

[22] Lee G, et al. Derivation of neural crest cells from human pluripotent stem cells. Nat Protoc 2010;5(4):688–701.

[23] Zhou Y, Snead ML. Derivation of cranial neural crest-like cells from human embryonic stem cells. Biochem Biophys Res Commun 2008;376(3):542–7.

[24] Doi D, et al. Isolation of human induced pluripotent stem cell-derived dopaminergic progenitors by cell sorting for successful transplantation. Stem Cell Rep 2014;2(3):337–50.

[25] DeQuach JA, et al. Decellularized porcine brain matrix for cell culture and tissue engineering scaffolds. Tissue Eng Part A 2011;17(21–22): 2583–92.

[26] Liu J, et al. Role of miRNAs in neuronal differentiation from human embryonic stem cell-derived neural stem cells. Stem Cell Rev 2012; 8(4):1129–37.

[27] Fu X, et al. Genetic approach to track neural cell fate decisions using human embryonic stem cells. Protein Cell 2014;5(1):69–79.

[28] Maroof AM, et al. Directed differentiation and functional maturation of cortical interneurons from human embryonic stem cells. Cell Stem Cell 2013;12(5):559–72.

[29] Israel MA, et al. Probing sporadic and familial Alzheimer's disease using induced pluripotent stem cells. Nature 2012;482(7384):216–20.

[30] Woodruff G, et al. The presenilin-1 DeltaE9 mutation results in reduced gamma-secretase activity, but not total loss of PS1 function, in isogenic human stem cells. Cell Rep 2013;5(4):974–85.

[31] Schondorf DC, et al. iPSC-derived neurons from GBA1-associated Parkinson's disease patients show autophagic defects and impaired calcium homeostasis. Nat Commun 2014;5:4028.

[32] Novarino G, et al. Mutations in BCKD-kinase lead to a potentially treatable form of autism with epilepsy. Science 2012;338(6105):394–97.

[33] Ordonez MP, et al. Disruption and therapeutic rescue of autophagy in a human neuronal model of Niemann Pick type C1. Hum Mol Genet 2012;21(12):2651–62.

[34] Lee G, et al. Modelling pathogenesis and treatment of familial dysautonomia using patient-specific iPSCs. Nature 2009;461(7262):402–6.

[35] Lee G, et al. Large-scale screening using familial dysautonomia induced pluripotent stem cells identifies compounds that rescue IKB-KAP expression. Nat Biotechnol 2012;30(12):1244–8.

[36] Kakinohana O, et al. Survival and differentiation of human embryonic stem cell-derived neural precursors grafted spinally in spinal ischemia-injured rats or in naive immunosuppressed minipigs: a qualitative and quantitative study. Cell Transpl 2012;21(12):2603–19.

[37] Hargus G, et al. Differentiated Parkinson patient-derived induced pluripotent stem cells grow in the adult rodent brain and reduce motor asymmetry in Parkinsonian rats. Proc Natl Acad Sci USA 2010;107(36):15921–6.

[38] Sevc J, et al. Effective long-term immunosuppression in rats by subcutaneously implanted sustained-release tacrolimus pellet: effect on spinally grafted human neural precursor survival. Exp Neurol 2013;248:85–99.

[39] Lu P, et al. Long-distance growth and connectivity of neural stem cells after severe spinal cord injury. Cell 2012;150(6):1264–73.

[40] Koester SK, Bolton WE. Intracellular markers. J Immunol Methods 2000;243(1–2):99–106.

[41] Jacobberger JW, Fogleman D, Lehman JM. Analysis of intracellular antigens by flow cytometry. Cytometry 1986;7(4):356–64.

[42] Jacobberger J. Flow cytometric analysis of intracellular protein epitopes. In: Carleton C. Stewart, Janet K.A. Nicholson, editors. Immunophenotyping. Wiley-Liss, Inc.; 2000. pp. 361–405.

[43] Koester SK, Bolton WE. Strategies for cell permeabilization and fixation in detecting surface and intracellular antigens. Methods Cell Biol 2001;63:253–68.

[44] Chambers SM, et al. Combined small-molecule inhibition accelerates developmental timing and converts human pluripotent stem cells into nociceptors. Nat Biotechnol 2012;30(7):715–20.

[45] Turac G, et al. Combined flow cytometric analysis of surface and intracellular antigens reveals surface molecule markers of human neuropoiesis. PLoS One 2013;8(6):e68519.

[46] Tang C, et al. An antibody against SSEA-5 glycan on human pluripotent stem cells enables removal of teratoma-forming cells. Nat Biotechnol 2011;29(9):829–34.

[47] Kelly OG, et al. Cell-surface markers for the isolation of pancreatic cell types derived from human embryonic stem cells. Nat Biotechnol 2011;29(8):750–6.

[48] Uosaki H, et al. Efficient and scalable purification of cardiomyocytes from human embryonic and induced pluripotent stem cells by VCAM1 surface expression. PLoS One 2011;6(8):e23657.

[49] Dubois NC, et al. SIRPA is a specific cell-surface marker for isolating cardiomyocytes derived from human pluripotent stem cells. Nat Biotechnol 2011;29(11):1011–8.

[50] Baer PC, et al. Comprehensive phenotypic characterization of human adipose-derived stromal/stem cells and their subsets by a high throughput technology. Stem Cells Dev 2013;22(2):330–9.

[51] Dominici M, et al. Minimal criteria for defining multipotent mesenchymal stromal cells. The International Society for Cellular Therapy position statement. Cytotherapy 2006;8(4):315–7.

[52] Ong WK, et al. Identification of specific cell-surface markers of adipose-derived stem cells from subcutaneous and visceral fat depots. Stem Cell Rep 2014;2(2):171–9.

[53] Rostovskaya M, Anastassiadis K. Differential expression of surface markers in mouse bone marrow mesenchymal stromal cell subpopulations with distinct lineage commitment. PLoS One 2012;7(12):e51221.

[54] Cheong YK, et al. Identification of cell surface markers glypican-4 and CD200 that differentiate human corneal endothelium from stromal fibroblasts. Invest Ophthalmol Vis Sci 2013;54(7):4538–47.

[55] Lathia JD, et al. High-throughput flow cytometry screening reveals a role for junctional adhesion molecule a as a cancer stem cell maintenance factor. Cell Rep 2014;6(1):117–29.

[56] Sukhdeo K, et al. Multiplex flow cytometry barcoding and antibody arrays identify surface antigen profiles of primary and metastatic colon cancer cell lines. PLoS One 2013;8(1):e53015.

[57] Krutzik PO, Nolan GP. Fluorescent cell barcoding in flow cytometry allows high-throughput drug screening and signaling profiling. Nat Methods 2006;3(5):361–8.

[58] Gemei M, et al. Surface proteomic analysis of differentiated versus stem-like osteosarcoma human cells. Proteomics 2013;13(22):3293–7.

Chapter 17

Flow-Cytometric Identification and Characterization of Neural Brain Tumor-Initiating Cells for Pathophysiological Study and Biomedical Applications

Sujeivan Mahendram[*,1,3], Minomi K. Subapanditha[*,1,2], Nicole McFarlane[1,3], Chitra Venugopal[1,3] and Sheila K. Singh[1,2,3]

[1]McMaster Stem Cell and Cancer Research Institute, McMaster University, Hamilton, Ontario, Canada; [2]Departments of Biochemistry and Biomedical Sciences, Faculty of Health Sciences, McMaster University, Hamilton, Ontario, Canada; [3]Departments of Biomedical Sciences and Surgery, Faculty of Health Sciences, McMaster University, Hamilton, Ontario, Canada

17.1 INTRODUCTION TO NEURAL STEM CELLS

Stem cells are a broad class of undifferentiated cells that have the ability to undergo multilineage differentiation and self-renewal [1]. Stem cells can be derived from both the embryo and the adult human, and are classified according to their developmental capacity [2]. The zygote and early blastomeres are comprised of totipotent cells capable of forming all lineages, whereas pluripotent cells, such as embryonic stem cells, are able to form all except the extra-embryonic lineage of the developing embryo. Most stem cells fall into the more restrictive category of multipotent stem cells, which are limited in giving rise to all cell types within a particular lineage. Conversely, unipotent cells are capable of forming one specific type of cell in a given lineage.

Neural stem cells (NSCs) are one such family of multipotent stem cells found in the central nervous system (CNS) that have the capability of generating neural tissue. Reynolds et al. (1992) were the first to definitively isolate NSCs from the CNS that demonstrated features of self-renewal and multipotency [3]. The team identified a particular class of cells with the ability to proliferate into a large cluster of undifferentiated neurospheres, in response to epidermal growth factors, which consisted of single clusters of neurons and glia, as well as cells with the same characteristics as the initial cell. The prospective identification of NSC populations requires the identification of proteins that are selectively expressed in NSCs. Glial fibrillary acidic protein (GFAP), first described in 1971 by Eng et al. (2000) is a member of the cytoskeletal protein family and is widely expressed in NSCs [4,5]; other common markers for NSCs include nestin, an intermediate filament; the cell-surface carbohydrate CD15, also known as stage-specific embryonic antigen-1 (SSEA-1) or Lewis X (LeX) [6,7]; and the cell-surface glycoprotein Prominin-1 (*PROM1*, also known as CD133) [8,9].

Subpopulations of brain tumor cells share similar characteristics with NSCs, and multiple hypotheses seek to explain the observation. One such explanation is the transformation of NSCs to brain tumor cells [10]. There is also evidence the brain tumor cells can result from transformation of more differentiated cells [10]. The microenvironment of stem cells has been shown to regulate the fate of cells and has implications in giving rise to tumors with high frequency, especially when in high numbers [11]. Thus, NSCs may have the capacity to be transformed into cancer cells, since their longevity permits accumulation of alterations that could perturb their controlled self-renewal machinery [12].

17.2 TUMORS OF THE CENTRAL NERVOUS SYSTEM

Different types of tumors in the CNS are named either by the type of cell they originate from or by the area in which they grow. The most studied primary tumors in the CNS are medulloblastoma (MB) and malignant glioma, which originate in the cerebellum or posterior fossa and from glial cells, respectively.

---

* Equal Contribution.

Neural Surface Antigens. http://dx.doi.org/10.1016/B978-0-12-800781-5.00017-7

These tumors are categorized on the basis of morphological and immunohistochemical features, such as cellular polymorphism and expression of key cell-surface antigens [13]. WHO grading is generally employed in the histological classification of tumors in order to predict their biological behavior [14]. The fourth edition of the WHO classification of tumors in neuroepithelial tissue is classified into nine sectors, with a grading of I to IV assigned to each tumor depending on the malignancy. Grade IV, considered "cytologically malignant, mitotically active, necrosis-prone neoplasms," is applied to both MB and glioblastoma multiforme (GBM), and can be associated with poor patient outcomes [14].

17.2.1 Glioblastoma Multiforme

Gliomas represent a broad diagnostic group that is classified into three subtypes by the WHO: astrocytomas, oligodendrogliomas, and mixed (oligoastrocytomas). Among the astrocytomas, GBM is the most common adult CNS tumor and the most malignant form of glioma (grade IV) as characterized by the WHO system [14]. GBM can exist in either primary or secondary form, with primary GBMs arising in the brain de novo and being highly aggressive and invasive [9]. Secondary GBMs are a result of progression from less aggressive tumors, usually arising within 5–10 years in patients initially diagnosed with low-grade astrocytomas [13]. Despite the differences in development, the two subtypes of GBM are largely indistinguishable in terms of patient survival and pathology. The mean age of patients with primary GBM is approximately 65 years, whereas the mean age of secondary GBM is approximately 45 years, although there is higher age distribution for secondary GBM [15]. Despite the advances in understanding GBM biology, overall survival of patients with newly diagnosed GBM is 17–30% at 1 year, and only 3–5% at 2 years [16]. Survival rates are observed to be higher in younger GBM patients, with a 13% survival rate in the first 5 years of diagnosis for patients aged 15–45 years, and only 1% for those aged ≥75 years.

Since neuroimaging technology and clinical examination of patients does not provide a definitive diagnosis, histopathological analysis is mandatory. The WHO system grades astrocytic tumors into four categories based on histological features [14]. GBM is identified as a grade IV tumor as it displays the criteria of cytological abnormalities: anaplasia, mitotic activity, microvascular proliferation, and necrosis. These astrocytic tumors are immunohistochemically characterized via standardized markers including GFAP, S-100, vimentin, and the nuclear marker Ki67 for proliferative index [17]. Positive staining of GBM with the three formerly mentioned lineage markers suggests heterogeneity within the tumor, which in turn advocates a stem cell origin for GBMs. Furthermore, in accordance with the cancer stem cell (CSC) hypothesis, the subpopulation of tumor cells in GBM that display stem-like cell properties

have been shown to express the surface antigens CD133 and CD15 (discussed below), and are implicated in the diagnosis of GBM [18,19].

Another classification method of GBM is via genomic analysis based on differential gene expression patterns, with distinct genetic alterations in primary and secondary GBMs [20–22]. Most genes that are implicated in GBM are associated with regulation of cell cycle, proliferation, and apoptosis. Genetic aberrations leading to amplification of epithelial growth factor receptor, low frequency of TP53 tumor suppressor, low expression of neurofibromatosis type 1, and high expression of genes associated with the tumor necrosis factor family are some of the few mutations associated with GBM diagnosis [21].

Given its poor prognosis, advances in molecular biology, cellular biology, and genomics have improved the pathogenic characterization and treatment of GBM [23,24]. However, the survival rate of the majority of GBM patients still remains less than 2 years following standard treatment using radiation and chemotherapy [25].

17.2.2 Medulloblastoma

MB is a CNS embryonal tumor commonly diagnosed in children, with high prevalence in the age group of 5–10 years [26]. MB is categorized as a grade IV tumor by the WHO system, and is considered the most malignant brain tumor in children [14]. MBs are suggested to be derived from primitive, pluripotent, neuroepithelial stem cells by in situ differentiation along glial and neuronal pathways [27]. The expression of Zic protein by tumor cells, which is a biomarker of the human cerebellar granule cell lineage, suggests ties of origin to the precursor cells of this lineage found in the maturing cerebellum [28]. The existence of multiple cellular origins of MB has placed its definitive origin under speculation [29,30].

The most common type of treatment is surgery, followed by radiation and chemotherapy, but these approaches are often associated with recurrence and metastasis. Only 50% of high-risk patients survive 5 years following diagnosis, and risk suffering from cognitive deficits and endocrine disorders as a consequence of therapy [31,32]. Predictably, the high malignancy grading of MB results in its poor prognosis, but diagnosis of the tumor state has improved over the past decade due to key discoveries in tumor biology. Molecular subtyping has associated subtypes of MB with the presence of biomarkers Zic1, Math1, Calbindin, and GFAP [28,33]. Furthermore, aberrations of signaling pathways have been heavily linked to the onset of MB. MB tumor cells often demonstrate abnormal activation of Sonic hedgehog or Wnt signaling pathways, both of which are related to embryonic development [30].

Similar to GBM, MB is a heterogeneous tumor that possesses a tumor initiating cell subpopulation [18], which has led to the identification of CD133, CD15, and CD271 as cell-surface markers for MB.

17.3 CANCER STEM-CELL HYPOTHESIS

About 150 years ago, pathologists Rudolph Virchow and Julius Cohnhem both observed histological similarities between primitive tumors and the developing fetus, which led to their postulation that cancer arises from cells with stem cell properties, or may derive from tissue-specific stem cells [34,35]. A variation of these postulations was put forth in the cancer stem cell (CSC) hypothesis, which suggests that a tumor consists of a fraction of cells known as CSCs that are capable of extensive proliferation and self-renewal.

The 1960s was a period of breakthrough research for CSCs, and multiple pieces of evidence helped to lay the foundation for the CSC hypothesis. First and foremost, Till and McCulloch (1964) demonstrated that the presence of a fraction of cells in normal hematopoietic cells could give rise to myeloerythroid colonies [36]. This subpopulation of cells demonstrated resistance to radiation, giving them the ability to self-renew in lethally irradiated mice. Around the same period, Bruce and Van Der Gaag (1963) demonstrated the proliferative ability of this subpopulation in mice [37]. A quantitative assay of the fraction of cells in murine lymphomas showed that only 1–4% of the transplanted murine lymphoma cells was able to form colonies. Research on tumor autotransplantation by Brunschwig et al. (1965) further clarified the presence of a tumorigenic subpopulation, as reinjection of bulk malignant tumors into the subcutaneous layer produced only a low frequency of tumor formation [38], suggesting that not every cell in a tumor was capable of initiating the tumor.

These studies eventually led to the first discovery of a tumor-initiating population in human acute myelogenous leukemia [39]. The subpopulation was identified upon regeneration of a heterogeneous leukemia cell population in a xenograft mouse after injection of a rare homogeneous population of leukemic cells that expressed hematopoietic stem cell (HSC) markers. Nonhomogeneity of tumor cells is evident from the fact that only this subpopulation of CSCs has the ability to initiate and maintain the cancerous growth by having properties similar to normal stem cells. Since these cancer cells are responsible for tumor initiation, successful isolation of these cells would provide insight into mechanisms of tumorigenesis. However, it is important to note that the CSC population is not entirely static; rather, CSC potential is dynamic in nature, with great influence from its microenvironment. Signals from the local environment may have the capacity to induce differentiated cells to revert to a stem cell-like phenotype, which may pose a significant challenge when targeting the CSC population [40].

17.4 BRAIN TUMOR INITIATING CELLS

Tumors with vastly different histology may have different prognoses. Moreover, brain tumors that share similar morphology and phenotype can also have drastically varying prognoses and responses to treatment, as well as display differential behaviors in patients of varying ages. The first study that applied the CSC hypothesis to human brain tumors identified a subpopulation of brain cancer cells that were phenotypically diverse and possessed a capacity for proliferation, self-renewal, and differentiation not evident in the bulk population of brain tumors [41]. This tumorigenic subpopulation consisted of cells called brain tumor initiating cells (BTICs), and represented a mere fraction of the total number of tumor cells. The BTIC has the phenotypic signature of the original tumor, such that these cells in isolation could replicate the original tumor, both in vitro and in vivo [18,41]. Singh et al. (2004) were able to isolate BTICs from both pediatric tumors, such as MB; adult low-grade tumors, such as pilocytic astrocytomas; and high-grade tumors, such as GBM [18].

The BTICs had phenotypic similarity to normal NSCs, in that they expressed markers like CD133 and nestin. CD133 is a cell-surface antigen, unlike nestin, lending itself to be used for effective BTIC live sorting by flow cytometry (for a comprehensive overview on CD133, please refer to Chapter 10). Only the CD133$^+$ subpopulation of the BTICs was capable of tumor initiation in vivo [18]. However, both CD133$^+$ cells and CD133$^-$ cells were found to co-exist in tumors generated from isolated CD133$^+$ cells, implicating a cellular hierarchy within the tumor.

17.5 IDENTIFICATION OF NEURAL SURFACE ANTIGENS

17.5.1 CD133

In order to gain a better understanding of neural and CSC biology, several groups have demonstrated the importance of the cell-surface glycoprotein Prominin-1/CD133. The discovery of Prominin-1 (hereon referred to as CD133) originates from two studies conducted simultaneously in 1997 that aimed to characterize novel antigens, one of which searched for unique murine NSC markers while the other involved human hematopoietic stem and progenitor cells [8,9]. While murine CD133 was identified on plasma membrane protrusions of neuroepithelial stem cells, the human homolog was observed in a subpopulation of CD34$^+$ HSCs.

The topology of CD133 consists of an extracellular N-terminal domain, five transmembrane domains, and an intracellular C-terminal domain. Two large extracellular loops formed by the transmembrane domains provide sites of glycosylation, although their precise location appears to vary among species [42–45]. In humans, the gene encoding CD133 is found on chromosome 4 and contains 37 exons, 28 of which are coding [46]. The gene is also under the control of 5 promoters, which may generate up to

12 splice variants in order to give rise to a 115–120 kDa protein [47,48]. Following its initial discovery, the presence of CD133 was later observed on neural precursors and NSCs [49,50]. Additionally, it appeared to concentrate on the apical plasma membrane, the surface that characteristically segregates with stem cells [8,51].

Within the last decade, CD133 expression has been attributed to the isolation and identification of brain tumor CSCs [41,52,53]. A milestone finding by Singh et al. (2004) arose when a CD133-specific antibody was used to sort CD133+ cells from patient GBM tumor samples. Strikingly, only the CD133+ cell population initiated brain tumors following the transplantation of 100 cells into the frontal cortex of NOD/SCID mice [18]. The CD133+ subpopulation has also been shown to exhibit stem-like properties including self-renewal and multipotency, while at the same time having increased resistance to ionizing radiation compared to CD133− GBM cells [41,52,54,55]. However, a growing controversy regarding the requirement of CD133 expression in CSCs has risen in recent years. A number of findings suggest there to exist CD133− glioma stem cells. First, the vast majority of GBM samples as well as established glioma cell lines lack CD133 positivity [56,57]. Second, cells that appear to display features of CSCs can be isolated from both CD133+ and CD133− tumors, albeit with observable phenotypic differences. An analysis of the transcriptional profiles of CD133+ and CD133− GBM CSCs demonstrated that those that were CD133 positive grew as neurospheres, whereas the latter typically grew adherently [58]. Lastly, CD133− cells isolated from a heterogeneous tumor can not only give rise to CD133+ cells in vitro and in vivo, but they can also self-renew and initiate tumor propagation upon xenotransplantation [59,60].

Discrepancies in the results obtained from CD133-related experiments may in large part be due to the use of antibodies recognizing different epitopes of the glycosylated protein. The two most commonly used antibodies are specific for either the AC133 or AC141 epitope. While both epitopes are glycosylated and are spatially distinct, their biochemical nature and precise locations have yet to be elucidated [9,61]. There have also been no reports showing a direct cross-comparison of both antibodies. Furthermore, the dependency of glycosylated CD133 epitopes also poses an issue when isolating CD133+ subpopulations. Exclusively using either of the aforementioned antibodies to isolate the positive fraction discounts differentially expressed or nonglycosylated CD133. Both the AC133 and AC141 epitopes can be downregulated without being influenced by CD133 mRNA levels [62,63]. In fact, the distribution of CD133 mRNA across tissues is much more widespread than either epitope. Given that flow-cytometric analysis (discussed below) depends on the use of such antibodies, further methodological improvements or complementary alternatives for identifying the bona fide stem cell subpopulation are warranted.

Although there is growing evidence supporting the association of CD133 with CSCs and stem-like properties, its true biological function and CD133+ exclusivity as a BTIC marker remain to be determined.

17.5.2 CD15

As controversies regarding the necessity of CD133 expression in CSCs have developed over recent years, many have resorted to using other alternative markers for identifying these tumor-initiating subpopulations. One marker that has gained particular interest is CD15, also known as stage-specific embryonic antigen-1 (SSEA-1) or Lewis X (LeX). (Please refer to Chapter 13 by Itokazu & Yu for more details). CD15 is a sugar moiety that is typically expressed in neural progenitors and NSCs, as well as embryonic stem cells during early development [6,7]. In murine brains, its expression is localized to the subventricular zone [6]. In vitro culturing of sorted populations showed CD15+ cells to readily form neurospheres, whereas CD15− cells did not [6,7].

Unlike CD133, the association of CD15 to stemness is relatively more defined. For instance, Imura et al. (2006) showed that CD15+ cells were capable of differentiating into multiple cells types of the neural lineage, including neurons, astrocytes, and oligodendrocytes [64]. In addition, its expression appears to correlate with key proteins involved in stemness, including FGF8 and Wnt-1 [7,65]. While FGF8 is heavily associated with hematopoiesis, Wnt-1 plays an important role in midbrain neural specification [66]. Of note, CD15 co-immunoprecipitates with Wnt-1, further suggesting a direct relationship between both proteins [67,68].

Moreover, in a mouse model of MB, it was demonstrated that the CD15+ subpopulation was more effective at propagating tumors in comparison to CD133+ cells [69]. In human GBMs, CD15 was shown to enrich for the stem cell compartment in CD133− tumors [70]. These cells enriched for CD15 positivity displayed the capability to form neurospheres and differentiate into various neural lineages. When transplanted into immunocompromised mice, CD15+ cells were at least 100-fold more tumorigenic when compared to CD15− cells. Unlike the CD15− population, the CD15+ cells were able to recapitulate the heterogeneic marker expression seen in the original tumor [70]. Thus, it is possible that combinations of CD133 and CD15 marker expression, along with other putative markers like CD271, may better refine the true functional BTIC population.

17.5.3 CD271

Nerve growth factor receptor (NGFR), also known as CD271 or p75 neurotrophin receptor, interacts with the neurotrophic factor NGF, which is heavily implicated in neuronal survival in vivo [71]. The 45 kDa receptor consists of a signal peptide, an extracellular domain containing four 40 amino acid repeats, a single transmembrane domain, as well as a cytoplasmic domain [71]. The ligand-binding site of CD271 is postulated to be the negatively charged region of the extracellular region.

CD271 is found on the neuronal cell surface, and is expressed on the surface of neuroblastoma and MB cells [72,73]. While the overexpression of CD271 is observed in the developing external granular layer of the fetal cerebellum, its expression has not been documented in the adult brain [73]. Recently, Morrison et al. (2013) identified the role of the cell-surface marker in selecting for MB cell types [74]. CD271, when used in conjunction with CD133, was shown to select for cells with high self-renewal capacity both in vitro and in vivo. The self-renewal ability of the CD271$^+$ cells creates a drug-resistant phenotype in this subpopulation. Furthermore, the marker is differentially expressed in subtypes of MB with higher expression in the fetal and Shh molecular variant cerebellum, and lower expression in the most invasive classifications [74].

The capacity of CD271 in marking tumor initiation and chemoresistance at low concentrations signifies its therapeutic potential, particularly in MB.

17.5.4 High Aldehyde Dehydrogenase Activity

While the use of cell-surface antigens is one method for identifying neural BTICs, an alternative approach takes advantage of their innate aldehyde dehydrogenase (ALDH) activity. Naturally, these enzymes work to detoxify aldehyde products produced by reactive oxygen species, as they play a key role in a number of biological activities including cell proliferation and drug resistance [75–77]. HSCs display higher levels of cytosolic ALDH compared to less primitive progenitor cells, and this feature has previously been used to select for HSCs [78,79]. Similarly, high ALDH activity has also been used to identify both mesenchymal and epithelial stem cells, as well as primitive NSCs [80,81]. Not only can this functional assay be used to identify normal stem cells, it is also effective in enriching for their malignant counterparts [82].

Jones and colleagues (1995) originally identified a method of assessing cytosolic ALDH activity in viable cells using a fluorescent aldhehyde substrate [83]. In this assay, cells are incubated with a fluorescent substrate (dansylaminoacetaldehyde (DAAA)) to allow for take-up by passive diffusion. The uncharged substrate is converted by intracellular ALDH into dansyl glycine, a negatively charged molecule. Cells that have high ALDH activity (ALDH$^+$) will display an overall accumulation of the negatively charged product, ultimately causing these cells to become more fluorescent than ALDH$^-$ cells. These cells can then be isolated

and enriched using flow cytometry (see below) without the use of antibodies. In present day, this method is commonly referred to now as the Aldefluor™ assay, in which the substrate (BODIPY®-acetaldehyde (BAAA)) has been optimized to rapidly select for viable cells across multiple cell lineages in addition to HSCs.

While the Aldefluor™ assay can be used on its own, it can also be coupled with cell-surface markers such as CD133 and CD15 in order to provide a more efficient way of enriching for the CSC population within heterogeneous tumors.

17.6 FLOW-CYTOMETRIC IDENTIFICATION AND CHARACTERIZATION

The first fluorescence-based flow cytometer was developed in 1968. Early flow cytometers were basic, single-excitation source instruments, able to analyze multiple tubes with single colors. The refined instruments of today have multiple excitation sources and are able to analyze single tubes with multiple colors. Hundreds of thousands of cells may be analyzed, making it an ideal technology for exploring the heterogeneous nature of brain tumor samples and isolating rare BTICs. For an overview of the fundamentals of flow cytometry, please refer to Chapter 2.

We stain our primary GBM BTICs with CD133/2-APC (clone 293C3, Miltenyi Biotec). Our enzymatically digested tissues and neurospheres often have nondistinct scatter patterns (Figure 17.1(A)). We also use the viability dye 7-AAD (7-aminoactinomycin D) to exclude dead cells (Figure 17.1(B)). Given that CD133$^+$ cells rarely exhibit distinct populations, we use a matched isotype as a negative control to assess nonantigen-related binding and position our analysis regions (Figure 17.1(C) and (D)). We have consistently found that patients with a higher CD133 proportion have significantly lower survival times (unpublished data). When CD133$^+$ and CD133$^-$ sorted populations are checked for purity (Figure 17.2(A) and (B)) and plated separately, CD133$^+$ cells form bigger neurospheres compared to CD133$^-$ cells (Figure 17.2(C) and (D)). We also found that CD133$^+$ cells showed an increase in self-renewal capacity (as assessed by secondary sphere formation) compared to its counterpart.

We have also shown GBM patient survival to be inversely correlated to secondary sphere formation. Conceivably, those tumors containing the most CSCs result in high secondary sphere formation and are clinically more aggressive [84].

Primary brain tumor samples naturally grow as spheres under NSC culture conditions and must be enzymatically dissociated prior to flow sorting. This process produces cell death, and DNA released by dead cells can in turn cause cells to clump. We use DNAse to reduce this aggregation and re-suspend our stained cells in phosphate buffered saline with 2 μm EDTA, which helps to prevent cation-dependent

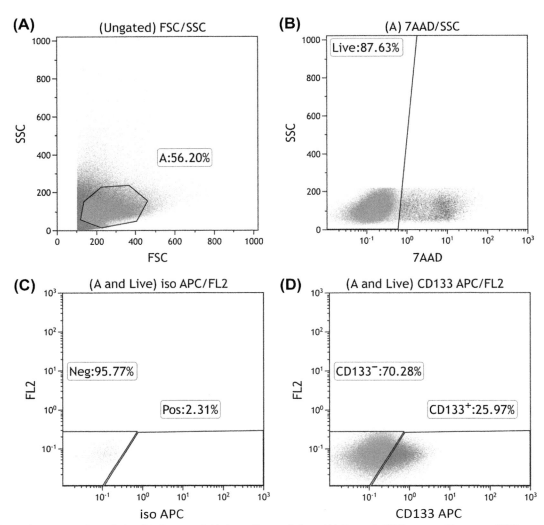

FIGURE 17.1 Flow-cytometric analysis of brain tumor-initiating cells populations. (A) Forward (FSC) versus side scatter (SSC) properties provide an image of all cells, including debris. (B) Cells that stain with the viability dye 7-AAD are excluded from analysis. (C) The position of the statistical quadrants is determined using appropriate isotype controls. (D) Tumor cells stained with surface marker CD133-APC.

cell–cell interactions. Additionally, cell suspensions are filtered through a 40 μm nylon mesh. All of these steps aid in maintaining the single cell suspension required for flow. As these manipulations already impart stress on the cells, it is important to reduce cell stress due to shear forces experienced during a sort. We have found that lower sheath pressure (30 psi) with a 100 μm nozzle improves cell survival. Moreover, we rarely sort at high data rates, as faster sort speeds are typically achieved by concentrating the cell suspensions, and this increases the tendency of our cells to clump. Our typical data rate is less than 6000 events per second.

The advent of multi-parameter fluorescent-activated cell sorting and monoclonal antibody technology has thus allowed for the purification of CSCs based on cell-surface markers. By using stem cell markers such as CD133 and CD15, we have been able to prospectively sort for specific stem cell-like populations and to study their self-renewal capacity.

17.7 LIMITATIONS OF CURRENT CSC MARKERS

17.7.1 Hypoxia

It is commonly recognized that the microenvironment plays a significant role in the regulation and maintenance of stem cells. Hypoxia is one important feature of the stem cell niche, as it has previously been shown to promote self-renewal and maintain an undifferentiated cell state [85]. In GBM, the hypoxic microenvironment is believed to induce BTIC proliferation and aid in promoting therapy resistance [86,87]. McCord et al. (2009) demonstrated that the expression of CD133 was upregulated when primary glioma

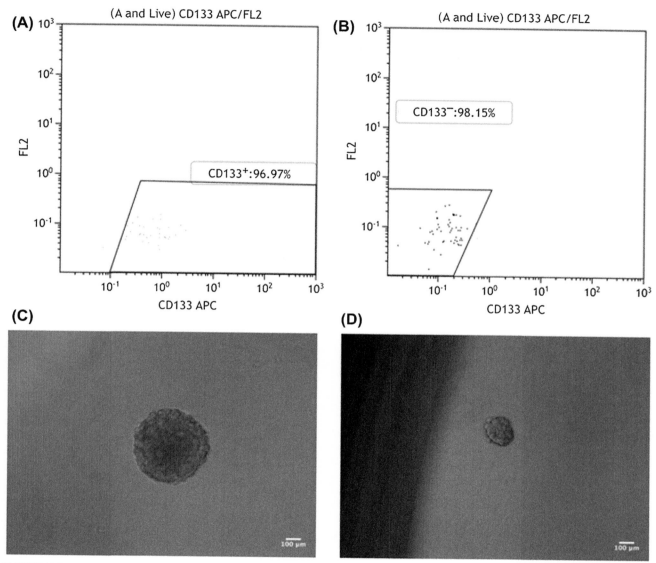

FIGURE 17.2 **Sorted cells in culture.** CD133+ cells form large neurospheres in culture. Purity check on (A) CD133+ and (B) CD133− sorted cells. (C) CD133+ cells form huge neurospheres when compared to (D) CD133− cells (10×magnification).

spheres were cultured under more physiological levels of oxygen (i.e., 7%) compared to normoxic (i.e., 21% O_2) culture conditions [88]. Key regulators of the hypoxic response include transcription factors of the hypoxia-inducible factor (HIF) family. In human lung cancer cells, it was shown that HIF-1α and HIF-2α induce *Oct4* and *Sox2* expression under hypoxic culture conditions, and that this in turn was necessary for the upregulation of CD133 [89].

While hypoxia-induced CD133 expression has been reported by a number of groups, these studies failed to assess whether glycosylation status was affected by such conditions. Lehnus et al. (2013) recently showed that indeed CD133 N-glycosylation in GBM cells is enhanced in response to hypoxia [90]. This finding also sheds light in providing an explanation for the existing controversies

between the tumorigenic potential of CD133+ and CD133− cells.

Thus, it is becoming increasingly necessary to culture cells under hypoxic or as close to physiological conditions as possible, in order to be able to attain a true isolation of the stem cell or CSC compartment when using external cell-surface markers like CD133.

17.7.2 Cell Dissociation Methods

The expression of CD133 and other cell-surface markers can also be affected by the method of cell dissociation used. It is important to take caution when using flow-cytometric analysis to isolate various subpopulations, as the type of cell dissociation reagent used and duration of treatment can significantly impact cell-surface marker expression. In particular, the use of the serine protease trypsin has been shown to cleave CD133 antigen expression, and consequently complicate assay interpretations [91]. Panchision et al. (2007) also demonstrated that disaggregation of neural precursor cells using trypsin, papain, or collagenase protease cocktails appeared to differentially influence the expression patterns of CD133 and CD15 [92]. As the need for refining cell fractions using flow cytometry increases, the requirement of using a standard cell dissociation reagent is also becoming important in order to help prevent conflicting results as reported in the literature.

17.7.3 Dynamic Growth

An important characteristic to consider when purifying populations by flow cytometry is that several commonly used markers differ in their expression patterns at varying stages of the cell cycle. As cells can shuttle between quiescent and activated states, as well as primitive and committed stages, it is possible that two phenotypically distinct populations may in fact be the same but are in different stages of the cell cycle [93,94]. One marker that appears to be influenced by cell cycle state is CD133. Using cell cycle profiling in human NSCs, Sun et al. (2009) showed that CD133⁻ cells predominate in G1/G0, whereas CD133⁺ cells are preferentially expressed during S, G2, or M phase [95]. As such, it is important to consider the specific stage of the cell cycle when using flow-cytometric methods to enrich for CD133 subpopulations, and also generally when analyzing cell-surface molecular profiles.

17.8 APPLICATIONS OF FUNCTIONAL BTIC ASSAYS

17.8.1 Neurosphere Formation

The neurosphere formation assay is a commonly used technique in neurogenesis and modeling early neural development. It presents an ideal clonal assay to quantitate the frequency of NSCs in a given heterogeneous cell population [96,97]. This standardized method, also referred to as the limiting dilution assay (LDA), is greatly facilitated by clonal sphere growth in vitro to test for self-renewal capacity (Figure 17.3). Briefly, tumorspheres are dissociated and serially diluted down to a single cell per well, after which the rate of subsequent sphere formation in culture over the next 7 days for example, is calculated. On day 7, the percentage of wells not containing spheres for each cell-plating density (F_0) is calculated and plotted against the number of cells per well (x). The number of cells required to form at least one tumorsphere in every well is determined from the point at which the line crosses the 0.37 threshold ($F_0 = e - x$). In a Poisson distribution of cells, $F_0 = 0.37$ corresponds to the dilution of one cell per well, so that this calculation (the 37% intersect) reflects the frequency of clonogenic stem cells in the entire cell population.

In vitro culture of neurospheres allows for the propagation of a heterogeneous population of NSCs and their progenitors at various stages throughout development. Since its original discovery approximately 20 years ago, the assay has undergone a number of advancements. Typically, cells exhibiting stem-like characteristics will proliferate and form clonal spheres when cultured under serum-free conditions,

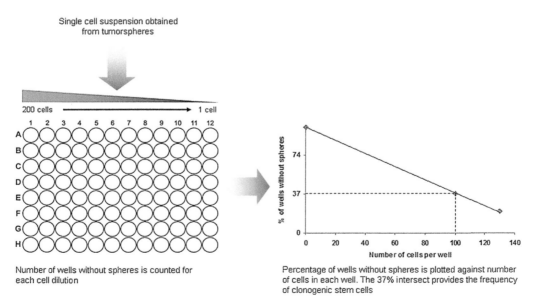

FIGURE 17.3 Schematic of limiting dilution assay. Tumorspheres are dissociated into single-cell suspension and are then plated into a 96-well plate at dilutions ranging from 200 cells/well to 1 cell/well. Following incubation for 3–7 days, the percentage of wells without spheres is calculated and plotted as a function of the number of cells per well. A line of best fit is then drawn, and the number of clonogenic cells is determined by the 37% intersect.

while those incapable of self-renewal die off following multiple passages [55,98,99]. In order to determine multipotency, it is now accepted that cells must be plated at clonal density and that the developing neurospheres must be able to give rise to neurons, astrocytes, and oligodendrocytes upon cues to differentiation.

However, certain drawbacks to the neurosphere formation assay do exist, which consequently limit its ability to exclusively identify the stem cell and CSC populations within these spheres [100]. First, while neurospheres can be formed by NSCs, they can also arise from progenitors and other less primitive cells. Second, cells within neurospheres can alternate between activated and quiescent states. More committed progenitors also have the ability to revert back to a stem cell-like state [101]. Third, neurosphere fusion appears to be a common and rapid event irrespective of the cell type [102]. These occurrences not only underestimate stem cell frequency calculated solely based on the number of neurospheres observed, but also complicate the notion that neurosphere size is an indicator of proliferative potential [102,103]. Lastly, despite the importance and wide application of the assay, it lacks a standardized protocol, as there is a high degree of variability between the types of growth factors and hormones used in the culture media [98]. This complicates data interpretation and comparison of assay results among research groups.

An alternative approach to help definitively prove the existence of stem-like populations is the use of molecular beacons [104–107]. This technology measures stemness by providing a fluorescent readout of the mRNA expression of key stemness markers and can be used in conjunction with neurosphere assays. Ilieva et al. (2013) showed that *Sox2* and *Oct4* molecular beacons localized to the center of neurospheres during differentiation [108]. In brief, hairpin molecular beacons tagged with a fluorophore and quencher hybridize to target mRNA [109]. Once hybridization occurs,

the fluorophore is released, allowing for the detection of the signal. This live-cell assay has recently been used to sort for undifferentiated NSCs and mouse embryonic stem cells by designing molecular beacons targeting *Sox2* mRNA [110].

17.8.2 In vivo Serial Transplantation

While in vitro assays provide a relatively quick method to help identify stem cell populations, a number of caveats exist as discussed above. The gold standard assay for testing the hallmark features of stem cells (i.e., self-renewal and pluripotency) lies in the serial transplantation of animal models. The use of such in vivo assays together with flow-cytometric cell sorting provides a powerful tool for testing the tumorigenic ability of prospectively sorted stem cell fractions, as it is the combination of these techniques from which informative data can be obtained. During in vivo serial transplantation, cells are orthotopically xenotransplanted into immunocompromised mice, after which the mice are monitored over several weeks for the formation of a tumor. In order to assess self-renewal, cells of the initial tumor must be isolated and grafted into a second immunocompromised mouse (Figure 17.4). As briefly mentioned earlier, Singh et al. (2004) demonstrated that xenotransplantation with as few as 100 CD133+ GBM cells into adult NOD/SCID mouse brains gave rise to tumors consisting of both CD133+ and CD133− cell populations [18]. Furthermore, when these primary tumors were serially transplanted into a second recipient mouse, the newly arising tumors were reminiscent of the initial tumor [18]. In contrast, injection with as many as 100,000 CD133− GBM cells did not have the same effect. Similarly, Harris et al. (2008) demonstrated the self-renewal capacity of glioma stem cells by transplanting glioma spheres into mouse brains and serially transplanting the xenografted tumors for more than six passages [111].

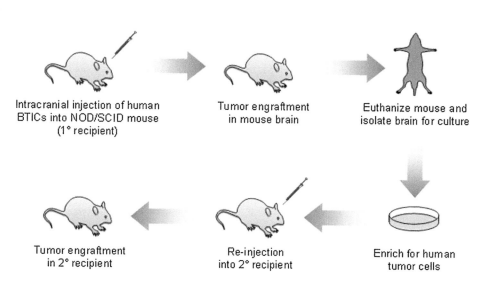

Intracranial injection of human BTICs into NOD/SCID mouse (1° recipient)

Tumor engraftment in mouse brain

Euthanize mouse and isolate brain for culture

Tumor engraftment in 2° recipient

Re-injection into 2° recipient

Enrich for human tumor cells

FIGURE 17.4 Schematic representation of in vivo serial xenotransplantation. Human BTICs are intracranially injected into an immunocompromised mouse. Once the tumor has engrafted, the mouse is euthanized and brain isolated for culture. The human tumor cells are then enriched and similarly re-injected into a second mouse to observe for secondary tumor engraftment.

Though it has generally been accepted that in vivo serial transplantation is the best functional assay to identify primitive stem cell populations, there are several issues worth noting. First, it is possible that the recipient grafting site can provide a tumor-proliferative environment. While it is commonly believed that stem cells require signals from the surrounding stroma for maintaining stemness, it is unclear whether separating tumor-initiating cells from the rest of a given population during the assay has any effect on the outcome. Orimo et al. (2005) used a tumor xenograft model to demonstrate that a population of breast carcinoma cells mixed with carcinoma-associated fibroblasts promotes tumor growth and uptake much more readily than those that have been mixed with normal mammary fibroblasts [112]. Second, nontumor cells engrafted next to tumor-associated stromal tissue can also become tumorigenic, likely due to the release of tumor-promoting cytokines and hormones [113]. It is also possible that non-CSCs can recapitulate the original tumor, as they may be able to give rise to a diverse array of cell types solely on the basis of undergoing a high level of genetic and epigenetic transformations. Lastly, in vivo assays like serial transplantations can be expensive and time-consuming, as they can take up to 6 months or more to complete.

17.9 CONCLUSION

Taken together, the identification of neural BTICs by flow-cytometric analysis requires a multitude of resources and factors to be considered. Despite existing controversies in the field, a considerable number of advancements have been made to help reasonably explain conflicting results. Based upon the aforementioned limitations, it is increasingly becoming imperative that standardized assays and reagents be utilized. Moreover, it is unlikely that a single marker of BTICs exists. Given that marker expression is heavily influenced by the culture conditions used and the cell cycle state, it is more plausible to suggest that several different markers make up a "blueprint" for CSC identification depending on the specific circumstances evaluated. As in the case for CD133, perhaps identifying its biological role in tumor development may be a more pertinent issue as opposed to it being a marker of CSCs. Nevertheless, it is crucial to explore the biology of purified BTIC populations in physiological conditions by complementing such in vitro assays with functionally based serial transplantations in vivo.

Therapeutic approaches in the future could potentially aim to treat cancers of the brain differently from others based solely on eradicating neural BTICs. Successful identification and characterization of this rare cell population will enable significant advancements towards personalized drug-targeting strategies, thereby reducing or even eliminating the need for radiation or chemotherapy while minimizing the risk of genetically altering healthy cells.

REFERENCES

[1] Till JE, McCulloch EA. A direct measurement of the radiation sensitivity of normal mouse bone marrow cells. Radiat Res 1961;14:213–22.

[2] Jaenisch R, Young R. Stem cells, the molecular circuitry of pluripotency and nuclear reprogramming. Cell 2008;132:567–82.

[3] Reynolds BA, Tetzlaff W, Weiss S. A multipotent EGF-responsive striatal embryonic progenitor cell produces neurons and astrocytes. J Neurosci 1992;12:4565–74.

[4] Eng LF, Vanderhaeghen JJ, Bignami A, Gerstl B. An acidic protein isolated from fibrous astrocytes. Brain Res 1971;28:351–4.

[5] Eng LF, Ghirnikar RS, Lee YL. Glial fibrillary acidic protein: GFAP-thirty-one years (1969–2000). Neurochem Res 2000;25:1439–51.

[6] Capela A, Temple S. LeX/ssea-1 is expressed by adult mouse CNS stem cells, identifying them as nonependymal. Neuron 2002;35:865–75.

[7] Capela A, Temple S. LeX is expressed by principle progenitor cells in the embryonic nervous system, is secreted into their environment and binds Wnt-1. Dev Biol 2006;291:300–13.

[8] Weigmann A, Corbeil D, Hellwig A, Huttner WB. Prominin, a novel microvilli-specific polytopic membrane protein of the apical surface of epithelial cells, is targeted to plasmalemmal protrusions of non-epithelial cells. Proc Natl Acad Sci USA 1997;94:12425–30.

[9] Yin AH, et al. AC133, a novel marker for human hematopoietic stem and progenitor cells. Blood 1997;90:5002–12.

[10] Oliver TG, Wechsler-Reya RJ. Getting at the root and stem of brain tumors. Neuron 2004;42:885–8.

[11] Moore KA, Lemischka IR. Stem cells and their niches. Science 2006;311:1880–5.

[12] Werbowetski-Ogilvie TE, et al. Evidence for the transmission of neoplastic properties from transformed to normal human stem cells. Oncogene 2011;30:4632–44.

[13] Maher EA, et al. Malignant glioma: genetics and biology of a grave matter. Genes Dev 2001;15:1311–33.

[14] Louis DN, et al. The 2007 WHO classification of tumours of the central nervous system. Acta Neuropathol 2007;114:97–109.

[15] Ohgaki H, et al. Genetic pathways to glioblastoma: a population-based study. Cancer Res 2004;64:6892–9.

[16] Adamson C, et al. Glioblastoma multiforme: a review of where we have been and where we are going. Expert Opin Invest Drugs 2009;18:1061–83.

[17] Niemiec J. Ki-67 labelling index in human brain tumours. Folia Histochem Cytobiol 2001;39:259–62.

[18] Singh SK, et al. Identification of human brain tumour initiating cells. Nature 2004;432:396–401.

[19] Mao XG, et al. Brain tumor stem-like cells identified by neural stem cell marker CD15. Transl Oncol 2009;2:247–57.

[20] Huse JT, Holland EC. Targeting brain cancer: advances in the molecular pathology of malignant glioma and medulloblastoma. Nat Rev Cancer 2010;10:319–31.

[21] Verhaak RG, et al. Integrated genomic analysis identifies clinically relevant subtypes of glioblastoma characterized by abnormalities in PDGFRA, IDH1, EGFR, and NF1. Cancer Cell 2010;17:98–110.

[22] Watanabe K, et al. Overexpression of the EGF receptor and p53 mutations are mutually exclusive in the evolution of primary and secondary glioblastomas. Brain Pathol 1996;6:217–23. discussion 223–214.

[23] Huse JT, Holland E, DeAngelis LM. Glioblastoma: molecular analysis and clinical implications. Annu Rev Med 2013;64:59–70.

[24] Tanaka S, Louis DN, Curry WT, Batchelor TT, Dietrich J. Diagnostic and therapeutic avenues for glioblastoma: no longer a dead end? Nat Rev Clin Oncol 2013;10:14–26.

[25] Grossman SA, et al. Survival of patients with newly diagnosed glioblastoma treated with radiation and temozolomide in research studies in the United States. Clin Cancer Res 2010;16:2443–9.

[26] Novakovic B. U.S. childhood cancer survival, 1973–1987. Med Pediatr Oncol 1994;23:480–6.

[27] Raffel C. Medulloblastoma: molecular genetics and animal models. Neoplasia 2004;6:310–22.

[28] Yokota N, et al. Predominant expression of human zic in cerebellar granule cell lineage and medulloblastoma. Cancer Res 1996;56:377–83.

[29] Read TA, Hegedus B, Wechsler-Reya R, Gutmann DH. The neurobiology of neurooncology. Ann Neurol 2006;60:3–11.

[30] Wang J, Wechsler-Reya RJ. The role of stem cells and progenitors in the genesis of medulloblastoma. Exp Neurol 2014;260C:69–73.

[31] Packer RJ, et al. Outcome for children with medulloblastoma treated with radiation and cisplatin, CCNU, and vincristine chemotherapy. J Neurosurg 1994;81:690–8.

[32] Mulhern RK, et al. Neurocognitive consequences of risk-adapted therapy for childhood medulloblastoma. J Clin Oncol 2005;23:5511–9.

[33] Salsano E, et al. Expression profile of frizzled receptors in human medulloblastomas. J Neuro-oncol 2012;106:271–80.

[34] Virchow R, Smith TP. Post-mortem examinations, with especial reference to medico-legal practice. Arch Pathol Anat Physiol Klin Med 1885;3:23.

[35] Cohnheim J. Ueber entzuendung und eiterung. Path Anat Physiol Klin Med 1867;40:1–79.

[36] McCulloch EA, Till JE. Proliferation of hemopoietic Colony-Forming cells transplanted into irradiated mice. Radiat Res 1964;22:383–97.

[37] Bruce WR, Van Der Gaag H. A quantitative assay for the number of murine lymphoma cells capable of proliferation in vivo. Nature 1963;199:79–80.

[38] Brunschwig A, Southam CM, Levin AG. Host resistance to cancer. clinical experiments by homotransplants, autotransplants and admixture of autologous leucocytes. Ann Surg 1965;162:416–25.

[39] Lapidot T, et al. A cell initiating human acute myeloid leukaemia after transplantation into SCID mice. Nature 1994;367:645–8.

[40] Vermeulen L, de Sousa e Melo F, Richel DJ, Medema JP. The developing cancer stem-cell model: clinical challenges and opportunities. Lancet Oncol 2012;13:e83–89.

[41] Singh SK, et al. Identification of a cancer stem cell in human brain tumors. Cancer Res 2003;63:5821–8.

[42] Corbeil D, Fargeas CA, Huttner WB. Rat prominin, like its mouse and human orthologues, is a pentaspan membrane glycoprotein. Biochem Biophy Res Commun 2001;285:939–44.

[43] McGrail M, et al. Expression of the zebrafish CD133/prominin1 genes in cellular proliferation zones in the embryonic central nervous system and sensory organs. Dev Dyn 2010;239:1849–57.

[44] Jaszai J, et al. Distinct and conserved prominin-1/CD133-positive retinal cell populations identified across species. PloS One 2011;6:e17590.

[45] Han Z, Papermaster DS. Identification of three prominin homologs and characterization of their messenger RNA expression in *Xenopus laevis* tissues. Mol Vision 2011;17:1381–96.

[46] Fargeas CA, et al. Identification of novel Prominin-1/CD133 splice variants with alternative C-termini and their expression in epididymis and testis. J Cell Sci 2004;117:4301–11.

[47] Shmelkov SV, et al. Alternative promoters regulate transcription of the gene that encodes stem cell surface protein AC133. Blood 2004;103:2055–61.

[48] Pleshkan VV, Vinogradova TV, Sverdlov ED. Methylation of the prominin 1 TATA-less main promoters and tissue specificity of their transcript content. Biochim Biophys Acta 2008;1779:599–605.

[49] Mizrak D, Brittan M, Alison M. CD133: molecule of the moment. J Pathol 2008;214:3–9.

[50] Uchida N, et al. Direct isolation of human central nervous system stem cells. Proc Natl Acad Sci USA 2000;97:14720–5.

[51] Fietz SA, et al. OSVZ progenitors of human and ferret neocortex are epithelial-like and expand by integrin signaling. Nat Neurosci 2010;13:690–9.

[52] Bao S, et al. Glioma stem cells promote radioresistance by preferential activation of the DNA damage response. Nature 2006;444:756–60.

[53] Singh SK, Clarke ID, Hide T, Dirks PB. Cancer stem cells in nervous system tumors. Oncogene 2004;23:7267–73.

[54] Galli R, et al. Isolation and characterization of tumorigenic, stem-like neural precursors from human glioblastoma. Cancer Res 2004;64:7011–21.

[55] Hemmati HD, et al. Cancerous stem cells can arise from pediatric brain tumors. Proc Natl Acad Sci USA 2003;100:15178–83.

[56] Joo KM, et al. Clinical and biological implications of CD133-positive and CD133-negative cells in glioblastomas. Lab Invest 2008;88:808–15.

[57] Beier D, et al. CD133(+) and CD133(−) glioblastoma-derived cancer stem cells show differential growth characteristics and molecular profiles. Cancer Res 2007;67:4010–5.

[58] Lottaz C, et al. Transcriptional profiles of CD133+ and CD133− glioblastoma-derived cancer stem cell lines suggest different cells of origin. Cancer Res 2010;70:2030–40.

[59] Wang J, et al. CD133 negative glioma cells form tumors in nude rats and give rise to CD133 positive cells. Int J Cancer 2008;122:761–8.

[60] Chen R, et al. A hierarchy of self-renewing tumor-initiating cell types in glioblastoma. Cancer Cell 2010;17:362–75.

[61] Miraglia S, et al. A novel five-transmembrane hematopoietic stem cell antigen: isolation, characterization, and molecular cloning. Blood 1997;90:5013–21.

[62] Corbeil D, et al. The human AC133 hematopoietic stem cell antigen is also expressed in epithelial cells and targeted to plasma membrane protrusions. J Biol Chem 2000;275:5512–20.

[63] Florek M, et al. Prominin-1/CD133, a neural and hematopoietic stem cell marker, is expressed in adult human differentiated cells and certain types of kidney cancer. Cell Tissue Res 2005;319:15–26.

[64] Imura T, Nakano I, Kornblum HI, Sofroniew MV. Phenotypic and functional heterogeneity of GFAP-expressing cells in vitro: differential expression of LeX/CD15 by GFAP-expressing multipotent neural stem cells and non-neurogenic astrocytes. Glia 2006;53:277–93.

[65] Crossley PH, Martinez S, Martin GR. Midbrain development induced by FGF8 in the chick embryo. Nature 1996;380:66–8.

[66] Prakash N, et al. A Wnt1-regulated genetic network controls the identity and fate of midbrain-dopaminergic progenitors in vivo. Development 2006;133:89–98.

[67] Reya T, et al. A role for Wnt signalling in self-renewal of haematopoietic stem cells. Nature 2003;423:409–14.

[68] Willert K, et al. Wnt proteins are lipid-modified and can act as stem cell growth factors. Nature 2003;423:448–52.

[69] Read TA, et al. Identification of CD15 as a marker for tumor-propagating cells in a mouse model of medulloblastoma. Cancer Cell 2009;15:135–47.

[70] Son MJ, Woolard K, Nam DH, Lee J, Fine HA. SSEA-1 is an enrichment marker for tumor-initiating cells in human glioblastoma. Cell Stem Cell 2009;4:440–52.

[71] Johnson D, et al. Expression and structure of the human NGF receptor. Cell 1986;47:545–54.

[72] Sonnenfeld KH, Ishii DN. Nerve growth factor effects and receptors in cultured human neuroblastoma cell lines. J Neurosci Res 1982;8:375–91.

[73] Barnes M, Eberhart CG, Collins R, Tihan T. Expression of p75NTR in fetal brain and medulloblastomas: evidence of a precursor cell marker and its persistence in neoplasia. J Neuro-oncol 2009;92:193–201.

[74] Morrison LC, et al. Deconstruction of medulloblastoma cellular heterogeneity reveals differences between the most highly invasive and self-renewing phenotypes. Neoplasia 2013;15:384–98.

[75] Huang EH, et al. Aldehyde dehydrogenase 1 is a marker for normal and malignant human colonic stem cells (SC) and tracks SC overpopulation during colon tumorigenesis. Cancer Res 2009;69:3382–9.

[76] Ginestier C, et al. ALDH1 is a marker of normal and malignant human mammary stem cells and a predictor of poor clinical outcome. Cell Stem Cell 2007;1:555–67.

[77] Douville J, Beaulieu R, Balicki D. ALDH1 as a functional marker of cancer stem and progenitor cells. Stem Cells Dev 2009;18:17–25.

[78] Sahovic EA, Colvin M, Hilton J, Ogawa M. Role for aldehyde dehydrogenase in survival of progenitors for murine blast cell colonies after treatment with 4-hydroperoxycyclophosphamide in vitro. Cancer Res 1988;48:1223–6.

[79] Gordon MY, Goldman JM, Gordon-Smith EC. 4-Hydroperoxy-cyclophosphamide inhibits proliferation by human granulocyte-macrophage colony-forming cells (GM-CFC) but spares more primitive progenitor cells. Leuk Res 1985;9:1017–21.

[80] Gentry T, et al. Simultaneous isolation of human BM hematopoietic, endothelial and mesenchymal progenitor cells by flow sorting based on aldehyde dehydrogenase activity: implications for cell therapy. Cytotherapy 2007;9:259–74.

[81] Corti S, et al. Identification of a primitive brain-derived neural stem cell population based on aldehyde dehydrogenase activity. Stem Cells 2006;24:975–85.

[82] Rasper M, et al. Aldehyde dehydrogenase 1 positive glioblastoma cells show brain tumor stem cell capacity. Neuro-oncology 2010;12:1024–33.

[83] Jones RJ, et al. Assessment of aldehyde dehydrogenase in viable cells. Blood 1995;85:2742–6.

[84] Venugopal C, et al. Bmi1 marks intermediate precursors during differentiation of human brain tumor initiating cells. Stem Cell Res 2012;8:141–53.

[85] Pistollato F, Chen HL, Schwartz PH, Basso G, Panchision DM. Oxygen tension controls the expansion of human CNS precursors and the generation of astrocytes and oligodendrocytes. Mol Cell Neurosci 2007;35:424–35.

[86] Evans SM, et al. Hypoxia is important in the biology and aggression of human glial brain tumors. Clin Cancer Res 2004;10:8177–84.

[87] Amberger-Murphy V. Hypoxia helps glioma to fight therapy. Curr Cancer Drug Targets 2009;9:381–90.

[88] McCord AM, Jamal M, Williams ES, Camphausen K, Tofilon PJ. CD133+ glioblastoma stem-like cells are radiosensitive with a defective DNA damage response compared with established cell lines. Clin Cancer Res 2009;15:5145–53.

[89] Iida H, Suzuki M, Goitsuka R, Ueno H. Hypoxia induces CD133 expression in human lung cancer cells by up-regulation of OCT3/4 and SOX2. Int J Oncol 2012;40:71–9.

[90] Lehnus KS, et al. CD133 glycosylation is enhanced by hypoxia in cultured glioma stem cells. Int J Oncol 2013;42:1011–7.

[91] Sakariassen PO, Immervoll H, Chekenya M. Cancer stem cells as mediators of treatment resistance in brain tumors: status and controversies. Neoplasia 2007;9:882–92.

[92] Panchision DM, et al. Optimized flow cytometric analysis of central nervous system tissue reveals novel functional relationships among cells expressing CD133, CD15, and CD24. Stem Cells 2007;25:1560–70.

[93] Li L, Clevers H. Coexistence of quiescent and active adult stem cells in mammals. Science 2010;327:542–5.

[94] Davies EL, Fuller MT. Regulation of self-renewal and differentiation in adult stem cell lineages: lessons from the Drosophila male germ line. Cold Spring Harb Symp Quant Biol 2008;73:137–45.

[95] Sun Y, et al. CD133 (Prominin) negative human neural stem cells are clonogenic and tripotent. PloS One 2009;4:e5498.

[96] Hill EJ, Woehrling EK, Prince M, Coleman MD. Differentiating human NT2/D1 neurospheres as a versatile in vitro 3D model system for developmental neurotoxicity testing. Toxicology 2008;249:243–50.

[97] Binello E, Qadeer ZA, Kothari HP, Emdad L, Germano IM. Stemness of the CT-2A immunocompetent mouse brain tumor model: characterization in vitro. J Cancer 2012;3:166–74.

[98] Chaichana K, Zamora-Berridi G, Camara-Quintana J, Quinones-Hinojosa A. Neurosphere assays: growth factors and hormone differences in tumor and nontumor studies. Stem Cells 2006;24:2851–7.

[99] Reynolds BA, Rietze RL. Neural stem cells and neurospheres–re-evaluating the relationship. Nat Methods 2005;2:333–6.

[100] Wan F, et al. The utility and limitations of neurosphere assay, CD133 immunophenotyping and side population assay in glioma stem cell research. Brain Pathol 2010;20:877–89.

[101] Pastrana E, Silva-Vargas V, Doetsch F. Eyes wide open: a critical review of sphere-formation as an assay for stem cells. Cell Stem Cell 2011;8:486–98.

[102] Singec I, et al. Defining the actual sensitivity and specificity of the neurosphere assay in stem cell biology. Nat Methods 2006;3:801–6.

[103] Louis SA, et al. Enumeration of neural stem and progenitor cells in the neural colony-forming cell assay. Stem Cells 2008;26:988–96.

[104] Bratu DP, Cha BJ, Mhlanga MM, Kramer FR, Tyagi S. Visualizing the distribution and transport of mRNAs in living cells. Proc Natl Acad Sci USA 2003;100:13308–13.

[105] Rhee WJ, Santangelo PJ, Jo H, Bao G. Target accessibility and signal specificity in live-cell detection of BMP-4 mRNA using molecular beacons. Nucleic Acids Res 2008;36:e30.

[106] Santangelo P, Nitin N, LaConte L, Woolums A, Bao G. Live-cell characterization and analysis of a clinical isolate of bovine respiratory syncytial virus, using molecular beacons. J Virol 2006;80:682–8.

[107] Rhee WJ, Bao G. Simultaneous detection of mRNA and protein stem cell markers in live cells. BMC Biotechnol 2009;9:30.

[108] Ilieva M, Dufva M. SOX2 and OCT4 mRNA-expressing cells, detected by molecular beacons, localize to the center of neurospheres during differentiation. PloS One 2013;8:e73669.

[109] Tyagi S, Kramer FR. Molecular beacons: probes that fluoresce upon hybridization. Nat Biotechnol 1996;14:303–8.

[110] Larsson HM, et al. Sorting live stem cells based on Sox2 mRNA expression. PloS One 2012;7:e49874.

[111] Harris MA, et al. Cancer stem cells are enriched in the side population cells in a mouse model of glioma. Cancer Res 2008;68:10051–9.

[112] Orimo A, et al. Stromal fibroblasts present in invasive human breast carcinomas promote tumor growth and angiogenesis through elevated SDF-1/CXCL12 secretion. Cell 2005;121:335–48.

[113] Tyan SW, et al. Breast cancer cells induce cancer-associated fibroblasts to secrete hepatocyte growth factor to enhance breast tumorigenesis. PloS One 2011;6:e15313.

Chapter 18

Using Cell Surface Signatures to Dissect Neoplastic Neural Cell Heterogeneity in Pediatric Brain Tumors

Tamra Werbowetski-Ogilvie

Regenerative Medicine Program, Department of Biochemistry & Medical Genetics and Physiology, University of Manitoba, Winnipeg, MB, Canada

ABBREVIATIONS

CSC Cancer stem cell
FACS Fluorescence-activated cell sorting
hESC Human embryonic stem cell
MB Medulloblastoma
Shh Sonic hedgehog
TPC Tumor-propagating cell

18.1 CELL SURFACE MARKERS TO DISTINGUISH HETEROGENEITY IN THE NEURAL LINEAGE HIERARCHY

There are many induction techniques used for generating neural stem cells and their differentiated progeny from both pluripotent stem cells and somatic tissue [1]. These methods lead to production of neuronal and glial cells that are then utilized in a variety of in vitro and in vivo assays. However, cell populations are typically a mixture of both differentiated and undifferentiated cells. For utilization in transplantation assays, one must generate pure and well-defined cell populations. Recent studies have begun to utilize flow cytometric and fluorescence-activated cell sorting (FACS) techniques to both identify and isolate various cell phenotypes along the neural lineage hierarchy. Although difficulties with dissociation methods, cell viability, and retention of cell surface marker expression have complicated these analyses, several markers have emerged that distinguish between multipotent neural stem cells, progenitors, and differentiated neurons and glia [2–5]. For example, CD15 (SSEA1), CD133, CD271 (p75NTR), CD146, and CD29 have been used to identify neural stem and precursor cells; whereas CD24 and CD56 (NCAM) can be used to isolate differentiated neurons derived from human embryonic stem cell cultures. Yuan et al. [5] utilized a slightly different combination of markers (CD184, CD271, CD44, CD24, and CD15) to distinguish neural stem cells from neurons and glia differentiated from human embryonic stem cells. Similar sets of cell surface markers, including CD133, CD15, and CD24, have been used to discriminate between fetal mouse multipotent cells and more differentiated progenitors and neurons [2]. In addition, CD326 has been utilized as an exclusion marker, in addition to the common pluripotent markers SSEA3, SSEA4, Tra-1-60, and Tra-1-80, to remove contaminating undifferentiated human embryonic stem cells from highly heterogeneous neural cell populations [6]. Identification of cell surface signatures that can be reproducibly utilized for isolation of highly pure and defined cell populations required for transplantation studies will lessen the risk of retention of primitive undifferentiated cells. Residual pluripotent stem cells are known to give rise to teratomas in transplantation assays, and these tumors consist of cell types derived from all three germ layers. Thus, the heterogeneous composition of cell types in neural cultures poses a substantial risk for future clinical applications. For a more comprehensive analysis of cell surface markers used during normal neural lineage specification from a variety of cell types, refer to Chapters 2, 14–16.

Immunophenotyping has emerged as a powerful tool for delineation of heterogeneity within normal neural cell populations. However, flow cytometry and FACS have also been widely utilized in neuro-oncology to identify and purify specific cell phenotypes in malignant brain tumors. As these tumors also exhibit extensive heterogeneity at the cellular level, it will be important to define cell surface signatures that identify not only putative tumor-propagating cells (TPCs) or cancer stem cell (CSC) populations, but also

Neural Surface Antigens. http://dx.doi.org/10.1016/B978-0-12-800781-5.00018-9

highly proliferative and invasive cells that exhibit variable responses to therapies.

18.2 MEDULLOBLASTOMA: AN EXAMPLE OF GENOMIC, MOLECULAR, AND CELLULAR HETEROGENEITY

Central nervous system tumors are among the most prevalent forms of childhood cancers, accounting for nearly 20% of all new cases (Canadian Cancer Society Statistics, 2014). Medulloblastoma (MB) is the most common malignant primary pediatric brain tumor [7,8]. Despite improved 5-year survival rates of 60–70%, MBs often recur as a consequence of tumor cell infiltration into normal tissue and frequent metastasis through the cerebral spinal fluid [7–9].

Advances in both genomic sequencing and microarray technologies have revolutionized our understanding of the extensive molecular and genetic heterogeneity that underlies MB. Multiple groups have demonstrated that MB comprises several molecular variants [10–13]. However, the current consensus is that MB consists of four distinct subtypes exhibiting different genomic alterations, gene expression profiles, and responses to treatment: a WNT variant, a Sonic hedgehog (Shh) variant, and the more highly aggressive Group 3 and Group 4 subgroups [14]. This has led to the identification of many subgroup-specific mutated genes [15–18]. In fact, studies have even demonstrated heterogeneity within a molecular variant, as Zhukova et al. revealed that p53 mutations are associated with poor outcome for Shh patients and that these may account for treatment failure within this subgroup [19]. This heterogeneity within the Shh variant emphasizes the need to identify additional biomarkers that may be linked with therapy-resistant cell phenotypes, relapse, and poor prognosis in these patients.

To date, most research on the four MB variants focuses on differential gene expression, copy number variations, and mutation analysis. However, the functional roles of these mutated and differentially expressed genes are poorly understood. After completion of the "omics" analyses, future studies must determine the roles of these genes in the maintenance and progression of MB subtypes. In addition, understanding how these genes may contribute to the heterogeneity at the cellular level will provide a more complete picture of the complexity of the disease. We must approach cellular heterogeneity in the same manner in which the genetic variation has been dissected, and not assume that the signaling pathways regulating stem cell/progenitor properties are applicable to all subtypes. Further complicating this issue is the finding that MB metastases are genetically divergent from the matched primary tumor [9], highlighting the need to evaluate cellular heterogeneity in both the primary and the metastatic compartment. Identifying variant-specific biomarkers that select for cellular phenotypes will inevitably provide a more comprehensive understanding of the MB subgroups and will shed light on the intertumoral and intratumoral heterogeneity that accompanies these malignancies.

18.3 THE CSC HYPOTHESIS AND BRAIN TUMORS

The CSC hypothesis emerged as a proposed explanation for tumor cell heterogeneity and recurrence. Current theory suggests that a subpopulation of cells within a tumor exhibit stem cell properties such as self-renewal capacity, or the ability to maintain themselves in the primitive "stem cell" state, and multilineage differentiation [20,21]. These CSCs or TPCs need not be rare but would either initiate or maintain tumor growth and must therefore be specifically targeted to prevent malignant progression and recurrence. Although this work originated in leukemia [22], a large number of studies have since been published on TPCs in solid tumors [20,21,23]. For example, Singh et al. demonstrated that isolation of a rare cell subpopulation based on CD133 cell surface expression selected for a brain tumor stem cell phenotype both in vitro and in vivo [24,25]. Also known as Prominin-1, CD133 is a glycoprotein with very little functional characterization [26]. For additional information on CD133 and its role in brain tumor-propagating/initiating cell populations, refer to Chapter 10. Although these studies demonstrated the existence of a brain tumor "stem-cell-like" phenotype, the use of CD133 as a brain TPC marker remains controversial. Most recently, several studies have shown that in addition to CD133+ cells, the CD133− subpopulation also exhibits self-renewal capacity and that CD133− cells can give rise to aggressive tumors in vivo [26–30]. This was observed for CD133− cells from both colon cancer [29] and brain tumors [27–30]. Future studies will probably further deconstruct CD133+/− populations to identify other markers, alone or in combination with CD133, that better select for self-renewing and highly invasive and metastatic phenotypes in a variety of cancers [23]. Given that CD133 is not exclusive to TPC populations and is also expressed in a variety of normal primitive and more differentiated cell types [26], these markers will probably have to be considered in a cell-context-dependent manner.

Recent studies have shown that CD15 or stage-specific embryonic antigen (SSEA-1) also selects for MB TPCs, particularly in mouse models of the Shh molecular variant [31,32]. Multiple murine models such as the Patched+/− (Ptc+/−) receptor mouse [33,34] and transgenic Smoothened (Smo) mouse [35] have become a mainstay in MB research for studying the developmental biology of the disease. Interestingly, Read et al. [31] showed that CD133+ cells from Ptc+/− mice were not amenable to tumorsphere formation in vitro and were not capable of forming tumors after cerebellar transplantation in vivo [31]. In this Shh variant mouse model, tumors are propagated by cells expressing CD15 (SSEA-1) and Math1 (neuronal progenitor marker) [31]. Ward et al. also demonstrated that CD15+ cells are tumor propagating in Ptc+/− mice; however, in this study, the authors suggested that the CD15+ population represents a more primitive stem cell fraction, as the cells were propagated long term under "stem-cell-enriched" conditions [32].

More research is necessary to identify additional cell surface markers that select for TPC populations irrespective of whether they are stem cells or progenitors. Given the molecular and genetic heterogeneity between and even within MB subtypes, the MB TPC signature may also differ between subgroups, highlighting heterogeneity at the cellular level. Moreover, different combinations of markers may select for putative stem cell, highly proliferative, or most invasive subtypes. Eradicating MB TPCs while leaving normal neural stem cells/progenitors intact will be a major challenge [36]. However, exhausting MB TPCs may serve as a concomitant treatment strategy that can be used together with or as a follow-up to standard chemotherapy or radiation aimed at reducing tumor burden [36].

18.4 IN SEARCH OF NEW MARKERS: THE CASE FOR CD271/p75NTR

The low-affinity neurotrophin receptor CD271/p75NTR has been linked with self-renewal properties of MB [37]. This was achieved by deconstructing Shh MB culture heterogeneity and generating subclones from a cell line that exhibited distinct cellular properties. Specifically, the subclones displayed vastly different self-renewal capacities when placed in ultralow attachment plates and propagated in suspension or tumorsphere culture or under stem-cell-enriched conditions. Tumorspheres from subclones displaying higher self-renewal vs lower self-renewal were comparatively screened for the presence of eight cell surface markers already known to play roles in neural lineage specification, brain TPCs, and/or tumor cell migration, invasion, and metastasis [37]. Analysis of higher vs lower self-renewing tumorspheres revealed significant differential expression of CD271 (but not CD133 or CD15) using flow cytometry [37].

Interestingly, CD271 expression was also higher in the tumor "core" or stationary cells versus actively migrating cells in a collagen matrix [37]. In this assay, hanging-drop aggregates were allowed to adhere to a plate, and the cells then migrated out from the aggregate over 48h. The core was manually dissected and separated from the actively migrating cells. CD271

was significantly higher in the core relative to the migrating MB cells. In contrast, CD133 showed opposing patterns, with levels significantly higher in the migrating cells. Cell sorting based on combinatory CD271 and CD133 expression provided support for these findings and demonstrated that CD271+/CD133− cells exhibit increased tumorsphere formation and self-renewal capacity; whereas CD271−/CD133+ cells exhibit a modest increase in cell motility [37]. This was validated by global gene expression analysis on higher versus lower self-renewing MB tumorspheres (N=3 for each) that revealed downregulation of a cell movement transcription program in higher self-renewing Shh MB cells [37]. Interestingly, when sorted subpopulations were recultured as tumorspheres and then subsequently analyzed for CD271 and CD133 expression, only the CD271+/CD133− fraction consistently recapitulated the overall distribution of cell surface markers in parental culture [37]. These results emphasized the heterogeneity in MB cultures and demonstrated that selection based on certain combinations of markers leads to reestablishment of phenotypic equilibrium.

Based on the expression patterns of CD271 in cell phenotypes in vitro, it was predicted that CD271 levels would be higher in the less aggressive Shh and Wnt variants relative to the more aggressive Groups 3 and 4 molecular subgroups that typically exhibit more extensive cell motility and metastasis [37]. To test this hypothesis, CD271 expression was assessed in a dataset derived from exon array profiling of 111 primary MB and 14 normal human cerebellar samples (nine fetal cerebellum and five adult cerebellum) [12]. Indeed, CD271 levels were highest in human fetal cerebellum and primary samples of the Shh MB molecular variant and lowest in Groups 3 and 4 [37]. Immunohistochemical studies have also demonstrated higher CD271 levels in the external granular layer and Purkinje layer in the developing human fetal cerebellum compared with undetectable expression in the adult cerebellum [38].

These studies revealed a previously unappreciated role for CD271 in selecting for MB self-renewing/TPC phenotypes and suggested that successful treatment of pediatric brain tumors requires concomitant targeting of a spectrum of transitioning self-renewing and highly aggressive cell subpopulations. This leads to a dynamic model in which a self-renewing CD271↑, CD133↓ cell in the primary MB mass may contribute to tumor propagation and maintenance (Figure 18.1(A)) [37]. Once a cell commits to entering a state of cell motility, the expression profiles change and a migrating/invading cell acquires a CD271↓, CD133↑ phenotype (Figure 18.1(B)) [37]. As the tumorsphere studies were conducted over only two passages and long-term self-renewal was not assessed, it is unknown whether the CD271↑, CD133↓ fraction represents a stem cell population or a more differentiated progenitor state [37].

FIGURE 18.1 (A) and (B) working model depicting the dynamic relationship between Shh MB cells exhibiting a higher self-renewal capacity (CD271↑, CD133↓ cells) over two passages in tumorsphere culture and those migrating/invading cells exhibiting a CD271↓, CD133↑ phenotype. (C) and (D) It is unclear whether these cell surface marker profiles represent a single TPC that shifts between cellular states (C) or multiple cell phenotypes exhibiting distinct properties (D). MB, medulloblastoma; Shh, sonic hedgehog; TPC, tumor-propagating cell.

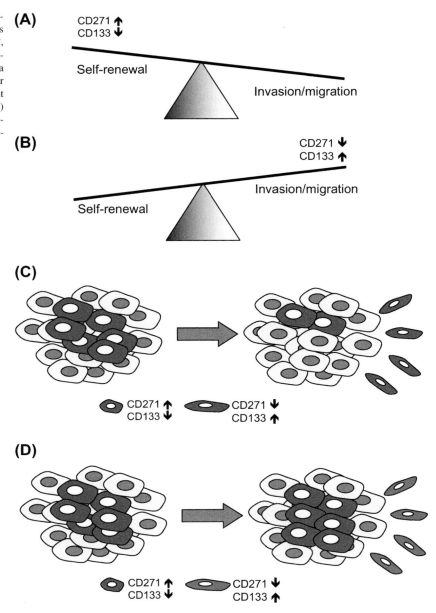

In addition, whether this represents a single TPC that shifts between cellular states or multiple TPC types exhibiting distinct phenotypes remains unclear (Figure 18.1(C) and (D)). In other cancers, evidence for both theories has been documented [39,40]; however, the most recent studies suggest that in breast cancer, TPC populations indeed transition between mesenchymal-like and epithelial-like states [41]. Interestingly, mesenchymal breast TPCs are mainly quiescent and are located at the invasive front; whereas epithelial TPCs are typically situated closer to the tumor core and are actively proliferating. Collectively, these results demonstrate that cell surface phenotyping can not only be used for identifying putative TPC or stem cell populations, but may also be utilized to select for other clinically relevant phenotypes including those cells that display a high invasive capacity and rapid proliferation.

18.5 CD271: ITS ROLE IN NEURODEVELOPMENT AND PROGENITOR/STEM CELL FUNCTION

CD271/p75NTR is a member of the tumor necrosis factor receptor family that plays crucial roles in nervous system

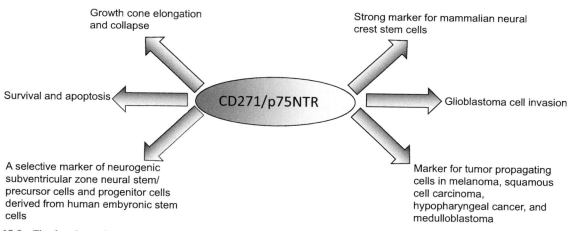

FIGURE 18.2 The functions of CD271/p75NTR in normal neurodevelopment, stem cell function, and cancer described in this chapter. Although CD271/p75NTR is primarily known for its roles during early neurodevelopment, more recent studies have linked this multifaceted cell surface marker to both normal and neoplastic stem cell and progenitor function.

development, including regulation of growth cone elongation and collapse as well as both cell survival and apoptosis [42,43]. CD271 is a 427-amino-acid transmembrane receptor that consists of an extracellular domain, a transmembrane domain, and an intracellular domain [42–44]. Within the extracellular domain, four cysteine-rich repeats help facilitate ligand binding [42,43]. CD271 has proapoptotic and prosurvival effects, depending on which (pro)neurotrophin ligand (NGF, BDNF, NT-3, or NT-4) is bound and whether CD271 binds to a coreceptor such as a member of the Trk family (TrkA, TrkB, and TrkC), sortilin (SORT1), or the Nogo receptor [42,43]. This diversity combined with the many signal transduction pathways targeted downstream leads to a wide array of cellular effects that depend on both cell type and cell state (i.e., more primitive or differentiated phenotypes).

In stem cell biology, CD271 has been shown to mark a neurogenic precursor or stem cell population in the subventricular zone (SVZ) from both rats and mice [45]. Using FACS to isolate CD271+ postnatal rat SVZ, the authors showed that sorted cells with the highest levels of CD271 generated the most neurospheres. They also demonstrated that CD271 regulates neurogenesis and the ongoing generation of olfactory bulb neurons in the SVZ [45]. Neuron production was particularly enhanced in CD271+ neural precursor cells after treatment with BDNF or NGF [45]. Moreover, following neural lineage specification from hESC's, CD271 expression was high at intermediate and later stages of neural differentiation; whereas CD133 expression was elevated in neural stem cell fractions and then decreased after extended differentiation [3]. Similar to the studies in MB self-renewing populations [37], the highest CD271 and CD133 levels are inversely correlated,

suggesting that they select for cells exhibiting distinct phenotypes. Finally, CD271 has also been shown to mark multipotent mammalian neural crest stem cells [46] and has since been employed to remove the neural crest population from heterogeneous neural cultures derived from hESC's [5].

The results from the MB studies [37] as well as those obtained for other tumors suggest that CD271 is an excellent candidate for further functional studies. For example, CD271 selects for putative TPCs in a variety of tumors, including melanoma [47,48], esophageal squamous cell carcinoma [49], and hypopharyngeal cancer [50]. A recent study also suggested that a combination of CD271 and the glycosphingolipid GD2 could be used to distinguish between primitive neuroectodermal tumors, another highly malignant brain tumor, and neuroblastoma, the most common extracranial neural tumor [51]. Incidentally, this combination has also been utilized to mark mesenchymal stem cells in the bone marrow [52,53]. Brain tumors and other malignancies that have taken on properties of epithelial-to-mesenchymal transition may consequently acquire the expression of markers associated with the mesenchymal stem cell lineage. CD271 has also been linked to invasion of adult glioblastoma brain tumor cells [54] and could be targeted with γ-secretase inhibitors [55]. Moreover, proteolysis of CD271 has been shown to regulate glioblastoma TPC proliferation and this was linked with Akt signaling [56]. Thus, CD271 appears to select for a range of phenotypes in a context-dependent manner. A summary of CD271's functions in neurodevelopment, stem cell biology, and cancer is presented in Figure 18.2. Given these findings, it will be essential to further delineate the functional role of CD271 in MB progression and whether it is specifically linked with a neural precursor or progenitor cell population in the Shh variant.

18.6 EXTENDING BEYOND CD133: USING HIGH-THROUGHPUT FLOW CYTOMETRY TO IDENTIFY NOVEL MARKERS OF NEURAL TUMOR CELL PHENOTYPES

It is likely that additional markers, perhaps in combination with CD271, will best select for putative TPCs in the Shh MB variant. Studies have utilized multiparameter flow cytometry to complement histopathology for diagnostic screening of a variety of pediatric tumors [51]. High-throughput flow cytometry screening platforms have been introduced as a mechanism for distinguishing cells at various stages of neural lineage specification from human embryonic stem cells [5] as well as for identification of primary vs metastatic colon cancer cell lines [57]. More recently, Lathia et al. have used this platform to identify adhesion receptors such as junctional adhesion molecule A that contribute to glioblastoma self-renewal and tumor growth [58]. These screening platforms can therefore be utilized to interrogate a wide variety of both normal and neoplastic cell phenotypes including cells exhibiting a high capacity for self-renewal, migration and/or invasion, and cell proliferation.

Characterization of the cell surface proteome enables researchers to design strategies for isolation of specific cell phenotypes that can then be utilized for additional functional, mechanistic, and preclinical studies both in vitro and in vivo. This technique has been adopted for analysis of higher vs lower self-renewing Shh MB tumorspheres to identify additional novel TPC biomarkers. Given the demonstration of heterogeneity within the well-characterized Shh subgroup [19] and the identification of a role for CD271 in Shh MB cells [37], it is reasonable to assume that within this molecular variant, there is a combination of markers that will select for various phenotypes.

To address this issue, MB cells representing the Shh variant were comparatively screened to identify cell surface markers that were differentially expressed between higher and lower self-renewing phenotypes (unpublished data). Human cell surface marker screening panels were utilized that consisted of 242 cell surface markers and nine isotype or negative controls for analysis. Daoy MB cells (representative of the Shh variant) continue to be utilized as a supplement to working with fresh patient tissue or minimally cultured samples for stem cell and progenitor studies [59–61], as it has been quite difficult

to establish cultures from primary MB tumors. Whereas several studies have utilized these screening platforms to identify novel markers associated with self-renewal or "stemness," these methods can also be used to identify novel cell surface signatures associated with other properties such as adhesion, proliferation, migration/invasion, and cell differentiation. Using standard cell lines, one can generate a list of candidate markers and then validate them using a variety of additional cell lines or fresh cultured patient samples, as well as fixed paraffin-embedded primary tissue.

To date, 25 cell surface markers have been identified that exhibit more than a twofold difference in expression by both frequency and mean fluorescence intensity between higher and lower self-renewing Shh MB phenotypes. Of note, several surface markers have been linked with neural cancers (GD2 [62–64], CD57 [65], CD97 [66], epidermal growth factor receptor (EGFR) [67], CD171 [68,69]), human stem cell/CSCs (SSEA4 [70,71], Tra-1-60 [71], GD2 [52,72], CD106 [73]), and neurodevelopment (CD108 [74]). For example, CD171, also called L1CAM, has been shown to play a role in glioblastoma radioresistance [75] as well as stimulation of migration and proliferation through fibroblast growth factor (FGF) in this tumor [68,69]. GD2 is a glycosphingolipid [63,64] that is highly expressed in human tumors of neuroectodermal origin, such as melanoma, glioma, and neuroblastoma [72], and has also been shown to identify CSCs in breast cancer [52].

Using high-throughput flow cytometry platforms or other discovery-based approaches, one can identify and validate a series of cell surface markers for long-term study both in vitro and in vivo and then move forward with the most promising candidate(s) for additional clinical studies.

18.7 CLINICAL IMPLICATIONS OF CELL SURFACE SIGNATURES IN MB AND OTHER BRAIN TUMOR PATHOLOGIES

Our knowledge of cellular heterogeneity between and within brain tumor subgroups is limited. Identification of TPC populations for specific MB molecular variants will enable isolation of a novel cell resource for design of next-generation targeted therapies. Delineating specific combinations of cell surface markers with known biological functions that "mark" these important cell populations could enable the design of antibody-based tumor eradication strategies. This would ultimately lessen the broad impact of toxic treatments such as radiation and chemotherapy on the child's developing nervous system and improve the quality of life for those children who survive long term.

Newly identified candidate biomarkers can be utilized not only for isolation of cell phenotypes within a brain tumor but also for stratification of tumors into a particular subgroup. Therefore, cell surface marker phenotyping or "cellular fingerprinting" combined with validations in primary patient samples are powerful tools that can be used for both diagnostic and prognostic purposes.

REFERENCES

[1] Schwartz PH, Brick DJ, Stover AE, Loring JF, Muller FJ. Differentiation of neural lineage cells from human pluripotent stem cells. Methods 2008;45:142–58.

[2] Panchision DM, Chen HL, Pistollato F, Papini D, Ni HT, et al. Optimized flow cytometric analysis of central nervous system tissue reveals novel functional relationships among cells expressing CD133, CD15, and CD24. Stem Cells 2007;25:1560–70.

[3] Pruszak J, Sonntag KC, Aung MH, Sanchez-Pernaute R, Isacson O. Markers and methods for cell sorting of human embryonic stem cell-derived neural cell populations. Stem Cells 2007;25:2257–68.

[4] Pruszak J, Ludwig W, Blak A, Alavian K, Isacson O. CD15, CD24, and CD29 define a surface biomarker code for neural lineage differentiation of stem cells. Stem Cells 2009;27:2928–40.

[5] Yuan SH, Martin J, Elia J, Flippin J, Paramban RI, et al. Cell-surface marker signatures for the isolation of neural stem cells, glia and neurons derived from human pluripotent stem cells. PLoS One 2011;6:e17540.

[6] Sundberg M, Jansson L, Ketolainen J, Pihlajamaki H, Suuronen R, et al. CD marker expression profiles of human embryonic stem cells and their neural derivatives, determined using flow-cytometric analysis, reveal a novel CD marker for exclusion of pluripotent stem cells. Stem Cell Res 2009;2:113–24.

[7] Louis D, Ohgaki H, Wiestler OD, Cavenee WK. WHO classification of tumours of the central nervous system. 4th ed. Switzerland: WHO Press; 2007.

[8] Mehta M, Chang S, Newton H, Guha A, Vogelbaum M. Principles and practice of neuro-oncology: a multidisciplinary approach (October 21 2010). 1st ed. Demos Medical Publishing; 2011. 951.

[9] Wu X, Northcott PA, Dubuc A, Dupuy AJ, Shih DJ, et al. Clonal selection drives genetic divergence of metastatic medulloblastoma. Nature 2012;482:529–33.

[10] Cho YJ, Tsherniak A, Tamayo P, Santagata S, Ligon A, et al. Integrative genomic analysis of medulloblastoma identifies a molecular subgroup that drives poor clinical outcome. J Clin Oncol 2011;29:1424–30.

[11] Kool M, Koster J, Bunt J, Hasselt NE, Lakeman A, et al. Integrated genomics identifies five medulloblastoma subtypes with distinct genetic profiles, pathway signatures and clinicopathological features. PLoS One 2008;3:e3088.

[12] Northcott PA, Korshunov A, Witt H, Hielscher T, Eberhart CG, et al. Medulloblastoma comprises four distinct molecular variants. J Clin Oncol 2011;29:1408–14.

[13] Thompson MC, Fuller C, Hogg TL, Dalton J, Finkelstein D, et al. Genomics identifies medulloblastoma subgroups that are enriched for specific genetic alterations. J Clin Oncol 2006;24:1924–31.

[14] Taylor MD, Northcott PA, Korshunov A, Remke M, Cho YJ, et al. Molecular subgroups of medulloblastoma: the current consensus. Acta Neuropathol 2012;123:465–72.

[15] Robinson G, Parker M, Kranenburg TA, Lu C, Chen X, et al. Novel mutations target distinct subgroups of medulloblastoma. Nature 2012;488:43–8.

[16] Northcott PA, Shih DJ, Peacock J, Garzia L, Morrissy AS, et al. Subgroup-specific structural variation across 1000 medulloblastoma genomes. Nature 2012;488:49–56.

[17] Jones DT, Jager N, Kool M, Zichner T, Hut-ter B, et al. Dissecting the genomic complexity underlying medulloblastoma. Nature 2012;488:100–5.

[18] Pugh TJ, Weeraratne SD, Archer TC, Pomeranz Krummel DA, Auclair D, et al. Medulloblastoma exome sequencing uncovers subtype-specific somatic mutations. Nature 2012;488:106–10.

[19] Zhukova N, Ramaswamy V, Remke M, Pfaff E, Shih DJ, et al. Subgroup-specific prognostic implications of TP53 mutation in medulloblastoma. J Clin Oncol 2013;31:2927–35.

[20] Rosen JM, Jordan CT. The increasing complexity of the cancer stem cell paradigm. Science 2009;324:1670–3.

[21] Visvader JE. Cells of origin in cancer. Nature 2011;469:314–22.

[22] Bonnet D, Dick JE. Human acute myeloid leukemia is organized as a hierarchy that originates from a primitive hematopoietic cell. Nat Med 1997;3:730–7.

[23] Aiken C, Werbowetski-Ogilvie T. Animal models of cancer stem cells: what are they really telling us? Curr Pathobiol Rep 2013;1:91–9.

[24] Singh SK, Clarke ID, Terasaki M, Bonn VE, Hawkins C, et al. Identification of a cancer stem cell in human brain tumors. Cancer Res 2003;63:5821–8.

[25] Singh SK, Hawkins C, Clarke ID, Squire JA, Bayani J, et al. Identification of human brain tumour initiating cells. Nature 2004;432:396–401.

[26] Wu Y, Wu PY. CD133 as a marker for cancer stem cells: progresses and concerns. Stem Cells Dev 2009;18:1127–34.

[27] Ogden AT, Waziri AE, Lochhead RA, Fusco D, Lopez K, et al. Identification of A2B5+CD133- tumor-initiating cells in adult human gliomas. Neurosurgery 2008;62:505–14. discussion 514–515.

[28] Wang J, Sakariassen PO, Tsinkalovsky O, Immervoll H, Boe SO, et al. CD133 negative glioma cells form tumors in nude rats and give rise to CD133 positive cells. Int J Cancer 2008;122:761–8.

[29] Shmelkov SV, Butler JM, Hooper AT, Hormigo A, Kushner J, et al. CD133 expression is not restricted to stem cells, and both CD133+ and CD133– metastatic colon cancer cells initiate tumors. J Clin Invest 2008;118:2111–20.

[30] Joo KM, Kim SY, Jin X, Song SY, Kong DS, et al. Clinical and biological implications of CD133-positive and CD133-negative cells in glioblastomas. Lab Invest 2008;88:808–15.

[31] Read TA, Fogarty MP, Markant SL, McLendon RE, Wei Z, et al. Identification of CD15 as a marker for tumor-propagating cells in a mouse model of medulloblastoma. Cancer Cell 2009;15:135–47.

[32] Ward RJ, Lee L, Graham K, Satkunendran T, Yoshikawa K, et al. Multipotent CD15+ cancer stem cells in patched-1-deficient mouse medulloblastoma. Cancer Res 2009;69:4682–90.

[33] Goodrich LV, Milenkovic L, Higgins KM, Scott MP. Altered neural cell fates and medulloblastoma in mouse patched mutants. Science 1997;277:1109–13.

[34] Wetmore C, Eberhart DE, Curran T. The normal patched allele is expressed in medulloblastomas from mice with heterozygous germ-line mutation of patched. Cancer Res 2000;60:2239–46.

[35] Hatton BA, Villavicencio EH, Tsuchiya KD, Pritchard JI, Ditzler S, et al. The Smo/Smo model: hedgehog-induced medulloblastoma with 90% incidence and leptomeningeal spread. Cancer Res 2008;68:1768–76.

[36] Castelo-Branco P, Tabori U. Promises and challenges of exhausting pediatric neural cancer stem cells. Pediatr Res 2012;71:523–8.

[37] Morrison LC, McClelland R, Aiken C, Bridges M, Liang L, et al. Deconstruction of medulloblastoma cellular heterogeneity reveals differences between the most highly invasive and self-renewing phenotypes. Neoplasia 2013;15:384–98.

[38] Barnes M, Eberhart CG, Collins R, Tihan T. Expression of p75NTR in fetal brain and medulloblastomas: evidence of a precursor cell marker and its persistence in neoplasia. J Neurooncol 2009;92:193–201.

[39] Mani SA, Guo W, Liao MJ, Eaton EN, Ayyanan A, et al. The epithelial-mesenchymal transition generates cells with properties of stem cells. Cell 2008;133:704–15.

[40] Tsuji T, Ibaragi S, Shima K, Hu MG, Katsurano M, et al. Epithelial-mesenchymal transition induced by growth suppressor p12CDK2-AP1 promotes tumor cell local invasion but suppresses distant colony growth. Cancer Res 2008;68:10377–86.

[41] Liu S, Cong Y, Wang D, Sun Y, Deng L, et al. Breast cancer stem cells transition between epithelial and mesenchymal states reflective of their normal counterparts. Stem Cell Rep 2014;2:78–91.

[42] Chen Y, Zeng J, Cen L, Wang X, Yao G, et al. Multiple roles of the p75 neurotrophin receptor in the nervous system. J Int Med Res 2009;37:281–8.

[43] Tomellini E, Lagadec C, Polakowska R, Le Bourhis X. Role of p75 neurotrophin receptor in stem cell biology: more than just a marker. Cell Mol Life Sci 2014.

[44] He XL, Garcia KC. Structure of nerve growth factor complexed with the shared neurotrophin receptor p75. Science 2004;304:870–5.

[45] Young KM, Merson TD, Sotthibundhu A, Coulson EJ, Bartlett PF. p75 neurotrophin receptor expression defines a population of BDNF-responsive neurogenic precursor cells. J Neurosci 2007;27:5146–55.

[46] Morrison SJ, White PM, Zock C, Anderson DJ. Prospective identification, isolation by flow cytometry, and in vivo self-renewal of multipotent mammalian neural crest stem cells. Cell 1999;96:737–49.

[47] Boiko AD, Razorenova OV, van de Rijn M, Swetter SM, Johnson DL, et al. Human melanoma-initiating cells express neural crest nerve growth factor receptor CD271. Nature 2010;466:133–7.

[48] Civenni G, Walter A, Kobert N, Mihic-Probst D, Zipser M, et al. Human CD271-positive melanoma stem cells associated with metastasis establish tumor heterogeneity and long-term growth. Cancer Res 2011;71:3098–109.

[49] Huang SD, Yuan Y, Liu XH, Gong DJ, Bai CG, et al. Self-renewal and chemotherapy resistance of p75NTR positive cells in esophageal squamous cell carcinomas. BMC Cancer 2009;9:9.

[50] Imai T, Tamai K, Oizumi S, Oyama K, Yamaguchi K, et al. CD271 defines a stem cell-like population in hypopharyngeal cancer. PLoS One 2013;8:e62002.

[51] Ferreira-Facio CS, Milito C, Botafogo V, Fontana M, Thiago LS, et al. Contribution of multiparameter flow cytometry immunophenotyping to the diagnostic screening and classification of pediatric cancer. PLoS One 2013;8:e55534.

[52] De Giorgi U, Cohen EN, Gao H, Mego M, Lee BN, et al. Mesenchymal stem cells expressing GD2 and CD271 correlate with breast cancer-initiating cells in bone marrow. Cancer Biol Ther 2011;11:812–5.

[53] Rasini V, Dominici M, Kluba T, Siegel G, Lusenti G, et al. Mesenchymal stromal/stem cells markers in the human bone marrow. Cytotherapy 2013;15:292–306.

[54] Johnston AL, Lun X, Rahn JJ, Liacini A, Wang L, et al. The p75 neurotrophin receptor is a central regulator of glioma invasion. PLoS Biol 2007;5:e212.

[55] Wang L, Rahn JJ, Lun X, Sun B, Kelly JJ, et al. Gamma-secretase represents a therapeutic target for the treatment of invasive glioma mediated by the p75 neurotrophin receptor. PLoS Biol 2008;6:e289.

[56] Forsyth PA, Krishna N, Lawn S, Valadez JG, Qu X, et al. p75 neurotrophin receptor cleavage by alpha- and gamma-secretases is required for neurotrophin mediated proliferation of brain tumor initiating cells. J Biol Chem 2014.

[57] Sukhdeo K, Paramban RI, Vidal JG, Elia J, Martin J, et al. Multiplex flow cytometry barcoding and antibody arrays identify surface antigen profiles of primary and metastatic colon cancer cell lines. PLoS One 2013;8:e53015.

[58] Lathia JD, Li M, Sinyuk M, Alvarado AG, Flavahan WA, et al. High-throughput flow cytometry screening reveals a role for junctional adhesion molecule a as a cancer stem cell maintenance factor. Cell Rep 2014;6:117–29.

[59] Vo DT, Subramaniam D, Remke M, Burton TL, Uren PJ, et al. The RNA-binding protein Musashi1 affects medulloblastoma growth via a network of cancer-related genes and is an indicator of poor prognosis. Am J Pathol 2012;181:1762–72.

[60] Wang X, Venugopal C, Manoranjan B, McFarlane N, O'Farrell E, et al. Sonic hedgehog regulates Bmi1 in human medulloblastoma brain tumor-initiating cells. Oncogene 2012;31:187–99.

[61] Manoranjan B, Wang X, Hallett RM, Venugopal C, Mack SC, et al. FoxG1 interacts with Bmi1 to regulate self-renewal and tumorigenicity of medulloblastoma stem cells. Stem Cells 2013;31:1266–77.

[62] Parsons K, Bernhardt B, Strickland B. Targeted immunotherapy for high-risk neuroblastoma–the role of monoclonal antibodies. Ann Pharmacother 2013;47:210–8.

[63] Lloyd KO, Old LJ. Human monoclonal antibodies to glycolipids and other carbohydrate antigens: dissection of the humoral immune response in cancer patients. Cancer Res 1989;49:3445–51.

[64] Hakomori S. Tumor malignancy defined by aberrant glycosylation and sphingo(glyco)lipid metabolism. Cancer Res 1996;56:5309–18.

[65] Schlitter AM, Dorneburg C, Barth TF, Wahl J, Schulte JH, et al. CD57(high) neuroblastoma cells have aggressive attributes ex situ and an undifferentiated phenotype in patients. PLoS One 2012;7:e42025.

[66] Safaee M, Clark AJ, Oh MC, Ivan ME, Bloch O, et al. Overexpression of CD97 confers an invasive phenotype in glioblastoma cells and is associated with decreased survival of glioblastoma patients. PLoS One 2013;8:e62765.

[67] Taylor TE, Furnari FB, Cavenee WK. Targeting EGFR for treatment of glioblastoma: molecular basis to overcome resistance. Curr Cancer Drug Targets 2012;12:197–209.

[68] Mohanan V, Temburni MK, Kappes JC, Galileo DS. L1CAM stimulates glioma cell motility and proliferation through the fibroblast growth factor receptor. Clin Exp Metastasis 2013;30:507–20.

[69] Yang M, Li Y, Chilukuri K, Brady OA, Boulos MI, et al. L1 stimulation of human glioma cell motility correlates with FAK activation. J Neurooncol 2011;105:27–44.

[70] Andrews PW. Human teratocarcinoma stem cells: glycolipid antigen expression and modulation during differentiation. J Cell Biochem 1987;35:321–32.

[71] Thomson JA, Itskovitz-Eldor J, Shapiro SS, Waknitz MA, Swiergiel JJ, et al. Embryonic stem cell lines derived from human blastocysts. Science 1998;282:1145–7.

[72] Battula VL, Shi Y, Evans KW, Wang RY, Spaeth EL, et al. Ganglioside GD2 identifies breast cancer stem cells and promotes tumorigenesis. J Clin Invest 2012;122:2066–78.

[73] Yang ZX, Han ZB, Ji YR, Wang YW, Liang L, et al. CD106 identifies a subpopulation of mesenchymal stem cells with unique immunomodulatory properties. PLoS One 2013;8:e59354.

[74] Pasterkamp RJ, Peschon JJ, Spriggs MK, Kolodkin AL. Semaphorin 7A promotes axon outgrowth through integrins and MAPKs. Nature 2003;424:398–405.

[75] Cheng L, Wu Q, Huang Z, Guryanova OA, Huang Q, et al. L1CAM regulates DNA damage checkpoint response of glioblastoma stem cells through NBS1. EMBO J 2011;30:800–13.

Chapter 19

Synopsis and Epilogue: Neural Surface Antigen Studies in Biology and Biomedicine—What We Have Learned and What the Future May Hold

Jan Pruszak

Institute of Anatomy and Cell Biology, University of Freiburg, Freiburg im Breisgau, Germany

19.1 INTRODUCTION

Stem cell in vitro differentiation systems can greatly profit from enhanced insights into the microenvironmental cues governing cell proliferation and phenotype establishment [1–4]. Neural growth factor signals, cell–matrix, and cell–cell interactions converge onto surface molecule-mediated downstream pathways, the specifics of which remain poorly elucidated [5]. Moreover, characteristic combinations of surface molecules can serve to identify and isolate specific cellular subsets by fluorescence-activated cell sorting (FACS) from viable cell suspensions [6]. Exploiting a range of complementary systems, including human embryonic stem cell lines, long-term expandable neural lines derived from induced pluripotent stem cells, and neural cancer lines, our research is aimed at a comprehensive and functional analysis of cluster of differentiation (CD) antigens expressed in human neural lineage differentiation (Figure 19.1). With the collection at hand, we attempt to illustrate current approaches to expand our knowledge of surface antigen expression patterns, their functional significance for cell–cell interactions, as well as their application as markers to isolate neural cell subpopulations.

19.2 PROGRESS IN NEURAL STEM CELL AND CANCER BIOLOGY HINGES UPON UNDERSTANDING CELL–CELL INTERACTIONS

Our insight into the complexities of nervous system development have substantially increased since the first description of the notion that the "neuron" represents its cellular unit [7–10], neatly embedded within its other main cellular components such as astroglia and oligodendrocytes [11,12]. Concepts of nervous system development and function have emerged, and been refined, and its ability to exhibit the phenomena of synaptic as well as cellular plasticity is now well-established [13]. The intricacies of cellular interactions at the synaptic junctions are under close investigation, but systematic study on the interaction via the broad array of other molecules expressed on the surface of neural cell types in development, tissue homeostasis, and neural disease has only begun to materialize. A particular need for these studies has arisen from two different focus areas: neural stem cell biology and neurooncology. The advent of pluripotent stem cells, cellular (re)programming [14], and gene editing tools [15] have raised justifiable hope for novel applications in cell-based therapies and regenerative and personalized medicine.

On the other hand, biomedical translation has been impeded by the difficulties of mimicking normal development in the dish [16,17]. Consequently, current differentiation protocols would greatly profit from a better understanding of how cell–cell interactions contribute to development of appropriate, physiological phenotypes [5]. Recent developments in the field, while acknowledging lack of insight into the underlying mechanisms, have reverted to three-dimensional differentiation systems or so-called organoid cultures to partially overcome the issues encountered in standard protocols [18,19]. A proper translation of physiological environments to biotechnological pipelines would benefit from a more complete picture of the surface molecular inputs a cell may need to encounter during neural lineage development toward a specific phenotype of biomedical interest. Please see Chapter 1 for fundamentals of neural stem cell biology and neural development.

One aspect of tumor biology and malignant transformation is that developmental pathways, or signaling reminiscent of those, appear to be reactivated [20]. The interaction

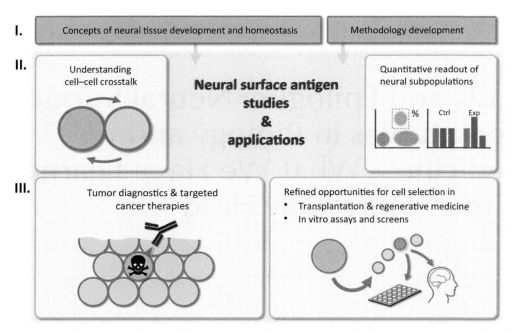

FIGURE 19.1 **Neural cell surface antigens—from basic biology toward biomedical applications.** Fundamental concepts of neural stem cell biology and cytometry-based analytic methods are covered by the first section of the book (Chapters 1 and 2, respectively). Section 2 (Chapters 3–13) focuses on particular surface molecules and surface molecule families mainly from a functional perspective, ranging from development to disease. The final section of the book (Chapters 14–19) illustrates how research on neural surface antigens may be translated to (multimarker) analytic and cell sorting paradigms for biomedical applications in neurooncology, disease modeling, and regenerative medicine.

of cancer cells and specific subpopulations—such as tumor-initiating or cancer stem cells with their microenvironment including vascular [21], immunological [22], and other nontransformed surrounding cell types—becomes increasingly important in devising novel approaches to neural cancer therapy and tumor eradication.

In both cases, the constituent subpopulations and developmental stages present in the dish or in the respective tumor tissues must be better defined. This latter point is of critical relevance from a pragmatic perspective: even without fully elucidating the underlying mechanisms of stem cell development or tumor malignancy that depend on surface molecule-mediated cell–cell interactions, characteristic surface molecular signatures can serve as markers to be exploited in cell sorting approaches or in antibody-based targeted therapies, respectively.

19.3 NEURAL FLOW CYTOMETRY AND CELL ISOLATION ARE TRICKY, BUT FEASIBLE

A technology that holds the proven, unrivaled capacity to accelerate research toward resolving surface molecule-mediated cell–cell interactions at a single cell level, as well as lineage analysis, is flow cytometry [6]. While translation from the hematological–oncological and immunological fields to neurobiology requires some methodological adaptations, this has been achieved and already fruitfully applied. Chapter 2 covers the "tricks of the trade" in neural

flow cytometry. Still, it is perceived as technically challenging and is not routinely used in a number of neurobiological laboratories. A rule of thumb may be that as long as a viable single-cell suspension can be generated, flow cytometry and FACS can be considered as analytical options complementary to, and in conjunction with, other readouts such as imaging or gene/protein expression assays. This holds true even for neural subpopulations considered to be rather delicate, including postmitotic mature neuronal subsets [23,24]. In addition to FACS, a number of other surface marker-based cell isolation methods exist, notably immunomagnetic cell selection [25], immunopanning [26], and more recently microfluidics-based cell isolation and expression analysis [27]. In other words, what limits further exploitation of subset isolation and flow cytometric methods in neurobiology may not be the limitations inherent to the technology itself, but rather the lack of markers and multiparametric marker signatures that could commonly be used to label particular well-defined subsets of neural cell populations.

19.4 NEURAL SURFACE ANTIGENS ARE CRITICAL MEDIATORS OF CELLULAR CROSSTALK IN NEUROBIOLOGY

One approach for identifying "novel" surface antigen molecules for functional study and as markers is careful neurohistochemical research, which can provide a clearer spatial picture of neural surface antigen expression. Please refer to Chapter 3 for neuroanatomical studies of

- Single molecule expression not exclusive to neural cell types or specific subsets
- Co-association with other molecules / heterodimer formation
- Intersection/integration of downstream signaling pathways
- Compensatory expression mechanisms
- Function dependent on baseline cellular state (epigenetic/transcriptional make-up)
- Membrane trafficking; localization (discrepancy of surface vs gene expression)
- Post-translational modifications (e.g., glycosylation)
- Alternative splicing mechanisms and diversity of isoforms
- Shedding and transfer of surface antigens
- Sensitivity of surface antigens to methods of analysis (harvesting, loss of neural processes; fixation and labeling procedures)

FIGURE 19.2 **Examples for complexities of neural surface antigen analysis and function.** In consequence, functional study of neural surface antigens may best be pursued through joint application of a variety of methods including flow cytometry, imaging, and other means of gene and protein expression analysis. When using surface antigens as markers, specificity may be enhanced by combinatorial labeling approaches.

CD36, CD44, and CD83 expression in the brain. Another promising lead toward resolving neural surface molecule expression is to more closely investigate in a neural context those molecules otherwise known to play a profound role in cellular crosstalk in other tissue systems. Neuroimmunological candidates have proven particularly fruitful in this regard, even without considering their immediate pathophysiological significance. Chapters 4 and 5 provide functional insights into the "death receptor" CD95 and other molecules (CD3, toll-like receptors) originally described in the context of the immune system, respectively. Beyond neuroimmunology, interaction of cells with their intermediate microenvironments, e.g., neural stem cell niches, is exemplified by tight neurovascular association and communication in development and in neurological pathologies. Please refer to Chapter 6 for an account of neuropilin/semaphorin signaling in this context as well as in neural migration, targeting, survival, and regeneration.

Another broad group of plasma membrane-associated functional modulators of cell state is represented by growth factor receptors. Chapter 7 not only introduces the neurotrophins as a pertinent example of growth factor receptor-mediated signaling, but also illustrates modes of cosignaling by association with other surface antigens at the membrane and/or cytoplasmic signaling pathway level. A common feature of neural cells, particularly, is the elaborate pattern of glycosylated epitopes expressed on their surface. Their synthesis, function, and utility as stage-specific markers is illustrated in Chapter 8, providing an overview on glycolipid antigens, as well as Chapter 9, focusing specifically on the cell surface proteoglycan NG2/Cspg4. A detailed summary of yet another prominent glycoprotein, CD133, and its conserved function, for example, in stem cells, neural development, and regeneration, is presented in Chapter 10.

While all of the above reports expertly illustrate the significant progress made in elucidating neural surface molecule-mediated signals, and while they now enable us to actually listen in to cellular crosstalk, any hubris would be unjustified (Figure 19.2).

19.5 EXPRESSION OF EVEN SOME OF THE BETTER-CHARACTERIZED NEURAL SURFACE ANTIGENS IS EXCEEDINGLY COMPLEX

A surface molecule commonly used as a well-accepted marker in the field of neurobiology is the neural cell adhesion molecule (NCAM/CD56). Throughout the book, we encounter it as a common marker for neural cell selection (Chapters 2, 14, 16, 18). Given its relatively widespread expression, it can be considered a nonexclusive "pan-neural" marker, rather than one to select particular subpopulations. It has proven useful for identification of neural cells, for example, in the quantification of the neural fraction in a mixed heterogeneous sample. However, the same surface molecule may be encountered in many different cellular scenarios or backgrounds. NCAM/CD56 is known to label natural killer cells in a hematological context, and mesenchymal or myogenic precursor cells, for instance [28,29]. Thus, the tissue and cellular context of the sample in question has to be taken into account. To further improve its utility for subset identification and isolation, it may best be used in conjunction with other markers.

NCAM is a prime example for glycosylation (PSA-NCAM) during development, and may exhibit other post-translational modifications. Chapter 11 adds yet another level of complexity when considering the fact that more than 20 isoforms of this molecule exist, enabled through alternative splicing. Of course, in light of the scenario drawn for protocadherins in the subsequent Chapter 12, this may seem trivial. Dozens of different loci, the propensity for complex DNA rearrangements, and the ability to form multimeric molecules for homophilic interactions provide a glimpse of the molecular complexity of nervous system network assembly and its specificity for cell–cell recognition. Just how critically important surface molecules can be for transcriptional regulation of neural development and function is exemplified by the role of integrins in enteric nervous system development and intestinal morphogenesis (see Chapter 13).

19.6 TOWARD AN INTEGRATED VIEW OF NEURAL SURFACE ANTIGEN SIGNALING

The hope for the collection at hand would be that the sum may become more than its parts for the reader: for example, we hear about NCAM's ability to act as a receptor in GDNF/GFRα1 receptor signaling, and that the 140-NCAM isoform, in particular, is able to interact with the RET receptor tyrosine kinase (Chapter 7). 140-NCAM is predominantly found on neural precursors (as opposed to more mature cell types) (Chapter 11). Moreover, enteric neural crest cells seem to express NCAM in conjunction with N-cadherin (CD325) during the phase of formation of enteric ganglia (Chapter 13). Interestingly, GDNF/GFRα1 signals also play an important role in that system (Chapter 13). We learn that NCAM/CD56 is present on a subset of CD271 cells (Chapter 11; CD271 being a marker covered in depth in Chapter 18 as a tumor subset-associated cell marker, and also used to identify enteric neural crest cells, Chapter 13). Similar trajectories across research area boundaries could be drawn for the CD133 (Chapters 1, 2, 4, 8, 10, 14–18) and the CD271 antigens (Chapters 3, 7, 16–18), to name a few, which we reencounter in different contexts throughout the book. An integrated synopsis of neural surface antigens from such varied perspectives may stimulate additional studies toward investigating analogous mechanisms encountered in a different context, beyond one's own specific field.

19.7 NEURAL SURFACE ANTIGENS SERVE AS VALUABLE MARKERS FOR CELL ANALYSIS AND CELL SELECTION

As mentioned above, stem cell applications suffer from the generation of unwanted cell populations. These can prevent the use of the cell types intended for biomedical purposes due to tumor risk in transplantation paradigms, or they may mask the effect to be determined by in vitro assay paradigms on a particular minor cell subset of interest. Thus, surface molecule-based cell analysis and/or isolation may need to be integrated into biomedical pipelines of cell generation, into neurodiagnostic clinical routines and as well as into devising better cell model systems.

Henry Klassen shares a personal account of his and others' seminal contributions to neural marker identification in Chapter 14. Buono et al. report on their complex seven-marker approach to analyze mouse subventricular zone neural stem and progenitor cells in Chapter 15, thereby defining critical stages of neural stem cell maturation and lineage.

As in hematology–oncology and immunology a dozen fluorescent markers are routinely applied in a combinatorial manner, fluorophore-based analyses of neural cells is by no means at its limit, and a number of screening tools are now available to catalog surface antigen expression in an efficient and cost-effective manner. Please see Chapters 16–18 for examples of neural surface antigen identification and biomedical applications, with particular focus on neural stem cells and tumor-initiating cells in nervous system cancers. To this end, multiwell high-throughput screening platforms have been exploited in order to comprehensively characterize surface antigen expression of neural stem [30] or neural tumor samples (glioblastoma, medulloblastoma, respectively) [31]. These works exemplify both that commonly established markers, such as CD271, can still be more closely resolved, or that markers such as CD321 (JAM-A) can be assigned to particular neural subpopulations and may be exploited as novel markers.

Our own research efforts have moved from single markers (FORSE-1; NCAM/CD56 [23]) to multiparametric codes to define particular neural subsets (e.g., CD15/CD24/CD29 [32]; CD49f/CD200 [33]). Recently, combining signatures from Yuan et al. [30] with our previous one [32], we were able to enrich dopaminergic neuronal cells from Parkinson patient-derived induced pluripotent stem cell differentiation systems via a CD15/CD24/CD29/CD44/CD184 code [24]. This collaborative study underlined the importance of performing such an enrichment step, as the resulting finding would largely have been masked by the otherwise predominant fraction of less mature or differently patterned cell types present in the dish. Glial cells, neural crest precursors, or other proliferative cell populations could be reduced by the sorting step, revealing a glycolipid signature consistent with adult brain. In contrast, unsorted fractions largely showed a fetal brainlike pattern, not necessarily a helpful feature when intending to study an age-associated neurodegenerative disease in a human in vitro model.

19.8 WHAT'S AROUND THE CORNER?

The future is bright for neural flow cytometry. Selecting increasingly better-defined and standardized samples for input into such screens may yield further interesting candidates for mechanism studies. For instance, surface molecule-based screens on patient-specific iPS cell-derived neural samples versus unaffected controls may yield candidates involved in the respective disease process at the cell surface level. Thus, potential disease-associated modifications in the surface molecular signature of particular neural or neuronal subpopulations may be uncovered. Such "surfaceome"-level studies at a single-cell resolution will greatly profit from opportunities established by novel combinatorial cytometric labeling and analytic approaches. In addition to antibody-based intracellular tags [33], intracellular barcoding using membrane-permeable dyes [34,35] or RNA-based probes or molecular beacons [36] offer opportunities for neural cell labeling to further resolve neural surface antigen pattern characteristics. In combinatorial neural flow cytometry, the currently limiting factor is, in fact, not the lack of compatible fluorophores, but rather the lack of identification of potential marker molecules themselves. Therefore, neural flow cytometry and neural stem cell biology in general are bound to greatly benefit from novel developments in single-cell surfaceome analytical methodology.

Mass cytometry is a technique in which the labeling fluorophore of the CD antigen antibody has been replaced with a heavy metal isotope. Free from limitations due to spectral fluorophore emission overlap, this approach offers the opportunity to combine the analytical readout of approximately 30 marker candidates per single cell [37]. While this approach requires specialized cytometry instruments adapted for mass spectrometric readout (cytometry by time-of-flight, CyTOF), such technology may become available at core facilities in a number of institutions in the near future. Moreover, surface marker readout paired with intracellular labels or expression profiling will further enhance exciting new options in single cell—as well as population-level analyses—of neural cell types [38–40]. Surface antigen-based readouts in neurobiology and stem cell biology can thereby immediately profit from recent advancements in more established flow cytometric domains in hematology, oncology, and immunology.

19.9 CONCLUSION

In summary, neural surface antigens represent critical tools for cell identification and isolation in biomedical approaches ranging from regenerative medicine to cancer. Further insights into the dynamics of their expression patterns in neural development and disease may emerge from more exhaustive global analytic platforms in conjunction with novel computational tools. Regardless of such opportunities for progress, the intricacies of surface molecule signaling may remain difficult to elucidate due to an extraordinarily complex, well-orchestrated arsenal of these molecules mediating cell interactions with their environment. It remains a source of amazement that this fine-tuned integration of signals resulting in an adequate cellular response actually and reliably works during physiological development and morphogenesis.

ACKNOWLEDGMENTS

I thank all members of my research team, and specifically those coworkers who have directly supported the work toward comprehensively characterizing neural cell types according to neural surface antigens over the past years: Harry Moe Hein Aung, Wesley Ludwig, Marcus Horl, Gizem Turaç, and Vishal Menon. I remain grateful to Ole Isacson who fostered and generously supported this endeavor at the onset. Funding by German Research Foundation Emmy Noether-grant PR1132/3-1, the Müller-Fahnenberg Foundation (Univ. Freiburg), and Neurex is gratefully acknowledged.

REFERENCES

[1] Williams CA, Lavik EB. Engineering the CNS stem cell microenvironment. Regen Med 2009;4(6):865–77.

[2] Lippmann ES, Azarin SM, Kay JE, Nessler RA, Wilson HK, Al-Ahmad A, et al. Derivation of blood–brain barrier endothelial cells from human pluripotent stem cells. Nat Biotechnol 2012;30(8): 783–91.

[3] Brafman DA. Constructing stem cell microenvironments using bioengineering approaches. Physiol Genomics 2013;45(23): 1123–35.

[4] Dalby MJ, Gadegaard N, Oreffo RO. Harnessing nanotopography and integrin-matrix interactions to influence stem cell fate. Nat Mater 2014;13(6):558–69.

[5] Solozobova V, Wyvekens N, Pruszak J. Lessons from the embryonic neural stem cell niche for neural lineage differentiation of pluripotent stem cells. Stem Cell Rev 2012;8(3):813–29.

[6] Perfetto SP, Chattopadhyay PK, Roederer M. Seventeen-colour flow cytometry: unravelling the immune system. Nat Rev Immunol 2004;4(8):648–55.

[7] Waldeyer W. Über einige neuere Forschungen im Gebiete der Anatomie des Centralnervensystems. Berl Klin Wochenschr 1891;28:691.

[8] Shepherd GM. Foundations of the neuron doctrine. New York: Oxford University Press; 1991.

[9] Fodstad H. The neuron theory. Stereotact Funct Neurosurg 2001;77(1–4):20–4.

[10] Bullock TH1, Bennett MV, Johnston D, Josephson R, Marder E, Fields RD. Neuroscience. The neuron doctrine, redux. Science 2005;310(5749):791–3.

[11] Somjen GG. Nervenkitt: notes on the history of the concept of neuroglia. Glia 1988;1(1):2–9.

[12] Oberheim NA, Goldman SA, Nedergaard M. Heterogeneity of astrocytic form and function. Methods Mol Biol 2012;814:23–45.

[13] Borodinsky LN, Belgacem YH, Swapna I, Visina O, Balashova OA, Sequerra EB, Tu MK, Levin JB, Spencer KA, Castro PA, Hamilton AM, Shim S. Spatiotemporal integration of developmental cues in neural development. Dev Neurobiol 2014. http://dx.doi.org/10.1002/dneu.22254.

[14] Okano H, Yamanaka S. iPS cell technologies: significance and applications to CNS regeneration and disease. Mol Brain 2014;7:22.

[15] M1 L, Suzuki K, Kim NY, Liu GH, Izpisua Belmonte JC. A cut above the rest: targeted genome editing technologies in human pluripotent stem cells. J Biol Chem 2014;289(8):4594–9.

[16] Irion S, Nostro MC, Kattman SJ, Keller GM. Directed differentiation of pluripotent stem cells: from developmental biology to therapeutic applications. Cold Spring Harb Symp Quant Biol 2008;73:101–10.

[17] Pruszak J, Isacson O. Molecular and cellular determinants for generating ES-cell derived dopamine neurons for cell therapy. Adv Exp Med Biol 2009;651:112–23.

[18] Lancaster MA, Knoblich JA. Organogenesis in a dish: modeling development and disease using organoid technologies. Science 2014;345(6194):1247125.

[19] Ader M, Tanaka EM. Modeling human development in 3D culture. Curr Opin Cell Biol 2014;31C:23–8.

[20] Micalizzi DS, Farabaugh SM, Ford HL. Epithelial-mesenchymal transition in cancer: parallels between normal development and tumor progression. J Mammary Gland Biol Neoplasia 2010;15(2):117–34.

[21] Hjelmeland AB, Lathia JD, Sathornsumetee S, Rich JN. Twisted tango: brain tumor neurovascular interactions. Nat Neurosci 2011; 14(11):1375–81.

[22] Sowers JL, Johnson KM, Conrad C, Patterson JT, Sowers LC. The role of inflammation in brain cancer. Adv Exp Med Biol 2014;816:75–105.

[23] Pruszak J, Sonntag KC, Aung MH, Sanchez-Pernaute R, Isacson O. Markers and methods for cell sorting of human embryonic stem cell-derived neural cell populations. Stem Cells 2007;25(9):2257–68.

[24] Schöndorf DC, Aureli M, McAllister FE, Hindley CJ, Mayer F, Schmid B, et al. iPSC-derived neurons from GBA1-associated Parkinson's disease patients show autophagic defects and impaired calcium homeostasis. Nat Commun 2014;5:4028.

[25] Grützkau A, Radbruch A. Small but mighty: how the MACS-technology based on nanosized superparamagnetic particles has helped to analyze the immune system within the last 20 years. Cytom A 2010;77(7):643–7.

[26] Barres BA. Designing and troubleshooting immunopanning protocols for purifying neural cells. Cold Spring Harb Protoc December 1, 2014;2014(12):pdb.ip073999.

[27] Citri A, Pang ZP, Südhof TC, Wernig M, Malenka RC. Comprehensive qPCR profiling of gene expression in single neuronal cells. Nat Protoc 2011;7(1):118–27.

[28] Battula VL, Treml S, Bareiss PM, Gieseke F, Roelofs H, de Zwart P, et al. Isolation of functionally distinct mesenchymal stem cell subsets using antibodies against CD56, CD271, and mesenchymal stem cell antigen-1. Haematologica 2009;94(2):173–84.

[29] Trapecar M, Kelc R, Gradisnik L, Vogrin M, Rupnik MS. Myogenic progenitors and imaging single-cell flow analysis: a model to study commitment of adult muscle stem cells. J Muscle Res Cell Motil 2014;35(5–6):249–57.

[30] Yuan SH, Martin J, Elia J, Flippin J, Paramban RI, Hefferan MP, et al. Cell-surface marker signatures for the isolation of neural stem cells, glia and neurons derived from human pluripotent stem cells. PLoS One 2011;6(3):e17540.

[31] Lathia JD, Li M, Sinyuk M, Alvarado AG, Flavahan WA, Stoltz K, et al. High-throughput flow cytometry screening reveals a role for junctional adhesion molecule a as a cancer stem cell maintenance factor. Cell Rep 2014;6(1):117–29.

[32] Pruszak J, Ludwig W, Blak A, Alavian K, Isacson O. CD15, CD24, and CD29 define a surface biomarker code for neural lineage differentiation of stem cells. Stem Cells 2009;27(12): 2928–40.

[33] Turaç G, Hindley CJ, Thomas R, Davis JA, Deleidi M, Gasser T, et al. Combined flow cytometric analysis of surface and intracellular antigens reveals surface molecule markers of human neuropoiesis. PLoS One 2013;8(6):e68519.

[34] Sukhdeo K, Paramban RI, Vidal JG, Elia J, Martin J, Rivera M, et al. Multiplex flow cytometry barcoding and antibody arrays identify surface antigen profiles of primary and metastatic colon cancer cell lines. PLoS One 2013;8(1):e53015.

[35] Menon V, Thomas R, Ghale AR, Reinhard C, Pruszak J. Flow cytometry protocols for surface and intracellular antigen analyses of neural cell types. J Vis Exp 2014;(94):e52241. http://dx.doi.org/10.3791/52241.

[36] Wile BM, Ban K, Yoon YS, Bao G. Molecular beacon-enabled purification of living cells by targeting cell type-specific mRNAs. Nat Protoc 2014;9(10):2411–24.

[37] Bodenmiller B, Zunder ER, Finck R, Chen TJ, Savig ES, Bruggner RV, et al. Multiplexed mass cytometry profiling of cellular states perturbed by small-molecule regulators. Nat Biotechnol 2012;30(9):858–67.

[38] Wang D, Bodovitz S. Single cell analysis: the new frontier in "omics". Trends Biotechnol 2010;28(6):281–90.

[39] Liberali P, Snijder B, Pelkmans L. Single-cell and multivariate approaches in genetic perturbation screens. Nat Rev Genet 2015;16(1):18–32. http://dx.doi.org/10.1038/nrg3768.

[40] Chattopadhyay PK, Gierahn TM, Roederer M, Love JC. Single-cell technologies for monitoring immune systems. Nat Immunol 2014;15(2):128–35.

Index

Note: Page numbers followed by "b", "f" and "t" indicate boxes, figures and tables respectively.

Printed and bound by CPI Group (UK) Ltd, Croydon, CR0 4YY

08/05/2025

01865030-0001